Statistics for Engineering
and Information Science

Series Editors
M. Jordan, S.L. Lauritzen, J.F. Lawless, V. Nair

Springer
New York
Berlin
Heidelberg
Barcelona
Hong Kong
London
Milan
Paris
Singapore
Tokyo

Statistics for Engineering and Information Science

Arnaud Doucet
Nando de Freitas
Neil Gordon
Editors

Sequential Monte Carlo Methods in Practice

Foreword by Adrian Smith

With 168 Illustrations

 Springer

Arnaud Doucet
Department of Electrical and Electronic
  Engineering
The University of Melbourne
Victoria 3010
Australia
doucet@ee.mu.oz.au

Nando de Freitas
Computer Science Division
387 Soda Hall
University of California
Berkeley, CA 94720-1776
USA
jfgf@cs.berkeley.edu

Neil Gordon
Pattern and Information Processing
Defence Evaluation and Research Agency
St. Andrews Road
Malvern, Worcs, WR14 3PS
UK
N.Gordon@signal.dera.gov.uk

Series Editors
Michael Jordan
Department of Computer Science
University of California, Berkeley
Berkeley, CA 94720
USA

Steffen L. Lauritzen
Department of Mathematical Sciences
Aalborg University
DK-9220 Aalborg
Denmark

Jerald F. Lawless
Department of Statistics
University of Waterloo
Waterloo, Ontario N2L 3G1
Canada

Vijay Nair
Department of Statistics
University of Michigan
Ann Arbor, MI 48109
USA

Library of Congress Cataloging-in-Publication Data
Doucet, Arnaud.
    Sequential Monte Carlo methods in practice / Arnaud Doucet, Nando de Freitas, Neil Gordon.
        p. cm. — (Statistics for engineering and information science)
    Includes bibliographical references and index.
    ISBN 0-387-95146-6 (alk. paper)
        1. Monte Carlo method.   I. de Freitas, Nando.   II. Gordon, Neil (Neil James), 1967–
    III. Title.   IV. Series.
    QA298 .D68 2001
    519.2′82—dc21   00-047093
Printed on acid-free paper.

Production managed by Jenny Wolkowicki; manufacturing supervised by Jeffrey Taub.
Camera-ready copy prepared from the authors' LaTeX files.
Printed and bound by Maple-Vail Book Manufacturing Group, York, PA.
Printed in the United States of America.

9 8 7 6 5 4 3 2 1

ISBN 0-387-95146-6   SPIN 10781886

Springer-Verlag   New York Berlin Heidelberg
A member of BertelsmannSpringer Science+Business Media GmbH

Foreword

It is a great personal pleasure to have the opportunity to contribute the foreword of this important volume.

Problems arising from new data arriving sequentially in time and requiring on-line decision-making responses are ubiquitous in modern communications and control systems, economic and financial data analysis, computer vision and many other fields in which real-time estimation and prediction are required.

As a beginning postgraduate student in Statistics, I remember being excited and charmed by the apparent simplicity and universality of Bayes' theorem as the key logical and computational mechanism for sequential learning. I also recall my growing sense of disappointment as I read Aoki's (1967) volume *Optimisation of Stochastic Systems* and found myself increasingly wondering how any of this wonderful updating machinery could be implemented to solve anything but the very simplest problems. This realisation was deeply frustrating. The elegance of the Kalman Filter provided some solace, but at the cost of pretending to live in a linear Gaussian world. Once into the nonlinear, non-Gaussian domain, we were without a universally effective approach and driven into a series of ingenious approximations, some based on flexible mixtures of tractable distributions to approximate and propagate uncertainty, or on local linearisations of nonlinear systems. In particular, Alspach and Sorenson's (1972) Gaussian sum approximations were developed into a systematic filtering approach in Anderson and Moore (1979); and the volume edited by Gelb (1974) provides an early overview of the use of the extended Kalman Filter.

In fact, the computational obstacles to turning the Bayesian handle in the sequential applications context were just as much present in complex, non-sequential applications and considerable efforts were expended in the 1970s and 1980s to provide workable computational strategies in general Bayesian Statistics.

Towards the end of the 1980s, the realisation emerged that the only universal salvation to hand was that of simulation. It involved combining increasingly sophisticated mathematical insight into, and control over, Monte Carlo techniques with the extraordinary increases we have witnessed in computer power. The end result was a powerful new toolkit for Bayesian computation.

Not surprisingly, those primarily concerned with sequential learning moved in swiftly to adapt and refine this simulation toolkit to the requirements of on-line estimation and prediction problems.

This volume provides a comprehensive overview of what has been achieved and documents the extraordinary progress that has been made in the past decade. A few days ago, I returned to Aoki's 1967 volume. This time as I turned the pages, there was no sense of disappointment. We now really can compute the things we want to!

Adrian Smith
Queen Mary and Westfield College
University of London
November 2000

Acknowledgments

During the course of editing this book, we were fortunate to be assisted by the contributors and several individuals. To address the challenge of obtaining a high quality publication each chapter was carefully reviewed by many authors and several external reviewers. In particular, we would like to thank David Fleet, David Forsyth, Simon Maskell and Antonietta Mira for their kind help. The advice and assistance of John Kimmel, Margaret Mitchell, Jenny Wolkowicki and Fred Bartlett from Springer-Verlag has eased the editing process and is much appreciated.

Special thanks to our loved ones for their support and patience during the editing process.

Arnaud Doucet
Electrical & Electronic Engineering Department
The University of Melbourne,
Victoria 3010, Australia
doucet@ee.mu.oz.au

Nando de Freitas
Computer Science Division,
387 Soda Hall,
University of California,
Berkeley, CA 94720-1776, USA
jfgf@cs.berkeley.edu

Neil Gordon
Pattern and Information Processing,
Defence Evaluation and Research Agency,
St Andrews Road,
Malvern, Worcs, WR14 3PS, UK.
N.Gordon@signal.dera.gov.uk

Contents

IV   Applications   319

Contributors

Christophe Andrieu   Department of Mathematics,
University of Bristol,
Bristol, BS8 1TW, England
C.Andrieu@bristol.ac.uk

Niclas Bergman   Data Fusion,
SaabTech Systems,
SE-175 88 Jarfalla, Sweden
ncbe@celsiustech.se

Carlo Berzuini   Dipartimento di Informatica e Sistemistica,
Via Ferrata 1, University of Pavia,
27100 Pavia, Italy
carlo@laplace.unipv.it

Andrew Blake   Microsoft Research Center,
1 Guildhall Street
Cambridge, CB2 3NH, England
ablake@microsoft.com

Erik Bølviken   Department of Mathematics,
University of Oslo,
N-0316 Oslo, Norway
erikb@math.uio.no

Wolfram Burgard   Institut für Informatik,
Albert-Ludwigs-Universität Freiburg,
D-79110 Freiburg, Germany
burgard@informatik.uni-freiburg.de

Rong Chen   Department of Statistics,
Texas A&M University,
College Station TX 77845, USA
chen@stat.tamu.edu

Tim Clapp   Astrium UK,
                      Gunnels Wood Road,
                      Stevenage, SG1 2AS, England

Dan Crisan   Department of Mathematics,
                      Huxley Building,
                      Imperial College, 180 Queens Gate,
                      London, SW7 2BZ, England
                      dcrisan@ic.ac.uk

Nando de Freitas   Computer Science Division
                      University of California, Berkeley
                      Berkeley, CA 94720-1776, USA
                      jfgf@cs.berkeley.edu

Pierre Del Moral   LSP, Bât. 1R1,
                      Université Paul Sabatier,
                      118 route de Narbonne,
                      31062 Toulouse, France
                      delmoral@cict.fr

Frank Dellaert   School of Computer Science,
                      Carnegie Mellon University,
                      Pittsburgh, PA 15213-3891, USA
                      dellaert@cs.cmu.edu

Petar Djurić   Department of Electrical and
                      Computer Engineering,
                      State University of New York,
                      Stony Brook, NY 11794, USA
                      djuric@ece.sunysb.edu

Arnaud Doucet   Department of Electrical & Electronic Engineering,
                      The University of Melbourne,
                      Victoria 3010, Australia
                      doucet@ee.mu.oz.au

Dieter Fox   Department of Computer Science and Engineering,
                      University of Washington,
                      Seattle, WA 98195-2350, USA
                      fox@cs.washington.edu

Andrew Gee   Speech, Vision and Robotics Group,
                      Department of Engineering, Trumpington Street,
                      Cambridge University,
                      Cambridge, CB2 1PZ, England
                      ahg@eng.cam.ac.uk

Walter Gilks   Medical Research Council,
                          Biostatistics Unit,
                          Robinson Way, Cambridge, CB2 2SR, England
                          wally.gilks@mrc-bsu.cam.ac.uk

Simon Godsill   Signal Processing Group,
                          Department of Engineering, Trumpington Street,
                          Cambridge University,
                          Cambridge, CB2 1PZ, England
                          sjg@eng.cam.ac.uk

Neil Gordon   Pattern and Information Processing,
                          Defence Evaluation and Research Agency,
                          St. Andrews Road, Malvern, WR14 3PS, England
                          N.Gordon@signal.dera.gov.uk

Ulf Grenander   Division of Applied Mathematics,
                          Brown University,
                          Providence, RI 02912, USA
                          ulf@brownvm.brown.edu

Tomoyuki Higuchi   The Institute of Statistical Mathematics,
                          4-6-7 Minami-Azabu, Minato-ku,
                          Tokyo 106-8569, Japan
                          higuchi@ism.ac.jp

Pedro Højen-Sørensen Section for Digital Signal Processing,
                          Department of Mathematical Modelling,
                          Technical University of Denmark,
                          DK-2800 Lyngby, Denmark
                          phs@eivind.imm.dtu.dk

Markus Hürzeler   UBS AG,
                          Nüschelerstrasse 22,
                          CH-8098, Zurich, Switzerland
                          markus.huerzeler@ubs.com

George Irwin   School of Electrical and Electronic Engineering,
                          The Queen's University of Belfast,
                          Belfast, BT9 5AH, Northern Ireland
                          g.irwin@ee.qub.ac.uk

Michael Isard   Computer Systems Research Center,
                          Palo Alto, CA 94301, USA
                          misard@pa.dec.com

Jean Jacod

Laboratoire de Probabilités,
Université de Paris 6, 4 place Jussieu,
75252 Paris, France
jj@cer.jussieu.fr

Genshiro Kitagawa

The Institute of Statistical Mathematics,
4-6-7 Minami-Azabu, Minato-ku,
Tokyo 106-8569, Japan
kitagawa@ism.ac.jp

Daphne Koller

Computer Science Department
Stanford University
Stanford, CA 94305-9010, USA
koller@cs.stanford.edu

Hans R. Künsch

Seminar für Statistik,
ETH Zentrum,
CH-8092 Zürich, Switzerland
kuensch@stat.math.ethz.ch

Aaron Lanterman

Coordinated Science Laboratory,
University of Illinois,
Urbana, IL 61801, USA
lanterma@ifp.uiuc.edu

François Le Gland

IRISA-INRIA, Campus de Beaulieu,
Campus de Beaulieu
35042 Rennes, France
legland@irisa.fr

Uri Lerner

Computer Science Department,
Stanford University,
Stanford, CA 94305-9010, USA
uri@cs.stanford.edu

Jane Liu

CDC Investment Management Corporation,
1251 Avenue of the Americas, 16th floor,
New York, NY 10020, USA

Jun Liu

Department of Statistics,
Harvard University,
1 Oxford Street,
Cambridge, MA 02138, USA
jliu@stat.harvard.edu

Tanya Logvinenko   Department of Statistics,
Stanford University,
Stanford, CA 94305, USA
tanyalog@stat.stanford.edu

Marc Loizeaux   Department of Statistics,
Florida State University,
Tallahassee, FL 32306-4330, USA

John MacCormick   Systems Research Center,
Compaq Computer Corporation,
Palo Alto, CA 94301, USA
jmac@pa.dec.com

Alan Marrs   Pattern and Information Processing,
Defence Evaluation and Research Agency,
St. Andrews Road, Malvern, WR14 3PS, England
A.Marrs@signal.dera.gov.uk

Shaun McGinnity   Openwave,
Charles House,
103-111 Donegall Street,
Belfast, BT1 2FJ, Northern Ireland
shaun.mcginnity@openwave.com

Michael Miller   Center for Imaging Science,
Johns Hopkins University,
Baltimore, MD 21218, USA
mim@cis.jhu.edu

Kevin Murphy   Computer Science Division,
University of California, Berkeley
Berkeley, CA 94720-1776, USA
murphyk@cs.berkeley.edu

Christian Musso   ONERA-DTIM-MCT,
92322 Chatillon, France
christian.musso@onera.fr

Mahesan Niranjan   Department of Computer Science,
University of Sheffield,
Sheffield, S1 4DP, England
M.Niranjan@dcs.shef.ac.uk

Nadia Oudjane   Universite de Paris 13
Department of Mathematics,
Avenue Jean-Baptiste Clement,
93430 Villetaneuse, France

Michael Pitt   Department of Economics,
University of Warwick,
Coventry, CV4 7AL, UK
M.K.Pitt@warwick.ac.uk

Elena Punskaya   Signal Processing Group,
Department of Engineering, Trumpington Street,
Cambridge University,
Cambridge, CB2 1PZ, England
op205@eng.cam.ac.uk

Stuart Russell   Computer Science Division
University of California, Berkeley
Berkeley, CA 94720-1776, USA
russell@cs.berkeley.edu

David Salmond   Guidance and Associated Processing,
Defence Evaluation and Research Agency,
Ively Road, Farnborough, GU14 0LX, England
djsalmond@dera.gov.uk

Seisho Sato   The Institute of Statistical Mathematics,
4-6-7 Minami-Azabu, Minato-ku,
Tokyo 106-8569, Japan
sato@ism.ac.jp

Neil Shephard   Nuffield College,
Oxford University,
Oxford, OX1 1NF, England
neil.shephard@nuffield.ox.ac.uk

Adrian Smith   Queen Mary and Westfield College,
Mile End Road, London, E1 4NS, England
principal@qmw.ac.uk

Anuj Srivastava   Department of Statistics,
Florida State University,
Tallahassee, FL 32306-4330, USA
anuj@stat.fsu.edu

Photis Stavropoulos   Department of Statistics,
University of Glasgow,
Glasgow G12 8QQ, Scotland

Geir Storvik   Department of Mathematics,
University of Oslo,
N-0316 Oslo, Norway
geirs@math.uio.no

Sebastian Thrun   School of Computer Science,
Carnegie Mellon University,
Pittsburgh, PA 15213-3891, USA
thrun@cs.cmu.edu

Mike Titterington   Department of Statistics,
University of Glasgow,
Glasgow G12 8QQ, Scotland
m.titerington@stats.gla.ac.uk

Mike West   Institute of Statistics & Decision Sciences,
Duke University,
Durham, NC 27708-0251, USA
mw@stat.duke.edu

Part I

Introduction

1

An Introduction to Sequential Monte Carlo Methods

Arnaud Doucet
Nando de Freitas
Neil Gordon

1.1 Motivation

Many real-world data analysis tasks involve estimating unknown quantities from some given observations. In most of these applications, prior knowledge about the phenomenon being modelled is available. This knowledge allows us to formulate Bayesian models, that is prior distributions for the unknown quantities and likelihood functions relating these quantities to the observations. Within this setting, all inference on the unknown quantities is based on the posterior distribution obtained from Bayes' theorem. Often, the observations arrive sequentially in time and one is interested in performing inference on-line. It is therefore necessary to update the posterior distribution as data become available. Examples include tracking an aircraft using radar measurements, estimating a digital communications signal using noisy measurements, or estimating the volatility of financial instruments using stock market data. Computational simplicity in the form of not having to store all the data might also be an additional motivating factor for sequential methods.

If the data are modelled by a linear Gaussian state-space model, it is possible to derive an exact analytical expression to compute the evolving sequence of posterior distributions. This recursion is the well known and widespread *Kalman filter*. If the data are modelled as a partially observed, finite state-space Markov chain, it is also possible to obtain an analytical solution, which is known as the hidden Markov model *HMM filter*. These two filters are the most ubiquitous and famous ones, yet there are a few other dynamic systems that admit finite dimensional filters (Vidoni 1999, West and Harrison 1997).

The aforementioned filters rely on various assumptions to ensure mathematical tractability. However, real data can be very complex, typically involving elements of non-Gaussianity, high dimensionality and nonlinearity, which conditions usually preclude analytic solution. This is a problem of fundamental importance that permeates most disciplines of science. According to the field of interest, the problem appears under many different names, including Bayesian filtering, optimal (nonlinear) filtering, stochastic filtering and on-line inference and learning. For over thirty years, many approximation schemes, such as the extended Kalman filter, Gaussian sum approximations and grid-based filters, have been proposed to surmount this problem. The first two methods fail to take into account all the salient statistical features of the processes under consideration, leading quite often to poor results. Grid-based filters, based on deterministic numerical integration methods, can lead to accurate results, but are difficult to implement and too computationally expensive to be of any practical use in high dimensions.

Sequential Monte Carlo (SMC) methods are a set of simulation-based methods which provide a convenient and attractive approach to computing the posterior distributions. Unlike grid-based methods, SMC methods are very flexible, easy to implement, parallelisable and applicable in very general settings. The advent of cheap and formidable computational power, in conjunction with some recent developments in applied statistics, engineering and probability, have stimulated many advancements in this field.

Over the last few years, there has been a proliferation of scientific papers on SMC methods and their applications. Several closely related algorithms, under the names of *bootstrap filters, condensation, particle filters, Monte Carlo filters, interacting particle approximations* and *survival of the fittest*, have appeared in several research fields. This book aims to bring together the main exponents of these algorithms with the goal of introducing the methods to a wider audience, presenting the latest algorithmic and theoretical developments and demonstrating their use in a wide range of complex practical applications. For lack of space, it has unfortunately not been possible to include all the leading researchers in the field, nor to address the theoretical and practical issues with the depth they deserve.

The chapters in the book are grouped in three parts. In the first part, a detailed theoretical treatment of various SMC algorithms is presented. The second part is mainly concerned with outlining various methods for improving the efficiency of the basic SMC algorithm. Finally, the third part discusses several applications in the areas of financial modelling and econometrics, target tracking and missile guidance, terrain navigation, computer vision, neural networks, time series analysis and forecasting, machine learning, robotics, industrial process control and population biology. In each of these parts, the chapters are arranged alphabetically by author.

The chapters are to a large extent self-contained and can be read independently. Yet, for completeness, we have added this introductory chapter to allow readers unfamiliar with the topic to understand the fundamentals and to be able to implement the basic algorithm. Here, we describe a general probabilistic model and the Bayesian inference objectives. After outlining the problems associated with the computation of the posterior distributions, we briefly mention standard approximation methods and point out some of their shortcomings. Subsequently, we introduce Monte Carlo methods, placing particular emphasis on describing the simplest – but still very useful – SMC method. This should enable the reader to start applying the basic algorithm in various contexts.

1.2   Problem statement

For sake of simplicity, we restrict ourselves here to signals modelled as Markovian, nonlinear, non-Gaussian state-space models, though SMC can be applied in a more general setting[1]. The unobserved signal (hidden states) $\{\mathbf{x}_t; t \in \mathbb{N}\}$, $\mathbf{x}_t \in \mathcal{X}$, is modelled as a *Markov process* of initial distribution $p(\mathbf{x}_0)$ and transition equation $p(\mathbf{x}_t|\mathbf{x}_{t-1})$. The observations $\{\mathbf{y}_t; t \in \mathbb{N}^*\}$, $\mathbf{y}_t \in \mathcal{Y}$, are assumed to be conditionally independent given the process $\{\mathbf{x}_t; t \in \mathbb{N}\}$ and of marginal distribution $p(\mathbf{y}_t|\mathbf{x}_t)$. To sum up, the model is described by

$$p(\mathbf{x}_0)$$
$$p(\mathbf{x}_t|\mathbf{x}_{t-1}) \quad \text{for } t \geq 1$$
$$p(\mathbf{y}_t|\mathbf{x}_t) \quad \text{for } t \geq 1.$$

We denote by $\mathbf{x}_{0:t} \triangleq \{\mathbf{x}_0, ..., \mathbf{x}_t\}$ and $\mathbf{y}_{1:t} \triangleq \{\mathbf{y}_1, ..., \mathbf{y}_t\}$, respectively, the signal and the observations up to time t.

Our aim is to estimate recursively in time the *posterior distribution* $p(\mathbf{x}_{0:t}|\mathbf{y}_{1:t})$, its associated features (including the marginal distribution $p(\mathbf{x}_t|\mathbf{y}_{1:t})$, known as the *filtering distribution*), and the expectations

$$I(f_t) = \mathbb{E}_{p(\mathbf{x}_{0:t}|\mathbf{y}_{1:t})}[f_t(\mathbf{x}_{0:t})] \triangleq \int f_t(\mathbf{x}_{0:t}) p(\mathbf{x}_{0:t}|\mathbf{y}_{1:t}) d\mathbf{x}_{0:t}$$

for some function of interest $f_t : \mathcal{X}^{(t+1)} \rightarrow \mathbb{R}^{n_{f_t}}$ integrable with respect to $p(\mathbf{x}_{0:t}|\mathbf{y}_{1:t})$. Examples of appropriate functions include the conditional mean, in which case $f_t(\mathbf{x}_{0:t}) = \mathbf{x}_{0:t}$, or the conditional covariance of $\mathbf{x}_t$ where $f_t(\mathbf{x}_{0:t}) = \mathbf{x}_t\mathbf{x}_t^{\mathrm{T}} - \mathbb{E}_{p(\mathbf{x}_t|\mathbf{y}_{1:t})}[\mathbf{x}_t]\mathbb{E}_{p(\mathbf{x}_t|\mathbf{y}_{1:t})}^{\mathrm{T}}[\mathbf{x}_t]$.

---

[1]For simplicity, we use $\mathbf{x}_t$ to denote both the random variable and its realisation. Consequently, we express continuous probability distributions using $p(d\mathbf{x}_t)$ instead of $\Pr(\mathbf{X}_t \in d\mathbf{x}_t)$ and discrete distributions using $p(\mathbf{x}_t)$ instead of $\Pr(\mathbf{X}_t = \mathbf{x}_t)$. If these distributions admit densities with respect to an underlying measure μ (usually counting or Lebesgue), we denote these densities by $p(\mathbf{x}_t)$. To make the material accessible to a wider audience, we shall allow for a slight abuse of terminology by sometimes referring to $p(\mathbf{x}_t)$ as a distribution.

At any time t, the posterior distribution is given by *Bayes' theorem*

$$p\left(\mathbf{x}_{0:t}|\mathbf{y}_{1:t}\right) = \frac{p\left(\mathbf{y}_{1:t}|\mathbf{x}_{0:t}\right)p\left(\mathbf{x}_{0:t}\right)}{\int p\left(\mathbf{y}_{1:t}|\mathbf{x}_{0:t}\right)p\left(\mathbf{x}_{0:t}\right)d\mathbf{x}_{0:t}}.$$

It is possible to obtain straightforwardly a recursive formula for this joint distribution $p\left(\mathbf{x}_{0:t}|\mathbf{y}_{1:t}\right)$,

$$p\left(\mathbf{x}_{0:t+1}|\mathbf{y}_{1:t+1}\right) = p\left(\mathbf{x}_{0:t}|\mathbf{y}_{1:t}\right)\frac{p\left(\mathbf{y}_{t+1}|\mathbf{x}_{t+1}\right)p\left(\mathbf{x}_{t+1}|\mathbf{x}_{t}\right)}{p\left(\mathbf{y}_{t+1}|\mathbf{y}_{1:t}\right)}. \qquad (1.1)$$

The marginal distribution $p\left(\mathbf{x}_{t}|\mathbf{y}_{1:t}\right)$ also satisfies the following recursion.

$$\textit{Prediction: } p\left(\mathbf{x}_{t}|\mathbf{y}_{1:t-1}\right) = \int p\left(\mathbf{x}_{t}|\mathbf{x}_{t-1}\right)p\left(\mathbf{x}_{t-1}|\mathbf{y}_{1:t-1}\right)d\mathbf{x}_{t-1}; \qquad (1.2)$$

$$\textit{Updating: } p\left(\mathbf{x}_{t}|\mathbf{y}_{1:t}\right) = \frac{p\left(\mathbf{y}_{t}|\mathbf{x}_{t}\right)p\left(\mathbf{x}_{t}|\mathbf{y}_{1:t-1}\right)}{\int p\left(\mathbf{y}_{t}|\mathbf{x}_{t}\right)p\left(\mathbf{x}_{t}|\mathbf{y}_{1:t-1}\right)d\mathbf{x}_{t}}. \qquad (1.3)$$

These expressions and recursions are deceptively simple because one cannot typically compute the normalising constant $p\left(\mathbf{y}_{1:t}\right)$, the marginals of the posterior $p\left(\mathbf{x}_{0:t}|\mathbf{y}_{1:t}\right)$, in particular $p\left(\mathbf{x}_{t}|\mathbf{y}_{t}\right)$, and $I\left(f_{t}\right)$ since they require the evaluation of complex high-dimensional integrals.

This is why, from the mid-1960s, a great many papers and books have been devoted to obtaining approximations for these distributions, including, as discussed in the previous section, the extended Kalman filter (Anderson and Moore 1979, Jazwinski 1970), the Gaussian sum filter (Sorenson and Alspach 1971) and grid-based methods (Bucy and Senne 1971). Other interesting work in automatic control was done during the 60s and 70s based on SMC integration methods (see (Handschin and Mayne 1969)). Most likely because of the modest computers available at the time, these last algorithms were overlooked and forgotten. In the late 1980s, the great increase of computational power made possible rapid advances in numerical integration methods for Bayesian filtering (Kitagawa 1987).

1.3   Monte Carlo methods

To address the problems described in the previous section, many scientific and engineering disciplines have recently devoted a considerable effort towards the study and development of *Monte Carlo* (MC) integration methods. These methods have the great advantage of not being subject to any linearity or Gaussianity constraints on the model, and they also have appealing convergence properties.

We start this section by showing that, when one has a large number of samples drawn from the required posterior distributions, it is not difficult to approximate the intractable integrals appearing in equations (1.1)-(1.3). It is, however, seldom possible to obtain samples from these distributions directly. One therefore has to resort to alternative MC methods, such as importance sampling. By making this general MC technique recursive, one obtains the *sequential importance sampling* (SIS) method. Unfortunately, it can easily be shown that SIS is guaranteed to fail as t increases. This problem can be surmounted by including an additional selection step. The introduction of this key step in (Gordon, Salmond and Smith 1993) led to the first operationally effective method. Since then, theoretical convergence results for this algorithm have been established. See, for example, (Del Moral 1996) and the chapters in this book by Crisan and Del Moral and Jacod.

1.3.1 Perfect Monte Carlo sampling

Let us assume that we are able to simulate N independent and identically distributed (i.i.d.) random samples, also named particles, $\left\{\mathbf{x}_{0:t}^{(i)}; i = 1, ..., N\right\}$ according to $p\left(\mathbf{x}_{0:t}|\mathbf{y}_{1:t}\right)$. An empirical estimate of this distribution is given by

$$P_N\left(d\mathbf{x}_{0:t}|\mathbf{y}_{0:t}\right) = \frac{1}{N}\sum_{i=1}^{N}\delta_{\mathbf{x}_{0:t}^{(i)}}\left(d\mathbf{x}_{0:t}\right),$$

where $\delta_{\mathbf{x}_{0:t}^{(i)}}\left(d\mathbf{x}_{0:t}\right)$ denotes the delta-Dirac mass located in $\mathbf{x}_{0:t}^{(i)}$. One obtains straightforwardly the following estimate of $I\left(f_t\right)$

$$I_N\left(f_t\right) = \int f_t\left(\mathbf{x}_{0:t}\right)P_N\left(d\mathbf{x}_{0:t}|\mathbf{y}_{1:t}\right) = \frac{1}{N}\sum_{i=1}^{N}f_t\left(\mathbf{x}_{0:t}^{(i)}\right).$$

This estimate is unbiased and, if the posterior variance of $f_t\left(\mathbf{x}_{0:t}\right)$ satisfies $\sigma_{f_t}^2 \triangleq \mathbb{E}_{p\left(\mathbf{x}_{0:t}|\mathbf{y}_{1:t}\right)}\left[f_t^2\left(\mathbf{x}_{0:t}\right)\right] - I^2\left(f_t\right) < +\infty$, then the variance of $I_N\left(f_t\right)$ is equal to $var\left(I_N\left(f_t\right)\right) = \frac{\sigma_{f_t}^2}{N}$. Clearly, from the strong law of large numbers,

$$I_N\left(f_t\right) \xrightarrow[N\to+\infty]{a.s.} I\left(f_t\right),$$

where $\xrightarrow{a.s.}$ denotes almost sure convergence. Moreover, if $\sigma_{f_t}^2 < +\infty$, then a central limit theorem holds

$$\sqrt{N}[I_N\left(f_t\right) - I\left(f_t\right)] \xRightarrow[N\to+\infty]{} \mathcal{N}\left(0, \sigma_{f_t}^2\right),$$

where $\Rightarrow$ denotes convergence in distribution. The advantage of this perfect MC method is clear. From the set of random samples $\left\{\mathbf{x}_{0:t}^{(i)}; i = 1, ..., N\right\}$, one can easily estimate any quantity $I\left(f_t\right)$ and the rate of convergence of this estimate is *independent of the dimension of the integrand*. In contrast, any deterministic numerical integration method has a rate of convergence that decreases as the dimension of the integrand increases.

Unfortunately, it is usually impossible to sample efficiently from the posterior distribution $p(\mathbf{x}_{0:t}|\mathbf{y}_{1:t})$ at any time t, $p(\mathbf{x}_{0:t}|\mathbf{y}_{1:t})$ being multivariate, non-standard, and only known up to a proportionality constant. In applied statistics, Markov chain Monte Carlo (MCMC) methods are a popular approach to sampling from such complex probability distributions (Gilks, Richardson and Spiegelhalter 1996, Robert and Casella 1999). However, MCMC methods are iterative algorithms unsuited to recursive estimation problems. So, alternative methods have to be developed.

1.3.2   Importance sampling

Importance sampling

An alternative classical solution consists of using the *importance sampling* method, see for example (Geweke 1989). Let us introduce an arbitrary so-called *importance sampling distribution* (also often referred to as the proposal distribution or the importance function) $\pi(\mathbf{x}_{0:t}|\mathbf{y}_{1:t})^2$. Assuming that we want to evaluate $I(f_t)$, and provided that the support of $\pi(\mathbf{x}_{0:t}|\mathbf{y}_{1:t})$ includes the support of $p(\mathbf{x}_{0:t}|\mathbf{y}_{1:t})$, we get the identity

$$I(f_t) = \frac{\int f_t(\mathbf{x}_{0:t}) w(\mathbf{x}_{0:t}) \pi(\mathbf{x}_{0:t}|\mathbf{y}_{1:t}) d\mathbf{x}_{0:t}}{\int w(\mathbf{x}_{0:t}) \pi(\mathbf{x}_{0:t}|\mathbf{y}_{1:t}) d\mathbf{x}_{0:t}},$$

where $w(\mathbf{x}_{0:t})$ is known as the *importance weight* ,

$$w(\mathbf{x}_{0:t}) = \frac{p(\mathbf{x}_{0:t}|\mathbf{y}_{1:t})}{\pi(\mathbf{x}_{0:t}|\mathbf{y}_{1:t})}.$$

Consequently, if one can simulate N i.i.d. particles $\{\mathbf{x}_{0:t}^{(i)}, i=1,...,N\}$ according to $\pi(\mathbf{x}_{0:t}|\mathbf{y}_{1:t})$, a possible Monte Carlo estimate of $I(f_t)$ is

$$\widehat{I}_N(f_t) = \frac{\frac{1}{N}\sum_{i=1}^N f_t(\mathbf{x}_{0:t}^{(i)}) w(\mathbf{x}_{0:t}^{(i)})}{\frac{1}{N}\sum_{j=1}^N w(\mathbf{x}_{0:t}^{(i)})} = \sum_{i=1}^N f_t(\mathbf{x}_{0:t}^{(i)}) \widetilde{w}_t^{(i)},$$

where the *normalised importance weights* $\widetilde{w}_t^{(i)}$ are given by

$$\widetilde{w}_t^{(i)} = \frac{w(\mathbf{x}_{0:t}^{(i)})}{\sum_{j=1}^N w(\mathbf{x}_{0:t}^{(j)})}. \tag{1.4}$$

For N finite, $\widehat{I}_N(f_t)$ is biased (ratio of two estimates) but asymptotically, under weak assumptions, the strong law of large numbers applies, that is, $\widehat{I}_N(f_t) \xrightarrow[N\to+\infty]{a.s.} I(f_t)$. Under additional assumptions, a central limit theorem with a convergence rate still independent of the dimension of the integrand

---

^{2}We underline the (possible) dependence of $\pi(\cdot)$ on $\mathbf{y}_{1:t}$ by writing $\pi(\mathbf{x}_{0:t}|\mathbf{y}_{1:t})$.

can be obtained (Geweke 1989). It is clear that this integration method can also be interpreted as a sampling method where the posterior distribution $p(\mathbf{x}_{0:t}|\mathbf{y}_{1:t})$ is approximated by

$$\widehat{P}_N(d\mathbf{x}_{0:t}|\mathbf{y}_{1:t}) = \sum_{i=1}^{N} \widetilde{w}_t^{(i)} \delta_{\mathbf{x}_{0:t}^{(i)}}(d\mathbf{x}_{0:t}), \qquad (1.5)$$

and $\widehat{I}_N(f_t)$ is nothing but the function $f_t(\mathbf{x}_{0:t})$ integrated with respect to the empirical measure $\widehat{P}_N(d\mathbf{x}_{0:t}|\mathbf{y}_{1:t})$:

$$\widehat{I}_N(f_t) = \int f_t(\mathbf{x}_{0:t}) \widehat{P}_N(d\mathbf{x}_{0:t}|\mathbf{y}_{1:t}).$$

Importance sampling is a general Monte Carlo integration method. However, in its simplest form, it is not adequate for *recursive estimation*. That is, one needs to get all the data $\mathbf{y}_{1:t}$ before estimating $p(\mathbf{x}_{0:t}|\mathbf{y}_{1:t})$. In general, each time new data $\mathbf{y}_{t+1}$ become available, one needs to recompute the importance weights over the entire state sequence. The computational complexity of this operation increases with time. In the following section, we present a strategy for overcoming this problem.

Sequential Importance Sampling

The importance sampling method can be modified so that it becomes possible to compute an estimate $\widehat{P}_N(d\mathbf{x}_{0:t}|\mathbf{y}_{1:t})$ of $p(\mathbf{x}_{0:t}|\mathbf{y}_{1:t})$ without modifying the past simulated trajectories $\left\{\mathbf{x}_{0:t-1}^{(i)}; i = 1, ..., N\right\}$. This means that the importance function $\pi(\mathbf{x}_{0:t}|\mathbf{y}_{1:t})$ at time t admits as marginal distribution at time $t-1$ the importance function $\pi(\mathbf{x}_{0:t-1}|\mathbf{y}_{1:t-1})$, that is

$$\pi(\mathbf{x}_{0:t}|\mathbf{y}_{1:t}) = \pi(\mathbf{x}_{0:t-1}|\mathbf{y}_{1:t-1}) \pi(\mathbf{x}_t|\mathbf{x}_{0:t-1},\mathbf{y}_{1:t}).$$

Iterating, one obtains

$$\pi(\mathbf{x}_{0:t}|\mathbf{y}_{1:t}) = \pi(\mathbf{x}_0) \prod_{k=1}^{t} \pi(\mathbf{x}_k|\mathbf{x}_{0:k-1},\mathbf{y}_{1:k}).$$

It is easy to see that this importance function allows us to evaluate recursively in time the importance weights (1.4). Indeed, one has

$$\widetilde{w}_t^{(i)} \propto \widetilde{w}_{t-1}^{(i)} \frac{p\left(\mathbf{y}_t|\mathbf{x}_t^{(i)}\right) p\left(\mathbf{x}_t^{(i)}\middle|\mathbf{x}_{t-1}^{(i)}\right)}{\pi\left(\mathbf{x}_t^{(i)}\middle|\mathbf{x}_{0:t-1}^{(i)},\mathbf{y}_{1:t}\right)}. \qquad (1.6)$$

An important particular case of this framework arises when we adopt the prior distribution as importance distribution

$$\pi(\mathbf{x}_{0:t}|\mathbf{y}_{1:t}) = p(\mathbf{x}_{0:t}) = p(\mathbf{x}_0) \prod_{k=1}^{t} p(\mathbf{x}_k|\mathbf{x}_{k-1}).$$

In this case, the importance weights satisfy $\widetilde{w}_t^{(i)} \propto \widetilde{w}_{t-1}^{(i)} p\left(\mathbf{y}_t | \mathbf{x}_t^{(i)}\right)$. In the following section, we restrict ourselves to the use of the prior distribution as importance sampling distribution. However, it is important to keep in mind that the method is far more general than this.

SIS is an attractive method, but it is nothing but a constrained version of importance sampling. Unfortunately, it is well known that importance sampling is usually inefficient in high-dimensional spaces (Gilks et al. 1996, Robert and Casella 1999). So, as t increases, this problem will arise in the SIS setting.

1.3.3   The Bootstrap filter

The problem encountered by the SIS method is that, as t increases, the distribution of the importance weights $\widetilde{w}_t^{(i)}$ becomes more and more skewed. Practically, after a few time steps, only one particle has a non-zero importance weight. The algorithm, consequently, fails to represent the posterior distributions of interest adequately. To avoid this degeneracy, one needs to introduce an additional selection step.

Notion

The key idea of the bootstrap filter is to eliminate the particles having low importance weights $\widetilde{w}_t^{(i)}$ and to multiply particles having high importance weights (Gordon et al. 1993). More formally, we replace the weighted empirical distribution $\widehat{P}_N\left(d\mathbf{x}_{0:t} | \mathbf{y}_{1:t}\right) = \sum_{i=1}^N \widetilde{w}_t^{(i)} \delta_{\mathbf{x}_{0:t}^{(i)}}\left(d\mathbf{x}_{0:t}\right)$ by the unweighted measure

$$P_N\left(d\mathbf{x}_{0:t} | \mathbf{y}_{1:t}\right) = \frac{1}{N} \sum_{i=1}^N N_t^{(i)} \delta_{\mathbf{x}_{0:t}^{(i)}}\left(d\mathbf{x}_{0:t}\right),$$

where $N_t^{(i)}$ is the number of offspring associated to particle $\mathbf{x}_{0:t}^{(i)}$; it is an integer number such that $\sum_{i=1}^N N_t^{(i)} = N$. If $N_t^{(j)} = 0$, then the particle $\mathbf{x}_{0:t}^{(j)}$ dies. The $N_t^{(i)}$ are chosen such that $P_N\left(d\mathbf{x}_{0:t} | \mathbf{y}_{0:t}\right)$ is close to $\widehat{P}_N\left(d\mathbf{x}_{0:t} | \mathbf{y}_{1:t}\right)$ in the sense that, for any function f_t,

$$\int f_t\left(\mathbf{x}_{0:t}\right) P_N\left(d\mathbf{x}_{0:t} | \mathbf{y}_{1:t}\right) \approx \int f_t\left(\mathbf{x}_{0:t}\right) \widehat{P}_N\left(d\mathbf{x}_{0:t} | \mathbf{y}_{1:t}\right). \tag{1.7}$$

After the selection step, the surviving particles $\mathbf{x}_{0:t}^{(i)}$, that is the ones with $N_t^{(i)} > 0$, are thus approximately distributed according to $p\left(\mathbf{x}_{0:t} | \mathbf{y}_{1:t}\right)$. There are many different ways to select the $N_t^{(i)}$, the most popular being the one introduced in (Gordon et al. 1993). Here, one obtains surviving particles by sampling N times from the (discrete) distribution $\widehat{P}_N\left(d\mathbf{x}_{0:t} | \mathbf{y}_{1:t}\right)$; this is equivalent to sampling the number of offspring $N_t^{(i)}$ according to a multinomial distribution of parameters $\widetilde{w}_t^{(i)}$. Equation (1.7) is satisfied in the sense that one can check easily that, for any bounded function f_t with $\|f_t\| = \sup_{\mathbf{x}_{0:t}} |f_t(\mathbf{x}_{0:t})|$, there exists C such that

$$\mathbb{E}\left[\left(\int f_t(\mathbf{x}_{0:t}) P_N(d\mathbf{x}_{0:t}|\mathbf{y}_{1:t}) - \int f_t(\mathbf{x}_{0:t}) \widehat{P}_N(d\mathbf{x}_{0:t}|\mathbf{y}_{1:t})\right)^2\right] \leq \frac{C\|f_t\|^2}{N}.$$

Algorithm description

We can now specify the algorithm in detail as follows.

Bootstrap Filter

1. *Initialisation*, $t = 0$.

   - For $i = 1, ..., N$, sample $\mathbf{x}_0^{(i)} \sim p(\mathbf{x}_0)$ and set $t = 1$.

2. *Importance sampling step*

   - For $i = 1, ..., N$, sample $\widetilde{\mathbf{x}}_t^{(i)} \sim p\left(\mathbf{x}_t | \mathbf{x}_{t-1}^{(i)}\right)$ and set $\widetilde{\mathbf{x}}_{0:t}^{(i)} = \left(\mathbf{x}_{0:t-1}^{(i)}, \widetilde{\mathbf{x}}_t^{(i)}\right)$.

   - For $i = 1, ..., N$, evaluate the importance weights

     $$\widetilde{w}_t^{(i)} = p\left(\mathbf{y}_t | \widetilde{\mathbf{x}}_t^{(i)}\right). \tag{1.8}$$

   - Normalise the importance weights.

3. *Selection step*

   - Resample with replacement N particles $\left(\mathbf{x}_{0:t}^{(i)}; i = 1, \ldots, N\right)$ from the set $\left(\widetilde{\mathbf{x}}_{0:t}^{(i)}; i = 1, \ldots, N\right)$ according to the importance weights.

   - Set $t \leftarrow t + 1$ and go to step 2.

Note that in equation (1.8), $\widetilde{w}_{t-1}^{(i)}$ does not appear because the propagated particles $\mathbf{x}_{0:t-1}^{(i)}$ have uniform weights after the resampling step at time $t-1$. Also, we do not need to store the paths of the particles from time 0 to time t if we are only interested in estimating $p(\mathbf{x}_t | \mathbf{y}_{1:t})$. A graphic representation of the algorithm is shown in Figure 1.1.

The bootstrap filter has several attractive properties. Firstly, it is very quick and easy to implement. Secondly, it is to a large extent modular. That is, when changing the problem one need only change the expressions for the importance distribution and the importance weights in the code. Thirdly, it can be straightforwardly implemented on a parallel computer. Finally, the resampling step is a black box routine that only requires as inputs the

i=1,...,N=10 particles

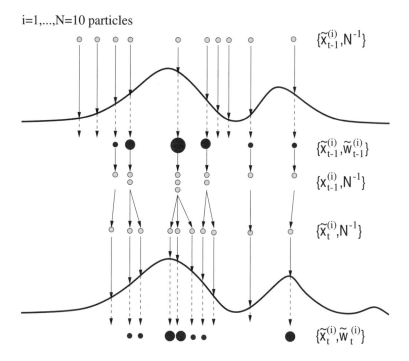

Figure 1.1. In this example, the bootstrap filter starts at time $t - 1$ with an unweighted measure $\{\widetilde{\mathbf{x}}_{t-1}^{(i)}, N^{-1}\}$, which provides an approximation of $p(\mathbf{x}_{t-1}|\mathbf{y}_{1:t-2})$. For each particle we compute the importance weights using the information at time $t - 1$. This results in the weighted measure $\{\widetilde{\mathbf{x}}_{t-1}^{(i)}, \widetilde{w}_{t-1}^{(i)}\}$, which yields an approximation $p(\mathbf{x}_{t-1}|\mathbf{y}_{1:t-1})$. Subsequently, the resampling step selects only the fittest particles to obtain the unweighted measure $\{\widetilde{\mathbf{x}}_{t-1}^{(i)}, N^{-1}\}$, which is still an approximation of $p(\mathbf{x}_{t-1}|\mathbf{y}_{1:t-1})$. Finally, the sampling (prediction) step introduces variety, resulting in the measure $\{\widetilde{\mathbf{x}}_t^{(i)}, N^{-1}\}$, which is an approximation of $p(\mathbf{x}_t|\mathbf{y}_{1:t-1})$.

importance weights and indices (both being one-dimensional quantities). This enables one to easily carry out sequential inference for very complex models.

An illustrative example

For demonstration purposes, we apply the bootstrap algorithm to the following nonlinear, non-Gaussian model (Gordon et al. 1993, Kitagawa 1996):

$$x_t = \frac{1}{2}x_{t-1} + 25\frac{x_{t-1}}{1 + x_{t-1}^2} + 8\cos(1.2t) + v_t$$

$$y_t = \frac{x_t^2}{20} + w_t,$$

where $x_1 \sim \mathcal{N}\left(0, \sigma_1^2\right)$, v_t and w_t are mutually independent white Gaussian noises, $v_k \sim \mathcal{N}\left(0, \sigma_v^2\right)$ and $w_k \sim \mathcal{N}\left(0, \sigma_w^2\right)$ with $\sigma_1^2 = 10$, $\sigma_v^2 = 10$ and $\sigma_w^2 = 1$. The estimated filtering distributions are shown in Figure 1.2.

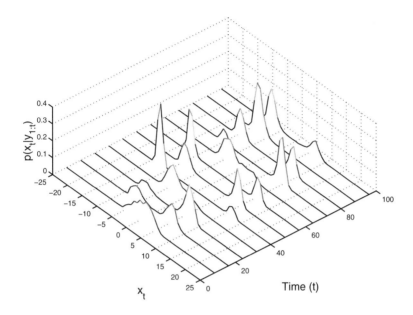

Figure 1.2. Estimated filtering distribution using 1000 particles.

Notice that for this model the minimum mean square estimates could be misleading because they do not provide enough information about the shape of the distribution. Indeed, one of the advantages of Monte Carlo methods is that they provide a complete description of the posterior distribution, not just a single point estimate.

1.4 Discussion

The aim of this chapter was to motivate the use of SMC methods to solve complex nonlinear, non-Gaussian on-line estimation problems. We also provided a brief introduction to SMC methods by describing one of the most basic particle filters. We hope we have convinced the reader of the enormous potential of SMC.

The algorithm we described is applicable to a very large class of models and is straightforward to implement. The price to pay for this simplicity is computationally inefficiency in some application domains.

The following chapters aim to develop and improve upon many of the ideas sketched here, and to propose new ones. Firstly, Chapters 2 and 3 provide a rigorous theoretical basis for SMC methods. Secondly, Chapters 4 to 14 describe numerous algorithmic developments which allow significant performance improvements over standard methods. Finally, Chapters 15 to 26 demonstrate the relevance of SMC to a wide range of complex practical applications.

The extended Kalman filter and related recursive sub-optimal estimation methods have been widely used for over 30 years in numerous applications. It is our belief that SMC methods can not only improve estimation performance in most of these applications, but also allow us to deal with more complex models that were out of reach a few years ago. We hope that this book constitutes a valuable and unified source of information on the topic.

Part II
Theoretical Issues

2

Particle Filters - A Theoretical Perspective

Dan Crisan

2.1 Introduction

The purpose of this chapter is to present a rigorous mathematical treatment of the convergence of particle filters. In general, we follow the notation and settings suggested by the editors, any extra notation being defined in the next section. Section 2.3.1 contains the main results of the paper: Theorems 2.3.1 and 2.3.2 provide *necessary and sufficient* conditions for the convergence of the particle filter to the posterior distribution of the signal. As an application of these results, we prove the convergence of a certain class of particle filters. This class includes several known filters (such as those presented in (Carpenter, Clifford and Fearnhead 1999b, Crisan, Del Moral and Lyons 1999, Gordon et al. 1993), but is by no means the most general one. Finally, we discuss some of the issues that are relevant in applications and which arise from the theoretical analysis of these methods.

We have tried to make the material as self-contained as possible, and therefore we have included in an appendix some useful definitions and results regarding conditional expectations/probabilities. We also include an elementary proof of the recurrence formula for the conditional distribution of the signal.

2.2 Notation and terminology

Let $\mathbb{R}^d$ be the d-dimensional Euclidean space and $\mathcal{B}\left(\mathbb{R}^d\right)$ be the σ-algebra of Borel subsets of $\mathbb{R}^d$. Hereafter we will use the following sets of functions:

$B\left(\mathbb{R}^d\right)$ - the set of bounded $\mathcal{B}\left(\mathbb{R}^d\right)$-measurable functions defined on $\mathbb{R}^d$.
$\mathcal{C}_b(\mathbb{R}^d)$ - the set of bounded continuous functions defined on $\mathbb{R}^d$.
$\mathcal{C}_k(\mathbb{R}^d)$ - the set of compactly supported continuous functions defined on

$\mathbb{R}^d$.

If $f \in \mathcal{C}_b(\mathbb{R}^d)$ we denote by $||f||$ the supremum norm of f

$$||f|| \triangleq \sup_{x \in \mathbb{R}^d} |f(x)|.$$

We also consider the following sets of measures:

$\mathcal{M}_F(\mathbb{R}^d)$ - the set of finite measures over $\mathcal{B}(\mathbb{R}^d)$.
$\mathcal{P}(\mathbb{R}^d)$- the set of probability measures over $\mathcal{B}(\mathbb{R}^d)$.

If $\mu \in \mathcal{M}_F(\mathbb{R}^d)$ (or $\mu \in \mathcal{P}(\mathbb{R}^d)$) and $f \in B(\mathbb{R}^{n_x})$, the integral of f with respect to μ is denoted by μf,

$$\mu f \triangleq \int_{\mathbb{R}^d} f(x) \mu(dx).$$

We endow $\mathcal{M}_F(\mathbb{R}^d)$, respectively, $\mathcal{P}(\mathbb{R}^d)$, with the weak topology. Let $(\mu_n)_{n=1}^\infty$ be a sequence of finite measures, then we say that μ_n converges (weakly) or in the weak topology to $\mu \in \mathcal{M}_F(\mathbb{R}^d)$ and write $\lim_{n \to \infty} \mu_n = \mu$ if

$$\lim_{n \to \infty} \mu_n f = \mu f, \text{ for all } f \in \mathcal{C}_b(\mathbb{R}^d),$$

and we have the same definition if $(\mu_n)_{n=1}^\infty$ is a sequence of probability measures. Also, we denote the Dirac measure concentrated at $a \in \mathbb{R}^d$ by δ_a and the constant function 1 by $\bar{1}$.

2.2.1   Markov chains and transition kernels

Let $(\Omega, \mathcal{F}, P)$ be a probability space and $X = \{X_t, t \in \mathbb{N}\}$ be a stochastic process defined on $(\Omega, \mathcal{F}, P)$ with values in $\mathbb{R}^{n_x}$. Let $\mathcal{F}_t^X$ be the σ-algebra generated by the process, i.e., $\mathcal{F}_t^X \triangleq \sigma(X_s, s \in [0, t])$. Then X is a *Markov chain* if, for all $t \in \mathbb{N}$ and $A \in \mathcal{B}(\mathbb{R}^{n_x})$,

$$P\left(X_{t+1} \in A \,|\, \mathcal{F}_t^X\right) = P\left(X_{t+1} \in A \,|\, X_t\right). \tag{2.1}$$

The transition kernel of the Markov chain X is the function $K_t(\cdot, \cdot)$ defined on $\mathbb{R}^{n_x} \times \mathcal{B}(\mathbb{R}^{n_x})$ such that, for all $t \in \mathbb{N}$ and $x \in \mathbb{R}^{n_x}$,

$$K_t(x, A) = P\left(X_{t+1} \in A \,|\, X_t = x\right). \tag{2.2}$$

The transition kernel K_t has the following properties:

○ $K_t(x, \cdot)$ is a probability measure on $\mathbb{R}^{n_x}$, for all $t \in \mathbb{N}$ and $x \in \mathbb{R}^{n_x}$.
○ $K_t(\cdot, A) \in B(\mathbb{R}^{n_x})$, for all $t \in \mathbb{N}$ and $A \in \mathcal{B}(\mathbb{R}^{n_x})$.

The distribution of X is uniquely determined by its initial distribution and its transition kernel. Let us denote by q_t the distribution of the random variable X_t,

$$q_t(A) \triangleq P(X_t \in A).$$

Then, from (2.2), we deduce that q_t satisfies the recurrence formula $q_{t+1} = q_t K_t$, where $q_t K_t$ is the measure defined as

$$(q_t K_t)(A) \triangleq \int_{\mathbb{R}^{n_x}} K_t(x, A) q_t(dx).$$

Hence, $q_t = q_0 K_0 K_1 ... K_{t-1}$. We say that the transition kernel K_t satisfies the Feller property if, for all $t > 0$, the function $K_t f : \mathbb{R}^{n_x} \to \mathbb{R}$ defined as

$$K_t f(x) \triangleq \int_{\mathbb{R}^{n_x}} f(y) K_t(x, dy)$$

is continuous for every $f \in C_b(\mathbb{R}^d)$. If K_t has the Feller property, then $K_t f \in C_b(\mathbb{R}^d)$ for all $f \in C_b(\mathbb{R}^d)$.

2.2.2   The filtering problem

Let $X = \{X_t, t \in \mathbb{N}\}$ be an $\mathbb{R}^{n_x}$-valued Markov process (called the signal process) with a Feller transition kernel $K_t(x, dy)$. Let also $Y = \{Y_t, t \in \mathbb{N}\}$ be an $\mathbb{R}^{n_y}$-valued stochastic process (called the observation process) defined as

$$Y_t \triangleq h(t, X_t) + W_t, \quad t > 0, \tag{2.3}$$

and $Y_0 = 0$. In (2.3), $h : \mathbb{N} \times \mathbb{R}^{n_x} \to \mathbb{R}^{n_y}$ is a Borel measurable function with the property that $h(t, \cdot)$ is continuous on $\mathbb{R}^{n_x}$ for all $t \in \mathbb{N}$ and, for all $t > 0$, $W_t : \Omega \to \mathbb{R}^{n_y}$ are independent random vectors and their laws are absolutely continuous with respect to the Lebesgue measure λ on $\mathbb{R}^{n_y}$. We denote by $g(t, \cdot)$ the density of W_t with respect to λ, and we further assume that $g(t, \cdot)$ is bounded and continuous. The filtering problem consists of computing the conditional distribution of the signal given the $\sigma-$algebra generated by the observation process from time 0 to the current time; in other words, one is interested in computing the (random) probability measure π_t, where

$$\pi_t(A) \triangleq P(X_t \in A \,|\, \sigma(Y_{0:t})), \quad \pi_t f = \mathbb{E}\left[f(X_t) \,|\, \sigma(Y_{0:t})\right] \tag{2.4}$$

for all $f \in B(\mathbb{R}^{n_x})$ and $A \in \mathcal{B}(\mathbb{R}^{n_x})$, where $Y_{0:t} \triangleq (Y_0, Y_1, \ldots, Y_t)$. Then $\pi_t = \pi_t^{Y_{0:t}}$, where

$$\pi_t^{y_{0:t}}(A) \triangleq P(X_t \in A \,|\, Y_{0:t} = y_{0:t}), \quad \pi_t^{y_{0:t}} f = \mathbb{E}\left[f(X_t) \,|\, Y_{0:t} = y_{0:t}\right] \tag{2.5}$$

and $y_{0:t} \triangleq (y_0, y_1, \ldots, y_t) \in (\mathbb{R}^{n_y})^{t+1}$. Notice that, while π_t is a random probability measure, $\pi_t^{y_{0:t}}$ is a deterministic probability measure. We also introduce p_t and $p_t^{y_{0:t-1}}$, the predicted conditional probability measures for $t > 0$, where $p_t = p_t^{Y_{0:t-1}}$ and

$$p_t^{y_{0:t-1}}(A) = P(X_t \in A \,|\, Y_{0:t-1} = y_{0:t-1}), \quad p_t^{y_{0:t-1}} f = \mathbb{E}\left[f(X_t) \,|\, Y_{0:t-1} = y_{0:t-1}\right].$$

In the appendix we include a short introduction to conditional probabilities/expectations.

We have the following recurrence relations for π_t, respectively, $\pi_t^{\mathbf{y}_{0:t}}$:

$$
\begin{cases}
\dfrac{d\pi_t}{dp_t} = \dfrac{g_t^{\mathbf{Y}_t}}{\int g_t^{\mathbf{Y}_t}(x)p_t(dx)} \\
p_{t+1} = \pi_t K_t
\end{cases}
\quad,\quad
\begin{cases}
\dfrac{d\pi_t^{\mathbf{y}_{0:t}}}{dp_t^{\mathbf{y}_{0:t-1}}} = \dfrac{g_t^{\mathbf{y}_t}}{\int g_t^{\mathbf{y}_t}(x)p_t^{\mathbf{y}_{0:t-1}}(dx)} \\
p_{t+1}^{\mathbf{y}_{0:t}} = \pi_t^{\mathbf{y}_{0:t}} K_t
\end{cases}
\quad,\quad (2.6)
$$

where $g_t^{\mathbf{y}_t} \in C_b(\mathbb{R}^{n_x})$ is defined by $g_t^{\mathbf{y}_t} := g(\mathbf{y}_t - h(t,\cdot))$ and, since $Y_0 = 0$, π_0 is the law of X. An elementary proof of (2.6) is included in the appendix. In the general case there is no closed solution for the system (2.6). In section 2.4.1 we present a generic class of particle filters that can be used to solve (2.6) numerically.

2.2.3 Convergence of measure-valued random variables

Essentially, the result of any algorithm to solve the filtering problem based on a sequential Monte Carlo method is a random measure which approximates π_t. In order to establish whether the algorithm is good or bad, we would need to define in what way a random measure or, more precisely, a sequence of random measures, can approximate another measure.

Let $(\Omega, \mathcal{F}, P)$ be a probability space and let $(\mu^n)_{n=1}^{\infty}$ be a sequence of random measures, $\mu^n : \Omega \to \mathcal{M}_F(\mathbb{R}^d)$ and $\mu \in \mathcal{M}_F(\mathbb{R}^d)$ be a deterministic finite measure. As we shall see, in the case of approximations obtained using particle filters, n represents the number of particles used in the approximating particle system. Below we study two types of convergence:

1. $\lim_{n\to\infty} \mathbb{E}\left[|\mu^n f - \mu f|\right] = 0$ for all $f \in C_b(\mathbb{R}^d)$;
2. $\lim_{n\to\infty} \mu^n = \mu$, P−a.s..

We will denote the first type of convergence by $\mathrm{e}\lim_{n\to\infty} \mu^n = \mu$. If there exists an integrable random variable $w : \Omega \to \mathbb{R}$ such that $\mu^n \mathbf{1} \leq w$ for all n, then $\lim_{n\to\infty} \mu^n = \mu$, P−a.s., implies $\mathrm{e}\lim_{n\to\infty} \mu^n = \mu$ by then Dominated Convergence Theorem. The extra condition is satisfied if $(\mu^n)_{n=1}^{\infty}$ is a sequence of random *probability* measures because, in this case, $\mu^n \mathbf{1} = 1$ for all n.

Remark 1. *If* $\mathrm{e}\lim_{n\to\infty} \mu^n = \mu$, *then sequences* $n(m)$ *exist such that* $\lim_{m\to\infty} \mu^{n(m)} = \mu$, P−*a.s..*

Proof. Since $\mathbb{R}^d$ is a locally compact separable metric space, there exists a countable set $\mathcal{M}$ which is dense in $C_k(\mathbb{R}^d)$. Then $\mathcal{M} \cup \{\bar{\mathbf{1}}\}$ is a convergence determining set, i.e., if ν_n, $n = 1, 2...$, and ν are finite measures and $\lim_{n\to\infty} \nu_n f = \nu f$ for all $f \in \mathcal{M} \cup \{\bar{\mathbf{1}}\}$, then $\lim_{n\to\infty} \nu_n = \nu$. Since $\lim_{n\to\infty} \mathbb{E}\left[|\mu^n f - \mu f|\right] = 0$ for all $f \in \mathcal{M} \cup \{\bar{\mathbf{1}}\}$ and $\mathcal{M}$ is countable, one can find a subsequence $n(m)$ such that, with probability 1, $\lim_{m\to\infty} \mu^{n(m)} f = \mu f$ for all $f \in \mathcal{M} \cup \{\bar{\mathbf{1}}\}$. □

Moreover, if the rate of convergence for $\mathbb{E}\left[|\mu^n f - \mu f|\right]$ is known, then these sequences can be explicitly specified. For instance, if for all $f \in \mathcal{M} \cup \{\bar{\mathbf{1}}\}$, $\mathbb{E}\left[|\mu^n f - \mu f|\right] \leq c_f n^{-\frac{1}{2}}$, for all $n > 0$, then using a Borel-Cantelli argument one can prove that $\lim_{m \to \infty} \mu^{m^4} = \mu$, P−a.s..

If $\mathcal{M} \in \mathcal{C}_k(\mathbb{R}^d)$ is the set defined above, then the following

$$d_{\mathcal{M}}(\mu, \nu) \triangleq |\mu \bar{\mathbf{1}} - \nu \bar{\mathbf{1}}| + \sum_{f_k \in \mathcal{M}} \frac{|\mu f_k - \nu f_k|}{2^k \|f_k\|}, \qquad (2.7)$$

is a distance on $\mathcal{M}_F(\mathbb{R}^d)$ (or $\mathcal{P}(\mathbb{R}^d)$), which generates the weak topology

$$\lim_{n \to \infty} \nu_n = \nu \iff \lim_{n \to \infty} d_{\mathcal{M}}(\nu_n, \nu) = 0.$$

Using $d_{\mathcal{M}}$, the almost sure convergence **2.** is equivalent to

2'. $\lim_{n \to \infty} d_{\mathcal{M}}(\mu^n, \mu) = 0$, P−a.s..

Also, if there exists an integrable random variable $w : \Omega \to \mathbb{R}$ such that $\mu^n \bar{\mathbf{1}} \leq w$ for all n, then, similarly, **1.** implies

1'. $\lim_{n \to \infty} \mathbb{E}\left[d_{\mathcal{M}}(\mu^n, \mu)\right] = 0$.

However, a stronger condition (such as tightness) would be needed in order to ensure that **1.** is equivalent to **1'.**

The same definitions are valid for the case when the limiting measure μ is not deterministic, but random, $\mu : \Omega \to \mathcal{M}_F(\mathbb{R}^d)$. If μ is random we have, just as before, that **2.** is equivalent to **2'.**. Also, **2.** implies **1.** and **1.** implies **1'.**, under the condition that there exists an integrable random variable $w : \Omega \to \mathbb{R}$ such that $\mu^n \bar{\mathbf{1}} \leq w$ and $\mu \bar{\mathbf{1}} \leq w$ for all n.

The limiting measures that we are interested in are $\pi_t^{\mathbf{y}_{0:t}}$ and π_t, hence deterministic and, random *probability* measures.

2.3 Convergence theorems

2.3.1 The fixed observation case

We shall look first at the case in which the observation process has an arbitrary but fixed value $\mathbf{y}_{0:T}$, where T is a finite but large time horizon. We assume that the recurrence formula (2.6) for $\pi_t^{\mathbf{y}_{0:t}}$ – the conditional distribution of the signal given the event $\{Y_{0:t} = \mathbf{y}_{0:t}\}$ – holds true for this particular value for all $0 \leq t \leq T$ (remember that (2.6) is valid $P_{Y_{0:t}}$-almost surely). Then (2.6) requires the use of an intermediate step, the predicted conditional probability measure $p_t^{\mathbf{y}_{0:t-1}}$:

$$\pi_{t-1}^{\mathbf{y}_{0:t-1}} \longrightarrow p_t^{\mathbf{y}_{0:t-1}} \longrightarrow \pi_t^{\mathbf{y}_{0:t}}.$$

Therefore it is natural to study algorithms that provide recursive approximations for $\pi_t^{\mathbf{y}_{0:t}}$ by using intermediate approximations for $p_t^{\mathbf{y}_{0:t-1}}$. We shall denote by $(\pi_t^n)_{n=1}^{\infty}$ and $(p_t^n)_{n=1}^{\infty}$ the approximating sequence for $\pi_t^{\mathbf{y}_{0:t}}$ and $p_t^{\mathbf{y}_{0:t-1}}$ and assume that π_t^n and p_t^n are random measures, not necessarily probabilities, such that $p_t^n \neq 0$, $\pi_t^n \neq 0$ (none of them is trivial) and $p_t^n g_t^{\mathbf{y}_t} > 0$ for all $n > 0$, $0 \leq t \leq T$. Let also $\bar{\pi}_t^n$ be defined as a (random) probability measure absolutely continuous with respect to p_t^n for $t \in \mathbb{N}$ and $n \geq 1$ such that

$$\frac{d\bar{\pi}_t^n}{dp_t^n} = \frac{g_t^{\mathbf{y}_t}}{p_t^n g_t^{\mathbf{y}_t}}. \tag{2.8}$$

The following theorems give us *necessary and sufficient* conditions for the convergence of p_t^n and π_t^n to $p_t^{\mathbf{y}_{0:t-1}}$ and, respectively, $\pi_t^{\mathbf{y}_{0:t}}$. In order to simplify notation, for the remainder of this subsection we suppress the dependence on $\mathbf{y}_{0:t}$ and we denote $\pi_t^{\mathbf{y}_{0:t}}$ by π_t, $p_t^{\mathbf{y}_{0:t-1}}$ by p_t, and $g_t^{\mathbf{y}_t}$ by g_t, but always keep in mind that the observation process is a given fixed path $\mathbf{y}_{0:T}$

Theorem 2.3.1. *The sequences p_t^n, π_t^n converge to p_t, respectively, to π_t with convergence taken to be of type **1**. if and only if the following three conditions are satisfied:*
a1. *For all $f \in \mathcal{C}_b(\mathbb{R}^{n_x})$, $\lim_{n\to\infty} \mathbb{E}\left[|\pi_0^n f - \pi_0 f|\right] = 0$*
b1. *For all $f \in \mathcal{C}_b(\mathbb{R}^{n_x})$, $\lim_{n\to\infty} \mathbb{E}\left[|p_t^n f - \pi_{t-1}^n K_{t-1} f|\right] = 0$*
c1. *For all $f \in \mathcal{C}_b(\mathbb{R}^{n_x})$, $\lim_{n\to\infty} \mathbb{E}\left[|\pi_t^n f - \bar{\pi}_t^n f|\right] = 0$.*

Proof. The sufficiency part is proved by mathematical induction. The theorem is true for $n = 0$ from the condition **a1**. We need to show that, if p_{t-1}^n, π_{t-1}^n converge to p_{t-1}, respectively, to π_{t-1} then p_t^n, π_t^n converge to p_t, respectively, to π_t. Since $p_t = \pi_{t-1} K_{t-1}$, we have, for all $f \in \mathcal{C}_b(\mathbb{R}^{n_x})$,

$$|p_t^n f - p_t f| \leq |p_t^n f - \pi_{t-1}^n K_{t-1} f| + |\pi_{t-1}^n K_{t-1} f - \pi_{t-1} K_{t-1} f|. \tag{2.9}$$

By taking the expectation of both sides of equation (2.9) we obtain the following $\lim_{n\to\infty} \mathbb{E}\left[|p_t^n f - p_t f|\right] = 0$, since the expected value of the first term on the right hand side of (2.9) converges to 0 from **b1.** and the expected value of the second one converges to 0 from the induction hypothesis as $K_{t-1} f \in \mathcal{C}_b(\mathbb{R}^{n_x})$ for all $f \in \mathcal{C}_b(\mathbb{R}^{n_x})$ – the Feller property of the transition kernel.
We then use

$$
\begin{aligned}
|\bar{\pi}_t^n f - \pi_t f| &= \left| \frac{p_t^n f g_t}{p_t^n g_t} - \frac{p_t f g_t}{p_t g_t} \right| \\
&\leq \left| \frac{p_t^n f g_t}{p_t^n g_t} - \frac{p_t^n f g_t}{p_t g_t} \right| + \left| \frac{p_t^n f g_t}{p_t g_t} - \frac{p_t f g_t}{p_t g_t} \right| \\
&\leq \frac{\|f\|}{p_t g_t} |p_t^n g_t - p_t g_t| + \frac{1}{p_t g_t} |p_t^n f g_t - p_t f g_t|, \tag{2.10}
\end{aligned}
$$

therefore

$$\mathbb{E}\left[|\bar{\pi}_t^n f - \pi_t f|\right] \leq \frac{\|f\|}{p_t g_t}\mathbb{E}\left[|p_t^n g_t - p_t g_t|\right] + \frac{1}{p_t g_t}\mathbb{E}\left[|p_t^n f g_t - p_t f g_t|\right], \quad (2.11)$$

and both terms on the right hand side of (2.11) converge to 0. Finally,

$$|\pi_t^n f - \pi_t f| \leq |\pi_t^n f - \bar{\pi}_t^n f| + |\bar{\pi}_t^n f - \pi_t f|. \quad (2.12)$$

As the expected value of the first term on the right hand side of (2.12) converges to 0 using **c1.** , and the expected value of the second term converges to 0 using (2.11), we find that $\lim_{n\to\infty}\mathbb{E}\left[|\pi_t^n f - \pi_t f|\right] = 0$.

We now prove the necessity part. Assume that for all $t \geq 0$ and for all $f \in \mathcal{C}_b(\mathbb{R}^{n_x})$, $\lim_{n\to\infty}\mathbb{E}\left[|p_t^n f - p_t f|\right] = 0$ and $\lim_{n\to\infty}\mathbb{E}\left[|\pi_t^n f - \pi_t f|\right] = 0$ (this, in particular implies **a1.**). From (2.11), we see that $\lim_{n\to\infty}\mathbb{E}\left[|\pi_t f - \bar{\pi}_t^n f|\right] = 0$, and since

$$\mathbb{E}\left[|\pi_t^n f - \bar{\pi}_t^n f|\right] \leq \mathbb{E}\left[|\pi_t^n f - \pi_t f|\right] + \mathbb{E}\left[|\pi_t f - \bar{\pi}_t^n f|\right], \quad (2.13)$$

we obtain **c1.**. Finally, since $p_t = \pi_{t-1}K_{t-1}$, we have, for all $f \in \mathcal{C}_b(\mathbb{R}^{n_x})$,

$$\mathbb{E}\left[|p_t^n f - \pi_{t-1}^n K_{t-1}f|\right] \leq \mathbb{E}\left[|p_t^n f - p_t f|\right] + \mathbb{E}\left[|\pi_{t-1}K_{t-1}f - \pi_{t-1}^n K_{t-1}f|\right], \quad (2.14)$$

which implies **b1.**. □

The corresponding theorem for the almost sure convergence of p_t^n, π_t^n to p_t, respectively, to π_t holds true.

Theorem 2.3.2. *The sequences p_t^n, π_t^n converge almost surely to p_t and π_t, i.e. with convergence of type **2.**, if and only if the following three conditions are satisfied;*
a2. $\lim_{n\to\infty}\pi_0^n = \pi_0$, $P-a.s.$
b2. $\lim_{n\to\infty}d_\mathcal{M}\left(p_t^n, \pi_{t-1}^n K_{t-1}\right) = 0$, $P-a.s.$
c2. $\lim_{n\to\infty}d_\mathcal{M}\left(\pi_t^n, \bar{\pi}_t^n\right) = 0$, $P-a.s.$.

Proof. The sufficiency part of the theorem is proved as above by mathematical induction using inequalities (2.9), (2.10) and (2.12). The necessity part of the theorem results from the following argument. Assume that for all $t \geq 0$ p_t^n and π_t^n converge almost surely to p_t and, respectively, to π_t. This implies that $\pi_{t-1}^n K_{t-1}$ converges almost surely to p_t $(= \pi_{t-1}K_{t-1})$ and, using (2.10), that $\bar{\pi}_t^n$ converges almost surely to π_t. Hence, almost surely $\lim_{n\to\infty}d_\mathcal{M}\left(p_t^n, p_t\right) = 0$, $\lim_{n\to\infty}d_\mathcal{M}\left(\pi_t^n, \pi_t\right) = 0$, $\lim_{n\to\infty}d_\mathcal{M}\left(\pi_{t-1}^n K_{t-1}, p_t\right) = 0$ and $\lim_{n\to\infty}d_\mathcal{M}\left(\bar{\pi}_t^n, \pi_t\right) = 0$, where $d_\mathcal{M}$ is the distance defined in (2.7). Finally, using the triangle inequalities

$$d_\mathcal{M}\left(p_t^n, \pi_{t-1}^n K_{t-1}\right) \leq d_\mathcal{M}\left(p_t^n, p_t\right) + d_\mathcal{M}\left(p_t, \pi_{t-1}^n K_{t-1}\right)$$
$$d_\mathcal{M}\left(\pi_t^n, \bar{\pi}_t^n\right) \leq d_\mathcal{M}\left(\pi_t^n, \pi_t\right) + d_\mathcal{M}\left(\pi_t, \bar{\pi}_t^n\right),$$

we get **b2.** and **c2.**. □

Remark 2. *Theorems 2.3.1 and 2.3.2 are very natural. They say that, we obtain approximations of p_t and π_t for all $t \geq 0$. if and only if we have to start from an approximation of π_0, and then 'follow closely' the recurrence formula (2.6) for p_t and π_t.*

Now, the question arises whether we can lift the results to the case when the observation process is random and not just fixed to a given observation path.

2.3.2   The random observation case

In the previous section both the converging sequences and the limiting measures depend on the fixed value of the observation. Let us make this explicit by writing that

$$\lim_{n\to\infty} p_t^{n,\mathbf{y}_{0:t-1}} = p_t^{\mathbf{y}_{0:t-1}}, \qquad \lim_{n\to\infty} \pi_t^{n,\mathbf{y}_{0:t}} = \pi_t^{\mathbf{y}_{0:t}}, \qquad (2.15)$$

where the limits in (2.15) are either of type **1.** or of type **2.**. Since $p_t = p_t^{Y_{0:t-1}}$ and $\pi_t = \pi_t^{Y_{0:t}}$, we expect that

$$\lim_{n\to\infty} p_t^{n,Y_{0:t-1}} = p_t, \qquad \lim_{n\to\infty} \pi_t^{n,Y_{0:t}} = \pi_t.$$

Let us look first at the convergence of type **1.**. We have

$$\mathbb{E}\left[\left|p_t^{n,Y_{0:t-1}}f - p_t f\right|\right] = \int_{(\mathbb{R}^{n_y})^t} \mathbb{E}\left[\left|p_t^{n,\mathbf{y}_{0:t-1}}f - p_t^{\mathbf{y}_{0:t}}f\right|\right] P_{Y_{0:t-1}}(d\mathbf{y}_{0:t-1})$$

$$\mathbb{E}\left[\left|\pi_t^{n,Y_{0:t}}f - p_t f\right|\right] = \int_{(\mathbb{R}^{n_y})^{t+1}} \mathbb{E}\left[|\pi_t^{n,\mathbf{y}_{0:t}}f - \pi_t^{\mathbf{y}_{0:t}}f|\right] P_{Y_{0:t}}(d\mathbf{y}_{0:t}).$$

Therefore, if $\mathrm{e}\lim_{n\to\infty} p_t^{n,\mathbf{y}_{0:t-1}} = p_t^{\mathbf{y}_{0:t-1}}$ for $P_{Y_{0:t-1}}$-almost all values $\mathbf{y}_{0:t-1}$ and $\mathrm{e}\lim_{n\to\infty} \pi_t^{n,\mathbf{y}_{0:t}} = \pi_t^{\mathbf{y}_{0:t}}$ for $P_{Y_{0:t}}$-almost all values $\mathbf{y}_{0:t}$ and we have two functions $v(\mathbf{y}_{0:t-1})$ and $w(\mathbf{y}_{0:t})$ such that, for all $n \geq 0$,

$$\mathbb{E}\left[|p_t^{n,\mathbf{y}_{0:t-1}}f - p_t^{\mathbf{y}_{0:t-1}}f|\right] \leq v_f(\mathbf{y}_{0:t-1}), \quad P_{Y_{0:t-1}}\text{-a.s.} \quad (2.16)$$

$$\mathbb{E}\left[|\pi_t^{n,\mathbf{y}_{0:t}}f - \pi_t^{\mathbf{y}_{0:t}}f|\right] \leq w_f(\mathbf{y}_{0:t}), \quad P_{Y_{0:t}}\text{-a.s.,} \quad (2.17)$$

then, by the Dominated Convergence Theorem, we have the following limits $\mathrm{e}\lim_{n\to\infty} p_t^{n,Y_{0:t-1}} = p_t$ and $\mathrm{e}\lim_{n\to\infty} \pi_t^{n,Y_{0:t}} = \pi_t$. Conditions (2.16) and (2.17) are trivially satisfied for probability approximations; in this case, $v_f = w_f = 2\|f\|$. However, we cannot obtain a necessary and sufficient condition in this way, since $\mathrm{e}\lim_{n\to\infty} p_t^{n,Y_{0:t}} = p_t$ and $\mathrm{e}\lim_{n\to\infty} \pi_t^{n,Y_{0:t}} = \pi_t$ doesn't imply $\mathrm{e}\lim_{n\to\infty} p_t^{n,\mathbf{y}_{0:t}} = p_t^{\mathbf{y}_{0:t}} P_{Y_{0:t-1}}$-a.s. and, respectively, $\mathrm{e}\lim_{n\to\infty} \pi_t^{n,\mathbf{y}_{0:t}} = \pi_t^{\mathbf{y}_{0:t}} P_{Y_{0:t}}$-a.s.. We have, though, the following proposition:

Proposition 2.3.3. *Provided for all $t \geq 0$, there exists a constant $c_t > 0$ such that $p_t g_t \geq c_t$, the sequences $p_t^{n,Y_{0:t-1}}, \pi_t^{n,Y_{0:t}}$ converge to p_t and to π_t*

with convergence taken to be of type 1. if and only, if for all $f \in C_b(\mathbb{R}^{n_x})$,

$$\lim_{n \to \infty} \mathbb{E}\left[|\pi_0^n f - \pi_0 f|\right] = 0$$

$$\lim_{n \to \infty} \mathbb{E}\left[\left|p_t^{n,Y_{0:t-1}} f - \pi_{t-1}^{n,Y_{0:t}} K_{t-1} f\right|\right] = 0$$

$$\lim_{n \to \infty} \mathbb{E}\left[\left|\pi_t^{n,Y_{0:t}} f - \bar{\pi}_t^{n,Y_{0:t}} f\right|\right] = 0.$$

Proof. The proof is identical to that of Theorem 2.3.1, so we omit it. □

We turn now to the almost sure convergence of $p_t^{n,Y_{0:t-1}}$ and $\pi_t^{n,Y_{0:t}}$ to p_t and, respectively, to π_t. In this case everything carries through smoothly.

Proposition 2.3.4. *The sequences $p_t^{n,Y_{0:t-1}}, \pi_t^{n,Y_{0:t}}$ converge almost surely to p_t, respectively, to π_t, for all $t \geq 0$ if and only if, for all $t \geq 0$,*

$$\lim_{n \to \infty} \pi_0^n = \pi_0, P\text{-}a.s..$$

$$\lim_{n \to \infty} d_{\mathcal{M}}\left(p_t^{n,Y_{0:t-1}}, \pi_{t-1}^{n,Y_{0:t-1}} K_{t-1}\right) = 0, \quad P\text{-}a.s..$$

$$\lim_{n \to \infty} d_{\mathcal{M}}\left(\pi_t^{n,Y_{0:t}}, \bar{\pi}_t^{n,Y_{0:t}}\right) = 0, \quad P\text{-}a.s..$$

Proof. Again, the proof for this theorem is similar to that of Theorem 2.3.2, the only difference being the proof that $\lim_{n \to \infty} p_t^{n,Y_{0:t-1}} = p_t$, P–a.s. implies $\lim_{n \to \infty} \bar{\pi}_t^{n,Y_{0:t}} = \pi_t$, $P - a.s.$ which is as follows. Let $\mathcal{M}$ be a convergence determining set of functions in $C_b(\mathbb{R}^{n_x})$, for instance the set we used to construct the distance $d_{\mathcal{M}}$. Then one can find a subset $\Omega' \subset \Omega$ such that $P(\Omega') = 1$ and for all $\omega \in \Omega'$ $\lim_{n \to \infty} p_t^{n,Y_{0:t-1}(\omega)} g_t = p_t g_t(\omega)$ and $\lim_{n \to \infty} p_t^{n,Y_{0:t-1}(\omega)} (g_t f) = p_t(g_t f)(\omega)$ for all $f \in \mathcal{M}$. Hence for all $\omega \in \Omega'$

$$\lim_{n \to \infty} \bar{\pi}_t^{n,Y_{0:t}} f(\omega) = \lim_{n \to \infty} \frac{p_t^{n,Y_{0:t}(\omega)}(g_t f)}{p_t^{n,Y_{0:t}(\omega)} g_t} = \frac{p_t(g_t f)}{p_t g_t}(\omega) = \pi_t f(\omega), \quad \forall f \in \mathcal{M},$$

which implies $\lim_{n \to \infty} \bar{\pi}_t^{n,Y_{0:t}} = \pi_t$, P-a.s. □

In the next section we shall present a class of particle filters which satisfies the conditions for these results. This is not the most general class of particle filter and the reader will find other particle filters in the remaining chapters of the book. Our intention is not to exhaust all classes of particle filters, but simply to exemplify the use of the results.

2.4 Examples of particle filters

2.4.1 Description of the particle filters

The algorithms presented below involve the use of a system of n particles which evolve (mutate) according to the law of a given Markov process and, at fixed times, give birth to a number of offspring. Several possible branching mechanisms are described and, after imposing some weak restrictions on those branching mechanisms, the empirical measure associated to the particle systems is proved to converge (as n tends to ∞) to the conditional distribution of the signal, given the observation.

Just as we did above, we shall denote by π_t^n the approximation to π_t and by p_t^n the approximation to p_t. The particle filter will have the following description.

Initialisation. The filter starts with π_0^n – the empirical measure associated to a sample of size n from π_0 – in other words, with n random particles of mass $\frac{1}{n}$, with positions $\mathbf{x}_0^{(i)}$, $i = 1, 2, ..., n$ such that for all i, $\mathbf{x}_0^{(i)}$ has distribution π_0:

$$\pi_0^n := \frac{1}{n}\sum_{i=1}^{n}\delta_{\left\{\mathbf{x}_0^{(i)}\right\}}.$$

Obviously, e $\lim_{n\to\infty}\pi_0^n = \pi_0$ and also $\lim_{n\to\infty}\pi_0^n = \pi_0$, P-a.s..

Iteration. We describe how to obtain π_t^n from π_{t-1}^n (hence this is a recursive algorithm). The approximation π_{t-1}^n will have the form $\frac{1}{n}\sum_{i=1}^{n}\delta_{\left\{\mathbf{x}_{t-1}^{(i)}\right\}}$, i.e., it will be the empirical measure associated to a system of n random particles. The first step is to move each particle using the transition kernel of the signal. If $\mathbf{x}_{t-1}^{(i)}$ is the position of the ith particle, then $\bar{\mathbf{x}}_t^{(i)}$ – the new position – has the distribution $K_{t-1}\left(\mathbf{x}_{t-1}^{(i)}, \cdot\right)$ and the particles move independent of each other, p_t^n will be the empirical distribution associated with the new cloud of particles

$$p_t^n = \frac{1}{n}\sum_{i=1}^{n}\delta_{\left\{\bar{\mathbf{x}}_t^{(i)}\right\}}.$$

For each particle, we compute the weight $w_t^{(i)} = \frac{ng_t(\bar{\mathbf{x}}_0^{(i)})}{\sum_{j=1}^{n}g_t(\bar{\mathbf{x}}_0^{(j)})}$ and obviously $\frac{1}{n}\sum_{i=1}^{n}w_t^{(i)}\delta_{\left\{\bar{\mathbf{x}}_t^{(i)}\right\}}$ is exactly the measure $\bar{\pi}_t^n$ as defined in (2.8),

$$\bar{\pi}_t^n = \frac{1}{n}\sum_{i=1}^{n}w_t^{(i)}\delta_{\left\{\bar{\mathbf{x}}_t^{(i)}\right\}}.$$

Each particle is then replaced by a number of offspring – $\xi_t^{(i)}$ – with the mean number of offspring being $w_t^{(i)}$ and of finite variance. We impose the condition that the total number of particles does not change, i.e., $\sum_{i=1}^{n}\xi_t^{(i)} = n$. We denote the positions of the newly obtained particles by $\mathbf{x}_t^{(i)}$, $i = 1, ..., n$ and define π_t^n as

$$\pi_t^n = \frac{1}{n} \sum_{i=1}^{n} \delta_{\{\mathbf{x}_t^{(i)}\}}.$$

Let A_t^n be the covariance matrix of the random vector $\xi_t := (\xi_t^{(i)})_{i=1}^{n}$, $A_t^n := \mathbb{E}[(\xi_t - w_t)^T (\xi_t - w_t)]$, where T denotes the transpose and let $w_t := (w_t^{(i)})_{i=1}^{n}$ be the vector of weights or means. Then we assume that there exists a constant c_t, such that

$$q^T A_t^n q \le n c_t, \text{ for all } q \in \mathbb{R}^{n_x}, q = \left(q^{(i)} \right)_{i=1}^{n}, \left| q^{(i)} \right| \le 1 \; i = 1, ..., n. \tag{2.18}$$

The condition that the total number of particles remains constant is not essential. Indeed in (Crisan et al. 1999) we present a similar algorithm with variable number of particles. Theorems 2.3.1 and 2.3.2 and their corollaries can be used in order to prove the convergence of the algorithm in (Crisan et al. 1999).

Remark 3. *An alternative way to obtain p_t^n from π_{t-1}^n is to sample n times from the measure $\pi_{t-1}^n K_{t-1}$ and define p_t^n to be the empirical measure associated with this sample.*

Remark 4. *Condition (2.18) is equivalent to*

$$q^T A_t^n q \le n \bar{c}_t, \text{ for all } q \in \mathbb{R}^{n_x}, q = \left(q^{(i)} \right)_{i=1}^{n}, 0 \le q^{(i)} \le 1 \; i = 1, ..., n \tag{2.19}$$

for a fixed constant $\bar{c}_t$.

Proof. Obviously (2.18) implies (2.19), so we need only show the inverse implication. Let $q \in \mathbb{R}^{n_x}$ be an arbitrary vector such that $q = \left(q^{(i)} \right)_{i=1}^{n}$, $\left| q^{(i)} \right| \le 1$, $i = 1, ..., n$. Let also $q_+ = \left(q_+^{(i)} \right)_{i=1}^{n}$, $q_- = \left(q_-^{(i)} \right)_{i=1}^{n}$, $0 \le q_+^{(i)}, q_-^{(i)} \le 1 \; i = 1, ..., n$, $q_+^{(i)} \triangleq \max \left(q^{(i)}, 0 \right)$, $q_-^{(i)} \triangleq \max \left(-q^{(i)}, 0 \right)$. We define $\|\cdot\|_A$ to be the semi-norm associated to the matrix A (if all its eigenvalues are positive, then $\|\cdot\|_A$ is a genuine norm), $\|q\|_A \triangleq \sqrt{q^T A q}$. Then $q = q_+ - q_-$ and, using the triangle inequality and (2.19),

$$\|q\|_{A_t^n} \le \|q_+\|_{A_t^n} + \|q_-\|_{A_t^n} \le 2\sqrt{n\bar{c}_t},$$

which implies that (2.18) holds with $c_t = 4\bar{c}_t$. □

The only part not yet specified is the branching mechanism ξ_t. Below we introduce two branching mechanisms that leave the total number of particles constant and satisfy (2.18).

2.4.2 Branching mechanisms

We introduce two possible branching mechanisms, the multinomial branching mechanism and the tree-based branching mechanism. Neither of the mechanisms is new; both can be found in the literature under various names (see, for instance, (Crisan et al. 1999) and the references therein).

The multinomial branching mechanism.

In this case, we choose $\xi_t = $ Multinomial $\left(n, \frac{w_t^{(1)}}{n}, \ldots, \frac{w_t^{(n)}}{n} \right)$. We have

$$\mathbb{E}\left[\xi_t^{(i)} \right] = w_t^{(i)}, \ \mathbb{E}\left[\left(\xi_t^{(i)} - w_t^{(i)} \right)^2 \right] = w_t^{(i)} \left(1 - \frac{w_t^{(i)}}{n} \right) \text{ and, in addition, we}$$

have $\mathbb{E}\left[\left(\xi_t^{(i)} - w_t^{(i)} \right) \left(\xi_t^{(j)} - w_t^{(j)} \right) \right] = -\frac{w_t^{(i)} w_t^{(j)}}{n}$. Then for all $q \in \mathbb{R}^{n_x}$, $q = \left(q^{(i)} \right)_{i=1}^n$, $\left| q^{(i)} \right| \leq 1$, $i = 1, ..., n$,

$$
\begin{aligned}
q^T A_t^n q &= \sum_{i=1}^n w_t^{(i)} \left(1 - \frac{w_t^{(i)}}{n} \right) \left(q^{(i)} \right)^2 - 2 \sum_{\substack{i,j=1 \\ i \neq j}}^n \frac{w_t^{(i)} w_t^{(j)}}{n} q^{(i)} q^{(j)} \\
&= \sum_{i=1}^n w_t^{(i)} \left(q^{(i)} \right)^2 - \frac{1}{n} \left(\sum_{i=1}^n w_t^{(i)} q^{(i)} \right)^2 \\
&\leq \sum_{i=1}^n w_t^{(i)}
\end{aligned}
$$

and, since $\sum_{i=1}^n w_t^{(i)} = n$, (2.18) holds with $c_t = 1$. The properties of the multinomial branching mechanism have been extensively studied in the literature (again, see, for instance, (Crisan et al. 1999) and the references therein).

The tree-based branching mechanism.

We consider now a binary tree with n leaves and denote by r the root of the tree (see Figure 2.1). To the ith particle we associate the ith leaf of the tree and attach to it the corresponding weight w_i. Let m be an arbitrary node of the tree which is not a leaf, m_1 and m_2 be its direct offspring and I_m (respectively, I_{m_1} and I_{m_2}) be the set of leaves/particles which are the descendants of m (respectively, m_1 and m_2). Let w_m be the weight attached to the node m defined as the sum of the weights of all the leaves which are its descendants

$$w_m \triangleq \sum_{i \in I_m} w_i.$$

Obviously, $w_r = n$ because I_r comprises the whole set of leaves/particles and $w_m = w_{m_1} + w_{m_2}$, since I_{m_1} and I_{m_2} represent a partition of I_m.

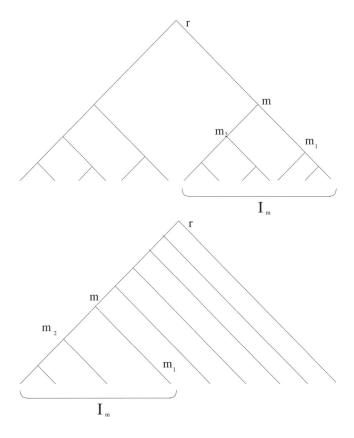

Figure 2.1. Branching binary trees.

We construct next a set of random variables ξ_m for all the nodes of the tree including the leaves such that

$$\xi_m = \begin{cases} [w_m] & \text{with probability} \quad 1 - \{w_m\} \\ [w_m] + 1 & \text{with probability} \quad \{w_m\} \end{cases}, \qquad (2.20)$$

where $[x]$ is the integer of part $x \in \mathbb{R}$ and $\{x\} \triangleq x - [x]$. The random variables associated to the leaves will constitute the branching mechanisms of the corresponding particles. The construction is done recursively starting from the root and finishing with the leaves such that *the sum of the random variables associated to the children nodes is equal to the random variable associated to the parent node.*

We define ξ_r to be identically equal to n. We show next the rule of recursion.

Suppose that we have constructed ξ_m with the distribution given by (2.20). First we assign $[w_{m_1}]$ to m_1 and $[w_{m_2}]$ to m_2 and observe that $\xi_m - [w_{m_1}] - [w_{m_2}]$ is equal to $0, 1$ or 2. These are the only possibilities since ξ_m can have only two values, $[w_m]$ or $[w_m] + 1$ and $[w_m] - [w_{m_1}] - [w_{m_2}]$ can be either 0 or 1. If $\xi_m - [w_{m_1}] - [w_{m_2}] = 2$ we assign one unit to each of the nodes, if $\xi_m - [w_{m_1}] - [w_{m_2}] = 0$ we do nothing, and if $\xi_m - [w_{m_1}] - [w_{m_2}] = 1$ we assign the unit randomly to one of the nodes such that the resulting random variables will have the right distribution. More precisely, if $[w_m] = [w_{m_1}] + [w_{m_2}]$, we define

$$\xi_{m_1} \triangleq [w_{m_1}] + (\xi_m - [w_m])\eta_m$$
$$\xi_{m_2} \triangleq [w_{m_2}] + (\xi_m - [w_m])(1 - \eta_m),$$

where

$$\eta_m = \begin{cases} 0 & \text{with probability} & \frac{\{w_{m_2}\}}{\{w_m\}} \\ 1 & \text{with probability} & \frac{\{w_{m_1}\}}{\{w_m\}} \end{cases}. \qquad (2.21)$$

The distribution of the auxiliary random variable is well defined because

$$\begin{aligned} w_m = w_{m_1} + w_{m_2} \\ [w_m] = [w_{m_1}] + [w_{m_2}] \end{aligned} \implies \frac{\{w_{m_1}\}}{\{w_m\}} + \frac{\{w_{m_2}\}}{\{w_m\}} = 1.$$

If $[w_m] = [w_{m_1}] + [w_{m_2}] + 1$, we define

$$\xi_{m_1} \triangleq [w_{m_1}] + 1 + (\xi_m - [w_m] - 1)\eta_m$$
$$\xi_{m_2} \triangleq [w_{m_2}] + 1 + (\xi_m - [w_m] - 1)(1 - \eta_m),$$

where

$$\eta_m = \begin{cases} 0 & \text{with probability} & \frac{1 - \{w_{m_2}\}}{1 - \{w_m\}} \\ 1 & \text{with probability} & \frac{1 - \{w_{m_1}\}}{1 - \{w_m\}} \end{cases}. \qquad (2.22)$$

The distribution of the auxiliary random variable is well defined because

$$\begin{aligned} w_m = w_{m_1} + w_{m_2} \\ [w_m] = [w_{m_1}] + [w_{m_2}] + 1 \end{aligned} \implies \frac{1 - \{w_{m_1}\}}{1 - \{w_m\}} + \frac{1 - \{w_{m_2}\}}{1 - \{w_m\}} = 1.$$

We take all the random variables η_m independent of each other. In either case

$$\xi_{m_1} \triangleq [w_{m_1}] + a_m + (\xi_m - b_m)\eta_m \qquad (2.23)$$
$$\xi_{m_2} \triangleq [w_{m_2}] + a_m + (\xi_m - b_m)(1 - \eta_m), \qquad (2.24)$$

where $a_m = [w_m] - [w_{m_1}] + [w_{m_2}] = \{w_{m_1}\} + \{w_{m_2}\} - \{w_m\}$ and $b_m = [w_m] - a_m$.

Let ξ_i $i = 1, ..., n$ be the resulting random variables corresponding to the leaves of the tree, and therefore to the particles of the system. From (2.20) we see that

1. $E[\xi_i] = w_i$
2. $\mathrm{Var}(\xi_i) = E[\xi_i^2] - E[\xi_i]^2 = \{w_i\}(1 - \{w_i\}) \le \frac{1}{4}$.

We also have

$$
\begin{aligned}
E\left[(\xi_{m_1} - E[\xi_{m_1}])\,|\xi_m\right] &= E\left[(a_m - \{w_{m_1}\} + (\xi_m - [w_m] - a_m)\eta_m)\,|\xi_m\right] \\
&= \frac{a_m - \{w_{m_1}\}}{a_m - \{w_m\}}(\xi_m - w_m),
\end{aligned}
$$

and

$$
\begin{aligned}
E&\left[(\xi_{m_1} - E[\xi_{m_1}])(\xi_{m_2} - E[\xi_{m_2}])\,|\xi_m\right] \\
&= E\left[\prod_{i=1}^{2}(a_m - \{w_{m_i}\} + (\xi_m - [w_m] - a_m)\eta_m)\,|\xi_m\right] \\
&= (a_m - \{w_{m_1}\})(a_m - \{w_{m_2}\})\left(\frac{2(\xi_m - w_m)}{a_m - \{w_m\}} - 1\right).
\end{aligned}
$$

Hence, if i and j are two different particles or leaves, then

$$
\begin{aligned}
E&\left[(\xi_i - E[\xi_{m_i}])(\xi_j - E[\xi_{m_j}])\right] \\
&= -\prod_{s=1}^{h_i-1}\frac{a_{m_s} - \{w_{m_{s-1}}\}}{a_{m_s} - \{w_{m_s}\}}\prod_{t=1}^{h_j-1}\frac{a_{n_t} - \{w_{n_{t-1}}\}}{a_{n_t} - \{w_{n_t}\}}\left(a_{m_{h_i}} - \left\{w_{m_{h_i-1}}\right\}\right)\left(a_{n_{h_j}} - \left\{w_{n_{h_j-1}}\right\}\right),
\end{aligned}
$$

where $m_{h_i} = m_{h_j}$ is the most recent common ancestor of i and j, h_i and h_j are the number of generations from the most common ancestor to i, respectively, j, $m_1, m_2, ..., m_{h_i}$ is the genealogical line of i, and $m_1, m_2, ..., m_{h_i}$ is the genealogical line of j. Now, since the following apply $\mathrm{Var}(\xi_i) \le \frac{1}{4}$ and $E\left[(\xi_i - E[\xi_{m_i}])(\xi_j - E[\xi_{m_j}])\right] \le 0$, we have that $q^T A_t^n q \le \frac{1}{4}n$, for all $q \in \mathbb{R}^{n_x}$, $q = (q^{(i)})_{i=1}^{n}$, $0 \le q^{(i)} \le 1$, $i = 1, .., n$. Hence by Remark 4, (2.18) holds with $c_t = \frac{1}{2}$.

2.4.3 Convergence of the algorithm

We fix first the observation process to an arbitrary value $\mathbf{y}_{0:T}$, where T is a finite time horizon. We prove first that the random measures resulting from the class of algorithms described above converge to $\pi_t^{\mathbf{y}_{0:t}}$ (respectively $p_t^{\mathbf{y}_{0:t-1}}$) for all $0 \le t \le T$. For this we introduce the following σ-algebras

$$
\begin{aligned}
\mathcal{F}_t &= \sigma\left(\bar{\mathbf{x}}_s^{(i)}, \mathbf{x}_s^{(i)}, s \le t, \quad i = 1, .., n\right) \\
\bar{\mathcal{F}}_t &= \sigma\left(\bar{\mathbf{x}}_s^{(i)}, \mathbf{x}_s^{(i)}, s < t, \quad \bar{\mathbf{x}}_t^{(i)}, \quad i = 1, .., n\right).
\end{aligned}
$$

Theorem 2.4.1. Let $(p_t^n)_{n=1}^{\infty}$ and $(\pi_t^n)_{n=1}^{\infty}$ be the measure valued sequences produced by the class of algorithms described above. Then, for all $0 \le t \le T$, we have

$$
\mathrm{e}\lim_{n\to\infty} p_t^n = p_t^{\mathbf{y}_{0:t-1}} \qquad \mathrm{e}\lim_{n\to\infty} \pi_t^n = \pi_t^{\mathbf{y}_{0:t}}.
$$

Proof. We apply Theorem 2.3.1. Since **a1.** is clearly satisfied, we only need to show that **b1.** and **c1.** hold true. If $f \in C_b(\mathbb{R}^{n_x})$, then

$$\mathbb{E}\left[f\left(\bar{\mathbf{x}}_t^{(i)}\right)|\mathcal{F}_{t-1}\right] = K_{t-1}f\left(\mathbf{x}_{t-1}^{(i)}\right) \quad \text{for all } i = 1, ..., n.$$

Hence, $\mathbb{E}\left[p_t^n f | \mathcal{F}_t\right] = \pi_{t-1}^n K_{t-1} f$ and, using the independence of the motion of the particles, we have

$$\mathbb{E}\left[\left(p_t^n f - \pi_{t-1}^n K_{t-1}f\right)^2 | \mathcal{F}_{t-1}\right] = \frac{1}{n}\pi_{t-1}^n\left(K_{t-1}f^2 - (K_{t-1}f)^2\right).$$

Therefore, $\mathbb{E}\left[\left(p_t^n f - \pi_{t-1}^n K_{t-1}f\right)^2\right] \leq \frac{||f||^2}{n}$ and **b1.** is satisfied. Now, since $\pi_t^n = \frac{1}{n}\sum_{i=1}^n \xi_t^{(i)}\delta_{\left\{\bar{\mathbf{x}}_t^{(i)}\right\}}$, we have

$$\mathbb{E}\left[\pi_t^n f | \bar{\mathcal{F}}_t\right] = \bar{\pi}_t^n f$$
$$\mathbb{E}\left[(\pi_t^n f - \bar{\pi}_t^n f)^2 | \bar{\mathcal{F}}_t\right] = \frac{1}{n^2}(q_t^n)^T A_t^n q_t^n, \tag{2.25}$$

where q_t^n is the vector with entries $(q_t^n)^{(i)} = f\left(\bar{\mathbf{x}}_t^{(i)}\right)$. From (2.18) and (2.25) we see that

$$\mathbb{E}\left[(\pi_t^n f - \bar{\pi}_t^n f)^2 | \bar{\mathcal{F}}_t\right] \leq \frac{c_t ||f||^2}{n}.$$

Therefore, $\mathbb{E}\left[(\pi_t^n f - \bar{\pi}_t^n f)^2\right] \leq c_t\frac{||f||^2}{n}$ and **c1.** is satisfied. □

Theorem 2.4.2. *Let $(p_t^n)_{n=1}^\infty$ and $(\pi_t^n)_{n=1}^\infty$ be the measure-valued sequences produced by the algorithm with multinomial branching mechanism described above. Then, for all $0 \leq t \leq T$, we have*

$$\lim_{n\to\infty} p_t^n = p_t^{\mathbf{y}_{0:t-1}} \qquad \lim_{n\to\infty} \pi_t^n = \pi_t^{\mathbf{y}_{0:t}} \quad P\text{-a.s.}$$

Proof. We apply Theorem 2.3.2. Let $\mathcal{M} \in C_b(\mathbb{R}^d)$ be the countable, convergence determining set of functions defined in the previous section. Since $\mathbb{E}\left[f\left(\bar{\mathbf{x}}_t^{(i)}\right)|\mathcal{F}_{t-1}\right] = K_{t-1}f\left(\mathbf{x}_{t-1}^{(i)}\right)$ and using the independence of $\bar{\mathbf{x}}_t^{(1)}$, $\bar{\mathbf{x}}_t^{(2)}$, ..., $\bar{\mathbf{x}}_t^{(n)}$ given $\mathcal{F}_t$, we have

$$\mathbb{E}\left[\left(p_t^n f - \pi_{t-1}^n K_{t-1}f\right)^4 | \mathcal{F}_{t-1}\right] \tag{2.26}$$
$$= \mathbb{E}\left[\left(\frac{1}{n}\sum_{i=1}^n \left(f\left(\bar{\mathbf{x}}_t^{(i)}\right) - K_{t-1}f\left(\mathbf{x}_{t-1}^{(i)}\right)\right)\right)^4 | \mathcal{F}_{t-1}\right]$$
$$= \frac{1}{n^4}\sum_{i=1}^n \mathbb{E}\left[\left(f\left(\bar{\mathbf{x}}_t^{(i)}\right) - K_{t-1}f\left(\mathbf{x}_{t-1}^{(i)}\right)\right)^4 | \mathcal{F}_{t-1}\right]$$
$$+ \frac{2}{n^4}\sum_{1\leq i<j\leq n} \mathbb{E}\left[\left(f\left(\bar{\mathbf{x}}_t^{(i)}\right)-K_{t-1}f\left(\mathbf{x}_{t-1}^{(i)}\right)\right)^2\left(f\left(\bar{\mathbf{x}}_t^{(j)}\right)-K_{t-1}f\left(\mathbf{x}_{t-1}^{(j)}\right)\right)^2 | \mathcal{F}_{t-1}\right].$$

Hence, by taking expectation in both terms of equation (2.26), we have $\mathbb{E}\left[\left(p_t^n f - \pi_{t-1}^n K_{t-1} f\right)^4\right] \leq \frac{16\|f\|^4}{n^2}$ and using a Borel Cantelli argument we get $\lim_{n\to\infty}\left|p_t^n f - \pi_{t-1}^n K_{t-1} f\right| = 0$ for all $f \in \mathcal{M}$, P-a.s.. Therefore $\lim_{n\to\infty} d_{\mathcal{M}}\left(p_t^n, \pi_{t-1}^n K_{t-1}\right) = 0$, and hence we have **b2.**. Similarly, one proves that, for all $f \in \mathcal{M}$,

$$\mathbb{E}\left[\left(\pi_t^n f - \bar{\pi}_t^n f\right)^4 \mid \mathcal{F}_t\right] \leq \frac{16\,\|f\|^4}{n^2}, \tag{2.27}$$

which implies, as above, that $\lim_{n\to\infty} d_{\mathcal{M}}\left(\pi_t^n, \bar{\pi}_t^n\right) = 0$, i.e., **c2.**. $\qquad\square$

The tree-based branching mechanism has more complex correlations between higher moments of the individual branching mechanisms, and we have not been able so far to prove that (2.27) holds true. However, we conjecture that this is the case and therefore we have almost sure convergence of for this algorithm as well.

We turn now to the case in which the observation process is random. With similar arguments one uses Propositions 2.3.3 and 2.3.4 to prove the following corollaries

Corollary 2.4.3. *Provided for all $t \geq 0$, there exists a constant $c_t > 0$ such that $p_t g_t \geq c_t$, we have*

$$\operatorname{e} \lim_{n\to\infty} p_t^{n,Y_{0:t-1}} = p_t \qquad \operatorname{e} \lim_{n\to\infty} \pi_t^{n,Y_{0:t}} = \pi_t.$$

Corollary 2.4.4. *Let $(p_t^n)_{n=1}^{\infty}$ and $(\pi_t^n)_{n=1}^{\infty}$ be the measure-valued sequences produced by the algorithm with multinomial branching mechanism described above. Then, for all $0 \leq t \leq T$, we have*

$$\lim_{n\to\infty} p_t^{n,Y_{0:t-1}} = p_t \qquad \lim_{n\to\infty} \pi_t^{n,Y_{0:t}} = \pi_t. \quad P--a.s..$$

2.5   Discussion

The remarkable feature of Theorems 2.3.1 and 2.3.2 is their provision of efficient techniques for proving convergence of particle algorithms. The *necessary and sufficient* conditions **a1(2).**, **b1(2).**, **c1(2).** are natural and easy to verify, as can be seen from the proofs of Theorems 2.4.1 and 2.4.2. Of course, these theorems can be applied when the algorithms studied provide not only π_t^n – the approximation to π_t – but also p_t^n – the intermediate approximation to p_t. However, this may not always be the case, and one can have algorithms where π_t^n is obtained from π_{t-1}^n without using intrinsically the intermediate step p_t^n. In this case one *defines* p_t^n to be $\pi_{t-1}^n K_{t-1}$, or indeed any measure which is 'close' to $\pi_{t-1}^n K_{t-1}$ (so that **b1(2).** is satisfied) and then tries to apply the theorems. This implies that one need only check **c1(2).**, i.e., whether π_t^n is closed to $\bar{\pi}_t^n$, where $\bar{\pi}_t^n$ is the probability measure defined in (2.8).

The results presented in this chapter are directed towards proving convergence of the particle filters, their rates of convergence are not discussed. One can easily prove that, for the class of algorithms presented here, we have

$$\mathbb{E}\left[|\pi_t^n f - \pi_t f|\right] \leq \frac{m_t}{\sqrt{n}} ||f||, \quad \text{for all } f \in \mathcal{C}_b(\mathbb{R}^d),$$

where m_t are constants independent of n, for all $t \geq 0$ (see (Crisan et al. 1999) for a proof of this result).

The two branching algorithms presented are both unbiased, in the sense that the mean of the offspring distribution of each particle is its corresponding weight. Also, the variance of the offspring distribution has the same order. However, of the two, the tree-based branching algorithm has minimal variance for individual particles. If ξ_i is the number of offspring of the particle i, then ξ_i has the prescribed mean w_i (the corresponding weight) and is an integer-valued random variable with the minimal variance among all the integer-valued random variables with the given mean w_i. So if, say, $w_i = 3.4$, then ξ_i takes only the values 3 (with probability 0.6) or 4 (with probability 0.4); hence, it stays as close as possible to the required w_i. By contrast, when one uses the multinomial branching mechanism, ξ_i can have any value between 0 and n with positive probability, meaning extra randomness is introduced in the system.

The minimal variance property for the tree-based branching algorithms holds true not only for individual particles, but also for groups of particles. If m is an arbitrary node of the tree and I_m is the set of particles corresponding to the offspring of m, then $\xi_m = \sum_{i \in I_m} \xi_i$ has the required mean $\sum_{i \in I_m} w_i$ and has minimal variance. Moreover, we prove in a forthcoming paper that in a discrete time/space set-up the tree-based branching algorithm (or, rather, a variant of it) gives the best approximation to π_t, when the measure of comparison is the relative entropy with respect to π_t.

Although we think that the tree-based branching mechanism is optimal for any correction/resampling procedure that we might undertake, we also believe that the updating procedure has to be greatly improved. The raw Monte Carlo method used in the class of algorithms presented is clearly not optimal because it introduces too much randomness. The answer may lie in applying a PDE approach – discretising the state-space and moving the particles using a quasi-deterministic law which approximates K_t.

Of course, there are other routes by which one may introduce vast improvements to these methods, such as resampling out of the current population using the new observation and then updating the position of the particles. Undoubtedly, there is virgin territory to be explored in this area and exciting developments await both practitioners and theoreticians.

2.6 Appendix

2.6.1 Conditional probabilities and conditional expectations

In this section, we state briefly some definitions and results that we need in the article. One can find details and proofs of these results in any monograph on probability theory, for example in (Shiryaev 1996) or (Williams 1991). We include this section for reasons of completeness, and also because we want to rigorously justify why π_t and $\pi_t^{\mathbf{y}_{0:t}}$ as defined in (2.4) and (2.5) are (random, respectively, deterministic) probability measures, and why we have $\pi_t f = \mathbb{E}\left[f\left(X_t\right)|\sigma\left(Y_{0:t}\right)\right]$ and $\pi_t^{\mathbf{y}_{0:t}} f = \mathbb{E}\left[f\left(X_t\right)|Y_{0:t} = \mathbf{y}_{0:t}\right]$, respectively. We also state the results on conditional probabilities and expectations needed to prove the recurrence formula (2.6) in the following subsection.

Let $(\Omega, \mathcal{F}, P)$ be a probability space and $\mathcal{G} \in \mathcal{F}$ be a sub-σ-algebra of $\mathcal{F}$. The conditional expectation of an integrable $\mathcal{F}$-measurable random variable ξ given $\mathcal{G}$ is defined as the integrable $\mathcal{G}$-measurable random variable, denoted by $\mathbb{E}\left[\xi\,|\mathcal{G}\right]$, with the property

$$\int_A \xi \, dP = \int_A \mathbb{E}\left[\xi\,|\mathcal{G}\right] dP, \qquad (2.28)$$

for all $A \in \mathcal{G}$. Then $\mathbb{E}\left[\xi\,|\mathcal{G}\right]$ exists and is *almost surely* unique. By this we mean that if $\bar{\xi}$ is another $\mathcal{G}$-measurable integrable random variable such that $\int_A \xi \, dP = \int_A \bar{\xi} \, dP$ for all $A \in \mathcal{G}$, then $\mathbb{E}\left[\xi\,|\mathcal{G}\right] = \bar{\xi}$, $P-$a.s..

Here are some of the properties of the conditional expectation.

a. If $\alpha_1, \alpha_2 \in \mathbb{R}$ and ξ_1, ξ_2 are $\mathcal{F}$-measurable, then

$$\mathbb{E}\left[\alpha_1\xi_1 + \alpha_2\xi_2\,|\mathcal{G}\right] = \alpha_1\mathbb{E}\left[\xi_1\,|\mathcal{G}\right] + \alpha_2\mathbb{E}\left[\xi_2\,|\mathcal{G}\right], \quad P\text{-a.s..}$$

b. If $\xi \geq 0$, then $\mathbb{E}\left[\xi\,|\mathcal{G}\right] \geq 0$, P-a.s..
c. If $0 \leq \xi_n \nearrow \xi$, then $\mathbb{E}\left[\xi_n\,|\mathcal{G}\right] \nearrow \mathbb{E}\left[\xi\,|\mathcal{G}\right]$, P-a.s..
d. If $\mathcal{H}$ is a sub-σ-algebra of $\mathcal{G}$, then $\mathbb{E}\left[\mathbb{E}\left[\xi\,|\mathcal{G}\right]|\mathcal{H}\right] = \mathbb{E}\left[\xi\,|\mathcal{H}\right]$, P-a.s..
e. If ξ is $\mathcal{G}$-measurable, then $\mathbb{E}\left[\xi\eta\,|\mathcal{G}\right] = \xi\mathbb{E}\left[\eta\,|\mathcal{G}\right]$, P-a.s..
f. If $\mathcal{H}$ is independent of $\sigma\left(\sigma\left(\xi\right), \mathcal{G}\right)$, then $\mathbb{E}\left[\xi\,|\sigma\left(\mathcal{G}, \mathcal{H}\right)\right] = \mathbb{E}\left[\xi\,|\mathcal{G}\right]$, P-a.s..

The conditional probability of a set $A \in \mathcal{F}$ with respect to the σ-algebra $\mathcal{G}$ is the random variable denoted by $P\left(A|\mathcal{G}\right)$ defined as $P\left(A|\mathcal{G}\right) \triangleq \mathbb{E}\left[I_A\,|\mathcal{G}\right]$, where I_A is the indicator function of the set A. From (2.28) we deduce that $P\left(A \cap B\right) = \int_B P\left(A|\mathcal{G}\right) dP$ for all $B \in \mathcal{G}$. Let $\eta_1, ..., \eta_k$ be $\mathcal{F}$-measurable random variables. Then $\mathbb{E}\left[\xi\,|\eta_1, ..., \eta_k\right]$ – the conditional expectation of ξ with respect to $\eta_1, ..., \eta_k$ – is the conditional expectation of ξ with respect to the σ-algebra generated by $\eta_1, ..., \eta_k$, i.e., $\mathbb{E}\left[\xi\,|\eta_1, ..., \eta_k\right] = \mathbb{E}\left[\xi\,|\sigma\left(\eta_1, ..., \eta_k\right)\right]$ and we have the analogue definition of $P\left(A|\eta_1, ..., \eta_k\right)$ – the conditional probability of A with respect to $\eta_1, ..., \eta_k$.

The fact that $P(A|\mathcal{G})$ is not pointwise uniquely defined may be troublesome. It implies that, for all $A \in \mathcal{B}(\mathbb{R}^{n_x})$, $\pi_t(A)$ is not pointwise uniquely defined but only almost sure. If $A_1, A_2, ... \in \mathcal{B}(\mathbb{R}^{n_x})$ is a sequence of pairwise disjoint sets, then, by properties a. and c.,

$$\pi_t\left(\bigcup_n A_n | \mathcal{G}\right) = \sum_n \pi_t(A_n|\mathcal{G}), \quad P - \text{a.s..}$$

However, for a given ω, the conditional probability $\pi_t(\cdot)(\omega)$ cannot be considered a-priori a measure on $\mathbb{R}^{n_x}$. One might suppose that $\pi_t(\cdot)(\omega)$ is a measure except for a set $\mathcal{N} \in \Omega$ of measure 0. This may not be the case for the following reason. Let $\mathcal{N}(A_1, A_2, ...)$ be the set of points ω such that the countable additivity property fail for these $A_1, A_2, ...$ Then the excluded set is $\mathcal{O}$, where $\mathcal{O} = \bigcup \mathcal{N}(A_1, A_2, ...)$, and the union is taken over all sequences $(A_i)_{i=1}^\infty$, such that for all $i > 0$, $A_i \in \mathcal{B}(\mathbb{R}^{n_x})$ and although the P-measure of $\mathcal{N}(A_1, A_2, ...)$ is zero, the P-measure of $\mathcal{N}$ may not be zero because of the uncountable union. However, it is natural to request that the conditional probability of the signal $\pi_t(\cdot)(\omega)$ be a (probability) measure.

Definition 2.6.1. *Let (Ω, F, P) be a probability, $(E, \mathcal{E})$ be a measurable space, $X : \Omega \to E$ be an $\mathcal{E}/\mathcal{F}$-measurable random element, and $\mathcal{G}$ a sub-σ-algebra of $\mathcal{F}$. A function $Q(\omega, B)$ defined for all $\omega \in \Omega$ and $B \in \mathcal{E}$ is a regular conditional distribution/probability of X with respect to $\mathcal{G}$ if*
(a) for each $\omega \in \Omega$, $Q(\omega, \cdot)$ is a probability measure on $(E, \mathcal{E})$.
(b) for each $B \in \mathcal{F}$, $Q(\cdot, B)$ is $\mathcal{G}$-measurable and $Q(\cdot, B) = P(X \in B|\mathcal{G})$, P−a.s..

Definition 2.6.2. *A measurable space $(E, \mathcal{E})$ is a Borel space if there exists a one-to-one mapping $f : (E, \mathcal{E}) \to (\mathbb{R}, \mathcal{B}(\mathbb{R}))$ such that $f(E) \in \mathcal{B}(\mathbb{R})$, f is $\mathcal{E}$-measurable and f^{-1} is $\mathcal{B}(\mathbb{R})/\mathcal{E}$- measurable.*

We state the following theorem without proof. The reader can find a proof in (Parthasarathy 1967) pp. 146-150.

Theorem 2.6.3. *Let $X = X(\omega)$ be a random element with values in a Borel space $(E, \mathcal{E})$. There then exists a regular conditional distribution of X with respect to $\mathcal{G}$.*

Since $(\mathbb{R}^{n_x}, \mathcal{B}(\mathbb{R}^{n_x}))$ is a Borel space, there exists a regular conditional distribution of X_t with respect to $\sigma(Y_{0:t})$. Therefore, if for all $A \in \mathcal{B}(\mathbb{R}^{n_x})$, we assign to $\pi_t(A)$ the value $Q(\cdot, A)$ (we can do this because $\pi_t(A)$ is defined only almost surely), where Q is the regular conditional distribution of X_t with respect to $\sigma(Y_{0:t})$, then π_t is a probability measure.

Remark 5. *If π_t is defined as above, then the identity in (2.4), $\pi_t f = \mathbb{E}[f(X_t)|\sigma(Y_{0:t})]$ holds true, P−a.s., for all $\mathcal{B}(\mathbb{R}^{n_x})$-measurable functions f.*

Proof. If $f = I_B$, where I_B is the characteristic function of an arbitrary set $B \in \mathcal{B}(\mathbb{R}^{n_x})$ the required formula holds by Definition 2.6.1(b). Consequently, it holds for simple functions. Let $\xi \geq 0$ be an arbitrary non-negative function and $0 \leq f_n \nearrow f$, where f_n are simple functions. Using property c. of the conditional expectation we have $\mathbb{E}[f(X_t) | \sigma(Y_{0:t})] = \lim_{n \to \infty} \mathbb{E}[f_n(X_t) | \sigma(Y_{0:t})]$, P-a.s.. But, since π_t is a probability measure for each $\omega \in \Omega$, we have, by Monotone Convergence Theorem, $\lim_{n \to \infty} \pi_t f_n = \pi_t f$. Hence, the identity holds for all non-negative measurable functions. The general case reduces to this one by using the representation $\xi = \xi_+ - \xi_-$. $\qquad\square$

Let ξ, η be $\mathcal{F}$-measurable functions. Since $\mathbb{E}[\xi | \eta]$ is a $\sigma(\eta)$-measurable random variable, there exists a function $m = m(y) : \mathbb{R} \to \mathbb{R}$ such that $m(\eta) = \mathbb{E}[\xi | \eta]$. We denote $m(y)$ by $\mathbb{E}[\xi | \eta = y]$ and call it the conditional expectation of ξ with respect to the event $\{\eta = y\}$. Therefore, via the change of variable formula, we have, for all $A \in \mathcal{B}(\mathbb{R})$,

$$\int_{\{\omega; \eta \in A\}} \xi(\omega) P(d\omega) = \int_{\{\omega; \eta \in A\}} m(\eta(\omega)) dP(d\omega) = \int_A m(y) P_\eta(dy),$$
(2.29)

where P_η is the probability distribution of η. We can use (2.29) as the defining formula for conditional expectation of ξ with respect to the event $\{\eta = y\}$. Indeed, the conditional expectation of ξ with respect to the event $\{\eta = y\}$, denoted by $\mathbb{E}[\xi | \eta = y]$ is the $\mathcal{B}(\mathbb{R})$-measurable random variable such that

$$\int_{\{\omega; \eta \in A\}} \xi dP = \int_A \mathbb{E}[\xi | \eta = y] P_\eta(dy)$$
(2.30)

holds true for all $A \in \mathcal{B}(\mathbb{R})$ (the definition has the obvious extension to ξ, η random vectors). Again, the function $y \to \mathbb{E}[\xi | \eta = y]$ is P_η-*almost surely* unique.

If we know $\mathbb{E}[\xi | \eta = y]$, *then we can deduce* $\mathbb{E}[\xi | \eta]$, *and conversely from* $\mathbb{E}[\xi | \eta]$ *we can reconstruct the conditional expectation* $\mathbb{E}[\xi | \eta = y]$. The expectation $\mathbb{E}[\xi | \eta = y]$ satisfies the following identity P_η−a.s.

$$\mathbb{E}[\xi f(\eta) | \eta = y] = f(y) \mathbb{E}[\xi | \eta = y]$$
(2.31)

for all $f \in \mathcal{B}(\mathbb{R})$. Moreover, if ξ and η are independent and $g \in \mathcal{B}(\mathbb{R}^2)$, then P_η−a.s.,

$$\mathbb{E}[\xi | \eta = y] = \mathbb{E}[\xi]$$
(2.32)
$$\mathbb{E}[g(\xi, \eta) | \eta = y] = \mathbb{E}[g(\xi, y)].$$
(2.33)

The conditional probability of the event given by $A \in \mathcal{F}$ with respect to the event/under the condition that $\{\eta = y\}$ (notation $P(A | \eta = y)$) is defined as $\mathbb{E}[I_A | \eta = y]$. $P(A | \eta = y)$ is the $\mathcal{B}(\mathbb{R})$-measurable random variable such that

$$P\left(A \cap \{\eta \in B\}\right) = \int_B P\left(A \,|\, \eta = y\right) P_\eta\left(dy\right) \tag{2.34}$$

for all $B \in \mathcal{B}\left(\mathbb{R}\right)$.

Now, if π_t is the regular conditional distribution of X_t with respect to $Y_{0:t}$ then, for all $A \in \mathcal{B}\left(\mathbb{R}^{n_x}\right)$, $\pi_t\left(A\right)$ is $Y_{0:t}$-measurable. Hence, there exists a function $m = m\left(A, \mathbf{y}_{0,t}\right) : \mathcal{B}\left(\mathbb{R}^{n_x}\right) \times \mathrm{Im}\left(Y_{0:t}\right) \to \mathbb{R}$, such that, *pointwise* (and not just almost surely),

$$\pi_t\left(A\right)\left(\omega\right) = m\left(A, Y_{0:t}\left(\omega\right)\right).$$

Since, for all $\omega \in \Omega$, $\pi_t\left(\cdot\right)\left(\omega\right)$ is a probability measure, it follows that for all $\mathbf{y}_{0,t} \in \mathrm{Im}\left(Y_{0:t}\right)$, $m\left(\cdot, \mathbf{y}_{0,t}\right)$ is a probability measure on $\mathcal{B}\left(\mathbb{R}^{n_x}\right)$, Then, just as above, we assign to $\pi_t^{\mathbf{y}_{0:t}}\left(A\right)$ the value $m\left(A, \mathbf{y}_{0,t}\right)$ and we have that $\pi_t^{\mathbf{y}_{0:t}}$ is a probability measure and $\pi_t^{\mathbf{y}_{0:t}} f = \mathbb{E}\left[f\left(X_t\right) | Y_{0:t} = \mathbf{y}_{0:t}\right]$ for all $f \in B\left(\mathbb{R}^{n_x}\right)$.

2.6.2   The recurrence formula for the conditional distribution of the signal

We first need to prove the following lemma

Lemma 2.6.4. *Let $P_{Y_{s:t}} \in \mathcal{P}\left(\left(\mathbb{R}^{n_y}\right)^{t-s+1}\right)$ be the probability distribution of $Y_{s:t}$ and λ be the Lebesgue measure λ on $\left(\left(\mathbb{R}^{n_y}\right)^{t-s+1}, \mathcal{B}\left(\left(\left(\mathbb{R}^{n_y}\right)^{t-s+1}\right)\right)\right)$. Then, for all $0 < s \le t < \infty$, $P_{Y_{s:t}}$ is absolutely continuous with respect to λ and its Radon-Nikodym derivative is*

$$\frac{dP_{Y_{s:t}}}{d\lambda}\left(\mathbf{y}_{s:t}\right) = \Upsilon\left(\mathbf{y}_{s:t}\right) \triangleq \int_{\left(\mathbf{R}^{n_x}\right)^{t-s+1}} \prod_{i=s}^{t} g_i(\mathbf{y}_i - h\left(i, \mathbf{x}_i\right)) P_{X_{s,t}}\left(d\mathbf{x}_{s:t}\right).$$

Proof. Let $C_{s:t} = C_s \times \ldots \times C_t$, where C_r are arbitrary Borel sets, $C_r \in \mathcal{B}\left(\mathbb{R}^{n_y}\right)$ for all $s \le r \le t$. We need to prove that

$$P_{Y_{s:t}}\left(C_{s:t}\right) = P\left(\{Y_{s:t} \in C_{s:t}\}\right) = \int_{C_{s:t}} \Upsilon\left(\mathbf{y}_{s:t}\right) d\mathbf{y}_s \ldots d\mathbf{y}_t. \tag{2.35}$$

By (2.34), or rather by its vector-valued analogue.

$$P\left(\{Y_{s:t} \in C_{s:t}\}\right) = \int_{\left(\mathbf{R}^{n_x}\right)^{t-s+1}} P\left(Y_{s:t} \in C_{s:t} \,|\, X_{s:t} = \mathbf{x}_{s:t}\right) P_{X_{s:t}}\left(d\mathbf{x}_{s:t}\right),$$

$$\tag{2.36}$$

and using the fact that, for all i, X_i and W_i are independent and also $W_s, ..., W_t$ are independent, we have from (2.33)

$$
\begin{aligned}
P\left(Y_{s:t} \in C_{s:t} \,|\, X_{s:t} = \mathbf{x}_{s:t}\right) &= \mathbb{E}\left[\prod_{i=s}^{t} I_{\{C_i\}}\left(h(i, X_i) + W_i\right) \,|\, X_{0,t} = \mathbf{x}_{0:t}\right] \\
&= \mathbb{E}\left[\prod_{i=s}^{t} I_{\{C_i\}}\left(h(i, \mathbf{x}_i) + W_i\right)\right] \\
&= \prod_{i=s}^{t} \mathbb{E}\left[I_{\{C_i\}}\left(h(i, \mathbf{x}_i) + W_i\right)\right] \\
&= \prod_{i=s}^{t} \int_{C_i} g_i(\mathbf{y}_i - h(i, \mathbf{x}_i)) d\mathbf{y}_i. \qquad (2.37)
\end{aligned}
$$

By combining (2.36) and (2.37) and applying Fubini, we obtain (2.35). $\quad\square$

Proposition 2.6.5. *The conditional distribution of the signal satisfies the following recurrence relations, for $t \geq 0$:*

$$
\left\{
\begin{aligned}
\frac{d\pi_t^{\mathbf{y}_{0:t}}}{dp_t^{\mathbf{y}_{0:t-1}}} &= \frac{g_t^{\mathbf{y}_t}}{\int_{\mathbb{R}^{n_x}} g_t^{\mathbf{y}_t}(\mathbf{x}_t) p_t^{\mathbf{y}_{0:t-1}}(d\mathbf{x}_t)} \\
p_{t+1}^{\mathbf{y}_{0:t}} &= \pi_t^{\mathbf{y}_{0:t}} K_t
\end{aligned}
\right. .
$$

where $g_t^{\mathbf{y}_t} \triangleq g(\mathbf{y}_t - h(t, \cdot))$ and the recurrence is satisfied $P_{Y_{0:t}}$-almost surely or, equivalently, λ-almost surely.

Proof. We prove first the second identity since it is the simpler of the two. For all $f \in B(\mathbb{R}^{n_x})$, we have, using the Markov property of X, $\mathbb{E}\left[f(X_{t+1}) \,|\, \mathcal{F}_t^X\right] = \mathbb{E}\left[f(X_{t+1}) \,|\, X_t\right] = K_t f(X_t)$. Now, since $W_{0:t}$ is independent of $X_{0:t+1}$, we have, using property f. of conditional expectations,

$$
\mathbb{E}\left[f(X_{t+1}) \,|\, \sigma\left(\mathcal{F}_t^X, \sigma(W_{0:t})\right)\right] = \mathbb{E}\left[f(X_{t+1}) \,|\, \mathcal{F}_t^X\right].
$$

Hence, using property d. of conditional expectations,

$$
\begin{aligned}
p_{t+1} f &= \mathbb{E}[f(X_{t+1}) \,|\, Y_{0:t}] \\
&= \mathbb{E}\left[\mathbb{E}\left[f(X_{t+1}) \,|\, \sigma\left(\mathcal{F}_t^X, \sigma(W_{0:t})\right)\right] \,|\, \sigma(Y_{0:t})\right] \\
&= \mathbb{E}\left[K_t f(X_t) \,|\, \sigma(Y_{0:t})\right] \\
&= \pi_t K_t f
\end{aligned}
$$

which implies $p_{t+1}^{\mathbf{y}_{0:t}} = \pi_t^{\mathbf{y}_{0:t}} K_t$. We prove now the first identity. Let $C_{0:t} = C_0 \times ... \times C_t$ where C_r are arbitrary Borel sets, $C_r \in \mathcal{B}(\mathbb{R}^{n_y})$, for all $0 \leq r \leq t$. We need to prove that

$$
\int_{C_{0:t}} \pi_t^{\mathbf{y}_{0:t}}(A) \, P_{Y_{0:t}}(d\mathbf{y}_{0:t}) = \int_{C_{0:t}} \frac{\int_A g_t^{\mathbf{y}_t}(\mathbf{x}_t) p_t^{\mathbf{y}_{0:t-1}}(d\mathbf{x}_t)}{\int_{\mathbb{R}^{n_x}} g_t^{\mathbf{y}_t}(\mathbf{x}_t) p_t^{\mathbf{y}_{0:t-1}}(d\mathbf{x}_t)} P_{Y_{0:t}}(d\mathbf{y}_{0:t}).
$$

$$(2.38)$$

By (2.34) the first term in (2.38) is equal to $P(\{X_t \in A\} \cap \{Y_{0:t} \in C_{0:t}\})$. So, we need to prove the same thing for the second term. Since we have $\sigma(X_{0:t}, W_{0:t-1}) \supset \sigma(X_t, Y_{0:t-1})$, one obtains, using property f. of conditional expectations,

$$P(Y_t \in A_t | X_t, Y_{0:t-1}) = P(P(Y_t \in A_t | X_{0:t}, W_{0:t-1}) | X_t, Y_{0:t-1}), \quad (2.39)$$

and using property d. of conditional expectations and (2.37),

$$
\begin{aligned}
P(Y_t \in A_t | X_{0:t}, W_{0:t-1}) &= P(Y_t \in A_t | X_{0:t}) \\
&= P\left(Y_{0:t} \in (\mathbb{R}^{n_y})^t \times A_t | X_{0:t}\right) \\
&= \int_{A_t} g_t(\mathbf{y}_t - h(t, X_t)) d\mathbf{y}_t. \quad (2.40)
\end{aligned}
$$

From (2.39) and (2.40), we derive $P(Y_t \in A_t | X_t, Y_{0:t-1}) = \int_{A_t} g_t(\mathbf{y}_t - h(t, X_t) d\mathbf{y}_t$, which gives us

$$P(Y_t \in A_t | X_t = \mathbf{x}_t, Y_{0:t-1} = \mathbf{y}_{0:t-1}) = \int_{A_t} g_t^{\mathbf{y}_t}(\mathbf{x}_t) d\mathbf{y}_t. \quad (2.41)$$

Hence,

$$
\begin{aligned}
P_{Y_{0:t}}(A_{0:t}) &= P(\{Y_t \in A_t\} \cap \{X_t \in \mathbf{R}^{n_x}\} \cap \{Y_{0:t-1} \in A_{0:t-1}\}) \\
&= \int_{\mathbf{R}^{n_x} \times A_{0:t-1}} P(Y_t \in A_t | X_t = \mathbf{x}_t, Y_{0:t-1} = \mathbf{y}_{0:t-1}) P_{X_t, Y_{0:t-1}}(d\mathbf{x}_t d\mathbf{y}_{0:t-1}) \\
&= \int_{\mathbf{R}^{n_x} \times A_{0:t-1}} \int_{A_t} g_t^{\mathbf{y}_t}(\mathbf{x}_t) d\mathbf{y}_t p_t^{\mathbf{y}_{0:t-1}}(d\mathbf{x}_t) P_{Y_{0:t-1}}(d\mathbf{y}_{0:t-1}) \\
&= \int_{A_{0:t}} \int_{\mathbf{R}^{n_x}} g_t^{\mathbf{y}_t}(\mathbf{x}_t) p_t^{\mathbf{y}_{0:t-1}}(d\mathbf{x}_t) d\mathbf{y}_t P_{Y_{0:t-1}}(d\mathbf{y}_{0:t-1}). \quad (2.42)
\end{aligned}
$$

In (2.42), we used the identity

$$P_{X_t, Y_{0:t-1}}(d\mathbf{x}_t d\mathbf{y}_{0:t-1}) = p_t^{\mathbf{y}_{0:t-1}}(d\mathbf{x}_t) P_{Y_{0:t-1}}(d\mathbf{y}_{0:t-1}), \quad (2.43)$$

which is, again, a consequence of the vector-valued equivalent of (2.34), since for all $A \in \mathcal{B}(\mathbb{R}^{n_x})$ we have

$$
\begin{aligned}
P((X_t, Y_{0:t-1}) \in A \times C_{0:t-1}) &= \int_{\mathbf{C}_{0:t-1}} P(X_t \in A | Y_{0:t-1} \in \mathbf{y}_{0:t-1}) P_{Y_{0:t-1}}(d\mathbf{y}_{0:t-1}) \\
&= \int_{A \times \mathbf{C}_{0:t-1}} p_t^{\mathbf{y}_{0:t-1}}(d\mathbf{x}_t) P_{Y_{0:t-1}}(d\mathbf{y}_{0:t-1}).
\end{aligned}
$$

From (2.42) we see that

$$P_{Y_{0:t}}(d\mathbf{y}_{0:t}) = \int_{\mathbf{R}^{n_x}} g_t^{\mathbf{y}_t}(\mathbf{x}_t) p_t^{\mathbf{y}_{0:t-1}}(d\mathbf{x}_t) d\mathbf{y}_t P_{Y_{0:t-1}}(d\mathbf{y}_{0:t-1}).$$

Hence, the second term in (2.38) is equal to

$$\Xi \triangleq \int_{C_{0:t}} \int_A g_t^{\mathbf{y}_t}(\mathbf{x}_t) p_t^{\mathbf{y}_{0:t-1}}(d\mathbf{x}_t) d\mathbf{y}_t P_{Y_{0:t-1}}(d\mathbf{y}_{0:t-1}),$$

which, in turn, using again (2.41) and (2.43), is equal to

$$\Xi = \int_{A \times C_{0:t-1}} \left(\int_{C_t} g_t^{\mathbf{y}_t}(\mathbf{x}_t) d\mathbf{y}_t \right) p_t^{\mathbf{y}_{0:t-1}} (d\mathbf{x}_t) P_{Y_{0:t-1}} (d\mathbf{y}_{0:t-1})$$

$$= \int_{A \times C_{0:t-1}} P\left(Y_t \in C_t | X_t = \mathbf{x}_t, Y_{0:t-1} = \mathbf{y}_{0:t-1}\right) P_{X_t, Y_{0:t-1}} (d\mathbf{x}_t d\mathbf{y}_{0:t-1})$$

$$= P\left(\{X_t \in A\} \cap \{Y_{0:t} \in C_{0:t}\}\right).$$

$\square$

Acknowledgment. The author would like to thank Jessica Gaines for her careful reading of the manuscript and many useful suggestions.

3

Interacting Particle Filtering With Discrete Observations

Pierre Del Moral
Jean Jacod

3.1 Introduction

We consider a pair of processes (X, Y), where X represents the state of a system (or signal) and Y represents the observation: X may take its values in an arbitrary measurable space $(E, \mathcal{E})$, but it is important for what follows that Y take its values in $\mathbb{R}^q$ for some $q \geq 1$.

A distinctive feature of our study is that, although the pair (X, Y) might be indexed by $\mathbb{R}_+$, the actual observations take place at **discrete times** only: this is not for mathematical convenience, but because discrete time observations arise in a natural way as soon as real data are implicated. We will shall assume that the pair (X, Y) is Markov (see below). So, up to a change of time, one can suppose that the observation times are the non-negative integers, and the basic assumption will then be that the process $(X_n, Y_n)_{n \in \mathbb{N}}$ is a (possibly non-homogeneous) Markov chain.

In this paper we are interested in the nonlinear filtering problem (in short, **NLF**). That is, we want to find the one step predictor conditional probability given for each $n \in \mathbb{N}$ and each measurable function f on E such that $f(X_n)$ is integrable by

$$\eta_{n,Y} f = \mathbb{E}(f(X_n)|Y_0, \ldots, Y_{n-1})$$

(with the convention that $\eta_{0,Y}$ is the law of X_0), and the so-called filter conditional distribution given by

$$\widehat{\eta}_{n,Y} f = \mathbb{E}(f(X_n)|Y_0, \ldots, Y_n).$$

Throughout the chapter, we fix the observations $Y_n = y_n$, $n \in \mathbb{N}$, **and instead of** $\eta_{n,Y}$ **and** $\widehat{\eta}_{n,Y}$ **we simply write** η_n **and** $\widehat{\eta}_n$.

Our aim is to design an interacting particle system approach (**IPS** in short) for numerical computation of the distributions η_n and $\widehat{\eta}_n$.

Let us now isolate the two types of NLF problems covered by our work.

- **Case A**

  The state signal $(X_n)_{n \in \mathbb{N}}$ is an E-valued non-homogeneous Markov chain with 1-step transition probabilities $(Q_n)_{n \geq 1}$ (i.e. Q_n is the transition from time $n - 1$ to time n) and initial law η_0. The observation sequence $(Y_n)_{n \in \mathbb{N}}$ takes the form

  $$Y_n = H_n(X_n, V_n) \tag{3.1.1}$$

  for some measurable functions H_n from $E \times F$ (with $(F, \mathcal{F})$ an auxiliary measurable space) into $\mathbb{R}^q$, and where the sequence (V_n) is i.i.d. with values in $(F, \mathcal{F})$. The basic assumption concerning H and the law of the V_n's is that for any $x \in E$ and any n the variable $H_n(x, V_n)$ admits a *strictly positive* and *bounded* density $y \mapsto \bar{g}_n(x, y)$ w.r.t. Lebesgue measure on $\mathbb{R}^q$.

- **Case B**

  The pair signal/observation process (X, Y) is an $E \times \mathbb{R}^q$-valued non-homogeneous Markov chain with 1-step transition probabilities equal to $(P_n)_{n \geq 1}$ and initial law μ_0 having the form (dy denotes Lebesgue measure)

  $$\left. \begin{aligned} \mu_0(dx, dy) &= \eta_0(dx)\bar{G}_0(x, y)dy \\ P_n(x, y; dx', dy') &= Q_n(x, dx')\bar{G}_n(x, y; x', y')dy' \end{aligned} \right\} \tag{3.1.2}$$

  where each Q_n is a transition probability and η_0 is a probability and each function $\bar{G}_n$ is non-negative measurable and *bounded*. For homogeneity of notation we can consider $\bar{G}_0$ to be a function on $E \times \mathbb{R}^q \times E \times \mathbb{R}^q$, with the convention $\bar{G}_0(x, y; x', y') = \bar{G}_0(x', y')$. Notice that necessarily $\int \bar{G}_n(x, y; x', y')dy' = 1$.

Clearly, Case A is a particular case of Case B, with

$$\bar{G}_n(x, y; x', y') = \bar{g}_n(x', y'). \tag{3.1.3}$$

However, we single out Case A: it is easier to handle, and in fact the filtering problem in Case B will be reduced to an associated Case A problem. Observe also that in Case B the process X itself is Markov.

A crucial practical advantage of the first category of NLF problems is that it leads to a natural IPS-approximating scheme which is easily implemented. The first idea is to consider the equations that sequentially update the flow of distributions $\{\eta_n \; ; \; n \geq 0\}$, which are of the form

$$\eta_n = \Phi_n(\eta_{n-1}), \qquad n \geq 1, \tag{3.1.4}$$

with continuous mappings Φ_n (to be defined later) on the set $\mathbf{M}_1(E)$ of all probability measures on E. In this formulation, the NLF problem is now reduced to the problem of solving a dynamical system taking values in the infinite dimensional state-space $\mathbf{M}_1(E)$. To overcome this problem, a natural idea is to approximate η_n for $n \geq 1$ by a sequence of empirical measures

$$\eta_n^N = \frac{1}{N} \sum_{i=1}^{N} \delta_{\xi_n^i}, \tag{3.1.5}$$

(δ_a stands for the Dirac measure at $a \in E$) associated with a system of N interacting particles $\xi_n = (\xi_n^1, \dots, \xi_n^N)$ moving in the set E. It is indeed natural, in view of $(3.1.4)$ above, to construct the sequence (ξ_n) as a Markov chain taking values in E^N, starting with the initial distribution $\widehat{\eta}_0$ and with 1-step transition probabilities $\widetilde{Q}_n$ given by

$$\widehat{\eta}_0(dx) = \prod_{p=1}^{N} \eta_0(dx^p), \qquad \widetilde{Q}_n(z, dx) = \prod_{p=1}^{N} \Phi_n(m(z))(dx^p), \tag{3.1.6}$$

where $dx \overset{\text{def}}{=} dx^1 \times \cdots \times dx^N$ is an infinitesimal neighborhood of the point $x = (x^1, \dots, x^N) \in E^N$, $z = (z^1, \dots, z^N) \in E^N$ and where $m(z)$ is the empirical distribution associated with z, i.e. $m(z) = \frac{1}{N} \sum_{i=1}^{N} \delta_{z^i}$.

The rationale behind this is that η_n^N is the empirical measure associated with N independent variables with common law $\Phi_n(\eta_{n-1}^N)$, so as soon as η_{n-1}^N is a good approximation of η_{n-1} then, in view of $(3.1.4)$, η_n^N should be a good approximation of η_n.

The chapter is organised as follows. Section 3.2 is devoted to recalling some facts about NLF problems. In Section 3.3 we present an IPS to solve Case A. Since such an IPS is based upon Monte Carlo simulations, we first study in detail in Subsection 3.3.1 the situation in which all laws are explicitly known and where the simulation of random variables according to the laws $Q_n(x, .)$ is possible. Then, in Subsection 3.3.2, we consider the case where the laws are only "approximately" known and the simulations can be carried out according to laws which are close to the $Q_n(x, .)$'s. In Section 3.4, we study Case B. Finally, we give some applications to a condition under which the state process X is the solution to a stochastic differential equation.

Most of the results presented here are taken from (Del Moral and Jacod 1998), (Del Moral and Jacod 1999), (Del Moral and Miclo 2000) and (Del Moral 1998). However, the presentation here is quite different, and some new results have been added, mainly in the way Subcases A2 and B2 are treated, and Case B is more general here than in the above-mentioned papers. The ideas behind the proofs are provided, as well as full proofs for the new results.

Notice that here we restrict our attention to bounded test functions. The analysis for unbounded test functions is more involved. See (Del Moral and Jacod 1999) and (Del Moral and Guionnet 1999).

3.2   Nonlinear filtering: general facts

1) In this section we remind the reader of some elementary facts about nonlinear filtering. In the sequel, we use the traditional notation for transition kernels: if P and Q are two transition kernels and μ is a measure and f is a measurable functions (all on $(E, \mathcal{E})$), then we have another transition kernel PQ (the usual product, or composition), and a function Pf, and a measure μP, and a number μPf. We also denote by $\mathbf{M}_1(E)$ the set of all probability measures on $(E, \mathcal{E})$.

First, we give some facts that hold under Case B, and thus under Case A. Recall that the observations $y_0, y_1, \cdots$ are given and fixed, and introduce the functions

$$G_n(x, x') = \bar{G}_n(x, y_{n-1}; x', y_n) \tag{3.2.1}$$

(recall that $\bar{G}_0(x, y; x', y')$ does not actually depend on y, so the above is well defined even for $n = 0$).

Below, $\mathbb{E}$ denotes the expectation on the original probability space, on which the chain (X, Y) is Markov with the prescribed initial condition. Then Bayes' formula gives an explicit expression for η_n and $\hat{\eta}_n$, namely

$$\eta_n(f) = \frac{\gamma_n(f)}{\gamma_n(1)}, \qquad \hat{\eta}_n(f) = \frac{\hat{\gamma}_n(f)}{\hat{\gamma}_n(1)} \tag{3.2.2}$$

for any measurable function f for which the following expressions make sense:

$$\left.\begin{array}{l} \gamma_n(f) = \mathbb{E}\left(f(X_n) \prod_{k=0}^{n-1} G_k(X_{k-1}, X_k)\right), \\[2mm] \hat{\gamma}_n(f) = \mathbb{E}\left(f(X_n) \prod_{k=0}^{n} G_k(X_{k-1}, X_k)\right), \end{array}\right\} \tag{3.2.3}$$

with the convention $\prod_\emptyset = 1$. This is because $G_0(x, x')$ does not depend on x, the value X_{-1} does not actually occur in the formula (3.2.3). Indeed, (3.2.2) comes from the following fact. Observing for example that $\hat{\gamma}_n(f)$ is a function of the observed values $y_0, \ldots, y_n$, we see that

$$(y_0, \ldots, y_n) \mapsto \hat{\gamma}_n(1)(y_0, \ldots, y_n)$$

is the density of the variable $(Y_0, \ldots, Y_n)$ w.r.t. Lebesgue measure on $(\mathbb{R}^q)^{n+1}$.

Observe that γ_n and $\hat{\gamma}_n$ can be considered finite positive measures (since each function G_n is bounded by hypothesis). These are often referred to as *Feynman-Kac formulae*. Notice that $\gamma_0 = \eta_0$, so (3.2.2) gives that the prediction measure at time 0 is indeed the initial measure η_0, since at this stage we have no observation at all.

2) Now we derive some (well known) formulae which allow us to compute the filter and the one-step prediction measure in a recursive way. First, the

Markov property in (3.2.9) gives for $n \geq 1$:

$$\widehat{\gamma}_n(f) = \widehat{\gamma}_{n-1}(\widehat{L}_n f), \qquad (3.2.4)$$

where

$$\widehat{L}_n f(x) = \int Q_n(x, dx') G_n(x, x') f(x').$$

Therefore, if we set for $0 \leq p \leq n$:

$$\widehat{L}_{p,n} = \widehat{L}_{p+1} \widehat{L}_{p+2} \ldots \widehat{L}_n, \qquad (3.2.5)$$

with the convention $\widehat{L}_{n,n} = Id$, we readily get

$$\widehat{\gamma}_n = \widehat{\gamma}_p \widehat{L}_{p,n}, \qquad \widehat{\gamma}_0(f) = \eta_0(fG_0).$$

In other words, we have

$$\widehat{\gamma}_n(f) = \eta_0(G_0 \widehat{L}_{0,n} f), \qquad (3.2.6)$$

and using (3.2.2) and (3.2.4), we get

$$\widehat{\eta}_n = \widehat{\Psi}_n(\widehat{\eta}_{n-1}), \qquad \text{where} \quad \widehat{\Psi}_n(\eta)(f) = \frac{\eta \widehat{L}_n f}{\eta \widehat{L}_n 1}.$$

Next, use again the Markov property to obtain for $n \geq 1$:

$$\gamma_n(f) = \widehat{\gamma}_{n-1}(Q_n f), \qquad (3.2.7)$$

and thus by (3.2.6),

$$\gamma_n(f) = \begin{cases} \eta_0(f) & \text{if } n = 0 \\ \eta_0(G_0 \widehat{L}_{0,n-1} Q_n f) & \text{if } n \geq 1. \end{cases} \qquad (3.2.8)$$

It also follows from (3.2.2) and (3.2.7) that

$$\eta_n(f) = \widehat{\eta}_{n-1}(Q_n f) \qquad \text{if } n \geq 1.$$

3) Let us now expand upon what we learned under Case A. First, all the foregoing remains true in Case A with $\bar{G}_n$ given by (3.1.3). So we set

$$g_n(x) = \bar{g}_n(x, y_n),$$

and (3.2.3) writes as

$$\gamma_n(f) = \mathbb{E}\left(f(X_n) \prod_{k=0}^{n-1} g_k(X_k) \right) \qquad (3.2.9)$$

and

$$\widehat{\gamma}_n(f) = \mathbb{E}\left(f(X_n) \prod_{k=0}^{n} g_k(X_k) \right).$$

Next, in (3.2.4), $\widehat{L}_n$ becomes $\widehat{L}_n(f) = Q_n(g_n f)$. Here, we also have a handy recursive formula for γ_n. If we set

$$L_{p,n} = L_{p+1}L_{p+2}\ldots L_n, \quad \text{with} \quad L_n f(x) = g_{n-1}(x)Q_n f(x) \quad (3.2.10)$$

for $0 \le p \le n$ and with the convention $L_{n,n} = Id$, we deduce from (3.2.8) that

$$\gamma_0 = \eta_0, \quad \gamma_n = \gamma_p L_{p,n}, \quad \widehat{\gamma}_n(f) = \gamma_n(f g_n). \quad (3.2.11)$$

If we next introduce the following mappings $(\Psi_n)_{n \ge 0}$ and $(\Phi_n)_{n \ge 1}$ from $\mathbf{M}_1(E)$ into itself by

$$\Psi_n(\eta)(f) = \frac{\eta(f g_n)}{\eta(g_n)},$$

$$\Phi_n(\eta)(f) = \Psi_{n-1}(\eta)Q_n f = \frac{\eta L_n f}{\eta L_n 1}, \quad (3.2.12)$$

we have

$$\widehat{\eta}_n = \Psi_n(\eta_n), \quad \eta_{n+1} = \Phi_{n+1}(\eta_n). \quad (3.2.13)$$

Finally, in view of (3.2.7) and (3.2.11) and (3.2.2) we get

$$\gamma_{n+1}(1) = \widehat{\gamma}_n(1) = \gamma_n(g_n) = \eta_n(g_n)\gamma_n(1),$$

and thus have for $n \ge 0$

$$\gamma_n(1) = \prod_{p=0}^{n-1} \eta_p(g_p), \quad \gamma_n(f) = \eta_n(f) \prod_{p=0}^{n-1} \eta_p(g_p), \quad (3.2.14)$$

with the usual convention $\prod_\emptyset = 1$.

3.3   An interacting particle system under Case A

3.3.1   Subcase A1

We label *Subcase A1* the Case A, when one knows explicitly the densities $\bar{g}_n$ (hence all functions g_n), as well as η_0 and Q_n, and one also knows how to simulate random variables exactly according to the laws η_0 and $Q_n(x,\,.)$ for all $x \in E$ and $n \ge 1$.

1) Let us now define our IPS for approximating the flows η_n and $\widehat{\eta}_n$. This is done according to (3.1.5) and (3.1.6), but we will be more precise in our description below.

In fact we have two nested particle systems, each with size N. First, N particles $\xi_n = (\xi_n^i)_{1 \le i \le N}$ at time n which are used for approximating η_n by means of the empirical measure η_n^N given by (3.1.5); next, N particles $\widehat{\xi}_n = (\widehat{\xi}_n^i)_{1 \le i \le N}$ at time n which are used as an intermediate step. The resulting motion of the particles decomposes into two separate mechanisms $\xi_n \longrightarrow \widehat{\xi}_n \longrightarrow \xi_{n+1}$, which can be modelled as follows.

The motion of these particles is defined on an auxiliary probability space, with respect to which the measure and expectation are denoted by $\widetilde{\mathbb{P}}$ and $\widetilde{\mathbb{E}}$, to differentiate them from the measure and expectation defined on the original space on which the basic process (X, Y) lives. We also denote by $\mathcal{G}_n$ the σ-field generated by the variables ξ_p for $p \leq n$ and $\widehat{\xi}_p$ for $p < n$, while $\widehat{\mathcal{G}}_n$ is the σ-field generated by the variables ξ_p and $\widehat{\xi}_p$ for $p \leq n$.

First, the variables ξ_0^i are drawn independently according to the initial law η_0. Next, the mechanism proceeds, by induction on n, according to the following two steps Markov rule.

Selection/Updating

$$\widetilde{\mathbb{P}}(\widehat{\xi}_n \in (dx_1, \dots, dx_N)|\mathcal{G}_n) = \prod_{p=1}^{N} \sum_{i=1}^{N} \frac{g_n(\xi_n^i)}{\sum_{j=1}^{N} g_n(\xi_n^j)} \delta_{\xi_n^i}(dx_p). \qquad (3.3.1)$$

Mutation/Prediction

$$\widetilde{\mathbb{P}}(\xi_{n+1} \in (dz_1, \dots, dz_N)|\widehat{\mathcal{G}}_n) = \prod_{p=1}^{N} Q_{n+1}(\widehat{\xi}_n^p, dz_p). \qquad (3.3.2)$$

Once again, the approximated measure η_n^N is defined by (3.1.5), and by analogy with equation (3.2.13) we approximate $\widehat{\eta}_n$ by

$$\widehat{\eta}_n^N = \Psi_n(\eta_n^N).$$

So in the selection transition, one updates the positions in accordance with the fitness functions $\{g_n; n \geq 1\}$ and the current configuration. More precisely, at each time $n \geq 0$, each particle examines the system of particles $\xi_n = (\xi_n^1, \dots, \xi_n^N)$ and chooses randomly a site ξ_n^i, $1 \leq i \leq N$, with a probability which depends on the entire configuration ξ_n and given by

$$\frac{g_n(\xi_n^i)}{\sum_{j=1}^{N} g_n(\xi_n^j)}.$$

This mechanism is called the Selection/Updating transition because the particles are selected for reproduction, the most fit individuals being more likely to be selected. In other words, this transition allows heavy particles to give birth to some particles at the expense of light particles, which die. The second mechanism is called Mutation/Prediction because at this step each particle evolves randomly according to a given transition probability kernel.

The preceding scheme is clearly a system of interacting particles undergoing adaptation in a time non-homogeneous environment represented by the fitness functions $\{g_n; n \in \mathbb{N}\}$.

In other words, we approximate the two-step transitions of the system

$$\eta_n \xrightarrow{\text{Updating}} \widehat{\eta}_n \xrightarrow{\text{Prediction}} \eta_{n+1}$$

by a two-step Markov chain taking values in the set of finitely discrete probability measures with atoms of size some integer multiple of $1/N$, namely,

$$\eta_n^N \stackrel{\text{def.}}{=} \frac{1}{N}\sum_{i=1}^N \delta_{\xi_n^i} \xrightarrow{\text{Selection}} \frac{1}{N}\sum_{i=1}^N \delta_{\widehat{\xi}_n^i} \xrightarrow{\text{Mutation}} \eta_{n+1}^N = \frac{1}{N}\sum_{i=1}^N \delta_{\xi_{n+1}^i}.$$

Remark: The above gives us good approximations of η_n and $\widehat{\eta}_n$ for all $n \geq 0$. But of course, for η_n^N, this really makes sense only when $n \geq 1$, since we are supposed to know exactly η_0. ∎

In view of (3.2.12) and (3.3.3), we have

$$\widehat{\eta}_n^N = \Psi_n(\eta_n^N) = \Psi_n\left(\frac{1}{N}\sum_{i=1}^N \delta_{\xi_n^i}\right) = \sum_{i=1}^N \frac{g_n(\xi_n^i)}{\sum_{j=1}^N g_n(\xi_n^j)}\delta_{\xi_n^i}. \qquad (3.3.3)$$

Hence, another way of stating the selection/updating step consists of saying that, conditional on $\mathcal{G}_n$, the variables $\widehat{\xi}_n^i$ for $i = 1,\ldots,N$ are i.i.d. with law $\Psi_n(\eta_n^N)$. Taking also into account the mutation/prediction step, and again conditional on $\mathcal{G}_n$, the variables ξ_{n+1}^i are also i.i.d., with law $\Phi_{n+1}(\eta_n^N)$. Hence, if

$$\delta_{n+1}^N f = \eta_{n+1}^N f - \Phi_{n+1}(\eta_n^N)f$$

we have for $n \geq 0$:

$$\widetilde{\mathbb{E}}(\delta_{n+1}^N f | \mathcal{G}_n) = 0,$$
$$\widetilde{\mathbb{E}}((\delta_{n+1}^N f)^2 | \mathcal{G}_n) = \frac{1}{N}\Phi_{n+1}(\eta_n^N)\left((f - \Phi_{n+1}(\eta_n^N)(f))^2\right). \quad (3.3.4)$$

Similarly, we also have if $\delta_0^N = \eta_0^N f - \eta_0 f$:

$$\widetilde{\mathbb{E}}(\delta_0^N f) = 0, \qquad \widetilde{\mathbb{E}}((\delta_0^N f)^2) = \frac{1}{N}\eta_0\left((f - \eta_0(f))^2\right). \qquad (3.3.5)$$

2) For studying the behavior of the IPS as $N \to \infty$, it is easier to begin by studying the approximations of the measures γ_n because of the linear structure of these measures. Starting with η_n^N above, we can introduce a natural approximation γ_n^N by mimicking the relations (3.2.14), which gives

$$\gamma_n^N(1) = \prod_{p=0}^{n-1} \eta_p^N(g_p), \qquad \gamma_n^N(f) = \eta_n^N(f)\,\gamma_n^N(1). \qquad (3.3.6)$$

Then, recalling (3.2.10), observe that for $q \geq 0$ and any bounded measurable φ, and with the conventions that $\gamma_{-1}^N = \gamma_0 = \eta_0$ and $L_0 = Id$ and $\Phi_0(\eta_{-1}^N) = \eta_0$, we see from (3.3.6) that

$$
\gamma_q^N(\varphi) - \gamma_q(\varphi) = \sum_{p=0}^{q} \left(\gamma_p^N(L_{p,q}\varphi) - \gamma_{p-1}^N(L_p L_{p,q}\varphi) \right)
$$

$$
= \sum_{p=0}^{q} \gamma_p^N(1) \left(\eta_p^N(L_{p,q}\varphi) - \Phi_p\left(\eta_{p-1}^N\right)(L_{p,q}\varphi) \right). \quad (3.3.7)
$$

By choosing $\varphi = L_{q,n}f$ with $q \leq n$ and by (3.3.4) and (3.3.5), we see that the process

$$
M_q^N(f) \overset{\text{def.}}{=} \gamma_q^N(L_{q,n}f) - \gamma_q(L_{q,n}f), \qquad 0 \leq q \leq n \qquad (3.3.8)
$$

is a $(\mathcal{G}_q)_{0 \leq q \leq n}$-martingale with the angle bracket given by

$$
\langle M^N(f) \rangle_q = \frac{1}{N} \sum_{p=0}^{q} \left(\gamma_p^N(1) \right)^2 \Phi_p(\eta_{p-1}^N) \left(\left(L_{p,n}f - \Phi_p(\eta_{p-1}^N)L_{p,n}f \right)^2 \right)
$$

$$
(3.3.9)
$$

We now derive some easy consequences of the martingale properties described above. Below, f is an arbitrary bounded measurable function on $(E, \mathcal{E})$, with sup-norm $\|f\|$. We also denote by α_n the sup-norm of the function g_n.

The first consequence follows from taking expectations in (3.3.8) and (3.3.9), with $p = n$ (hence $M_n^N(f) = \gamma_n^N(f) - \gamma_n(f)$), giving

$$
\widetilde{E}\left(\gamma_n^N(f) \right) = \gamma_n(f),
$$

$$
\widetilde{E}\left(\left(\gamma_n^N(f) - \gamma_n(f) \right)^2 \right)
$$
$$
= \frac{1}{N} \sum_{p=0}^{n} \mathbb{E}\left(\left(\gamma_p^N(1) \right)^2 \Phi_p(\eta_{p-1}^N) \left(\left(L_{p,n}f - \Phi_p(\eta_{p-1}^N)L_{p,n}f \right)^2 \right) \right). \quad (3.3.10)
$$

Now, (3.2.10) gives

$$
\|L_n f\| \leq \alpha_{n-1}\|f\|,
$$

while (3.2.14) gives

$$
\gamma_n^N(1) \leq \prod_{p=1}^{n} \alpha_p.
$$

Hence, (3.3.10) gives

$$
\widetilde{E}\left(\left(\gamma_n^N(f) - \gamma_n(f) \right)^2 \right)^{\frac{1}{2}} \leq C_{n,2} \frac{\|f\|}{\sqrt{N}}, \quad \text{where } C_{n,2} = \sqrt{n+1} \left(\prod_{k=1}^{n} \alpha_k \right). \quad (3.3.11)
$$

Other classical estimates on martingales give an analogue to (3.3.11) for the pth moment, with a constant $C_{n,p}$.

Exponential estimates can also be deduced from (3.3.9) using Bernstein-type martingales inequalities or (Del Moral 1998) for an alternative proof based on (3.3.7) and Hoeffding's inequality. More precisely, there exist some finite constants $C_1(n)$ and $C_2(n)$ such that for any $\epsilon > 0$,

$$IP\left(\sup_{0\leq p\leq n} \left|\gamma_p^N(L_{p,n}f) - \gamma_p(L_{p,n}f)\right| > \epsilon\right) \leq C_1(n) \ \exp - \frac{N\epsilon^2}{C_2(n)\,\|f\|^2}. \quad (3.3.12)$$

3) In fact, the measures γ_n^N are just tools for studying η_n^N and $\widehat{\eta}_n^N$. But in view of (3.2.14), and with a bit of work (see e.g. (Del Moral and Jacod 1998)) one can deduce from (3.3.11) and (3.3.12) similar results for the difference $\eta_n^N f - \eta_n f$. Some more work gives in turn the result for $\widehat{\eta}_n^N f - \widehat{\eta}_n f$. At the end, one can prove the following result.

Proposition 3.3.1. *There exist constants* $C(n)$, $C_1(n)$, $C_2(n)$, *which depend on* n *and on the observed values* $y_0, \dots, y_n$, *such that*

$$\widetilde{E}\left(\left|\eta_n^N(f) - \eta_n(f)\right|\right) \leq C(n)\frac{\|f\|}{\sqrt{N}}, \quad \widetilde{E}\left(\left|\widehat{\eta}_n^N(f) - \widehat{\eta}_n(f)\right|\right) \leq C(n)\frac{\|f\|}{\sqrt{N}},$$
$$(3.3.13)$$

$$\widetilde{P}\left(\left|\eta_n^N(f) - \eta_n(f)\right| > \epsilon\right) \leq C_1(n) \ \exp - \frac{N\epsilon^2}{C_2(n)\,\|f\|^2},$$

and

$$\widetilde{P}\left(\left|\widehat{\eta}_n^N(f) - \widehat{\eta}_n(f)\right| > \epsilon\right) \leq C_1(n) \ \exp - \frac{N\epsilon^2}{C_2(n)\,\|f\|^2}.$$

More precise estimates of the exponential rates above are given in (Del Moral and Guionnet 1998a).

The conclusion for all these estimates is that the rate of convergence of η_n^N and $\widehat{\eta}_n^N$ to η_n and $\widehat{\eta}_n$ as $N \to \infty$ is $1/\sqrt{N}$. The same holds for the convergence of γ_n^N and $\widehat{\gamma}_n^N$ to γ_n and $\widehat{\gamma}_n$. Further exponential bounds for Zolotarev's semi-norms can be found in (Del Moral and Ledoux 2000).

It is also useful to have an estimate for the total variation distance between η_n and the laws of the ξ_n^i's. Observing that

$$\eta_n(f) = \frac{\gamma_n(f)}{\gamma_n(1)}, \quad \eta_n^N(f) = \frac{\gamma_n^N(f)}{\gamma_n^N(1)} \quad \text{and} \quad \widetilde{E}(\gamma_n^N(f)) = \gamma_n(f),$$

we get

$$\widetilde{E}(\eta_n^N(f)) - \eta_n(f) = \widetilde{E}\left(\left(\eta_n^N(f) - \eta_n(f)\right)\left(1 - \frac{\gamma_n^N(1)}{\gamma_n(1)}\right)\right).$$

Using Cauchy-Schwarz's inequality, this clearly implies

$$\left| \widetilde{\mathbb{E}}(\eta_n^N(f)) - \eta_n(f) \right| \leq C(n)\, \frac{\|f\|}{N} \tag{3.3.14}$$

for some constant $C(n)$, that depends on the time parameter n and on the observed values $y_0, \ldots, y_n$. Since the variables ξ_n^i for $i = 1, \ldots, N$ have the same law, the expectation of a bounded function f against this law is

$$\widetilde{\mathbb{E}}(f(\xi_n^i)) = \widetilde{\mathbb{E}}(\eta_n^N(f)).$$

Below is a consequence of (3.3.14), where $\|\cdot\|_{\mathrm{tv}}$ denotes the total variation distance:

$$\|\mathcal{L}aw(\xi_n^i) - \eta_n\|_{\mathrm{tv}} \leq \frac{C(n)}{N}. \tag{3.3.15}$$

In much the same way, for any $i = 1, \ldots, N$ we can show that

$$\widetilde{\mathbb{E}}(f(\widehat{\xi}_n^i)) = \widetilde{\mathbb{E}}\left(\widehat{\eta}_n^N(f)\right)$$

and

$$\widetilde{\mathbb{E}}(\widehat{\eta}_n^N(f)) - \widehat{\eta}_n(f)) = \widetilde{\mathbb{E}}\left(\left(\widehat{\eta}_n^N(f) - \widehat{\eta}_n(f)\right) \left(1 - \frac{\gamma_n^N(g_n)}{\gamma_n(g_n)}\right) \right),$$

from which the conclusion is that, for a constant $C(n)$ as before,

$$\|\mathcal{L}aw(\widehat{\xi}_n^i) - \widehat{\eta}_n\|_{\mathrm{tv}} \leq \frac{C(n)}{N}. \tag{3.3.16}$$

4) Now, if rates of convergence are important, the constants above such as $C(n)$, etc..., are also crucial. In (3.3.11) we have a precise estimate for $C_{n,2}$, which is already quite bad, but things are even worse for the constants in Proposition 3.3.1: one can see (in (Del Moral and Jacod 1998) for example) that standard estimates give constants $C(n)$ and $C_i(n)$ that grow typically as e^{Cn^2} for some $C > 0$ when $n \to \infty$.

But these estimates are obtained using very coarse majorations, as in (3.3.11) above. The precise magnitude of the error is better given by central limit theorems. A full discussion of these fluctuations is beyond the scope of this chapter, but the form of the angle bracket (3.3.9) suggests that one can prove that the sequence of random variables

$$U_n^N(f) = \sqrt{N}\,\left(\gamma_n^N(f) - \gamma_n(f)\right)$$

converges in law, as $N \to \infty$, to a centered Gaussian variable $U_n(f)$ with variance

$$\widetilde{\mathbb{E}}\left(U_n(f)^2\right) = \sum_{p=0}^{n} (\gamma_p(1))^2 \, \eta_p\left((L_{p,n}f - \eta_p L_{p,n}f)^2 \right) \tag{3.3.17}$$

for any $f \in \mathcal{B}_b(E)$ (the set of all bounded measurable functions on $(E, \mathcal{E})$). We can even prove that the processes $(U_n^N(f) : n \geq 0, f \in \mathcal{B}_b(E))$ converge finite-dimensionally in law to $(U_n(f) : n \geq 0, f \in \mathcal{B}_b(E))$, where the later is a centered Gaussian field whose covariance may be explicitly computed. Notice that in (3.3.17) we have $\eta_p L_{p,n} f = \eta_n f$; use (3.2.2) and (3.2.11).

However, we are more interested in the errors $\eta_n^N(f) - \eta_n(f)$ and $\widehat{\eta}_n^N(f) - \widehat{\eta}_n(f)$. To this effect, we observe that by (3.2.2) and (3.2.14) on the one hand and by (3.2.12), (3.2.13) and (3.3.3) on the other hand, we get the decompositions

$$W_n^N(f) \overset{\text{def.}}{=} \sqrt{N}\left(\eta_n^N(f) - \eta_n(f)\right) = \frac{1}{\gamma_n^N(1)} U_n^N\left(f - \eta_n(f)\right),$$

$$\widehat{W}_n^N(f) \overset{\text{def.}}{=} \sqrt{N}\left(\widehat{\eta}_n^N(f) - \widehat{\eta}_n(f)\right) = \frac{1}{\eta_n^N(1)} W_n^N(g_n(f - \widehat{\eta}_n(f))).$$

Since $\gamma_n^N(1) \to \gamma_n(1)$ and $\eta_n^N(g_n) \to \eta_n(g_n)$ in probability, it is easy to deduce from the above convergence of the $U_n^N(f)$'s the following central limit theorem.

Theorem 3.3.2. *For any bounded measurable function f, the sequence of variables $W_n^N(f)$ converges in law to a centered Gaussian variable $W_n(f)$ whose variance is given by*

$$\widetilde{\mathbb{E}}\left(W_n(f)^2\right) = \sum_{p=0}^{n} \left(\frac{\gamma_p(1)}{\gamma_n(1)}\right)^2 \eta_p\left((L_{p,n}(f - \eta_n(f)))^2\right). \tag{3.3.18}$$

Furthermore, the sequence of variables $\widehat{W}_n^N(f)$ converges in law to the variable

$$\widehat{W}_n(f) = \frac{1}{\eta_n(g_n)} W_n(g_n(f - \widehat{\eta}_n f)). \tag{3.3.19}$$

We can even show that the random fields $W_n^N(f)$ and $\widehat{W}_n^N(f)$, indexed by n and $f \in \mathcal{B}_b(E)$, converge finite-dimensionally in law to a centered Gaussian field. Furthermore, this convergence also holds, under suitable hypotheses, for measurable functions f which are not necessarily bounded; see (Del Moral and Jacod 1999) for details. Glivenko-Cantelli and Donsker theorems can also be found in (Del Moral and Ledoux 2000).

5) Let us present an alternative, but somewhat more tractable, description of the variances of $W_n(f)$ and $\widehat{W}_n(f)$ in Theorem 3.3.2. To do this, we recall (3.2.10). First, by (3.2.2) and (3.2.11) we have for $0 \leq p \leq n$:

$$\frac{\gamma_n(1)}{\gamma_p(1)} = \frac{\gamma_p L_{p,n}(1)}{\gamma_p(1)} = \eta_p L_{p,n} 1.$$

Thus, we deduce from (3.3.18) that

$$\widetilde{E}\left(W_n(f)^2\right) = \sum_{p=0}^{n} \frac{\eta_p\left((L_{p,n}(f - \eta_n f))^2\right)}{(\eta_p L_{p,n} 1)^2}. \tag{3.3.20}$$

Second, (3.2.12) and (3.2.13) yield $\eta_n\left(g_n(f - \widehat{\eta}_n f)\right) = 0$. We also have $L_{p,n}(g_n f) = g_p \widehat{L}_{p,n}(f)$ by (3.2.4) and (3.2.10), whence

$$\widehat{\eta}_p \widehat{L}_{p,n} 1 = \frac{\eta_p L_{p,n}(g_n)}{\eta_p(g_p)} = \frac{\gamma_n(g_n)}{\gamma_p(g_p)} = \frac{\eta_n(g_n)\gamma_n(1)}{\eta_p(g_p)\gamma_p(1)} = \frac{\eta_n(g_n)}{\eta_p(g_p)} \eta_p L_{p,n} 1.$$

Therefore, we deduce from (3.3.19) and (3.3.20) that

$$\widetilde{E}\left(\widehat{W}_n(f)^2\right) = \sum_{p=0}^{n} \frac{\eta_p\left(\left(g_p \widehat{L}_{p,n}(f - \widehat{\eta}_n f)\right)^2\right)}{(\eta_p(g_p))^2 \left(\widehat{\eta}_p \widehat{L}_{p,n} 1\right)^2}. \tag{3.3.21}$$

3.3.2   Subcase A2

1) We label *Subcase A2* the Case A, when the three main ingredients: the density $\bar{g}_n$, the initial law η_0 and the transitions Q_n are not known, and/or when one cannot simulate random variables exactly according to the laws η_0 or $Q_n(x,.)$.

One will replace $\bar{g}_n$, η_0 and Q_n by approximate quantities $\bar{g}_n^{(m)}$, $\eta_0^{(m)}$ and $Q_n^{(m)}$, in such a way that the functions $g_n^{(m)}(x) = \bar{g}_n^{(m)}(x, y_n)$ are known, and that one can simulate random variables exactly according to the laws $\eta_0^{(m)}$ and $Q_n^{(m)}(x,.)$ for all x. The index m (an integer) is a measure of the quality of the approximation, and in the IPS to be defined below, with N particles, we will make m depend on the number N.

We have thus two filtering schemes: one is relative to the original setting $(\bar{g}_n, \eta_0, Q_n)$, and we have the prediction and filtering measures η_n and $\widehat{\eta}_n$ together with all quantities defined in Section 3.2; the other one is relative to the approximate setting $(\bar{g}_n^{(m)}, \eta_0^{(m)}, Q_n^{(m)})$, and we have the corresponding prediction and filtering measures $\eta_n^{(m)}$ and $\widehat{\eta}_n^{(m)}$ together with all quantities defined also in Section 3.2, but with a superscript (m): for example $\Psi_n^{(m)}$, $\Phi_n^{(m)}$, $\gamma_n^{(m)}$, $\widehat{\gamma}_n^{(m)}$, $L_{p,n}^{(m)}$, and so on.

As for the IPS, it is conducted according to the setting

$$(\bar{g}_n^{(m)}, \eta_0^{(m)}, Q_n^{(m)}),$$

which is the only one available for explicit computations. So in (3.3.1), (3.3.2) and (3.3.3) we replace g_n, Q_{n+1} and Ψ_n by $g_n^{(m)}$, $Q_{n+1}^{(m)}$ and $\Psi_n^{(m)}$. The approximate prediction and filtering measures are still denoted by η_n^N and $\widehat{\eta}_n^N$, and the approximation for γ_n is γ_n^N, as given by (3.3.6) with $g_p^{(m)}$ instead of g_p. These approximate measures are good (for big N's) for approximating $\eta_n^{(m)}$, $\widehat{\eta}_n^{(m)}$ and $\gamma_n^{(m)}$, so we have first to evaluate the differences between these and η_n, $\widehat{\eta}_n$ and γ_n in terms of differences between $(\bar{g}_n^{(m)}, \eta_0^{(m)}, Q_n^{(m)})$ and $(\bar{g}_n, \eta_0, Q_n)$.

To this effect, we introduce a number of assumptions. Recall that $\|\mu\|_{\mathrm{tv}}$ denotes the total variation of the signed finite measure μ on $(E, \mathcal{E})$.

Assumption HE There exists a signed finite measure η_0' on $(E, \mathcal{E})$ and a constant C such that

$$\|m(\eta_0^{(m)} - \eta_0) - \eta_0'\|_{\mathrm{tv}} \le \frac{C}{m}.$$

Assumption HG For each n there exists a measurable bounded function g_n' on $(E, \mathcal{E})$ and a constant C_n such that

$$|m(g_n^{(m)}(x) - g_n(x)) - g_n'(x)| \le \frac{C_n}{m}.$$

Assumption HQ For all n there exists a signed finite transition measure Q_n' from $(E, \mathcal{E})$ into itself and a constant C_n such that

$$\|m(Q_n^{(m)}(x, .) - Q_n(x, .)) - Q_n'(x, .)\|_{\mathrm{tv}} \le \frac{C_n}{m}.$$

Proposition 3.3.3. *Assume that HE, HG and HQ are satisfied, and denote below by C_n a constant that may vary from line to line, and that depends only on n and on the observed values $y_0, \dots, y_n$.*

a) If for $0 \le p \le n$ we define the signed transition measure $L_{p,n}'$ by

$$L_{p,n}' = \sum_{q=p+1}^{n} L_{p,q-1} L_q' L_{q,n} \qquad (3.3.22)$$

with

$$L_q'(f)(x) = g_{q-1}(x)Q_q'(f)(x) + g_{q-1}'(x)Q_q(f)(x)$$

(with $\sum_\emptyset = 0$), we have for $f \in \mathcal{B}_b(E)$:

$$\left\| m\left(L_{p,n}^{(m)}f - L_{p,n}f\right) - L_{p,n}'f \right\| \le \frac{C_n}{m} \|f\|. \qquad (3.3.23)$$

b) Further, if we define the signed finite measures γ_n', η_n' and $\widehat{\eta}_n'$ by

$$
\left.
\begin{aligned}
\gamma_n'(f) &= \eta_0 L_{0,n}'f + \eta_0' L_{0,n}f, \\
\eta_n'(f) &= \frac{1}{\gamma_n(1)}\, \gamma_n'(f - \eta_n f), \\
\widehat{\eta}_n'(f) &= \tfrac{1}{\widehat{\gamma}_n(1)}\, (\gamma_n'(g_n(f - \widehat{\eta}_n f) + \gamma_n(g_n'(f - \widehat{\eta}_n f))),
\end{aligned}
\right\} \qquad (3.3.24)
$$

we then have

$$\left\| m\left(\gamma_n^{(m)} - \gamma_n\right) - \gamma_n' \right\|_{tv} \le \frac{C_n}{m}, \qquad (3.3.25)$$

$$\left\| m\left(\eta_n^{(m)} - \eta_n\right) - \eta_n' \right\|_{tv} \le \frac{C_n}{m}, \qquad (3.3.26)$$

$$\left\| m\left(\widehat{\eta}_n^{(m)} - \widehat{\eta}_n\right) - \widetilde{\eta}_n' \right\|_{tv} \le \frac{C_n}{m}. \qquad (3.3.27)$$

Proof. a) If R_m is a sequence of signed finite transition measures on $(E, \mathcal{E})$, we write $R_m = O(1/m^2)$ as soon as the sequence $\|m^2 R_m(x,.)\|_{tv}$ is bounded uniformly in x.

To prove (3.3.23) we use a downward induction on the parameter p. For $p = n$, $L_{n,n}^{(m)} = L_{n,n} = Id$ and the result is obvious. Suppose that (3.3.23) holds at rank $(p+1)$, that is

$$L_{p+1,n}^{(m)} f = L_{p+1,n} f + \frac{1}{m} L_{p+1,n}' f + O\left(\frac{1}{m^2}\right) \qquad (3.3.28)$$

with $L_{p+1,n}'$ given by (3.3.22). HG and HQ and (3.2.10) now yield

$$L_{p+1}^{(m)} = L_p + (g_p^{(m)} - g_p) \cdot Q_{p+1} + g_p \cdot (Q_{p+1}^{(m)} - Q_{p+1}) + (g_p^{(m)} - g_p) \cdot (Q_{p+1}^{(m)} - Q_{p+1}).$$

Since $L_{p+1,n}^{(m)} = 0(1)$ by (3.3.28), we get by (3.2.10) and (3.3.22):

$$L_{p,n}^{(m)} = L_{p+1}^{(m)} \, L_{p+1,n}^{(m)} = L_{p+1} L_{p+1,n}^{(m)} + \frac{1}{m} L_{p+1}' L_{p+1,n}^{(m)} + O\left(\frac{1}{m^2}\right).$$

A simple calculation shows that (3.3.23) holds for p, with $L_{p,n}'$ given again by (3.3.22).

b) Recalling that $\gamma_n = \eta_0 L_{0,n}$ and

$$\gamma_n^{(m)} = \eta_0^{(m)} L_{0,n}^{(m)},$$

we get

$$\gamma_n^{(m)} = \gamma_n + (\eta_0^{(m)} - \eta_0) L_{0,n} + \eta_0 (L_{0,n}^{(m)} - L_{0,n}) + (\eta_0^{(m)} - \eta_0)(L_{0,n}^{(m)} - L_{0,n}),$$

thus (3.3.25) follows from HE and (3.3.23). Similarly, we have

$$\widehat{\gamma}_n^{(m)}(f) = \gamma_n^{(m)}(f g_n^{(m)})$$

(see (3.2.11)), thus

$$\widehat{\gamma}_n^{(m)}(f) = \widehat{\gamma}_n(f) + (\gamma_n^{(m)} - \gamma_n)(g_n f) + \gamma_n((g_n^{(m)} - g_n)f) + (\gamma_n^{(m)} - \gamma_n)((g_n^{(m)} - g_n)f),$$

and we deduce from (3.3.25) and HG that

$$\left\| m\left(\widehat{\gamma}_n^{(m)} - \widehat{\gamma}_n\right) - \widehat{\gamma}_n' \right\|_{tv} \le \frac{C_n}{m}, \qquad (3.3.29)$$

where

$$\widehat{\gamma}_n'(f) = \gamma_n'(g_n f) + \gamma_n(g_n' f).$$

For (3.3.26), let us first recall the properties

$$\eta_n^{(m)}(f) = \frac{\gamma_n^{(m)}(f)}{\gamma_n^{(m)}(1)}, \qquad \eta_n(f) = \frac{\gamma_n(f)}{\gamma_n(1)}, \qquad \gamma_n(1) > 0.$$

We observe that it is sufficient to prove the result for m large enough and (3.3.25) implies $\gamma_n^{(m)}(1) \to \gamma_n(1) > 0$ as $m \to \infty$. In view of the foregoing, we can assume that $\gamma_n^{(m)}(1) \geq \varepsilon$ for some $\varepsilon_n > 0$ and all m. Next, a simple computation yields

$$m(\eta_n^{(m)}f - \eta_n f) - \eta_n' f = \frac{1}{(\gamma_n 1)^2 \gamma_n^{(m)} 1} \left((\gamma_n 1)^2 \left(m(\gamma_n^{(m)}f - \gamma_n f) - \gamma_n' f \right) \right.$$

$$\left. - \gamma_n 1 \gamma_n f \left(m(\gamma_n^{(m)} 1 - \gamma_n 1) - \gamma_n' 1 \right) - (\gamma_n' 1 \gamma_n f - \gamma_n 1 \gamma_n' f)(\gamma_n 1 - \gamma_n^{(m)} 1) \right).$$

At this stage, (3.3.26) readily follows from (3.3.25).

Finally, we prove (3.3.27) in exactly the same way, using (3.3.29) instead of (3.3.25).   ∎

2) The Interacting Particle System.

As seen in the previous proposition, the rates of convergence of $\eta_0^{(m)}$, $g_n^{(m)}$ and $Q_n^{(m)}$ to η_0, g_n and Q_n as $m \to \infty$, as stated in Hypotheses HE, HG and HQ, are adjusted to give the rate for the convergence of $\eta_n^{(m)}$ and $\widehat{\eta}_n^{(m)}$ to η_n and $\widehat{\eta}_n$.

Now, the IPS is conducted according to the scheme $(\eta_0^{(m)}, \bar{g}_n^{(m)}, Q_n^{(m)})$. Thus, if one wants to evaluate, for example, the difference $\eta_n^N f - \eta_n f$ there are two different errors: one is $\eta_n^N f - \eta_n^{(m)} f$, which is taken care of by (3.3.13) and is of order of magnitude $1/\sqrt{N}$. The other one is $\eta_n^{(m)} f - \eta_n f$, and is of order of magnitude $1/m$ by (3.3.26). These two errors should be balanced, so we are led to choose $m = m(N)$ to depend on N in such a way that $m(N)$ and $\sqrt{N}$ are equivalent as $N \to \infty$. We can take, for example, $m(N)$ to be the integer part $[\sqrt{N}]$ of $\sqrt{N}$.

In order to have estimates as in Proposition 3.3.1, we should also prove that the estimates of $\eta_n^N f - \eta_n^{(m)} f$ in this proposition are uniform in m. This can be proved, giving the result below.

Proposition 3.3.4. *Assume HE, HG and HG. If we conduct the IPS with the approximate quantities* $(\eta_0^{(m(N))}, \bar{g}_n^{(m(N))}, Q_n^{(m(N))})$ *where* $m(N) = [\sqrt{N}]$, *then there exist constants* $C(n)$, $C_1(n)$, $C_2(n)$, *which depend on* n *and on the observed values* $y_0, \ldots, y_n$, *such that*

$$\widetilde{I\!E}\left(|\eta_n^N(f) - \eta_n(f)| \right) \leq C(n) \frac{\|f\|}{\sqrt{N}}, \qquad \widetilde{I\!E}\left(|\widehat{\eta}_n^N(f) - \widehat{\eta}_n(f)| \right) \leq C(n) \frac{\|f\|}{\sqrt{N}},$$

$$\widetilde{I\!P}\left(\left|\eta_n^N(f) - \eta_n(f)\right| > \epsilon\right) \le C_1(n) \ \exp -\frac{N\epsilon^2}{C_2(n)\,\|f\|^2},$$

$$\widetilde{I\!P}\left(\left|\widehat{\eta}_n^N(f) - \widehat{\eta}_n(f)\right| > \epsilon\right) \le C_1(n) \ \exp -\frac{N\epsilon^2}{C_2(n)\,\|f\|^2}.$$

Arguing, as in Section 3.3.1, that the IPS is conducted according to the scheme $(\eta_0^{(m)}, \bar{g}_n^{(m)}, Q_n^{(m)})$ we can check that for any bounded function f such that $\|f\| \le 1$ and giving

$$\left|\widetilde{I\!E}(\eta_n^N(f)) - \eta_n^{(m)}(f)\right| \le \frac{C(n)}{N} \quad \text{and} \quad \left|\widetilde{I\!E}(\widehat{\eta}_n^N(f)) - \widehat{\eta}_n^{(m)}(f)\right| \le \frac{C(n)}{N}$$

for some constant $C(n)$, that only depends on the time parameter n and on the observed values $y_0, \dots, y_n$. Therefore (3.3.26) and (3.3.27) imply that

$$\left|\widetilde{I\!E}(\eta_n^N(f)) - \eta_n(f)\right| \le \frac{C(n)}{\sqrt{N}} \quad \text{and} \quad \left|\widetilde{I\!E}(\widehat{\eta}_n^N(f)) - \widehat{\eta}_n(f)\right| \le \frac{C(n)}{\sqrt{N}}$$

for any bounded function f such that $\|f\| \le 1$ and for some other constant $C(n)$. Since the variables ξ_n^i (respectively $\widehat{\xi}_n^i$) for $i = 1, \dots, N$ have the same law, we see that

$$\|\mathcal{L}\mathrm{aw}(\xi_n^i) - \eta_n\|_{\mathrm{tv}} \le \frac{C(n)}{\sqrt{N}} \quad \text{and} \quad \|\mathcal{L}\mathrm{aw}(\widehat{\xi}_n^i) - \widehat{\eta}_n\|_{\mathrm{tv}} \le \frac{C(n)}{\sqrt{N}}$$

(this should be compared with (3.3.15) and (3.3.16)).

More interesting is the central limit theorem 3.3.2. The same line of argument allows us to deduce from Theorem 3.3.2 the following result.

Theorem 3.3.5. *Assume HE, HG and HG, and suppose that we conduct the IPS with the approximate quantities*

$$(\eta_0^{(m(N))}, \bar{g}_n^{(m(N))}, Q_n^{(m(N))}) \quad \text{where} \quad m(N) = [\sqrt{N}].$$

a) For any bounded measurable function f, the sequence of random variables

$$W_n^N(f) = \sqrt{N}(\eta_n^N f - \eta_n f)$$

converges in law to a Gaussian variable with mean $\eta_n'(f)$ (see (3.3.24)) and variance given by (3.3.18) or equivalently (3.3.20).

b) For any bounded measurable function f, the sequence of random variables

$$\widehat{W}_n^N(f) = \sqrt{N}(\widehat{\eta}_n^N f - \widehat{\eta}_n f)$$

converges in law to a Gaussian variable with mean $\widehat{\eta}_n'(f)$ (see (3.3.24)) and variance given by (3.3.21).

3.4   An interacting particle system under Case B

3.4.1   Subcase B1

As we said in the introduction, the situation becomes much more involved in Case B, since the noise sources may be related to the signal. Here again we begin with Subcase B1, which means that we know the densities $\bar{G}_n$ (hence all functions G_n of (3.2.1)), as well as the initial conditions μ_0 and η_0 (connected by (3.1.2)) and the transitions P_n and Q_n (also connected by (3.1.2)), and further one can simulate random variables exactly according to the laws μ_0 and $P_n(x, y; .)$ for all $x \in E$, $y \in \mathbb{R}^d$ and $n \geq 1$.

1) We shall modify the setting in a controlled way so that the new setting fits in Case A. To this effect, our first objective is to enlarge the state-space of the signal and to construct an approximating model that has the same Feynman-Kac formulation as the one in (3.2.3). To describe precisely this approximating model, we need to introduce some additional notation.

- We denote by $\mathcal{X}_n = (\widehat{X}_n, \widehat{Y}_n)$, $n \geq 0$ the time in-homogeneous Markov chain with product state-space $E \times \mathbb{R}^q$, with initial law μ_0 and transition probability kernels $\{Q_n \; ; \; n \geq 1\}$ given by

$$\left. \begin{array}{l} \mu_0(dx, dy) = \eta_0(dx)\bar{G}_0(x, y)dy, \qquad \text{(see (3.1.2))} \\[2mm] \mathcal{Q}_n\left((x, y), d(x', y')\right) = Q_n(x, dx')\bar{G}_n(x, y_{n-1}, x', y')dy'. \end{array} \right\} \quad (3.4.1)$$

  The probability measure and the expectation corresponding to this Markov chain are denoted by $\widehat{P}$ and $\widehat{E}$. Observe that $\mathcal{Q}_n(x, y; .)$ does not depend on y.

- We also choose a Borel-bounded function θ from $\mathbb{R}^q$ into $(0, \infty)$ such that

$$\int \theta(y)dy=1, \qquad \int y\theta(y)dy=0, \qquad \int |y|^3\theta(y)\,dy<\infty. \quad (3.4.2)$$

  Then we set for any $\alpha \in (0, \infty)$ and $(x, y) \in E \times \mathbb{R}^q$

$$g_n^{(\alpha)}(x, y) = \alpha^{-q}\,\theta\left((y - y_n)/\alpha\right). \quad (3.4.3)$$

By using classical approximation arguments, and under appropriate regularity conditions on the functions G_n, a natural choice of α-approximating measures for γ_n and $\widehat{\gamma}_n$ in (3.2.9) (as $\alpha \to 0$) are the marginals $\gamma_n^{(\alpha)}$ and $\widehat{\gamma}_n^{(\alpha)}$ on the first component of the measures $\nu_n^{(\alpha)}$ and $\widehat{\nu}_n^{(\alpha)}$ on $E \times \mathbb{R}^q$ defined for any $\varphi \in \mathcal{B}_b(E \times \mathbb{R}^q)$ by the Feynman-Kac type formulae

$$\nu_n^{(\alpha)}(\varphi)= \widehat{E}\left(\varphi(\mathcal{X}_n)\prod_{k=0}^{n-1} g_k^{(\alpha)}(\mathcal{X}_k)\right), \quad \widehat{\nu}_n^{(\alpha)}(\varphi)= \widehat{E}\left(\varphi(\mathcal{X}_n)\prod_{k=0}^{n} g_k^{(\alpha)}(\mathcal{X}_k)\right),$$

$$(3.4.4)$$

with the convention $\prod_\emptyset = 1$. In other words, for any $f \in \mathcal{B}_b(E)$ we have

$$\gamma_n^{(\alpha)}(f) = \nu_n^{(\alpha)}(f \otimes 1), \qquad \widehat{\gamma}_n^{(\alpha)}(f) = \widehat{\nu}_n^{(\alpha)}(f \otimes 1). \qquad (3.4.5)$$

Another way of interpreting (3.4.4) is as Feynman-Kac formulae (3.2.3) for a filtering scheme in Case A1:

$$\left. \begin{array}{l} \text{the state process is} \quad \mathcal{X}_n = (\widehat{X}_n, \widehat{Y}_n) \\[2mm] \text{the observation is} \quad \mathcal{Y}_n^{(\alpha)} = \widehat{Y}_n + \alpha \, V_n, \end{array} \right\}, \qquad (3.4.6)$$

where the V_n's are i.i.d. q-dimensional variables, independent of $\mathcal{X}$ and with distribution $\theta(dz)dz$.

2) In view of this interpretation, we introduce the prediction and filtering measures associated with the scheme (3.4.6), and with respect to the observations $y_0, \dots, y_n$ for the true model. Namely, for any measurable function φ on $E \times \mathbb{R}^q$, we set

$$\left. \begin{array}{l} \mu_n^{(\alpha)}(\varphi) = \widehat{E}\Big(\varphi\left(\mathcal{X}_n\right) \Big| \mathcal{Y}_0^{(\alpha)} = y_0, \dots, \mathcal{Y}_{n-1}^{(\alpha)} = y_{n-1} \Big) = \dfrac{\nu_n^{(\alpha)}(\varphi)}{\nu_n^{(\alpha)}(1)} \\[4mm] \widehat{\mu}_n^{(\alpha)}(\varphi) = \widehat{E}\Big(\varphi\left(\mathcal{X}_n\right) \Big| \mathcal{Y}_0^{(\alpha)} = y_0, \dots, \mathcal{Y}_{n-1}^{(\alpha)} = y_{n-1}, \mathcal{Y}_n^{(\alpha)} = y_n \Big) = \dfrac{\widehat{\nu}_n^{(\alpha)}(\varphi)}{\widehat{\nu}_n^{(\alpha)}(1)}. \end{array} \right\} \qquad (3.4.7)$$

The first marginals of these measures, denoted by $\eta_n^{(\alpha)}$ and $\widehat{\eta}_n^{(\alpha)}$, are thus defined for any $f \in \mathcal{B}_b(E)$ by

$$\eta_n^{(\alpha)}(f) = \mu_n^{(\alpha)}(f \otimes 1) = \dfrac{\gamma_n^{(\alpha)}(f)}{\gamma_n^{(\alpha)}(1)}, \qquad \widehat{\eta}_n^{(\alpha)}(f) = \widehat{\mu}_n^{(\alpha)}(f \otimes 1) = \dfrac{\widehat{\gamma}_n^{(\alpha)}(f)}{\widehat{\gamma}_n^{(\alpha)}(1)}. \qquad (3.4.8)$$

Observe that $\nu_0^{(\alpha)} = \mu_0^{(\alpha)} = \mu_0$ and $\eta_0^{(\alpha)} = \eta_0$.

At this stage, we define quantities associated with the filtering scheme (3.4.6) as in Section 3.2, except that we only consider the marginals of E of the various transition kernels or measures on $(E \times \mathbb{R}^q)$. Notice once again that if φ is a function on $E \times \mathbb{R}^q$, then $Q_n\varphi$ can be considered a function on E, as well as on $E \times \mathbb{R}^q$.

First, as in (3.2.4), we define the kernels $\widehat{L}_n^{(\alpha)}$ on $(E, \mathcal{E})$ and their iterates $\widehat{L}_{p,n}^{(\alpha)}$ for $0 \le p \le n$ by

$$\widehat{L}_n^{(\alpha)} f = Q_n(g_n^{(\alpha)}(f \otimes 1)), \qquad \widehat{L}_{p,n}^{(\alpha)} = \widehat{L}_{p+1}^{(\alpha)} \dots \widehat{L}_n^{(\alpha)} \quad \text{with} \quad \widehat{L}_{n,n}^{(\alpha)} = Id. \qquad (3.4.9)$$

Next, we set

$$\widehat{G}_0^{(\alpha)}(x) = \int \bar{G}_0(x, y) g_0^{(\alpha)}(x, y) dy. \qquad (3.4.10)$$

Observing that the function G_0 of Section 3.2 is replaced here by the function $g_0^{(\alpha)}$, and taking into account that $\nu_0(g_0^{(\alpha)}(f \otimes 1)) = \eta_0(\widehat{G}_0 f)$, we can thus deduce from (3.2.6) and (3.2.8) that

$$\left.\begin{aligned}
&\nu_0^{(\alpha)}=\mu_0, &&\widehat{\nu}_0^{(\alpha)}(\varphi)=\mu_0(g_0^{(\alpha)}\varphi),\\
&\nu_n^{(\alpha)}(\varphi)=\eta_0(\widehat{G}_0^{(\alpha)}\widehat{L}_{0,n-1}^{(\alpha)}Q_n\varphi), &&\widehat{\nu}_n^{(\alpha)}(\varphi)=\eta_0(\widehat{G}_0^{(\alpha)}\widehat{L}_{0,n}Q_n(g_n^{(\alpha)}\varphi)), && n\geq 1.
\end{aligned}\right\}$$
$$(3.4.11)$$

This in turn gives the marginals on E:

$$\left.\begin{aligned}
&\gamma_0^{(\alpha)}=\eta_0, &&\widehat{\gamma}_0^{(\alpha)}(f)=\eta_0(\widehat{G}_0^{(\alpha)}f),\\
&\gamma_n^{(\alpha)}(f)=\eta_0(\widehat{G}_0^{(\alpha)}\widehat{L}_{0,n-1}^{(\alpha)}Q_nf), &&\widehat{\gamma}_n^{(\alpha)}(f)=\eta_0(\widehat{G}_0^{(\alpha)}\widehat{L}_{0,n}f), && n\geq 1.
\end{aligned}\right\}$$
$$(3.4.12)$$

Now, we evaluate the errors involved by replacing the original scheme by the scheme (3.4.6), in terms of α. For this, we need a regularity assumption on the functions $\bar{G}_n$, which is as follows.

Hypothesis R

a) The function $y \mapsto \bar{G}_0(x,y)$ is three times differentiable, with partial derivatives of order 1, 2 and 3 uniformly bounded in (x,y).

b) Setting $\widetilde{L}_n f(x,y) = \int Q_n(x,dx')f(x')\bar{G}_n(x,y_{n-1};x',y)$, for each $n \geq 1$ and each bounded measurable function f on E, $\|f\| \leq 1$, the function

$$y \mapsto \widetilde{L}_n f(x,y)$$

is three times differentiable, with partial derivatives of order 1, 2 and 3 uniformly bounded in (x,y).

Observe that, if the functions $y \mapsto \bar{G}_n(x,y_{n-1};x',y)$ are three times differentiable with partial derivatives of order up to 3 bounded uniformly in (x,x',y) for each $n \geq 1$, then R-b is satisfied.

Below, we denote by $\bar{G}''_{0,jk}(x,y)$ and by $\widetilde{L}''_{n,jk}f(x,y)$ the second order partial derivatives of the functions $y \mapsto \bar{G}_0(x,y)$ and $y \mapsto \widetilde{L}_n f(x,y)$ w.r.t. the components y^j and y^k. We can thus define the finite kernels $\widehat{L}_n^*$ and the function G_0^* on E by

$$\widehat{L}_n^* f(x) = \frac{1}{2}\sum_{j,\ell=1}^{q}\widetilde{L}''_{n,j\ell}f(x,y_n)\int\theta(z)z^j z^\ell dz, \qquad (3.4.13)$$

$$G_0^*(x) = \frac{1}{2}\sum_{j,\ell=1}^{q}\bar{G}''_{0,j\ell}(x,y_0)\int\theta(z)z^j z^\ell dz. \qquad (3.4.14)$$

Then, recalling that $\widehat{L}_n$ is defined by (3.2.4), the key lemma is:

Lemma 3.4.1. *Under R we have the following estimates, where C_n denotes a constant that depends only on n and on the observations $y_0, \ldots, y_n$,* *and f is a bounded measurable function on E:*

$$|\widehat{L}_n^{(\alpha)} f(x) - \widehat{L}_n f(x) - \alpha^2 \widehat{L}_n^* f(x)| \leq C_n \alpha^3 \|f\|, \tag{3.4.15}$$

$$|\widehat{G}_0^{(\alpha)}(x) - G_0(x) - \alpha^2 G_0^*(x)| \leq C_0 \alpha^3. \tag{3.4.16}$$

$$|\alpha^q \mathcal{Q}_n((g_n^{(\alpha)})^2 f)(x) - u\widehat{L}_n f(x)| \leq C_n \alpha \|f\|, \tag{3.4.17}$$

where

$$u = \int \theta(y)^2 dy.$$

Proof. Since $\int \theta(y)dy = 1$ and $\widehat{L}_n f(x) = \widetilde{L}_n f(x, y_n)$, we have

$$\widehat{L}_n^{(\alpha)} f(x) - \widehat{L}_n f(x) = \int \Big(\widetilde{L}_n f(x, y_n + \alpha y) - \widetilde{L}_n f(x, y_n) \Big) \theta(y)dy.$$

Then, taking R into account and using a third-order Taylor expansion for the function $y \mapsto \widetilde{L}_n f(x, y)$ around y_n, we readily deduce (3.4.15) from (3.4.2).

Since $G_0(x) = \bar{G}_0(x, y_0)$, the proof of (3.4.16) is similar. Finally,

$$\alpha^q \mathcal{Q}_n((g_n^{(\alpha)})^2 f)(x) \quad u\widehat{L}_n f(x) = \int \Big(\widetilde{L}_n f(x, y_n + \alpha y) - \widetilde{L}_n f(x, y_n) \Big) \theta(y)^2 dy.$$

So (3.4.17) follows again from R, and from a first-order Taylor expansion for the function $y \mapsto \widetilde{L}_n f(x, y)$ around y_n. ∎

We define the kernels $\widehat{L}_{p,n}^*$ for $0 \leq p \leq n$ and the measures γ_n^*, $\widehat{\gamma}_n^*$, η_n^* and $\widehat{\eta}_n^*$ on $(E, \mathcal{E})$ for $n \geq 0$ by

$$\widehat{L}_{p,n}^* = \sum_{q=p+1}^n \widehat{L}_{p,q-1} \widehat{L}_q^* \widehat{L}_{q,n}; \qquad \text{hence} \quad \widehat{L}_{n-1,n}^* = \widehat{L}_n^*,$$

$$\gamma_n^*(f) = \eta_0 \Big((G_0^* \widehat{L}_{0,n-1} + G_0 \widehat{L}_{0,n-1}^*) \mathcal{Q}_n f \Big), \quad \eta_n^*(f) = \frac{\gamma_n^*(f)\gamma_n(1) - \gamma_n^*(1)\gamma_n(f)}{(\gamma_n 1)^2}, \tag{3.4.18}$$

for $n \geq 1$ and $\gamma_0^* = 0$ and $\eta_0^* = 0$, and for $n \geq 0$:

$$\widehat{\gamma}_n^*(f) = \eta_0 \Big((G_0^* \widehat{L}_{0,n} + G_0 \widehat{L}_{0,n}^*) f \Big), \quad \widehat{\eta}_n^*(f) = \frac{\widehat{\gamma}_n^*(f)\widehat{\gamma}_n(1) - \widehat{\gamma}_n^*(1)\widehat{\gamma}_n(f)}{(\widehat{\gamma}_n 1)^2}. \tag{3.4.19}$$

Now we have an analogue to Proposition 3.3.3

Proposition 3.4.2. *Under R we have the following estimates, where C_n denotes a constant that depends only on n and on the observations $y_0, \ldots, y_n$, and f is a bounded measurable function on E:*

$$|\widehat{L}_{p,n}^{(\alpha)} f(x) - \widehat{L}_{p,n} f(x) - \alpha^2 \widehat{L}_{p,n}^* f(x)| \leq C_n \alpha^3 \|f\|, \qquad 0 \leq p \leq n, \quad (3.4.20)$$

$$\|\gamma_n^{(\alpha)} - \gamma_n - \alpha^2 \gamma_n^*\|_{tv} \leq C_n \alpha^3, \quad \|\widehat{\gamma}_n^{(\alpha)} - \widehat{\gamma}_n - \alpha^2 \widehat{\gamma}_n^*\|_{tv} \leq C_n \alpha^3, \quad (3.4.21)$$

$$\|\eta_n^{(\alpha)} - \eta_n - \alpha^2 \eta_n^*\|_{tv} \leq C_n \alpha^3, \quad \|\widehat{\eta}_n^{(\alpha)} - \widehat{\eta}_n - \alpha^2 \widehat{\eta}_n^*\|_{tv} \leq C_n \alpha^3. \quad (3.4.22)$$

Proof. In view of (3.4.15), the proof of (3.4.20) is exactly the same as the proof of (3.3.23) in Proposition 3.3.3, once observed that $\widehat{L}_{p+1,n}$ is a positive bounded kernel for all $p < n$ and

$$\widehat{L}_{p+1}^{(\alpha)} 1 \leq \widehat{L}_{p+1} 1 + C_{p+1} \alpha^2$$

by (3.4.15).

The first estimate in (3.4.21) is trivial for $n = 0$. If $n \geq 1$, comparing (3.2.8) with (3.4.12), we get

$$\gamma_n^{(\alpha)}(f) - \gamma_n(f) = \eta_0 \left((\widehat{G}_0^{(\alpha)} - G_0) \widehat{L}_{0,n-1}^{(\alpha)} Q_n f \right) + \eta_0 \left(G_0 (\widehat{L}_{0,n-1}^{(\alpha)} - \widehat{L}_{0,n-1}) Q_n f \right).$$

The first estimate in (3.4.21) follows from (3.4.16) and (3.4.20). The second estimate is proved similarly, using (3.2.6) instead of (3.2.8), for $n \geq 0$.

Finally, for (3.4.22) it is enough to prove the result for α small enough. In view of our standing assumption H, and of (3.4.21), we can assume that $\gamma_n(1)$, $\gamma_n^{(\alpha)}(1)$, $\widehat{\gamma}_n(1)$ and $\widehat{\gamma}_n^{(\alpha)}(1)$ are all bigger than some $\varepsilon_n > 0$. In view of (3.2.2) and (3.4.8), the two estimates in (3.4.22) are deduced from (3.4.21) exactly as (3.3.26) and (3.3.27) were deduced from (3.3.25) and (3.3.29). ∎

3) The Interacting Particle System.

For any given N and $\alpha > 0$ we can conduct an IPS of size N for the filtering scheme (3.4.6). We shall choose $\alpha = \alpha(N)$ depending on N, in a way prescribed later. Let us briefly describe how it works.

We have two nested particle systems. First, there are N particles $\kappa_n = (\kappa_n^i)_{1 \leq i \leq N}$ at time n which are used for approximating $\mu_n^{(\alpha)}$ using the empirical measure μ_n^N; next we have N particles $\widehat{\kappa}_n = (\widehat{\kappa}_n^i)_{1 \leq i \leq N}$ at time n, which are used as an intermediate step. All κ_n^i and $\widehat{\kappa}_n^i$ take their values in $E \times \mathbb{R}^q$, and we single out their components ξ_n^i and ζ_n^i for κ_n^i and $\widehat{\xi}_n^i$ and $\widehat{\zeta}_n^i$ for $\widehat{\kappa}_n^i$, taking values respectively in E and in $\mathbb{R}^q$.

The motion of these particles is again defined on an auxiliary probability space, on which we denote by $\mathcal{G}_n$ the σ-field generated by the variables κ_p for $p \leq n$ and $\widehat{\kappa}_p$ for $p < n$, and by $\widehat{\mathcal{G}}_n$ the σ-field generated by the variables κ_p and $\widehat{\kappa}_p$ for $p \leq n$.

First, the variables κ_0^i are drawn independently according to the initial law μ_0 of (3.4.1). Next, the mechanism goes, by induction on n, according to the following two steps Markov rule.

Selection/Updating

$$\widetilde{\mathbb{P}}(\widehat{\kappa}_n \in (dz_1, \dots, dz_N)|\mathcal{G}_n) = \prod_{p=1}^{N} \sum_{i=1}^{N} \frac{g_n^{(\alpha)}(\kappa_n^i)}{\sum_{j=1}^{N} g_n^{(\alpha)}(\kappa_n^j)} \delta_{\kappa_n^i}(dz_p). \quad (3.4.23)$$

Mutation/Prediction

$$\widetilde{\mathbb{P}}(\kappa_{n+1} \in (dz_1, \dots, dz_N)|\widehat{\mathcal{G}}_n) = \prod_{p=1}^{N} \mathcal{Q}_{n+1}(\widehat{\kappa}_n^p, dz_p). \quad (3.4.24)$$

The approximating measures μ_n^N and η_n^N for $\mu_n^{(\alpha)}$ and $\eta_n^{(\alpha)}$ (see (3.4.8)) are thus

$$\mu_n^N = \frac{1}{N}\sum_{i=1}^{N} \delta_{(\xi_n^i, \zeta_n^i)}, \qquad \eta_n^N = \frac{1}{N}\sum_{i=1}^{N} \delta_{\xi_n^i}. \quad (3.4.25)$$

Next, comparing with (3.3.3), we set as approximating measures for $\widehat{\mu}_n^{(\alpha)}$ and $\widehat{\eta}_n^{(\alpha)}$:

$$\widehat{\mu}_n^N = \sum_{i=1}^{N} \frac{g_n^{(\alpha)}(\kappa_n^i)}{\sum_{j=1}^{N} g_n^{(\alpha)}(\kappa_n^j)} \delta_{\kappa_n^i}, \qquad \widehat{\eta}_n^N = \sum_{i=1}^{N} \frac{g_n^{(\alpha)}(\kappa_n^i)}{\sum_{j=1}^{N} g_n^{(\alpha)}(\kappa_n^j)} \delta_{\xi_n^i}. \quad (3.4.26)$$

4) We will assert the quality of our IPS through a central limit theorem only. For each $\alpha > 0$ we have Theorem 3.3.2, which we recall in our setting for the marginals η_n^N and $\widehat{\eta}_n^N$ only, since these are the quantities of interest. Set

$$W_n^{N,(\alpha)}(f) = \sqrt{N}\left(\eta_n^N(f) - \eta_n^{(\alpha)}(f)\right),$$

and

$$\widehat{W}_n^{N,(\alpha)}(f) = \sqrt{N}\left(\widehat{\eta}_n^N(f) - \widehat{\eta}_n^{(\alpha)}(f)\right).$$

If $\alpha > 0$ is fixed, Theorem 3.3.2 yields that, as $N \to \infty$, the sequences $(W_n^{N,(\alpha)}(f))$ and $(\widehat{W}_n^{N,(\alpha)}(f))$ converge in law to centered Gaussian variables $W_n^{(\alpha)}(f)$ and $\widehat{W}_n^{(\alpha)}(f)$ with variances given by (recall (3.3.20) and (3.3.21))

$$\widetilde{\mathbb{E}}\left(W_n^{(\alpha)}(f)^2\right) = \sum_{p=0}^{n} \frac{\mu_p^{(\alpha)}\left(\left(g_p^{(\alpha)}\widehat{L}_{p,n-1}^{(\alpha)}Q_n(f - \eta_n^{(\alpha)}f)\right)^2\right)}{\left(\mu_p^{(\alpha)}(g_p^{(\alpha)}\widehat{L}_{p,n-1}^{(\alpha)}1)\right)^2} \quad (3.4.27)$$

and

$$\widetilde{E}\left(\widehat{W}_n^{(\alpha)}(f)^2\right) = \sum_{p=0}^n \frac{\mu_p^{(\alpha)}\left(\left(g_p^{(\alpha)}\widehat{L}_{p,n}^{(\alpha)}(f - \widehat{\eta}_n^{(\alpha)}f)\right)^2\right)}{\left(\mu_p^{(\alpha)}(g_p^{(\alpha)})\right)^2\left(\widehat{\eta}_p^{(\alpha)}\widehat{L}_{p,n}^{(\alpha)}1\right)^2}. \tag{3.4.28}$$

By Letting $\alpha \to 0$, both quantities (3.4.27) and (3.4.28) increase in a way that is controlled by using Lemma 3.4.1 and Proposition 3.4.2. Take for example the summand number p in (3.4.27). For the denominator, we may write by virtue of (3.4.7), (3.4.9), (3.4.11), (3.4.12) and (3.4.21)

$$\mu_p^{(\alpha)}(g_p^{(\alpha)}\widehat{L}_{p,n-1}^{(\alpha)}1) = \frac{\nu_p^{(\alpha)}(g_p^{(\alpha)}\widehat{L}_{p,n-1}^{(\alpha)}1)}{\nu_p^{(\alpha)}(1)} = \frac{\gamma_n^{(\alpha)}(1)}{\gamma_p^{(\alpha)}(1)} = \frac{\gamma_n(1)}{\gamma_p(1)} + O(\alpha^2)$$

As for the numerator, we can write it first as

$$\frac{1}{\nu_p^{(\alpha)}(1)}\,\eta_0\left(\widehat{G}_0^{(\alpha)}\widehat{L}_{0,p-1}^{(\alpha)}Q_p\left(\left(g_p^{(\alpha)}\widehat{L}_{p,n-1}Q_n(f - \eta_n^{(\alpha)}f)\right)^2\right)\right).$$

Next, we can replace $\widehat{G}_0^{(\alpha)}$ and $\widehat{L}_{0,p-1}^{(\alpha)}$ and $\widehat{L}_{p,n-1}^{(\alpha)}$ and $\eta_n^{(\alpha)}$ by G_0 and $\widehat{L}_{0,p-1}$ and $\widehat{L}_{p,n-1}$ and η_n, up to a relative error of size $O(\alpha^2)$, by using (3.4.15) and (3.4.16). Then, if we use (3.4.17), we see readily that

$$\alpha^q\,\mu_p^{(\alpha)}\left(\left(g_p^{(\alpha)}\widehat{L}_{p,n-1}^{(\alpha)}Q_n(f - \eta_n^{(\alpha)}f)\right)^2\right)$$

converges to

$$\frac{u}{\gamma_p(1)}\,\eta_0\left(G_0\widehat{L}_{0,p}\left((\widehat{L}_{p,n-1}Q_n(f - \eta_nf))^2\right)\right)$$

as $\alpha \to 0$. Then, using (3.2.2), (3.2.6) and (3.2.7), we finally see that

$$\lim_{\alpha\to0}\alpha^q E(W_n^{N,(\alpha)}(f)^2) = u\sum_{p=0}^n\frac{\gamma_p(1)\gamma_{p+1}(1)}{\gamma_n(1)^2}\widehat{\eta}_p\left((\widehat{L}_{p,n-1}Q_n(f - \eta_nf))^2\right). \tag{3.4.29}$$

In much the same way, we also obtain that

$$\lim_{\alpha\to0}\alpha^q E(\widehat{W}_n^{N,(\alpha)}(f)^2) = u\sum_{p=0}^n\frac{\gamma_p(1)\gamma_{p+1}(1)}{\widehat{\gamma}_n(1)^2}\widehat{\eta}_p\left((\widehat{L}_{p,n}(f - \widehat{\eta}_nf))^2\right) \tag{3.4.30}$$

(recall that $\gamma_{p+1}(1) = \widehat{\gamma}_p(1)$).

In view of these two results, the way $\alpha = \alpha(N)$ should depend on N is clear. Indeed, using (3.4.22), we see

$$\eta_n^N f - \eta_n f = \frac{1}{\sqrt{N}}W_n^{N,(\alpha(N))}(f) + \alpha(N)^2\eta_n^*(f) + O(\alpha(N)^3)$$

and a similar expression for $\widehat{\eta}_n^N f - \widehat{\eta}_n f$. The first term in the right side above is of order $1/\sqrt{N\alpha(N)^q}$ by (3.4.29). Optimizing the choice of $\alpha(N)$ leads to

$$\alpha(N) = \frac{1}{N^{\frac{1}{4+q}}}. \tag{3.4.31}$$

Therefore it is not surprising to have the following result, which is proved in (Del Moral and Jacod 1999) for $\widehat{\eta}_n$, and whose proof for η_n is quite similar.

Theorem 3.4.3. *Assume R, and take $\alpha = \alpha(N)$ as given by (3.4.31) in the procedure given by (3.4.23) and (3.4.24). Let f be any bounded measurable function on E.*
   a) *The sequence of variables*

$$W_n^N(f) = N^{\frac{2}{4+q}}\left(\eta_n^N f - \eta_n f\right)$$

converges in law to a Gaussian variable with mean $\eta_n^* f$ given by (3.4.18) and variance given by the right side of (3.4.29).
   b) *The sequence of variables*

$$\widehat{W}_n^N(f) = N^{\frac{2}{4+q}}\left(\widehat{\eta}_n^N f - \widehat{\eta}_n f\right)$$

converges in law to a Gaussian variable with mean $\widehat{\eta}_n^* f$ given by (3.4.19) and variance given by the right side of (3.4.30).

3.4.2 Subcase B2

1) Exactly as in Subsection 3.3.2, we label *Subcase B2* the Case B when the initial law η_0 and the transitions P_n are not explicitly known, or when one cannot simulate random variables exactly according to the laws μ_0 or $P_n(x, y; .)$. So, we replace μ_0 by

$$\mu_0^{(m)}(dx, dz) = \eta_0^{(m)}(dx)\bar{G}_0^{(m)}(x, y)dy$$

(where $\eta_0^{(m)}$ is a probability measure) and P_n by

$$P_n^{(m)}(x, y; dx', dy') = Q_n^{(m)}(x, dx')\bar{G}_n^{(m)}(x, y; x', y')dy'$$

(where $Q_n^{(m)}$ is a transition probability), as in (3.1.2). These are chosen in such a way that we can simulate variables according to $\mu_0^{(m)}$ and $P_n^{(m)}(x, y; .)$ for all (x, y). As in Subcase A2, the index m (an integer) measures the quality of the approximation, and will eventually depend on N.

For measuring the quality of this approximation, let us introduce some additional assumptions:

Assumption HG_n

There is a measurable bounded function $\bar{G}_n^\#$ and a constant C_n such that

$$|m(\bar{G}_n^{(m)}(x,y;x',y') - \bar{G}_n(x,y;x',y')) - \bar{G}_n^\#(x,y;x',y')| \leq \frac{C_n}{m}.$$

Furthermore, the functions $y' \mapsto \bar{G}_n'(x,y;x',y')$ are equi-continuous; we shall use the notation $G_n^\#(x,x') = \bar{G}_n^\#(x,y_{n-1};x',y_n)$ (if $n = 0$ we obviously have $\bar{G}_0^\#(x,y;x',y') = \bar{G}_0^\#(x',y')$ and $G_0^\#(x) = \bar{G}_0^\#(x,y_0)$).

Assumption HP

Setting

$$\tilde{L}_n^{(m)} f(x,y) = \int Q_n^{(m)}(x,dx') f(x') \bar{G}_n^{(m)}(x,y_{n-1};x',y)$$

(as in Assumption R), we suppose that for each $n \geq 1$ there is a finite kernel $\hat{L}_n^\#$ from $(E \times \mathbb{R}^q, \mathcal{E} \otimes \mathcal{R}^q)$ into $(E,\mathcal{E})$ and a constant C_n depending only on n and on y_{n-1}, such that

$$\|m(\tilde{L}_n^{(m)}(x,y;.) - \tilde{L}_n(x,y;.)) - \hat{L}_n^\#(x,y;.)\|_{tv} \leq \frac{C_n}{m}.$$

Furthermore, the functions $y \mapsto \tilde{L}_n^\# f(x,y)$ are equi-continuous for any bounded measurable function f.

Lemma 3.4.4. *The assumptions R, HQ and HG$_n$ for all $n \geq 1$ imply HP with*

$$\hat{L}_n^\# f(x,y)$$
$$= \int Q_n(x,dx') f(x') \bar{G}_n^\#(x,y_{n-1};x',y) + \int Q_n'(x,dx') f(x') G_n(x,y_{n-1};x',y). \tag{3.4.32}$$

Proof. Upon observing that, as with the notation (3.4.32),

$$m(\tilde{L}_n^{(m)} f(x,y) - \tilde{L}_n f(x,y)) - \hat{L}_n^\# f(x,y)$$
$$= \int \left(m(Q_n^{(m)}(x,dx') - Q_n(x,dx')) - Q_n'(x,dx') \right) f(x') \bar{G}_n(x,y_{n-1};x',y)$$
$$+ \int Q_n^{(m)}(x,dx') f(x')$$
$$\times \left\{ m\left((\bar{G}_n^{(m)}(x,y_{n-1};x',y) - \bar{G}_n^{(m)}(x,y_{n-1};x',y)) \right) - \bar{G}_n^\#(x,y_{n-1};x',y) \right\}$$
$$+ \int (Q_n^{(m)}(x,dx') - Q_n(x,dx')) f(x') \bar{G}_n^\#(x,y_{n-1};x',y),$$

the first part of HP is immediate (recall that $\bar{G}_n$ is bounded). The second part follows from the second part of HG$_n$, and from R. ∎

In the sequel, we assume HE, HG$_0$, R, and HP. Using the notation (3.4.13) and (3.4.14), we define a finite kernel $\widehat{L}'_n$ and a bounded function G'_0 by

$$\widehat{L}'_n f(x) = \widehat{L}^*_n f(x) + \widehat{L}^\#_n f(x, y_n), \qquad G'_0(x) = G^*_0(x) + G^\#_0(x).$$

Now, let $\widehat{L}_n^{(\alpha),(m)}$ and $\widehat{L}_{p,n}^{(\alpha),(m)}$ be the analogue of $\widehat{L}_n^{(\alpha)}$ and $\widehat{L}_{p,n}^{(\alpha)}$ with $\mathcal{Q}_n^{(m)}$ instead of $\mathcal{Q}_n$, and similarly $\widehat{G}_0^{(\alpha),(m)}$ is associated with $\bar{G}_0^{(m)}$ by (3.4.10). Similarly, we define $\gamma_n^{(\alpha),(m)}$ and $\widehat{\gamma}_n^{(\alpha),(m)}$ and $\eta_n^{(\alpha),(m)}$ and $\widehat{\eta}_n^{(\alpha),(m)}$.

Lemma 3.4.1 thus becomes

Lemma 3.4.5. *Assume R, HG$_0$ and HP. Let $\alpha \to 0$ and $m \to \infty$ in such a way that*

$$m\alpha^2 = 1 + o(\alpha). \tag{3.4.33}$$

Then there are constants $\varepsilon(n, \alpha)$ depending only on n and on the observations $y_0, \dots, y_n$, and on α, with $\varepsilon(n, \alpha) \to 0$ as $\alpha \to 0$, such that for any bounded measurable function f on E and any $x \in E$ we have:

$$|\widehat{L}_n^{(\alpha)} f(x) - \widehat{L}_n f(x) - \alpha^2 \widehat{L}'_n f(x)| \le \varepsilon(n, \alpha)\alpha^2 \|f\|, \tag{3.4.34}$$

$$|\widehat{G}_0^{(\alpha)}(x) - G_0(x) - \alpha^2 G'_0(x)| \le \varepsilon(0, \alpha)\alpha^2. \tag{3.4.35}$$

$$|\alpha^q \mathcal{Q}_n^{(m)}((g_n^{(\alpha)})^2 f)(x) - u\widehat{L}_n f(x)| \le \varepsilon(n, \alpha)\|f\|, \tag{3.4.36}$$

where

$$u = \int \theta(y)^2 dy.$$

Proof. We have

$$\widehat{L}_n^{(\alpha),(m)} f(x) - \widehat{L}_n f(x) = \int \left(\widetilde{L}_n^{(m)} f(x, y_n + \alpha y) - \widetilde{L}_n f(x, y_n + \alpha y) \right) \theta(y)dy$$
$$+ \int \left(\widetilde{L}_n f(x, y_n + \alpha y) - \widetilde{L}_n f(x, y_n) \right) \theta(y)dy.$$

The first term in the right side is

$$\frac{1}{m} \int \widehat{L}_n^\# f(x, y_n + \alpha y)\theta(y)dy$$

plus a remainder smaller than $\varepsilon(n, \alpha)\alpha^2 \|f\|$ in view of HP and (3.4.33); the last part of HP and (3.4.33) yield that it is also $\alpha^2 \widehat{L}_n^\# f(x, y_n)$ plus a remainder of the same form.

Next, the second term in the right side is $\alpha^2 \widehat{L}^*_n f(x)$, plus a remainder smaller than $C_n \alpha^3 \|f\|$, by the proof of Lemma 3.4.1. Hence (3.4.34) holds. The proofs of (3.4.35) and (3.4.36) are similar – and simpler. ∎

Continuing, we define the kernels $\widehat{L}'_{p,n}$ for $0 \le p \le n$ and the measures γ'_n, $\widehat{\gamma}'_n$, η'_n and $\widehat{\eta}'_n$ on $(E, \mathcal{E})$ for $n \ge 0$ by

$$\widehat{L}'_{p,n} = \sum_{q=p+1}^{n} \widehat{L}_{p,q-1} \widehat{L}'_q \widehat{L}_{q,n},$$

$$\gamma'_0(f) = \eta'_0(f),$$

$$\gamma'_{n+1}(f)$$
$$= \eta'_{n+1}(G_0 \widehat{L}_{0,n} Q_{n+1} f) + \eta_0 \Big(G'_0 \widehat{L}_{0,n} Q_{n+1} f + G_0 \widehat{L}'_{0,n} Q_{n+1} f + G_0 \widehat{L}_{0,n} Q'_{n+1} f \Big),$$

and

$$\eta'_n(f) = \frac{\gamma'_n(f)\gamma_n(1) - \gamma'_n(1)\gamma_n(f)}{(\gamma_n 1)^2}, \tag{3.4.37}$$

$$\widehat{\gamma}'_n(f) = \eta'_0(G_0 \widehat{L}_{0,n} f) + \eta_0 \Big((G'_0 \widehat{L}_{0,n} + G_0 \widehat{L}'_{0,n}) f \Big),$$

$$\widehat{\eta}'_n(f) = \frac{\widehat{\gamma}'_n(f)\widehat{\gamma}_n(1) - \widehat{\gamma}'_n(1)\widehat{\gamma}_n(f)}{(\widehat{\gamma}_n 1)^2}. \tag{3.4.38}$$

We can now deduce from Lemma 3.4.5 the following result, in exactly the same way Proposition 3.4.2 was deduced from Lemma 3.4.1.

Proposition 3.4.6. *Assume R, HE, HG$_0$ and HP. Let $\alpha \to 0$ and $m \to \infty$ in such a way that (3.4.33) holds.*
There are constants $\varepsilon(n, \alpha)$ depending only on n and on the observations $y_0, \dots, y_n$, and on α, with $\varepsilon(n, \alpha) \to 0$ as $\alpha \to 0$, such that for any bounded measurable function f on E we have

$$|\widehat{L}_{p,n}^{(\alpha),(m)} f(x) - \widehat{L}_{p,n} f(x) - \alpha^2 \widehat{L}'_{p,n} f(x)| \le \varepsilon(n, \alpha)\alpha^2 \|f\|, \quad 0 \le p \le n, \tag{3.4.39}$$

$$\|\gamma_n^{(\alpha),(m)} - \gamma_n - \alpha^2 \gamma'_n\|_{tv} \le \varepsilon(n, \alpha)\alpha^2, \quad \|\widehat{\gamma}_n^{(\alpha),(m)} - \widehat{\gamma}_n - \alpha^2 \widehat{\gamma}'_n\|_{tv} \le \varepsilon(n, \alpha)\alpha^2, \tag{3.4.40}$$

$$\|\eta_n^{(\alpha),(m)} - \eta_n - \alpha^2 \eta'_n\|_{tv} \le \varepsilon(n, \alpha)\alpha^2, \quad \|\widehat{\eta}_n^{(\alpha),(m)} - \widehat{\eta}_n - \alpha^2 \widehat{\eta}'_n\|_{tv} \le \varepsilon(n, \alpha)\alpha^2. \tag{3.4.41}$$

2) The Interacting Particle System.

Our IPS is of course conducted according to the setting $(\bar{G}_n^{(m)}, \eta_0^{(m)}, Q_n^{(m)})$, which is the only setting available for explicit computations, and using the recipe describe in Subsection 3.4.1. That is, we start with the initial law $\mu_0^{(m)}$, and apply (3.4.23), and also (3.4.24) with $Q_n^{(m)}$ instead of Q_n.

Notice that $g_n^{(\alpha)}$ is again given by (3.4.3). Notice also that $\alpha = \alpha(N)$ will be given by (3.4.31), while $m = m(N)$ will be taken to be equivalent to $1/\alpha(N)^2$. For example, we can take

$$m(N) = \left[N^{\frac{2}{4+q}} \right].$$
(3.4.42)

The IPS gives rise to the approximating measures η_n^N and $\widehat{\eta}_n^N$, as given by (3.4.25) and (3.4.26).

Notice that, with our choice of $\alpha(N)$ and $m(N)$ (see (3.4.31) and (3.4.42)), the relation (3.4.33) is satisfied as $N \to \infty$.

Then, the same computations as found in Subsection 3.4.1 give the following theorem.

Theorem 3.4.7. *Assume R, HE, HG$_0$ and HP, and take $\alpha = \alpha(N)$ and $m = m(N)$ as in (3.4.31) and (3.4.42). Let f be any bounded measurable function on E.*

a) *The sequence of variables*

$$W_n^N(f) = N^{\frac{2}{4+q}} \left(\eta_n^N f - \eta_n f \right)$$

converges in law to a Gaussian variable with mean $\eta_n' f$ given by (3.4.37) and variance given by the right side of (3.4.29).

b) *The sequence of variables*

$$\widehat{W}_n^N(f) = N^{\frac{2}{4+q}} \left(\widehat{\eta}_n^N f - \widehat{\eta}_n f \right)$$

converges in law to a Gaussian variable with mean $\widehat{\eta}_n' f$ given by (3.4.38) and variance given by the right side of (3.4.30).

3.5   Discretely observed stochastic differential equations

In this Section, we show how the previous results can be put to use when the state process $X = (X_t)_{t \in \mathbb{R}_+}$ is d-dimensional and is the solution to a stochastic differential equation of the form

$$dX_t = a(X_t)\, dt + b(X_t)\, dW_t, \qquad \mathcal{L}aw(X_0) = \eta, \qquad (3.5.1)$$

where η is the initial distribution on $\mathbb{R}^d$, and a, b are known functions, and W is a d-dimensional Wiener process. We suppose that a and b are such that there is a unique solution to (3.5.1) for every initial distribution. The process X is thus Markov, with a transition semi-group $(\bar{Q}_t)_{t \geq 0}$.

3.5.1   Case A

We suppose here that the observations occur at times $0, 1, \ldots$, according to the scheme (3.1.1), with the assumptions of Case A about H_n, V_n and $\bar{g}_n$. Then with the notation of this case we have $Q_n = \bar{Q}_1$ for all n.

We are in Subcase A1 when η_0 and $\bar{g}_n$ and $\bar{Q}_1$ are explicitly known, and when one can simulate variables exactly according to the laws η_0 and $\bar{Q}_1(x,.)$ for any x. Then Proposition 3.3.1 and Theorem 3.3.2 are valid (see (Del Moral and Jacod 1998) and (Del Moral and Jacod 1999)).

Such a situation is very rare. In practice, one may know a and b, and perhaps η_0 and the functions $\bar{g}_n$, but $\bar{Q}_1$ is usually not known explicitly. So we have to resort on an approximating scheme for the simulation. An appropriate way consists of using an Euler approximation scheme for the SDE (3.5.1).

In order to keep things simple, assume that we know $\bar{g}_n$ and η_0, and we also know how to simulate a variable according to the law η_0. We then can apply Theorem 3.3.5 as soon as HQ holds (since HG and HE are trivially met with $\eta_0' = 0$ and $g_n' = 0$). The approximate transitions $Q_n^{(m)} = Q_1^{(m)}$ will be the transitions at time 1 of the Euler approximating scheme with stepsize $1/m$ for equation (3.5.1).

In other words, for any starting point $x \in \mathbb{R}^d$ we define (and can easily simulate) by induction on i the following variables.

$$X(x)_0^{(m)} = x,$$
$$X(x)_{i+1}^{(m)} = X(x)_i^{(m)} + a(X(x)_i^{(m)})\frac{1}{m} + b(X(x)_i^{(m)})\frac{U_{i+1}}{\sqrt{m}}, \quad (3.5.2)$$

where the U_i's are i.i.d. centered Gaussian d-dimensional vectors with covariance matrix equal to the identity, and above we use a vector notation. Then $Q_1^{(m)}(x,.)$ is the law of $X(x)_m^{(m)}$.

Let us introduce the following assumption.

Assumption A1

In (3.5.1) the functions a and b are twice differentiable with bounded partial derivatives of first and second order, and the matrix $(bb^*)(x)$ is uniformly non-degenerate.

Then, according to (Bally and Talay 1996), the transitions $\bar{Q}_1(x,.)$ and $Q_1^{(m)}(x,.)$ have densities $y \mapsto q(x,y)$ and $y \mapsto q^{(m)}(x,y)$ and there are constants C, C' depending only on the coefficients a and b, such that

$$q(x,x') + q^{(m)}(x,x') \leq Ce^{-C'|x-x'|^2}, \quad (3.5.3)$$

$$\left.\begin{array}{c} q^{(m)}(x,x') - q(x,x') = \dfrac{\nu(x,x')}{m} + \dfrac{R_m(x,x')}{m^2}, \\[2mm] \text{with}\quad |\nu(x,x')| + |R_m(x,x')| \leq Ce^{-C'|x-x'|^2}. \end{array}\right\}$$

Then HQ holds trivially with

$$Q'_n f(x) = \int \nu(x, x') f(x') dx' \qquad (3.5.4)$$

for all n.

In other words, we have the following result (see (Del Moral and Jacod 1999)).

Theorem 3.5.1. *Assume (3.5.1) and A1, and suppose that we conduct the IPS with the approximate transitions $Q_n^{(m(N))}$, where $m(N) = [\sqrt{N}]$. Define η'_n and $\widehat{\eta}'_n$ by (3.3.24) with $\eta'_0 = 0$ and $g'_n = 0$ and Q'_n given by (3.5.4). Let f be any bounded measurable function on E.*

a) The sequence of random variables

$$W_n^N(f) = \sqrt{N}(\eta_n^N f - \eta_n f)$$

converges in law to a Gaussian variable with mean $\eta'_n(f)$ and variance given by (3.3.18) or equivalently (3.3.20).

b) The sequence of random variables

$$\widehat{W}_n^N(f) = \sqrt{N}(\widehat{\eta}_n^N f - \widehat{\eta}_n f)$$

converges in law to a Gaussian variable with mean $\widehat{\eta}'_n(f)$ and variance given by (3.3.21).

3.5.2 Case B

Here again, the process X is the solution to (3.5.1), while we observe at times $0, 1, \ldots$ the values of the q-dimensional process Y which is the solution to the equation

$$dY_t = a'(X_t, Y_t)dt + b'(X_t, Y_t)dW'_t, \qquad Y_0 = 0, \qquad (3.5.5)$$

where a' and b' are known functions and W' is a q-dimensional Wiener process, independent of W.

We assume that the initial law of (X, Y) is of the form μ_0, as given in (3.1.2), and we make the following assumptions on the coefficients of the two equations (3.5.1) and (3.5.5).

Assumption C1

In (3.5.1) and (3.5.5) the functions a, b, a' and b' are 3 times differentiable with bounded partial derivatives of orders 1, 2 and 3, and the matrices $(bb^*)(x)$ and $(b'b'^*)(x, y)$ are uniformly non-degenerate.

Then the process (X, Y) is homogeneous Markov with transition semigroup $(\bar{P}_t)_{t \geq 0}$, and the transition $\bar{P}_1(x, y; .)$ admits a positive density (x', y') $\mapsto p(x, y; x', y')$. Further $\bar{P}(x, y; dx' \times \mathbb{R}^q)$ does not depend on y and equals

$\bar{Q}_1(x,.)$ (the transition of X at time 1). So we are in exactly the situation of Case B, with $Q_n = \bar{Q}_1$ and $P_n = \bar{P}_1$ and

$$\bar{G}_n(x,y;x',y') = \frac{p(x,y;x',y')}{q(x,x')}, \qquad \text{where} \quad q(x,x') = \int p(x,y;x',y')dy'$$

(3.5.6)

for all $n \geq 1$ (notice that the integral above does not depend on y).

Here again, to keep things simple, we suppose that we know η_0 and $\bar{G}_0$, which we assume satisfy R-a. We further suppose that we can simulate exactly random variables with law μ_0. But we do not know explicitly $\bar{P}_1$ and we have to resort on an Euler scheme to approximately simulate a variable whose law is $\bar{P}_1(x,y;.)$.

More precisely, if the stepsize is $1/m$, we use the scheme (3.5.2) for (3.5.1), and similarly define the variables

$$Y(x,y)_0^{(m)} = y,$$

$$Y(x,y)_{i+1}^{(m)} = Y(x,y)_i^{(m)}$$
$$+a'(X(x)_i^{(m)}, Y(x,y)_i^{(m)}) \frac{1}{m} \; + \; b'(X(x)_i^{(m)}, Y(x,y)_i^{(m)}) \frac{U'_{i+1}}{\sqrt{m}}$$

for approximating (3.5.5). Here the U_i''s are i.i.d. centered Gaussian vectors with covariance equal to the identity, and independent of the sequence (U_i). Then $P_1^{(m)}(x,y;.)$ denotes the law of the pair $(X(x)_m^{(m)}, Y(x,y)_m^{(m)})$, and it is easy to simulate variables according to this law.

Now, under C1, the law $P_1^{(m)}(x,y;.)$ admits a density

$$(x',y') \mapsto p^{(m)}(x,y;x',y'),$$

such that $\int p^{(m)}(x,y;x',y')dy'$ does not depend on y and equals the function $q^{(m)}(x,x')$ of (3.5.3). Furthermore, the functions p and $p^{(m)}$ are three times differentiable, and according again to (Bally and Talay 1996) there exist constants C, C' depending only on the coefficients a, a', b and b', and such that

$$|\nabla_{y'}^j p(x,y;x',y')| + |\nabla_{y'}^j p^{(m)}(x,y;x',y')| \; \leq \; Ce^{-C'(|x-x'|^2+|y-y'|^2)},$$

(3.5.7)

for $j = 0,1,2,3$ (where $\nabla_{y'}^0 p = p$ and $\nabla_{y'}^j p$ denotes the vector of all partial derivatives of p, of order j, w.r.t. the components of y') and

$$\left. \begin{array}{l} p^{(m)}(x,y;x',y') - p(x,y;x',y') \; = \; \frac{\tilde{\nu}(x,y;x',y')}{m} + \frac{\tilde{R}_m(x,y;x',y')}{m^2}, \\[2mm] \text{with} \quad |\tilde{\nu}(x,y;x',y')| + |\tilde{R}_m(x,y;x',y')| \; \leq \; Ce^{-C'(|x-x'|^2+|y-y'|^2)}, \end{array} \right\}$$

(3.5.8)

and further the function $\tilde{\nu}$ is uniformly continuous.

Observe also that we have

$$P_1^{(m)}(x, y; dx', dy') = Q_1^{(m)}(x, dx')\bar{G}_n^{(m)}(x, y; x', y')dy'$$

for all $n \geq 1$, where, as in (3.5.6),

$$\bar{G}_n^{(m)}(x, y; x', y') = \frac{p^{(m)}(x, y; x', y')}{q^{(m)}(x, x')}.$$

This is exactly what obtains under Subcase B2 because we have HE with $\eta_0' = 0$, and R-a and HG$_0$ with $\bar{G}_0^{\#} = 0$ by hypothesis. Furthermore,

$$\tilde{L}_n f(x, y) = \int p(x, y_{n-1}, x', y) f(x') dx',$$

$$\tilde{L}_n^{(m)} f(x, y) = \int p^{(m)}(x, y_{n-1}, x', y) f(x') dx'.$$

Then R-b follows from (3.5.7), and HP follows from (3.5.8), with

$$\hat{L}_n^{\#} f x, y) = \int \nu(x, y_{n-1}; x', y) f(x') dx'. \tag{3.5.9}$$

Hence, we have the following result:

Theorem 3.5.2. *Assume (3.5.1) and (3.5.5) and C1 and R-a, and take $\alpha = \alpha(N)$ and $m = m(N)$ as given by (3.4.31) and (3.4.42). Suppose that we conduct the IPS with the approximate transitions $P_n^{m(N)}$ and the level $\alpha(N)$, and define η_n' and $\hat{\eta}_n'$ by (3.4.37) and (3.4.38) with $\eta_0' = 0$ and $G_0^{\#} = 0$ and $\hat{L}_n^{\#}$ given by (3.5.9). Let f be any bounded measurable function on E.*
  a) *The sequence of variables*

$$W_n^N(f) = N^{\frac{2}{4+q}} \left(\eta_n^N f - \eta_n f\right)$$

converges in law to a Gaussian variable with mean $\eta_n' f$ and variance given by the right side of (3.4.29).
  b) *The sequence of variables*

$$\widehat{W}_n^N(f) = N^{\frac{2}{4+q}} \left(\hat{\eta}_n^N f - \hat{\eta}_n f\right)$$

converges in law to a Gaussian variable with mean $\hat{\eta}_n' f$ and variance given by the right side of (3.4.30).

Part III

Strategies for Improving Sequential Monte Carlo Methods

4

Sequential Monte Carlo Methods for Optimal Filtering

Christophe Andrieu
Arnaud Doucet
Elena Punskaya

4.1 Introduction

Estimating the state of a nonlinear dynamic model sequentially in time is of paramount importance in applied science. Except in a few simple cases, there is no closed-form solution to this problem. It is therefore necessary to adopt numerical techniques in order to compute reasonable approximations. Sequential Monte Carlo (SMC) methods are powerful tools that allow us to accomplish this goal.

We describe here a generic SMC method that includes and generalises many algorithms presented recently in the literature (Doucet, Godsill and Andrieu 2000, Liu and Chen 1998). This algorithm combines importance sampling, a selection scheme and Markov chain Monte Carlo techniques. We demonstrate the efficiency of such algorithms on a problem arising in digital communications, namely demodulation of digital signals in flat fading channels.

This chapter is organised as follows. In Section 4.2, we introduce the Bayesian filtering problem. In Section 4.3, we present a generic SMC algorithm. Finally, in Section 4.4 we discuss the application of the method to digital communications.

4.2 Bayesian filtering and sequential estimation

4.2.1 Dynamic modelling and Bayesian filtering

Let t be a discrete time index. We are interested in estimating a hidden process $\{x_t; t \in \mathbb{N}\}$ based on some observations $\{y_t; t \in \mathbb{N}^*\}$, where $x_t \in$

$X \subseteq \mathbb{R}^{n_x}$ and $y_t \in Y \subseteq \mathbb{R}^{n_y}$. Let us introduce $x_{i:j} \triangleq (x_i, x_{i+1}, \dots, x_j)$ and $y_{i:j} \triangleq (y_i, \dots, y_j)$. Then $\{x_t; t \in \mathbb{N}\}$ is defined by

$$x_0 \sim \mu(dx_0), \quad x_t | (x_{0:t-1}, y_{1:t-1}) \sim K_t(dx_t | x_{0:t-1}, y_{1:t-1}), \qquad (4.2.1)$$

where $\mu(dx_0)$ is the distribution of x_0 and $\int_A K_t(dx_t | x_{0:t-1}, y_{1:t-1})$ is the probability for x_t to be in a (measurable) set A given $(x_{0:t-1}, y_{1:t-1})$. The observations are assumed to be distributed according to

$$y_t | (x_{0:t}, y_{1:t-1}) \sim g_t(dy_t | x_{0:t}, y_{1:t-1}). \qquad (4.2.2)$$

To simplify, we use the same notation for random variables and their realisations, and assume that

$$g_t(dy_t | x_{0:t}, y_{1:t-1}) = g_t(y_t | x_{0:t}, y_{1:t-1}) dy_t.$$

Equations (4.2.1) and (4.2.2) define a very general dynamic model, as it embodies hidden Markov models or nonlinear non-Gaussian state-space models, among others.

At time t, any inference of the hidden process $x_{0:t}$ is based on the posterior distribution $P(dx_{0:t} | y_{1:t})$. We are interested in estimating this distribution sequentially in time. It satisfies the following two-step recursion

$$\textit{Predict: } P(dx_{0:t} | y_{1:t-1}) = P(dx_{0:t-1} | y_{1:t-1}) K_t(dx_t | x_{0:t-1}, y_{1:t-1}) \quad (4.2.3)$$

$$\textit{Update: } P(dx_{0:t} | y_{1:t}) = \frac{g_t(y_t | x_{0:t}, y_{1:t-1}) P(dx_{0:t} | y_{1:t-1})}{\int_{X^{t+1}} g_t(y_t | x_{0:t}, y_{1:t-1}) P(dx_{0:t} | y_{1:t-1})}, \quad (4.2.4)$$

that is

$$P(dx_{0:t} | y_{1:t}) \propto \mu(dx_0) \prod_{i=1}^{t} g_i(y_i | x_{0:i}, y_{1:i-1}) K_i(dx_i | x_{0:i-1}, y_{1:i-1}). \quad (4.2.5)$$

This recursion appears to be very simple, yet one cannot compute the distribution $P(dx_{0:t} | y_{1:t})$ and its features of interest, including $P(dx_t | y_{1:t})$ and $\mathbb{E}(x_t | y_{1:t})$, in closed-form except in some very simple cases, such as linear Gaussian state-space models and finite state-space Markov chains.

4.2.2   Alternative dynamic models

The aim of this subsection is to underline that, although the dynamic model given by (4.2.1)-(4.2.2) defines the joint target distribution $P(dx_{0:t} | y_{1:t})$, one can find an infinite number of dynamic models that have the same joint distribution. One can, as a result, construct dynamic models that have more interesting theoretical properties than the original one. In particular, one can construct dynamic models that exhibit better mixing properties, while still having the same target distribution.

Example 4.2.1. *Consider the following dynamic model*

$$x_0 \quad \sim \quad \mu\,(dx_0)$$
$$x_t|\,(x_{0:t-1}, y_{1:t-1}) \quad \sim \quad \widetilde{K}_t\,(dx_t|\,x_{0:t-1}, y_{1:t-1}) \tag{4.2.6}$$
$$y_t|\,(x_{0:t}, y_{1:t-1}) \quad \sim \quad \widetilde{g}_t\,(dy_t|\,x_{0:t}, y_{1:t-1}) \tag{4.2.7}$$

where for any $(x_{0:t-1}, y_{1:t-1})$

$$\widetilde{g}_t\,(y_t|\,x_{0:t}, y_{1:t-1})\,\widetilde{K}_t\,(dx_t|\,x_{0:t-1}, y_{1:t-1})$$
$$\propto g_t\,(y_t|\,x_{0:t}, y_{1:t-1})\,K_t\,(dx_t|\,x_{0:t-1}, y_{1:t-1}). \tag{4.2.8}$$

Then the prediction and update equations (4.2.3)-(4.2.4) are modified to

$$\widetilde{P}\,(dx_{0:t}|\,y_{1:t-1}) = \widetilde{P}\,(dx_{0:t-1}|\,y_{1:t-1})\,\widetilde{K}_t\,(dx_t|\,x_{0:t-1}, y_{1:t-1}) \tag{4.2.9}$$

$$\widetilde{P}\,(dx_{0:t}|\,y_{1:t}) = \frac{\widetilde{g}_t\,(y_t|\,x_{0:t}, y_{1:t-1})\,\widetilde{P}\,(dx_{0:t}|\,y_{1:t-1})}{\int_{X^{t+1}} \widetilde{g}_t\,(y_t|\,x_{0:t}, y_{1:t-1})\,\widetilde{P}\,(dx_{0:t}|\,y_{1:t-1})}. \tag{4.2.10}$$

Although $\widetilde{P}\,(dx_{0:t}|\,y_{1:t-1}) \neq P\,(dx_{0:t}|\,y_{1:t-1})$, *it is straightforward to see that* $\widetilde{P}\,(dx_{0:t}|\,y_{1:t}) = P\,(dx_{0:t}|\,y_{1:t})$ *because of (4.2.8).*

Example 4.2.2. *Assume (4.2.1) and (4.2.2) depend on an unknown random parameter* θ. *That is,*

$$(x_0, \theta) \quad \sim \quad \mu\,(dx_0, d\theta)$$
$$x_t|\,(x_{0:t-1}, \theta, y_{1:t-1}) \quad \sim \quad K_t\,(dx_t|\,x_{0:t-1}, \theta, y_{1:t-1}) \tag{4.2.11}$$
$$y_t|\,(x_{0:t}, \theta, y_{1:t-1}) \quad \sim \quad g_t\,(y_t|\,x_{0:t}, \theta, y_{1:t-1}). \tag{4.2.12}$$

These equations define a probability distribution $P\,(dx_{0:t}, d\theta|\,y_{1:t})$. *Clearly, this system is not ergodic because, for different initial marginal distributions of* θ, *say* $\delta_{\theta_0}\,(d\theta)$ *and* $\delta_{\theta_1}\,(d\theta)$, *the asymptotic behavior of the system is likely to be different if the dependence of (4.2.11) and (4.2.12) on* θ *is not trivial. Now, consider the following alternative model based on the introduction of a time-varying parameter* θ_t

$$(x_0, \theta_0) \quad \sim \quad \mu\,(dx_0, d\theta)$$
$$x_t|\,(x_{0:t-1}, \theta_{t-1}, y_{1:t-1}) \quad \sim \quad K_t\,(dx_t|\,x_{0:t-1}, \theta_{t-1}, y_{1:t-1})$$
$$y_t|\,(x_{0:t}, \theta_{t-1}, y_{1:t-1}) \quad \sim \quad g_t\,(y_t|\,x_{0:t}, \theta_{t-1}, y_{1:t-1})$$

and

$$\theta_t|\,(x_{0:t}, \theta_{t-1}, y_{1:t}) \sim \Xi_t\,(d\theta_t|\,x_{0:t}, \theta_{t-1}, y_{1:t}),$$

where $\Xi_t\,(d\theta_t|\,x_{0:t}, \theta_{t-1}, y_{1:t}) \neq \delta_{\theta_{t-1}}\,(d\theta_t)$ *is a kernel with invariant distribution* $P\,(d\theta_t|\,x_{0:t}, y_{1:t})$, *say a Gibbs kernel or a Metropolis-Hastings kernel. The introduction of this kernel might lead to an ergodic model and, by construction, we know that this new model is such that* $P\,(dx_{0:t}, d\theta_t|\,y_{1:t})$ *and* $P\,(dx_{0:t}, d\theta|\,y_{1:t})$ *are similar.*

Defining an alternative dynamic model does not mean that one can now evaluate $P(dx_{0:t}| y_{1:t})$ with a closed-form expression. However, it allows us to design more efficient sampling algorithms. Recall that SMC methods are numerical techniques for computing an approximation of $P(dx_{0:t}| y_{1:t})$ by following the prediction and update equations corresponding to a dynamic model. One can then, for example, either use equations (4.2.3) and (4.2.4) if one adopts the dynamic model given by (4.2.1) and (4.2.2), or (4.2.9) and (4.2.10) if one uses the model given by (4.2.6) and (4.2.7). It can be expected that the performance of SMC methods will be closely related to the theoretical properties of the model, which is indeed the case.

4.3   Sequential Monte Carlo Methods

Bayesian filtering can be formulated as a particular case of the following general sequential estimation problem: estimate sequentially in time a sequence of probability distributions $\pi_t(dx_{0:t})$ (where it is assumed that it is possible to evaluate $\pi_t(dx_{0:t})$ pointwise up to a normalising constant). In the Bayesian filtering case, one has $\pi_t(dx_{0:t}) \triangleq P(dx_{0:t}| y_{1:t})$.

Assume that at time $t-1$ we have N paths (particles) $x_{0:t-1}^{(i)}$, $i = 1, \ldots, N$, distributed according to $\pi_{t-1}(dx_{0:t-1})$. Then, we can approximate $\pi_{t-1}(dx_{0:t-1})$ with the following empirical distribution

$$\pi_{t-1}^N(dx_{0:t-1}) = \frac{1}{N} \sum_{i=1}^{N} \delta_{x_{0:t-1}^{(i)}}(dx_{0:t-1}).$$

Particle filtering methods are a set of computationally efficient methods to obtain N new paths $x_{0:t}^{\prime(i)}$, $i = 1, \ldots, N$, distributed approximately according to $\pi_t(dx_{0:t}^\prime)$. These algorithms are a mixture of importance sampling, a selection scheme and MCMC.

4.3.1   Methodology

Importance sampling

The joint distributions $\pi_{t-1}(dx_{0:t-1})$ and $\pi_t(dx_{0:t})$ are of different dimension. We first modify and extend the current paths $x_{0:t-1}^{(i)}$ to obtain new paths $\tilde{x}_{0:t}^{(i)}$ using a kernel $q_t(d\tilde{x}_{0:t}| x_{0:t-1})$; in the Bayesian filtering case, this kernel will typically depend on $y_{1:t}$. After having applied the kernel, the new paths are distributed according to

$$q_t(d\tilde{x}_{0:t}) = \int_{X^t} q_t(d\tilde{x}_{0:t}| x_{0:t-1}) \pi_{t-1}(dx_{0:t-1}). \qquad (4.3.13)$$

The aim of this kernel is to obtain new paths whose distribution $q_t(d\tilde{x}_{0:t})$ is as close as possible to $\pi_t(d\tilde{x}_{0:t})$. Practically, it is not possible to have

$q_t\left(d\widetilde{x}_{0:t}\right) = \pi_t\left(d\widetilde{x}_{0:t}\right)$ so it is necessary to weight the new paths to obtain consistent estimates.

It is possible to carry out this correction through importance sampling. Assume that the following ratio (Radon-Nikodym derivative) is defined

$$w\left(\widetilde{x}_{0:t}\right) = \frac{\pi_t\left(d\widetilde{x}_{0:t}\right)}{q_t\left(d\widetilde{x}_{0:t}\right)}. \tag{4.3.14}$$

Using the importance sampling identity

$$\pi_t\left(d\widetilde{x}_{0:t}\right) = \frac{w\left(\widetilde{x}_{0:t}\right)q_t\left(d\widetilde{x}_{0:t}\right)}{\int_{X^{t+1}} w\left(\widetilde{x}_{0:t}\right)q_t\left(d\widetilde{x}_{0:t}\right)},$$

a Monte Carlo approximation of $\pi_t\left(dx_{0:t}\right)$ is given by

$$\widetilde{\pi}_t^N\left(dx_{0:t}\right) = \sum_{i=1}^{N} w_{0:t}^{(i)}\delta_{\widetilde{x}_{0:t}^{(i)}}\left(dx_{0:t}\right), \tag{4.3.15}$$

where $w_{0:t}^{(i)} \propto w_{0:t}\left(\widetilde{x}_{0:t}^{(i)}\right)$ and $\sum_{i=1}^{N} w_{0:t}^{(i)} = 1$. For the method to be efficient, it is important for the weights $w_{0:t}^{(i)}$ to be as uniform as possible. Refined importance sampling methods can also be used to reduce the bias and variance of the estimates

$$\int_{X^{t+1}} f\left(x_{0:t}\right)\widetilde{\pi}_t^N\left(dx_{0:t}\right)$$

of

$$\int_{X^{t+1}} f\left(x_{0:t}\right)\pi_t\left(dx_{0:t}\right).$$

See for example (Hesterberg 1995).

This strategy is simple, but it requires the pointwise evaluation of the weights $w_{0:t}\left(\widetilde{x}_{0:t}\right)$, and thus of $q_t\left(d\widetilde{x}_{0:t}\right)$, up to a normalising constant. That is, it requires that we evaluate (4.3.13) in closed-form. It is, however, impossible to carry out this task in most cases. An important exception is

$$q_t\left(d\widetilde{x}_{0:t}|\,x_{0:t-1}\right) = \delta_{x_{0:t-1}}\left(d\widetilde{x}_{0:t-1}\right)q_t\left(d\widetilde{x}_t|\,x_{0:t-1}\right),$$

that is, $\widetilde{x}_{0:t} = (x_{0:t-1}, \widetilde{x}_t)$. In this case, (4.3.14) satisfies

$$w\left(\widetilde{x}_{0:t}\right) = \frac{\pi_t\left(dx_{0:t-1}\right)}{\pi_{t-1}\left(dx_{0:t-1}\right)}\frac{\pi_t\left(d\widetilde{x}_t|\,x_{0:t-1}\right)}{q_t\left(d\widetilde{x}_t|\,x_{0:t-1}\right)}. \tag{4.3.16}$$

It is easy to see that the optimal importance distribution, according to the criterion which consists of minimising the variance of the importance weight $w\left(\widetilde{x}_{0:t}\right)$ *conditional upon* not modifying the path $x_{0:t-1}$, is given by

$$q_t\left(d\widetilde{x}_t|\,x_{0:t-1}\right) = \pi_t\left(d\widetilde{x}_t|\,x_{0:t-1}\right).$$

Thus, $w\left(\widetilde{x}_{0:t}\right) = \pi_t\left(dx_{0:t-1}\right)/\pi_{t-1}\left(dx_{0:t-1}\right)$ (Doucet 1998, Doucet, Godsill and Andrieu 2000). This method requires us to be able to compute

$\pi_t(dx_{0:t-1})$ pointwise up to a normalising constant, which is not always possible. Even when one can apply it, however, it is likely to be inefficient if there is a large discrepancy between $\pi_t(dx_{0:t-1})$ and $\pi_{t-1}(dx_{0:t-1})$. In this case, it is of interest to rejuvenate a part of the past of the path, that is having $\widetilde{x}_{0:t-1}^{(i)} \neq x_{0:t-1}^{(i)}$. Of course, if the goal is to keep a sequential method, it is impossible to rejuvenate the whole path in practice! The auxiliary particle filter introduced in (Pitt and Shephard 1999b) (see also Pitt and Shephard (2001: this volume)), tries to solve this problem by setting $\widetilde{x}_{0:t-1}^{(i)} = x_{0:t-1}^{(j)}$ where $j \in \{1, \dots, N\}$, according to a simulation-based rule.

Selection scheme

The aim of the selection step is to obtain an unweighted approximate empirical distribution $\ddot{\pi}_t^N(dx_{0:t})$ of the weighted measure $\widetilde{\pi}_t^N(dx_{0:t})$. The idea is to discard the samples with small weights and multiply those with large weights. The introduction of this key step in (Gordon et al. 1993) led to the first operational SMC method.

A selection procedure associates with each particle $\widetilde{x}_{0:t}^{(i)}$ a number of offspring $N_t^{(i)} \in \mathbb{N}$ ($i = 1, \dots, N$), such that $\sum_{i=1}^N N_t^{(i)} = N$, to obtain N new particles $\ddot{x}_{0:t}^{(i)}$ with associated empirical distribution $\ddot{\pi}_t^N$, where

$$\ddot{\pi}_t^N(dx_{0:t}) = N^{-1} \sum_{i=1}^N N_t^{(i)} \delta_{\widetilde{x}_{0:t}^{(i)}}(dx_{0:t}) = N^{-1} \sum_{i=1}^N \delta_{\ddot{x}_{0:t}^{(i)}}(dx_{0:t}). \qquad (4.3.17)$$

For the surviving particles $\widetilde{x}_{0:t}^{(i)}$, the weights are reset to $1/N$, and these particles are distributed approximately according to $\pi_t(dx_{0:t})$.

Many schemes have been presented in the literature to perform selection. This includes the multinomial sampling originally proposed in (Gordon et al. 1993), residual resampling (Higuchi 1997, Liu and Chen 1998) and minimum variance sampling (Carpenter, Clifford and Fearnhead 1999a, Crisan and Lyons 1999, Kitagawa 1996). All these algorithms ensure that $\mathbb{E}\left[N_t^{(i)}\right] = Nw_{0:t}^{(i)}$ while they differ in terms of $var\left[N_t^{(i)}\right]$. In practice, one must favour algorithms with small Monte Carlo variations. As demonstrated in (Crisan and Doucet 2000), the unbiasedness assumption can be lifted, that is we can have $\mathbb{E}\left[N_t^{(i)}\right] \neq Nw_{0:t}^{(i)}$, if another weak constraint on the selection scheme is valid. An example of a theoretically valid biased selection scheme is that of deterministic selection proposed in (Kitagawa 1996).

As mentioned in (Kitagawa 1996, Doucet, Andrieu and Fitzgerald 1998) and in Liu, Chen and Logvinenko (2001: this volume), results might be improved provided the weights are not reset to $1/N$ as in (4.3.17). One can carry out partial resampling (Carpenter et al. 1999a), or keep information about the weights using the splitting procedure proposed in (Doucet et al. 1998). According to this procedure, one uses a stochastic selection algo-

rithm to compute $N_t^{(i)}$, then, for the surviving particles, the new weights
are set to

$$\ddot{w}_{0:t}^{(i)} \propto \frac{w_{0:t}^{(i)}}{\left(N_t^{(i)} \Pr\left(N_t^{(i)} > 0 \right) \right)}.$$

All of these procedures can be implemented in $O\left(N\right)$ iterations.

Whatever the adopted selection scheme, it will introduce some Monte
Carlo errors. Hence, selection schemes should not necessarily be used at
each time step. A criterion to determine whether it is necessary to resample
or not has been proposed in (Liu and Chen 1995).

Markov transition kernel

When the distribution of the importance weights $w_{0:t}^{(i)}$ is skewed, the particles $\tilde{x}_{0:t}^{(i)}$, $i = 1, ..., N$ which have high importance weights are selected
many times. This results in a depletion of samples (Doucet, Godsill and
Andrieu 2000) as numerous particles $\ddot{x}_{0:t}^{(i)}$ and $\ddot{x}_{0:t}^{(j)}$ are in fact equal for
$i \neq j$. It is possible to carry out sample regeneration using a mixture
approximation (Gordon et al. 1993), see also Liu and West (2001: this
volume). This strategy might be difficult to apply when the state is high-
dimensional. An alternative clever approach based on Markov chain Monte
Carlo (MCMC) methods has been recently proposed in (Gilks and Berzuini
1999). The rationale behind the use of MCMC moves is based on the following remark. Assume that the particles $\ddot{x}_{0:t}^{(i)}$ are distributed marginally
according to $\pi_t\left(dx_{0:t}\right)$. If we apply to each particle a Markov transition ker-
nel $\Xi_t(dx_{0:t}|\ddot{x}_{0:t}^{(i)})$ of invariant distribution $\pi_t(dx_{0:t})$, then the new particles
$x_{0:t}^{(i)}$ are still distributed according to the posterior distribution of interest. It
is, therefore, possible to use all the standard MCMC methods; see (Robert
and Casella 1999). Yet, we do not require the Markov kernel to be ergodic.
Sequential algorithms, incorporating MCMC techniques, have been used
to carry out fixed-parameter estimation in (Gilks and Berzuini 1999), and
model selection in (Andrieu, de Freitas and Doucet 1999b) (using reversible
jump MCMC (Green 1995)).

A different interpretation of this approach is the following. One can think
of the transition kernel, not as a new algorithmic trick to improve the
results, but as simulating the dynamics of a modified probabilistic model.
Example 4.2.2 is a particular instance of this.

4.3.2 A generic algorithm

To sum up, a generic SMC algorithm proceeds as follows.

A Sequential Monte Carlo Algorithm

At time $t = 0$,

Step 0: *Initialisation*

- For $i = 1, ..., N$, sample $x_0^{(i)} \sim \pi_0 (dx_0)$ and set $t = 1$.

At time $t \geq 1$,

Step 1: *Importance Sampling step*

- For $i = 1, ..., N$, sample $\widetilde{x}_{0:t}^{(i)} \sim q_t \left(d\widetilde{x}_{0:t} \middle| x_{0:t-1}^{(i)} \right)$.

- For $i = 1, ..., N$, evaluate the importance weights $w_{0:t}^{(i)}$ using (4.3.14).

Step 2: *Selection step*

- Multiply/Discard particles $\left(\widetilde{x}_{0:t}^{(i)}; i = 1, \dots, N \right)$ with respect to high/low normalised importance weights $w_{0:t}^{(i)}$ to obtain N equally weighted particles $\left(\ddot{x}_{0:t}^{(i)}; i = 1, \dots, N \right)$.

Step 3: *Markov transition step*

- For $i = 1, ..., N$, sample $x_{0:t}^{(i)} \sim \Xi_t \left(dx_{0:t} \middle| \ddot{x}_{0:t}^{(i)} \right)$ where $\Xi_t (\cdot | \cdot)$ is a transition kernel of invariant distribution $\pi_t (dx_{0:t})$.

- Set $t \leftarrow t + 1$ and go to **Step 1**.

There are numerous possible extensions to this generic algorithm. We quickly list some of them here. The particles can be sampled from a kernel $q_t \left(d\widetilde{x}_{0:t} \middle| x_{0:t-1}, \pi_t^N (dx_{0:t-1}) \right)$ which depends on the previous empirical measure $\pi_t^N (dx_{0:t-1})$ (provided this dependence is smooth enough) or even jointly instead of evolving independently. Similarly, one can introduce a transition kernel dependent on the whole set of particles, in other words $\Xi_t (dx_{0:t} | \ddot{x}_{0:t}^{(1)}, \dots, \ddot{x}_{0:t}^{(N)})$, so as to design, for example, an efficient proposal for a Metropolis-Hastings sampler. This allows us to use the global information provided by the population of particles; see (Crisan and Doucet 2000) for details.

4.3.3   Convergence results

For any n, let $B (\mathbb{R}^n)$ be the space of bounded, Borel-measurable functions on $\mathbb{R}^n$, and $f \in B (\mathbb{R}^n)$: $\|f\| \triangleq \sup_{z \in \mathbb{R}^n} |f (z)|$. One can show that under very weak assumptions (importance weights upper bounded and regularity assumptions on the selection scheme) the following theorem holds. This theorem is an extension of previous results obtained in (Crisan et al. 1999).

Theorem 4.3.1. *(Crisan and Doucet 2000) For all $t \geq 0$, there exists c_t independent of N such that for any $f_t \in B\left((\mathbb{R}^{n_x})^{t+1}\right)$*

$$\mathbb{E}\left[\left(\frac{1}{N}\sum_{i=1}^{N} f_t\left(x_{0:t}^{(i)}\right) - \int_{X^{t+1}} f_t\left(x_{0:t}\right)\pi_t\left(dx_{0:t}\right)\right)^2\right] \leq c_t \frac{\|f_t\|^2}{N}. \quad (4.3.18)$$

Under minimal assumptions, this convergence result states that SMC methods are theoretically valid. The sequence c_t increases exponentially with time, however, so one must also increase the number of particles exponentially with time in order to ensure a given precision at time t. Nevertheless, it has been shown in (Del Moral and Guionnet 1998b) that, under additional assumptions on the theoretical optimal filter, one can obtain a uniform convergence result on the marginal distribution at time t, which is a bound independent of time.

Theorem 4.3.2. *(Del Moral and Guionnet 1998b) There exists c independent of N such that, for all $t \geq 0$ and for any $f \in B\left(\mathbb{R}^{n_x}\right)$*

$$\mathbb{E}\left[\left(\frac{1}{N}\sum_{i=1}^{N} f\left(x_t^{(i)}\right) - \int_{X} f\left(x_t\right)\pi_t\left(dx_t\right)\right)^2\right] \leq c\frac{\|f\|^2}{N^\alpha} \quad (4.3.19)$$

where $\alpha \leq 1$.

This convergence result is based on the fact that the theoretical optimal filter, that is $P\left(dx_t|y_{1:t}\right)$, exponentially forgets its initial condition $\mu\left(dx_0\right)$. This fact is closely linked to the ergodicity of the underlying dynamic model. See (Del Moral and Guionnet 1998b) for details. This result must not be misinterpreted. In (Del Moral and Guionnet 1998b), the algorithm under study is the standard bootstrap filter (Gordon et al. 1993), which is based on an exploration of the space according to the natural dynamics of the system

$$q_t\left(d\tilde{x}_{0:t}|x_{0:t-1}^{(i)}\right) = \delta_{x_{0:t-1}}\left(d\tilde{x}_{0:t-1}\right)K_t\left(dx_t|x_{0:t-1},y_{1:t-1}\right)$$

and this is why it appears linked to the ergodicity properties of the recursion (4.2.3)-(4.2.4). Now, assume that one uses

$$q_t\left(d\tilde{x}_{0:t}|x_{0:t-1}^{(i)}\right) = \delta_{x_{0:t-1}}\left(d\tilde{x}_{0:t-1}\right)\tilde{K}_t\left(dx_t|x_{0:t-1},y_{1:t-1}\right),$$

that is, that the particle filtering method is based on the dynamic model (4.2.6)-(4.2.7). Uniform convergence will then be achieved if the optimal filter associated with the recursion (4.2.9)-(4.2.10) has exponential forgetting properties. Detailed convergence including central limit theorems can be found in (Del Moral and Miclo 2000).

To sum up, we mention an obvious but useful aspect of SMC methods: the convergence properties of the SMC method one uses are linked to the possibly artificial dynamic model one is trying to mimic, not to the natural real model (4.2.1)-(4.2.2).

4.4   Application to digital communications

The generic SMC algorithm developed in the previous section can be particularly useful for solving a number of problems arising in digital communications. One of these problems is the demodulation of signals in noisy fading channels. Such channels produce random amplitude and phase variations on the signal, and recovering the original message requires recursive estimation of both the signal and channel distortions given the available data. The model of this dynamic system facilitates a state-space representation, and within a sequential framework general recursive expressions for the filtering distributions can be derived from which the estimates of the states can be obtained. However, the exact computation of these estimates involves a prohibitive computational cost, exponential in the (growing) number of observations, and thus approximate methods must be employed. Schemes proposed in the literature for different modulation methods include extended Kalman filtering (EKF) (Haeb and Meyr 1989, Lodge and Moher 1990), or coupled EKF and hidden Markov model (HMM) approaches (Collings and Moore 1994, Collings and Moore 1995). Other recent techniques involve the pilot symbol-aided schemes (PSAM) (Cavers 1991, Sampei and Sunaga 1993, Torrance and Hanzo 1995), decision-feedback (Kam and Ching 1992, Liu and Blostein 1995), and per-survivor processing (Vitetta and Taylor 1995).

We consider here a general case of M-ary modulated signals under conditions of frequency-nonselective (flat) Rayleigh fading channels. Our approach to demodulation based on a SMC algorithm allows us to compute the optimal estimates of the symbols from the observations. The tracking of the channel is naturally incorporated into this estimation scheme, and the case of a possibly non-Gaussian additive channel noise can be easily treated.

This section is organised as follows. First, the model specification and estimation objectives are stated. Next, an application of the generic SMC algorithm to the problem of demodulation is considered, and, finally, an extensive simulation study for the case of M-ary phase shift keyed (PSK) signals is carried out. The results show that the algorithm outperforms the current methods routinely used in most communication applications.

4.4.1 Model specification and estimation objectives

Representation of digitally modulated signals

In the general case of M-ary modulation, the information sequence is subdivided into blocks of k binary digits, which are mapped into $M = 2^k$ deterministic, finite energy waveforms $\{s_{\text{trans},m}(t), m = 1, 2, \ldots, M\}$. Let r_t be an indicator variable associated with one of M possible k-bit sequences $r_t \in \mathcal{R} = \{1, 2, \ldots, M\}$, $t = 0, 1, \ldots$, where t is a discrete time index. The signal waveform transmitted in the signaling interval of duration T may be represented as $s_{\text{trans}}(\tau) = \text{Re}[s_t(r_{0:t})h(\tau)\exp(j2\pi f_c \tau)]$, where f_c is a known carrier frequency, $h(\tau)$ is a real-valued signal pulse whose shape influences the spectrum of the transmitted signal $r_{0:t} \triangleq (r_0, \ldots r_t)^{\text{T}}$. $s_t(\cdot)$ performs the mapping from the digital sequence to waveforms. In a general case of a modulator with memory, $s_t(\cdot)$ depends on one or more previously transmitted symbols.

In the absence of an encoder, the message symbols are assumed to be *independent identically distributed* (i.i.d.). However, if error correcting coding, such as convolutional coding (Proakis 1995) or trellis coding (Du and Vucetic 1991), has been employed, the signals produced contain symbols which are not i.i.d. and as such display Markov properties. Thus, (to a first approximation) it is reasonable to assume that r_t is a first-order, time-homogeneous, M-state, Markov process with known transition probabilities $p_{ij} = \Pr\{r_{t+1} = j \mid r_t = i\}, i, j \in \mathcal{R}$, such that $p_{ij} \geq 0$, $\sum_{j=1}^{M} p_{ij} = 1$ for each i, and initial probability distribution $p_i = \Pr\{r_1 = i\}$, $p_i \geq 0$, $\sum_{i=1}^{M} p_i = 1$ for $i \in \mathcal{R}$ (see (Collings and Moore 1994, Collings and Moore 1995), for the same approach).

Channel model

The signal is passed through a noisy fading channel that can cause time-varying amplitude gain and phase shift. The channel can be described by a multiplicative discrete time disturbance l_t, which, in the case of Rayleigh fading, is a complex low-pass filtered Gaussian process with zero-mean real and imaginary components. (The time variations in l_t are realistically assumed to be slow in comparison to the message rate.) Ideally, in order to approximate the power spectral density of the fading process, a high-order low-pass pole-zero filter is required , (Stuber 1996). Thus, l_t can be modelled as the following ARMA(q, q) process (Butterworth filter of order q)

$$l_t = a^{\text{T}} l_{t-1:t-q} + b^{\text{T}} v_{t:t-q}, \tag{4.4.20}$$

where $l_{t-1:t-q} \triangleq (l_{t-1}, l_{t-2}, \ldots l_{t-q})^{\text{T}}$, $v_{t:t-q} \triangleq (v_t, v_{t-1}, \ldots v_{t-q})^{\text{T}}$, and v_t is a complex i.i.d. Gaussian noise , $v_t \overset{i.i.d.}{\sim} \mathcal{N}_c(0, 1)$. The coefficients of the filter $a \triangleq (a_1, a_2, \ldots, a_q)^{\text{T}}$, $b \triangleq (b_0, b_1, \ldots, b_q)^{\text{T}}$ are chosen so that the cut-off frequency of the filter matches the normalised channel Doppler frequency

$f_d T$ (T is the symbol rate), and the variance of the fading envelop is equal to 1. We assume here that $f_d T$, and thus a and b, are known.

Observations

At the demodulator, the received signal is passed through a filter whose impulse response is matched to the waveform $h(\tau)$ and is sampled at rate T^{-1}. The output of the filter is assumed to be corrupted by an additive noise with mutually independent real and imaginary parts, which are i.i.d. random sequences distributed as a mixture of zero-mean Gaussians with a known number of components[1] K. This assumption allows us to model non-Gaussian noise, and in particular outliers. In order to identify the parameters (the variance in our case) of the distribution from which the noise samples are drawn, it is convenient to introduce a latent allocation variable z_t, $z_t \in \mathcal{R}_z = \{1, 2, \dots, K\}$, $t = 1, 2, \dots$, such that $\Pr(z_t = j) = \lambda_j$, for $j = 1, \dots, K$, $\sum_{j=1}^{K} \lambda_j = 1$. Given the values of z_t, the additive noise component at time t is drawn from the complex Gaussian distribution with variance $\sigma_{z_t}^2$ correspondingly. Thus,

$$y_t = s_t(r_{0:t})l_t + \sigma_{z_t} w_t, \quad w_t \overset{i.i.d.}{\sim} \mathcal{N}_c(0,1),$$

where y_t and w_t are respectively the complex output of the matched filter and the additive complex zero-mean Gaussian noise.

State-space representation

In order to express the system in a state-space form, we define a state α_t such that $l_t = b^{\mathsf{T}} \alpha_{t:t-q}$. Then, from (4.4.20),

$$\alpha_t = a^{\mathsf{T}} \alpha_{t-1:t-q} + v_t,$$

and, thus, conditional upon the symbols and allocation variables, the problem can be formulated in the linear state-space form

$$\begin{aligned}
\alpha_{t:t-q} &= A\alpha_{t-1:t-q-1} + Bv_t \\
y_t &= C(r_{0:t})\alpha_{t:t-q} + D(z_t)w_t,
\end{aligned}$$

where $B = (1, 0, \dots, 0)^{\mathsf{T}}$, $C(r_{0:t}) = s_t(r_{0:t})b^{\mathsf{T}}$, $D(z_t) = \sigma_{z_t}$ and

$$A = \begin{pmatrix} a_1 & a_2 & \cdots & a_q & 0 \\ 1 & 0 & \cdots & 0 & 0 \\ 0 & 1 & \cdots & 0 & 0 \\ \vdots & \vdots & \ddots & \vdots & \vdots \\ 0 & 0 & \cdots & 1 & 0 \end{pmatrix}.$$

---

[1]The extension to non-zero mean Gaussian components is straightforward.

We assume $\alpha_{0:1-q} \sim \mathcal{N}_c\,(\widehat{\alpha}_0, P_0)$, where $P_0 > 0$, and let v_t, w_t be mutually independent for all $t > 0$. The message symbols r_t, the channel characteristics α_t and the latent allocation variable z_t are unknown for $t > 0$, whereas the model parameters a, b, $C\,(r_{0:t})$, $D(z_t)$, $\widehat{\alpha}_0$, P_0 are known for all $r_t \in \mathcal{R}$, $z_t \in \mathcal{R}_z$.

Estimation objectives

We are interested in obtaining the MMAP (marginal maximum a-posteriori) estimate of the symbols $\arg\max_{r_t} p(r_t \mid y_{1:t})$, and, if one is interested, the MMSE (conditional mean) estimate of the fading coefficients α_t, given by $\mathbb{E}\,(\alpha_t| y_{1:t})$. Obtaining these estimates requires a prohibitive computational cost exponential in the (growing) number of observations. It is thus necessary to develop some numerical approximations.

4.4.2   SMC applied to demodulation

Variance reduction via Rao-Blackwellisation

A generic SMC method developed previously can be directly applied to obtain the estimate of the joint posterior distribution $p\,(r_{0:t}, z_{0:t}, \alpha_{0:t}| y_{1:t})$. However, in our case, the states $\alpha_{0:t}$ can actually be integrated out, leading to a more efficient simulation algorithm with reduced variance of the estimates. This is an application of the so-called Rao-Blackwellisation method (Robert and Casella 1999). Indeed, $p\,(r_{0:t}, z_{0:t}, \alpha_{0:t}| y_{1:t})$ can be factorised in the following way:

$$p\,(r_{0:t}, z_{0:t}, \alpha_{0:t}| y_{1:t}) = p\,(\alpha_{0:t}| r_{0:t}, z_{0:t}, y_{1:t})\,p\,(r_{0:t}, z_{0:t}| y_{1:t}),$$

where, given $r_{0:t}$, $z_{0:t}$, $p\,(\alpha_{0:t}| r_{0:t}, z_{0:t}, y_{1:t})$ is a Gaussian distribution whose parameters can be computed using the Kalman filter. Thus, the problem of estimating $p\,(r_{0:t}, z_{0:t}, \alpha_{1:t}| y_{1:t})$ can be reduced to one of sampling from a lower-dimensional distribution $p\,(r_{0:t}, z_{0:t}| y_{1:t})$, which intuitively requires a reduced number of samples N in order to reach a given precision. See (Doucet, Gordon and Krishnamurthy 1999).

Let us define $x_t \triangleq (r_t, z_t)$. In this example, x_t is a finite state-space process which satisfies

$$K_t\,(x_t = (i, j)| x_{0:t-1}, y_{1:t-1}) = K\,(x_t = (i, j)| x_{t-1}) = p_{r_{t-1}i}\lambda_j,$$

and $g_t\,(y_t| x_{0:t}, y_{1:t-1}) = g\,(y_t| x_{0:t}, y_{1:t-1})$ is the likelihood that can be evaluated pointwise using the Kalman filter. See (Doucet 1998, Doucet, Godsill and Andrieu 2000, Doucet et al. 1999) for further details. We are in the framework of (4.2.1)-(4.2.2) where $\pi_t\,(x_{0:t}) = p\,(r_{0:t}, z_{0:t}| y_{1:t})$ is the discrete probability distribution to estimate at time t. We propose a SMC method to approximate it.

Algorithm

We describe briefly the implementation of this algorithm.

Sequential importance sampling step. We use a sampling distribution of the form

$$q\left(\widetilde{x}_{0:t}\middle|\, x_{0:t-1}, y_{1:t}\right) = \delta_{x_{0:t-1}}\left(\widetilde{x}_{0:t-1}\right) q\left(\widetilde{x}_t\middle|\, \widetilde{x}_{0:t-1}, y_{1:t}\right)$$

(we emphasise here that this kernel is dependent on the observations $y_{1:t}$). The optimal distribution satisfies

$$
\begin{aligned}
q\left(\widetilde{x}_t\middle|\, \widetilde{x}_{0:t-1}, y_{1:t}\right) &= \pi_t\left(\widetilde{x}_t\middle|\, \widetilde{x}_{0:t-1}\right) = p\left(\widetilde{x}_t\middle|\, \widetilde{x}_{0:t-1}, y_{1:t}\right) \\
&= \frac{g\left(y_t\middle|\, y_{1:t-1}, \widetilde{x}_{0:t}\right) K\left(\widetilde{x}_t\middle|\, \widetilde{x}_{t-1}\right)}{g\left(y_t\middle|\, y_{1:t-1}, \widetilde{x}_{0:t-1}\right)}
\end{aligned}
$$

then

$$
\begin{aligned}
w\left(\widetilde{x}_{0:t}\right) &= \frac{\pi_t\left(\widetilde{x}_{0:t-1}\right)}{\pi_{t-1}\left(\widetilde{x}_{0:t-1}\right)} = \frac{p\left(\widetilde{x}_{0:t-1}\middle|\, y_{1:t}\right)}{p\left(\widetilde{x}_{0:t-1}\middle|\, y_{1:t-1}\right)} \\
&\propto g\left(y_t\middle|\, y_{1:t-1}, \widetilde{x}_{0:t-1}\right)
\end{aligned}
$$

where

$$
\begin{aligned}
g\left(y_t\middle|\, y_{1:t-1}, \widetilde{x}_{0:t-1}\right) = &\sum_{m=1}^{M}\sum_{k=1}^{K}\left[g\left(y_t\middle|\, y_{1:t-1}, \widetilde{x}_{0:t-1}, r_t = m, z_t = k\right)\right. \\
&\left.\times\, p(r_t = m\middle|\, r_{t-1})p(z_t = k)\right].
\end{aligned}
$$

The distribution $g\left(y_t\middle|\, y_{1:t-1}, \widetilde{x}_{0:t-1}, r_t = m, z_t = k\right)$ is evaluated using the Kalman filter.

Thus, sampling from $p\left(\widetilde{x}_t\middle|\, \widetilde{x}_{0:t-1}, y_{1:t}\right)$ requires the evaluation of MK Kalman filter steps, which may be computationally intensive if MK is large. In this case, the *prior* distribution $K\left(\widetilde{x}_t\middle|\, \widetilde{x}_{t-1}\right) = p_{\widetilde{r}_{t-1}\widetilde{r}_t}\lambda_{\widetilde{z}_t}$ can be recommended as the importance distribution, for which the associated importance weights are proportional to $g\left(y_t\middle|\, y_{1:t-1}, \widetilde{x}_{0:t}\right)$. In this case, only one step of the Kalman filter for each particle has to be computed.

Selection step/Markov transition step. The selection step is done according to a stratified sampling scheme (Kitagawa 1996).

MCMC step. We use a Gibbs sampling method to update at each time t the path $x_{0:t}$ between $t - L + 1$ and t, where $L \in \mathbb{N}$ is the length of a given lag, see (Doucet et al. 1999, Punskaya, Andrieu, Doucet and Fitzgerald 2000).

Implementation issues

Clearly, the computational complexity of the proposed algorithm at each iteration is $O\left(N\right)$. Moreover, since both the optimal and prior importance distributions $q\left(\widetilde{x}_t\middle|\, \widetilde{x}_{0:t-1}, y_{1:t}\right)$ and the associated importance weights depend on $\widetilde{x}_{0:t-1}$ via a set of low-dimensional sufficient statistics (the mean

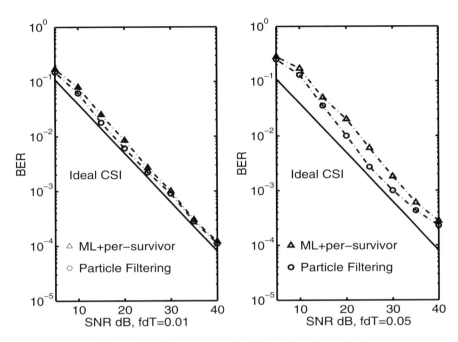

Figure 4.1. Bit error rate (additive Gaussian noise).

and covariance at time $t-1$ of the state α_{t-1} given by the Kalman filter associated to the path $\tilde{x}_{0:t-1}$), only these values need be kept in memory and, thus, the storage requirements for the proposed algorithm are also $O(N)$, and do not increase over time.

4.4.3 Simulations

Gaussian additive noise, $K = 1$

To assess the performance of the method proposed above, we applied it to the case of demodulation of 4-PSK ($s_t = A_c \exp(j\theta_t)$, $\theta_t = \frac{2\pi r_t}{M}$) signals which has been previously examined in the literature (Vitetta and Taylor 1995). The channel was generated from the low-pass filtered (3rd order Butterworth filter) zero-mean complex Gaussian noise with the bandwidth of the filter $f_d T$ times the bit rate, $f_d T$ being different in each example.

To prevent cycle slipping, known pilot symbols are periodically inserted into the transmitted symbol stream at rate $1/20$ (see (Georghiades and Han 1997, Gertsman and Lodge 1997, Seymour and Fitz 1995) for the same approach). The results for $N = 50$ compared with those obtained by Maximum Likelihood (ML) receiver employing per-survivor processing, and the one with ideal channel state information (CSI), are shown in Figure 4.1

for different signal-to-noise ratio (SNR) associated with the observations. Increasing the number of particles does not sensibly modify the results for this problem. As can be seen, the proposed algorithm outperforms the existing methods, and even in the case of fast fading ($f_d T = 0.05$) it performs well.

Mixture of Gaussians, $K = 2$

In the second experiment we applied the proposed algorithm to the case in which the additive noise is distributed as a two-component mixture of zero-mean Gaussians. The overall variance of the noise in this case is $\lambda_1 \sigma_1^2 + (1 - \lambda_1) \sigma_2^2$ with $\lambda_1 = 0.2$. σ_1^2 was chosen so that the SNR would be equal to 5 dB if $\lambda_1 = 1$ and in the absence of fading. The characteristics of the signal and channel are the same as for the first experiment, and the results are shown in Figure 4.2. It can be seen that our algorithm still performs well whereas other standard methods fail in this case.

The bit error rate performance of the proposed simulation-based algorithm has been extensively studied for different modulation schemes, and the complete results are available in (Punskaya et al. 2000). It is also shown in that study that these results can be significantly improved by obtaining the fixed-lag smoothed estimates of the symbols (with a small fixed delay

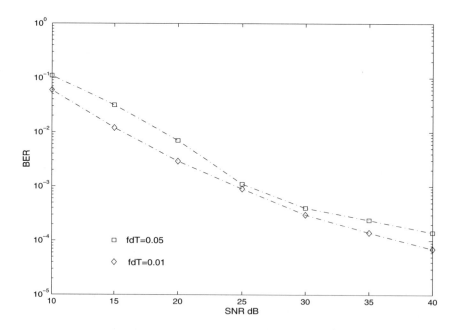

Figure 4.2. Bit error rate (additive non-Gaussian noise).

L), and when channel coding has been employed a joint signal detection and decoding can be easily carried out.

5

Deterministic and Stochastic Particle Filters in State-Space Models

Erik Bølviken
Geir Storvik

5.1 Introduction

Optimal or Bayesian filtering in state-space models is a question of computing series of linked numerical integrals where output from one is input to the other (Bucy and Senne 1971). Particle filtering can be regarded as comprising techniques for solving these integrals by replacing the complicated posterior densities involved by *discrete* approximations, based on *particles* (Kitagawa 1996). There is evidence that the numerical errors as the process is iterated often stabilise, or at least do not accumulate sharply (see section 5.2.5). Filters of this type can be constructed in many ways. Most of the contributions to this volume employ Monte Carlo designs (see also (Doucet 1998) and the references therein). Particles are then random drawings of state vectors under the current posterior. This amounts to Monte Carlo evaluations of integrals. Numerically inaccurate, but often practical and easy to implement, general methods to run the sampling have been developed. Alternatively, particles can be laid out through a deterministic plan, using more sophisticated and more accurate numerical integration techniques. Such an approach has been discussed in (Kitagawa 1987), (Pole and West 1988) and (Pole and West 1990), but recently most work has been based on Monte Carlo methods. To some extent, Monte Carlo and deterministic particle filters are complementary approaches, and one may also wonder whether they may be usefully combined (see (Monahan and Genz 1997) for such a combination in a non-dynamic setting). Emphasis in this paper is on deterministic filtering. A general framework can be found in the above-mentioned references and in (West and Harrison 1997)[Section 13.5]. We shall present a common perspective in the next section, where our contribution will be on design issues.

Deterministic filtering could be the method of choice when the state process has low dimension and high numerical accuracy is wanted. Our applications, which are a group of statistical problems from population biology, are of that type. The biological signals in some of the examples in Section 5.3 are extremely weak, yet governed by well-defined stochastic models derived from basic theory on fluctuations of animal populations. The scientific objective is in understanding the underlying mechanisms. This means model identification, and in a frequency context, computation and maximization of likelihood functions. Filtering is then a mean, to compute such quantities, rather than an end in itself, but the computational problems remain much the same as in engineering, where the interest is usually in the filtering estimates. Even though our statistical applications are off-line, computer speed is still highly relevant because the computations are often repeated many times to estimate uncertainty through bootstrapping. One difference from engineering, however, may be the attitude towards Monte Carlo error. In ad-hoc situations in which a filtered estimate is used for, say, guidance or control, added uncertainty that is a small fraction of the total may not matter much. In a context of basic science, as in Section 5.3, this could be different, and Monte Carlo error in likelihood evaluations can be awkward if the function is to be maximised. Various types of computing likelihoods will be compared in Section 5.3.

5.2   General issues

5.2.1   Model and exact filter

We are dealing with a discrete time vectorial process $\{\mathbf{x}_t\}$ in $\mathcal{R}^{n_x}$, observed indirectly through another process $\{\mathbf{y}_t\}$ in $\mathcal{R}^{n_y}$. The framework is a Markov model for $\{\mathbf{x}_t\}$ based on (possibly time-varying) transition densities $p(\mathbf{x}_t|\mathbf{x}_{t-1})$ and conditionally independent observations $\{\mathbf{y}_t\}$ given $\{\mathbf{x}_t\}$. The latter usually means that the conditional density of the vector $\mathbf{y}_{1:t} \stackrel{\Delta}{=} (\mathbf{y}_1,\ldots,\mathbf{y}_t)$ given the corresponding $\mathbf{x}_{1:t} \stackrel{\Delta}{=} (\mathbf{x}_1,\ldots,\mathbf{x}_t)$ factors into $p(\mathbf{y}_{1:t}|\mathbf{x}_{1:t}) = \prod_{s=1}^{t} p(\mathbf{y}_s|\mathbf{x}_s)$, where $p(\mathbf{y}_s|\mathbf{x}_s)$ is the density of $\mathbf{y}_s$ given $\mathbf{x}_s$. The symbol p will throughout be used to designate density functions of various kinds, for example in addition to those above, $p(\mathbf{y}_{1:t})$ for the density function of $\mathbf{y}_{1:t}$. These functions may depend on t themselves, say $p_t(\mathbf{y}_{1:t})$ or $p_s(\mathbf{y}_s|\mathbf{x}_s)$ rather than $p(\mathbf{y}_{1:t})$ or $p(\mathbf{y}_s|\mathbf{x}_s)$, but it is not necessary to include this in the notation.

The exact filter for $\{\mathbf{x}_t\}$ can be written as a set of recursive integration equations (Bucy and Senne 1971). Start with $p(\mathbf{x}_0) \stackrel{\Delta}{=} p(\mathbf{x}_0|\mathbf{y}_0)$ as prior for

$\mathbf{x}_0$ and calculate recursively

$$p(\mathbf{x}_t|\mathbf{y}_{1:t-1}) = \int p(\mathbf{x}_t|\mathbf{x}_{t-1})p(\mathbf{x}_{t-1}|\mathbf{y}_{1:t-1})d\mathbf{x}_{t-1} \qquad (5.2.1)$$

$$C_t = \int p(\mathbf{y}_t|\mathbf{x}_t)p(\mathbf{x}_t|\mathbf{y}_{1:t-1})d\mathbf{x}_t \qquad (5.2.2)$$

$$p(\mathbf{x}_t|\mathbf{y}_{1:t}) = C_t^{-1}p(\mathbf{y}_t|\mathbf{x}_t)p(\mathbf{x}_t|\mathbf{y}_{1:t-1}), \qquad (5.2.3)$$

for $t = 1, \ldots, T$. The normalisation constants $\{C_t\}$ then produce the log-likelihood function of observations $\mathbf{y}_{1:T}$ through

$$\log\{p(\mathbf{y}_{1:T})\} = \sum_{t=1}^{T} \log(C_t). \qquad (5.2.4)$$

The relationships (5.2.1-5.2.4) are well known and are consequences of $\{\mathbf{x}_t\}$ being a Markov process and $\{\mathbf{y}_t\}$ being conditionally independent given $\{\mathbf{x}_t\}$. The proof is an elementary application of Bayes' formula.

5.2.2 Particle filters

Computation of the posterior densities $p(\mathbf{x}_t|\mathbf{y}_{1:t})$ and the log likelihood function $p(\mathbf{y}_{1:T})$ is an exercise in high-dimensional numerical integration. General purpose integration methods are bound to be inefficient, and it seems better to design iterative schemes imitating the exact one (5.2.1-5.2.3). Particle filters proceed in this manner through *discrete* approximations to the exact posterior distributions. Let $\hat{p}(\mathbf{x}_t|\mathbf{y}_{1:t})$ be some discrete analogue to the exact density $p(\mathbf{x}_t|\mathbf{y}_{1:t})$. The points $\mathbf{x}_t^{(i)}$ on which $\hat{p}(\mathbf{x}_t)$ assigns positive probabilities are known as the *particles*. Their number N_t may vary. Suppose a reasonable approximation $\hat{p}_{t-1}(\mathbf{x}_{t-1}|\mathbf{y}_{1:t-1})$ is available at time $t - 1$. When inserted for the exact density $p(\mathbf{x}_{t-1}|\mathbf{y}_{1:t-1})$ on the right in (5.2.1), we obtain

$$\hat{p}(\mathbf{x}_t|\mathbf{y}_{1:t-1}) = \sum_{j=1}^{N_{t-1}} p(\mathbf{x}_t|\mathbf{x}_{t-1}^{(j)})\hat{p}(\mathbf{x}_{t-1}^{(j)}|\mathbf{y}_{1:t-1}). \qquad (5.2.5)$$

The main point of the design is to ensure a good approximation to the exact predictive density $p(\mathbf{x}_t|\mathbf{y}_{1:t-1})$. When (5.2.5) replaces its exact counterpart in (5.2.3), we immediately have

$$\tilde{p}(\mathbf{x}_t|\mathbf{y}_{1:t}) = \tilde{C}_t^{-1}p(\mathbf{y}_t|\mathbf{x}_t)\hat{p}(\mathbf{x}_t|\mathbf{y}_{1:t-1}), \qquad (\tilde{C}_t \text{ a constant}), \qquad (5.2.6)$$

as an approximate update density. To complete the recursion, (5.2.6) must be replaced by a particle approximation $\hat{p}(\mathbf{x}_t|\mathbf{y}_{1:t})$. Filters proposed in the literature vary as to how this step should be carried out. Most of the papers in this volume use Monte Carlo sampling. If the particles $\mathbf{x}_t^{(i)}$ are drawn randomly from (5.2.6), the probabilities become $\hat{p}(\mathbf{x}_t^{(i)}|\mathbf{y}_{1:t}) = N_t^{-1}$. We may enhance numerical accuracy by importance sampling (Doucet 1998), the probabilities are then the importance weights. In recent years, many tricks have been invented to obtain computationally fast sampling.

Stochastic generation of particles corresponds to Monte Carlo evaluations of integrals. An alternative is to use other, more accurate integration methods. This raises, as we shall see in Section 5.3, various issues regarding design, but to understand this we shall first review elements of the theory of numerical integration.

5.2.3 Gaussian quadrature

Consider the case of a *scalar* process x_t, denoted x in this sequel. The integrands in (5.2.1) and (5.2.2) are of the form $h(x)p(x)$ with h some function and p a density. Evaluations of integrals are often efficiently carried out by Gaussian quadrature. The simplest among these rules is the Gauss-Legendre method. The standard form is

$$\int_{-1}^{1} h(x)p(x)dx = \sum_{i=1}^{m} \gamma^{(i)}h(\xi^{(i)})p(\xi^{(i)}) + \mathcal{E}_0, \qquad (5.2.7)$$

where the abscissas $\xi^{(1)},\ldots,\xi^{(m)}$ and the positive weights $\gamma^{(1)},\ldots,\gamma^{(m)}$ are tabulated in numerical literature; see also (Press, Teukolsky, Vetterling and Flannery 1992) for a simple computer program. For a general interval (5.2.7) changes into

$$\int_{A}^{B} h(x)p(x)dx = \sum_{i=1}^{m} w^{(i)}h(x^{(i)})p(x^{(i)}) + \mathcal{E}, \qquad (5.2.8)$$

where

$$x^{(i)} = \frac{1}{2}(A + B + (B - A)\xi^{(i)}), \qquad w^{(i)} = \frac{1}{2}(B - A)\gamma^{(i)}. \qquad (5.2.9)$$

The error term $\mathcal{E}$ is

$$\mathcal{E} = \frac{1}{2}(B - A)^{2m+1}\frac{(m!)^4}{\{(2m)!\}^3(2m + 1)}(ph)^{(2m)}(\omega), \qquad (5.2.10)$$

where $(ph)^{(2m)}(\omega)$ is the derivative of order $2m$ of ph at some $\omega \in (A, B)$. The error term thus vanishes if the integrand is a polynomial of degree $< 2m$, and the accuracy is almost startling otherwise. Assuming $B-A = 10$, $\mathcal{E}$ being of order 10^{-10} times the derivative for $m = 10$ and of 10^{-32} for $m = 20$ for $B - A = 10$ (but higher order derivatives often grow!).

The integral approximation (5.2.8-5.2.9) has a probabilistic interpretation. Suppose the probability mass of p is negligible outside the interval (A, B). Define

$$\hat{p}(x^i) = w^{(i)}p(x^{(i)}) \qquad (5.2.11)$$

and insert $h \equiv 1$ in (5.2.8). Then

$$\sum_{i=1}^{m} \hat{p}(x^{(i)}) \approx \int_{A}^{B} p(x)dx \approx 1,$$

so that $\hat{p}(x^{(1)}), \ldots, \hat{p}(x^{(m)})$ is almost a proper probability vector. Normalisation is usually not worth the bother. Let $\hat{p}$ be the distribution assigning these probabilities to $x^{(i)}$, $i = 1, \ldots, m$. Then (5.2.8) expresses that $\hat{p}$ is an approximation to p in the sense that the expectation of $h(x)$ for any *smooth* function h is almost equal for the two distributions. Note that $\hat{p}(x^{(i)})$ and $p(x^{(i)})$ deviate.

Other quadrature rules produce similar particle approximations through an idea similar to importance sampling. Choose some density ψ and note that $\int hp = \int hR\psi$ where $R = p/\psi$. Quadrature with ψ as a weight function yields

$$\int h(x)p(x)dx = \int h(x)R(x)\psi(x)dx = \sum_{i=1}^{m} \gamma^{(i)} h(\xi^{(i)})R(\xi^{(i)}) + \mathcal{E}, \quad (5.2.12)$$

where abscissas $\xi^{(i)}$ and weights $\gamma^{(i)}$, which depend on ψ, differ from those in (5.2.7). The error term now vanishes if hR is polynomial of degree less that $2m$. The particle approximation (similar to (5.2.11)) becomes $x^{(i)} = \xi^{(i)}$ and $\hat{p}(x^{(i)}) = \gamma^{(i)} p(x^{(i)})/\psi(x^{(i)})$.

5.2.4   Quadrature filters

Quadrature filters are constructed by replacing the density $\tilde{p}(\mathbf{x}_t | \mathbf{y}_{1:t-1})$ in (5.2.6) by a particle approximation based on the quadrature formulas in Section 5.2.3. This leads to the following recursive scheme, where

$$\hat{p}(\mathbf{x}_t^{(i)} | \mathbf{y}_{1:t-1}) = \sum_{j=1}^{N_{t-1}} p(\mathbf{x}_t^{(i)} | \mathbf{x}_{t-1}^{(j)}) \hat{p}(\mathbf{x}_{t-1}^{(j)} | \mathbf{y}_{1:t-1}) \qquad (5.2.13)$$

$$\widehat{C}_t = \sum_{i=1}^{N_t} p(\mathbf{y}_t | \mathbf{x}_t^{(i)}) w_t^{(i)} \hat{p}(\mathbf{x}_t^{(i)} | \mathbf{y}_{1:t-1}) \qquad (5.2.14)$$

$$\hat{p}(\mathbf{x}_t^{(i)} | \mathbf{y}_{1:t}) = \widehat{C}_t^{-1} p(\mathbf{y}_t | \mathbf{x}_t^{(i)}) w_t^{(i)} \hat{p}(\mathbf{x}_t^{(i)} | \mathbf{y}_{1:t-1}), \qquad (5.2.15)$$

for $i = 1, \ldots, N_t$. Implementing the scheme (5.2.13-5.2.15) when $n_x > 1$ raises a number of problems. The mathematical theory of Gaussian quadrature described above is inherently one-dimensional and must be applied sequentially, one state variable at a time. The weights $w^{(i)}$ in (5.2.14) and (5.2.15) will then be *products* of weights from each of the n_x variables; see Section 5.3.2 for details. This also suggests that the computational requirements will grow rapidly with n_x. The number of operations in quadrature filtering is proportional to N_t^2 for each step of the recursion.

When $n_x = 1$, as in the example, N_t equaled the order m of the quadrature rule. But if n_x abscissas are used for each variable of the state vector, then $N_t = m^{n_x}$, easily a huge number. A particle representation of n_x *indepen-dent* variables actually needs such a large N_t to be accurate, but if there are correlations it may be possible reduce it.

5.2.5 Numerical error

It is possible to gain some insight into how numerical error propagates by elementary methods. Let

$$\delta_t = \log\{\hat{p}(\mathbf{y}_{1:t})\} - \log\{p(\mathbf{y}_{1:t})\} = \sum_{s=1}^{t} \log \widehat{C}_s - \sum_{s=1}^{t} \log C_s \qquad (5.2.16)$$

be the accumulated error in the log-likelihood function, and define

$$\varepsilon(\mathbf{x}_t|\mathbf{y}_{1:t-1}) = \exp(\delta_{t-1})\hat{p}(\mathbf{x}_t|\mathbf{y}_{1:t-1}) - p(\mathbf{x}_t|\mathbf{y}_{1:t-1}) \qquad (5.2.17)$$

as an indicator of numerical error present in the distribution $\hat{p}(\mathbf{x}_t|\mathbf{y}_{1:t-1})$. The factor $\exp(\delta_{t-1})$ is to avoid normalisations in the recursion (5.2.18) below.

By elementary manipulations it is easily proved (see appendix) that

$$\varepsilon(\mathbf{x}_{t+1}|\mathbf{y}_{1:t}) = C_t^{-1}\int p(\mathbf{x}_{t+1}|\mathbf{x}_t)p(\mathbf{y}_t|\mathbf{x}_t)\varepsilon(\mathbf{x}_t|\mathbf{y}_{1:t-1})d\mathbf{x}_t + \eta(\mathbf{x}_{t+1}), \qquad (5.2.18)$$

where

$$\eta(\mathbf{x}_{t+1}) = \exp(\delta_t)[\sum_i p(\mathbf{x}_{t+1}|\mathbf{x}_t^{(i)})\hat{p}(\mathbf{x}_t^{(i)}|\mathbf{y}_t) - \int p(\mathbf{x}_{t+1}|\mathbf{x}_t)\tilde{p}(\mathbf{x}_t|\mathbf{y}_{1:t})d\mathbf{x}_t].$$

$$(5.2.19)$$

If we insert (5.2.3) into (5.2.1) it emerges that the sequence of posterior den-sities $\{p(\mathbf{x}_t|\mathbf{y}_{t-1})\}$ themselves satisfy exactly the same recursion (5.2.18) but for the term $\eta(\mathbf{x}_{t+1})$, which signifies the contribution of the numerical error at time t. Suppose now that r_t is the smallest real number for which

$$|\varepsilon(\mathbf{x}_t|\mathbf{y}_{1:t-1})| \leq r_t p(\mathbf{x}_t|\mathbf{y}_{1:t-1}) \qquad (5.2.20)$$

so that r_t controls the *relative error* in $\hat{p}(\mathbf{x}_t|\mathbf{y}_{1:t-1})$. Then, by insert-ing (5.2.20) in the integrand (5.2.18) it follows that

$$r_{t+1} \leq r_t + \tilde{\eta}(\mathbf{x}_{t+1}), \qquad (5.2.21)$$

where

$$\tilde{\eta}(\mathbf{x}_{t+1}) = \eta(\mathbf{x}_{t+1})/p(\mathbf{x}_{t+1}|\mathbf{y}_{1:t}) \qquad (5.2.22)$$

is the relative numerical error introduced at time t.

Relative error in the approximations diminishes through the next step of the scheme. New error is brought, as well, but the old relative error does not blow up as the recursion is progressed.

It is possible for Gaussian quadrature filters to give a rigorous error statement. Suppose we have been able to find approximations so that

$$\left| \sum_{j=1}^{N_{t-1}} w_t^{(j)} p(\mathbf{x}_t^{(i)}|\mathbf{x}_{t-1}^{(j)}) p(\mathbf{x}_{t-1}^{(j)}|\mathbf{y}_{1:t-1}) - p(\mathbf{x}_t^{(i)}|\mathbf{y}_{t-1}) \right| \le \varepsilon p(\mathbf{x}_t^{(i)}|\mathbf{y}_{1:t-1})$$

$$(5.2.23)$$

for all i, t. Notice that this condition is for $p(\mathbf{x}_t^{(j)}|\mathbf{y}_{1:t})$ known and for a finite set of $\mathbf{x}_{t+1}^{(i)}$ values, which is always possible to obtain by setting N_t large enough. It is then proved (in the Appendix) that

$$r_{t+1} \le r_t (1+\varepsilon)^t + \varepsilon, \qquad (5.2.24)$$

which implies that

$$r_t \le (1+\varepsilon)^t - 1. \qquad (5.2.25)$$

By defining

$$\varepsilon(\mathbf{y}_t|\mathbf{y}_{1:t-1}) = \exp(\delta_{t-1})\hat{p}(\mathbf{y}_t|\mathbf{y}_{1:t-1}) - p(\mathbf{y}_t|\mathbf{y}_{1:t-1}) \qquad (5.2.26)$$

and replacing condition (5.2.23) with

$$\left| \sum_{i=1}^{N_t} w_t^{(i)} p(\mathbf{y}_t|\mathbf{x}_t^{(i)}) p(\mathbf{x}_t^{(i)}|\mathbf{y}_{1:t-1}) - p(\mathbf{y}_t|\mathbf{y}_{t-1}) \right| \le \varepsilon p(\mathbf{x}_t^{(i)}|\mathbf{y}_{1:t-1}) \quad (5.2.27)$$

for all t, it is by similar reasoning possible to show that

$$|\varepsilon(\mathbf{y}_t|\mathbf{y}_{1:t-1})| \le [r_t(1+\varepsilon) + \varepsilon] p(\mathbf{y}_t|\mathbf{y}_{1:t-1}). \qquad (5.2.28)$$

However,

$$\varepsilon(\mathbf{y}_t|\mathbf{y}_{1:t-1}) = \exp(\delta_{t-1})\widehat{C}_t - C_t = [\exp(\delta_t) - 1] p(\mathbf{y}_t|\mathbf{y}_{1:t-1}),$$

which results in

$$|\exp(\delta_t) - 1| \le r_t(1+\varepsilon) + \varepsilon \le (1+\varepsilon)^{t+1} - 1, \qquad (5.2.29)$$

and consequently that

$$\delta_t \le (t+1)\log(1+\varepsilon), \qquad (5.2.30)$$

meaning the error is approximately linear in t. This shows that, although the error will increase with time, it will not accumulate sharply. Notice that, in general, a better limit for the error in the likelihood evaluation is not possible. This can be seen by considering a situation in which all observations are independent, such that it is straightforward to show that the error involved in the computation of the log-likelihood is the sum of errors at each time point.

5.2.6   A small illustrative example

In order to compare the quadrature filter with the Monte Carlo filter, we shall consider the one-dimensional nonlinear reference model (Doucet 1998)

$$x_t = \frac{1}{2}x_{t-1} + 25\frac{x_{t-1}}{1 + x_{t-1}^2} + 8\cos(1.2t) + e_t \qquad (5.2.31)$$

$$z_t = \frac{x_t^2}{20} + v_t, \qquad (5.2.32)$$

where $x_0 \sim \mathcal{N}(0,5)$, e_t and v_t are mutually independent white Gaussian noises, $e_t \sim \mathcal{N}(0,\sigma_e^2)$ and $e_t \sim \mathcal{N}(0,\sigma_v^2)$ with $\sigma_e^2 = 10$ and $\sigma_v^2 = 1$. Data were simulated according to this model for $t = 1, ..., T = 100$. The quadrature filter was implemented as described in Section 5.2.4, while rejection sampling with the predictive distribution as a proposal was used for the Monte Carlo filter. Although more effective Monte Carlo filters could be used, our main concern here is to examine the variability in the result, making the choice of Monte Carlo filter less important.

The filters were first run with a large number of particles ($N_t = N = 500$ for the quadrature filter and $N_t = N = 2000$ for the Monte Carlo filter). The results from the two filters agreed in this case, giving us the true posterior as a reference. In Figure 5.1, the estimates of the predictive density $p(x_{t+1}|y_{1:t})$ for $t = 100$ are displayed. For both filters, $N = 50$ were used. The true density is given by the solid curve. The Monte Carlo filter (long dashed curve) is able to find the two modes of the density, but is somewhat inaccurate on the weights on these modes. The quadrature filter (dotted) curve, on the other hand, is clearly able to yield the true density on almost all parts.

5.3   Case studies from ecology

5.3.1   Problem area and models

Population biology is a field in which optimal filtering is likely to yield substantial benefits, and we shall in this section demonstrate what can be achieved with such a technique. It is generally recognised that, in the stationary regime, animal populations oscillate due to feedback links between predator and prey and interactions with environment. A reasonable mathematical framework, see (Royama 1992), (Stenseth, Bjørnstad and Falck 1996), imposes nonlinear autoregressive models of low order on logarithms of abundance. The parameters of the models, influenced by vegetation and climatic conditions, might vary between sites. There is a dearth of data. The usual data reflect the number of animals caught in traps annually or seasonally, sometimes gathered by means of scientific experiment. Other data is arrived at deductively, e.g., from the number of skins delivered to

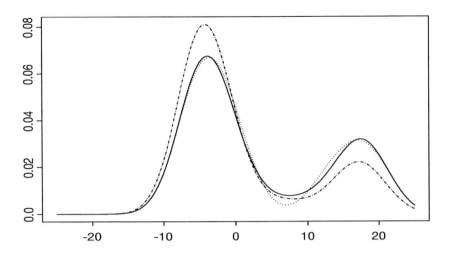

Figure 5.1. True and estimated predictive density $p(x_{101}|y_{1:100})$ for the reference model. The true density is given as a solid curve, the Monte Carlo filter estimate is given as a long dashed curve, and the quadrature estimate is given by dotted curve.

a company. The famous lynx series data, used in countless textbooks on statistical time series analysis (for example (Tong 1990)), are of the latter type.

The traditional approach in ecology (Stenseth, Bjørnstad and Saitoh 1996, Stenseth, Falck, Bjørnstad and Krebs 1997, Stenseth, Falck, Chain, Bjørnstad, Donoghue and Tong 1998)) is to identify the data with the animal population, but it is really more correct to regard data and population as different entities and employ a state model. This generates a more realistic picture of statistical uncertainties, deals with zero observations and opens for utilisation sources of data that have traditionally *not* been used scientifically; see below. Populations are, on first approximation, described by linear autoregressive models on log-scale, e.g.,

$$x_t = a_1 x_{t-1} + \ldots a_r x_{t-q} + e_t, \qquad (5.3.1)$$

where $\{e_t\}$ is independent noise with zero mean and constant variance. The mean of x_t is not estimable under the measurement models introduced below, so has been subtracted out. Thus, in (5.3.1) x_t is the difference from the long-term average. Interest is directed towards the period (i.e. the location of the maximum of the spectral density), the autoregressive coefficients (which have ecological interpretations) and the order q of the series (which relates to ecological hypotheses).

The linear model (5.3.1) is useful, but fails to deal with the hypothesis that the built-up phase of animal abundance tends to be faster for a population in decline. There is for the lynx empirical evidence supporting such a hypothesis; see (Stenseth et al. 1997). One possibility, utilised by these authors, is to impose a TAR (truncated autoregressive) model of the type outlined in (Tong 1990). The single autoregressive relationship is then split into two separate ones, i.e.

$$x_t = \begin{cases} a_1 x_{t-1} + \ldots a_r x_{t-q} + e_t, & \text{if } x_{t-d} \le \theta; \\ b_1 x_{t-1} + \ldots b_r x_{t-q} + e_t, & \text{if } x_{t-d} > \theta. \end{cases} \quad (5.3.2)$$

This split expresses that the population changes into a different phase (i.e. decline) d time units after the threshold θ has been passed. Mathematical properties of these models are summarised in (Tong 1990).

We shall consider three different measurement regimes $\{y_t\}$. The most common ones are *counts* (based on trappings). The natural model for y_t given x_t is then

$$p(y_t|x_t) = \frac{\lambda_t^{y_t}}{y_t!} \exp(-\lambda_t), \qquad \log(\lambda_t) = \alpha + \beta x_t. \quad (5.3.3)$$

Notice that the parameter β influences the distribution of $\{y_t\}$ through $\beta\sigma_e$, where σ_e is the standard deviation of the series $\{e_t\}$ in (5.3.1) and (5.3.2). It is therefore not possible to estimate β and σ_e jointly. The same defect occurs in the other observation models (5.3.4) and (5.3.5) below.

Other types of data are categorical, based on judgments by experienced field observers. These sources of information have never been utilised scientifically, but would be a welcome additional source of information in population biology. One type of assessment is binary, y_t being 0 or 1, $y_t = 0$ signifying a population below average and $y_t = 1$ above. Such judgments are with errors. There can be no precise definition as to the precise meaning of these categories. A plausible model is

$$p(y_t|x_t) = p_t^{y_t}(1 - p_t)^{y_t}, \qquad p_t = \frac{\exp(\alpha + \beta x_t)}{1 + \exp(\alpha + \beta x_t)}. \quad (5.3.4)$$

Here β is close to zero for bad observers and very large for good ones. The other parameter α captures *bias* in observer evaluation (upwards if $\alpha > 0$ and downwards if $\alpha < 0$). Other observations may be in terms of *differences* $x_t - x_{t-1}$ (*ratios* on the original scale). We shall in the next section discuss an example where $y_t \in \{-1, 0, 1\}$, where $y_t = -1$ means decline from one time point to the next, $y_t = 0$ signifies (at best) moderate changes and $y_t = 1$ growth. A possible model is now

$$p(y_t|x_t) = \begin{cases} c\exp(-\alpha_n - \beta(x_t - x_{t-1})) & \text{if } y_t = -1 \\ c & \text{if } y = 0 \\ c\exp(-\alpha_p + \beta(x_t - x_{t-1})) & \text{if } y_t = -1, \end{cases} \quad (5.3.5)$$

where c is a constant forcing the three probabilities to sum to one. Interpretations of the three parameters α_n, α_p and β are similar to those in (5.3.4).

5.3.2 Quadrature filters in practice

Constructing quadrature filters is a question of selecting particles $x_t^{(i)}$ and their associated weights $w_t^{(i)}$ in the scheme (5.2.13-5.2.15) above. In the present instance, several issues are involved in this selection. How should joint particle grids in several variables be defined? How is the discontinuity in the TAR model (5.3.2) handled? The process $\{x_t\}$ describing the population is stationary and varies within a well-defined, and in practice limited, region. Should that process be utilised?

It is most convenient to answer these questions when the autoregressive processes (5.3.1) and (5.3.2) have order $q = 2$. The state vector at t is then (x_t, x_{t-1}) with state representation

$$\begin{pmatrix} x_t \\ x_{t-1} \end{pmatrix} = \begin{pmatrix} a_1 & a_2 \\ 1 & 0 \end{pmatrix} \begin{pmatrix} x_{t-1} \\ x_{t-2} \end{pmatrix} + \begin{pmatrix} e_t \\ 0 \end{pmatrix}. \tag{5.3.6}$$

Notice the zero noise in the second component. The relation (5.2.1) of the exact filter becomes

$$p(x_t, x_{t-1}|y_{1:t-1}) = \int_{-\infty}^{\infty} p(x_t|x_{t-1}, x_{t-2})p(x_{t-1}, x_{t-2}|y_{1:t-1})dx_{t-2}, \tag{5.3.7}$$

which is a uni-dimensional integral despite the state vector's having two variables, it would have been univariate even for general q. Autoregressive processes are special because only one (of q) relations of the state representation has non-zero noise. This is responsible for the univariate integral (5.3.7) and is actually desirable. A general quadrature filter requires $O(N_t^2)$ operations for each step t of the recursion, but the constructions below are much faster.

However, the degenerate noise in (5.3.6) means that it is less clear how the joint grid for (x_t, x_{t-1}) should be constructed. At first glance, it may seem natural to use the version on the left of Figure 5.2, which utilises the correlation structure between x_t and x_{t-1}. Skew grids, as in Figure 5.2 left, follow by applying univariate particle approximations to the two factors sequentially. First, quadrature abscissas $x_{t-1}^{(j)}$ (needed for later integration) are laid out for x_{t-1} and the second approximation is conditioned on the first. The problem with this design is that the number of particles will grow explosively. Suppose there are N_{t-1} different particles for x_{t-1}. Each of them is unchanged as it moves ahead and each gives birth to m different particles for x_t, making $N_t = mN_{t-1}$. One way to keep this under control is to change the grid with fewer particles before progressing, but this approach requires the posterior densities to be interpolated numerically, adding to the error. Further it is not cheap computationally. Interpolation from skew grids, such as that in Figure 5.2 left, is also a cumbersome process, especially in higher dimensions.

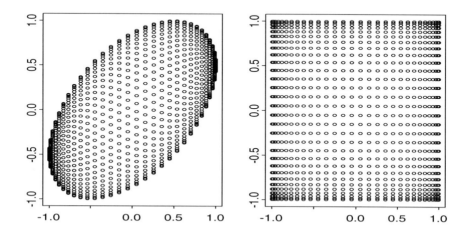

Figure 5.2. Particles for (x_t, x_{t-1}) as Gauss-Legendre abscissas in each dimension. On the left: particles constructed by conditioning x_t on x_{t-1}, on the right: particles constructed independently in each direction.

We have tested versions such as these, and they certainly work, but for stationary autoregressive processes we tend to prefer the simpler solution in Figure 5.2 right where the particles have been laid out as quadrature abscissas for the *marginal* densities of the two variables. This is not so parsimonious as approximations to the joint distribution, but now the number of particles can be kept constant as the algorithm progresses without interpolation. In detail, these methods work as described below when based on Gauss-Legendre quadrature.

First, a particle approximation to the density $p(x_0, x_{-1})$ is needed for initialisation. In the linear, Gaussian case the corresponding distribution has zero means and an easily computable covariance matrix; see (Priestley 1981). If σ_x is the standard deviation, we may apply (5.2.9) to each variable with $A = -B_0\sigma_x$ and $B = B_0\sigma_x$ for, say $B_0 = 5$. The square with edges $(\pm B_0\sigma_x, \pm B_0\sigma_x)$ then contains virtually the whole probability mass of the initial density $p(x_0, x_{-1})$, and the bivariate particle approximation becomes:

Initialisation

$$x_0^{(i)} = x_{-1}^{(i)} = B_0 \sigma_x \xi^{(i)}$$

$$\hat{p}(x_0^{(i)}, x_{-1}^{(j)}) = p(x_0^{(i)}, x_{-1}^{(j)}) B_0^2 \gamma^{(i)} \gamma^{(j)}.$$

$$(5.3.8)$$

Closed-form expressions for the joint density $p(x_0, x_{-1})$ are rarely available apart from in the Gaussian, linear case. A technique that works well (and which was used for the TAR model) is that of running the main recursion below a suitable number of steps *without data* from a start such as in (5.3.8). Since $\{x_t\}$ by definition is stationary, the recursion will settle at a steady state. For the TAR model, where simple necessary and sufficient conditions for stationarity are not available (see (Tong 1990)), this is actually a way of checking for stationarity.

Next comes the main recursive step. By adapting the general scheme (5.2.13-5.2.15) by means of Gauss-Legendre integration, we obtain the following method (with $N_t = N$ for all t).

Algorithm 5.3.1 (Main recursion). *Choose A_t and B_t (see text). Put*

$$x_t^i = \frac{1}{2}(A_t + B_t + (B_t - A_t)\xi^{(i)}), \qquad (5.3.9)$$

$$w_t^i = \frac{1}{2}(B_t - A_t)\gamma^{(i)}. \qquad (5.3.10)$$

Update the probabilities by

$$\hat{p}(x_t^{(i)}, x_{t-1}^{(j)}|y_{1:t-1}) = \sum_{k=1}^{N} p(x_t^{(i)}|x_{t-1}^{(j)}, x_{t-2}^{(k)})\hat{p}(x_{t-1}^{(j)}, x_{t-2}^{(k)}), \qquad (5.3.11)$$

$$\hat{q}(x_t^{(i)}, x_{t-1}^{(j)}|y_{1:t}) = p(y_t|x_t^{(i)}, x_{t-1}^{(j)})\hat{p}(x_t^{(i)}, x_{t-1}^{(j)}|y_{1:t-1})w_t^{(i)} w_{t-1}^{(j)} \qquad (5.3.12)$$

$$\hat{C}_t = \sum_{i=1}^{N} \sum_{j=1}^{N} \hat{q}(x_t^{(i)}, x_{t-1}^{(j)}|y_{1:t}) \qquad (5.3.13)$$

$$\hat{p}(x_t^{(i)}, x_{t-1}^{(j)}|y_{1:t}) = \hat{C}_t^{-1}\hat{q}(x^{(i)}, x^{(j)}|y_{1:t}), \qquad (5.3.14)$$

for $i, j = 1, \ldots, N$.

The form of the density functions $p(y_t|x_t, x_{t-1})$ (equation (5.3.12)) will vary for the different measurements models. Extension to higher-order processes is straightforward. A modification is needed for the TAR model to cope with the discontinuity at θ. The interval (A_t, B_t) must now be split into two pieces, i.e. (A_t, θ) and (θ, B_t), and the Gauss-Legendre abscissas applied to each part (if $\theta \notin (A_t, B_t)$ only one of the pieces is needed). The algorithm is in all other respects as above.

Several strategies can be used for selecting A_t and B_t. The simplest is to keep them unchanged, i.e. $A_t = A_0$ and $B_t = B_0$. The grids are then exactly the same everywhere, and the first two lines of the algorithm disappear. One advantage is that the transition probabilities $p(x_t^{(i)}|x_{t-1}^{(j)}, x_{t-2}^{(k)})$ do not change with t and can be pre-stored. This is our preferred method when dealing with the very noisy observations (5.3.4) and (5.3.5), but it is a possibility only because $\{x_t\}$ is stationary. In reality, many of the particles $(x_t^{(i)}, x_{t-1}^{(j)})$ have very low probabilities a-priori and can be removed with no noticeable effect. When $q = 3$, this trick can lead to huge reductions, up to several hundred times. When the data are more informative, as under the Poisson model (5.3.3), this strategy of fixed particles is not effective, and it is better to adapt (A_t, B_t) for each t. There are many ways to accomplish this. A rough way is first to compute, from the observed y_t, an interval where x_t is highly likely to be located, then a similar interval from the prediction density $p(x_t|y_{1:t-1})$ and use their intersection as (A_t, B_t). This worked well in all our examples. Notice that this must be done before to finding updated means and variances of x_t, and it is inadvisable to allow the computer to spend too much time on this detail.

5.3.3   Numerical experiments

Different methods will be used in this section to evaluate likelihoods according to the models in Section 5.3.1. Results produced by quadrature filters and Monte Carlo filters will be compared, and we shall for linear autoregression (5.3.1) also investigate a simple approximate technique proposed in (Schnatter 1992), which is a special case of the so-called Gaussian sum filter in control engineering (Sorenson and Alspach 1971). Instead of working with particle approximations in the scheme described in Section 5.2.2, we use Gaussian ones instead. See (Schnatter 1992) for details.

In Figure 5.3 we have computed likelihood functions from simulated data based on the *linear* model (5.3.1) under the three scenarios shown in Table 5.1, all of them taken from estimates obtained from real data. All data are annual. Scenarios I and II represent lemmings or other small rodents. The period of the population models (corresponding to the maximum of the spectral density function) is a little over three years and the length of the series (50 years) representative for real data. The third scenario is constructed from real observations for the Canadian snow shoe hare;

Scenario	Population			Observations				
	a_1	a_2	σ_ε	Type	α	β	T	S
I	−0.5	−0.5	1.0	Poisson (5.3.3)	1.0	1.0	50	1
II	−0.5	−0.5	1.0	Logistic (5.3.4)	0.0	2.0	50	1
III	1.5	−0.9	1.0	Categorical (5.3.5)	−0.8	−0.8	17	100

Table 5.1. Scenarios for the linear auto-regression used in Figure 5.3. S is the number of time series and T is the length of each time series.

see (Bølviken, Glöckner and Stenseth 1999). The period is now much longer (about 9.6 years), and the research question is whether it is possible to recover the dynamics of the animal fluctuations from a large number of short series of 17 years.

It is fairly clear that the categorical data must contain information about the frequencies of oscillations of the underlying populations, but perhaps not so obvious that these data can reflect other information about the dynamics. We have therefore in Figure 5.3 plotted likelihood profiles along curves of constant periods in the (a_1, a_2)-space. The values of the nuisance parameters in the measurement models (5.3.3-5.3.5) are the correct ones, except on the lower right, where the horizontal axis represents variations of β. The results are encouraging. In scenario III (lower half of Figure 5.3), where much data support the computations, it is shown that the parameters can be consistently estimated, in accordance with the theoretical results in (Bickel, Ritov and Ryden 1998). The Monte Carlo effects there are much dampened as averages of 100 independent likelihoods, one for each 17−year series. The similarity between methods is great in all cases considered, but the Monte Carlo versions (based on 1000 replications) for scenario I and II are too erratic to optimise by numerical software. By contrast, the likelihood surfaces for the two other methods are smooth. The closeness in shape between the results from the Schnatter approximation and the more accurate quadrature filter is noteworthy, and most impressive because Schnatter's method is extremely fast. However, there are numerical errors in the *level* of the likelihood curves produced by her method. If model fit is to be judged through AIC or some other likelihood-based criterion, her approximation is not good enough. The quadrature filter was run with $m = 20$ abscissas in each direction in scenarios I and II, but in scenario III, where the parameters are much closer to the boundary of the stationary region, it was necessary to raise the number to $m = 50$. Computing time for the Poisson case was about 14 seconds for each likelihood evaluation on an Intel Pentium 200MHz MMX running on Linux 2.0.

A second round of experiments is presented for the TAR model (5.3.2) in Figure 5.4 under the scenarios shown in Table 5.2. Notice that the two regimes are defined at lag $d = 2$ in accordance with empirical studies (Stenseth, Bjørnstad and Falck 1996). For the snow shoe hare (Scenario III'), the model is the superposition of two stationary regimes, correspond-

Scenario	Population				
	a_1	a_2	b_1	b_2	σ_ε
I'	−0.5	0.0	−0.5	−0.7	1.0
III'	1.0	−0.4	1.0	−0.9	1.0

Table 5.2. Scenarios for the TAR auto-regression used in Figure 5.4. S is the number of time series and T is the length of each time series. The observation models are equal to scenarios I and III in Table 5.1.

ing to periods of about 12 (a-regime) and 6 (b-regime) years. This reflects prevailing ecological theory, deduced from studies of the lynx, of a slow growth phase and a faster decline, but no empirical studies of this nature have as yet been undertaken for the hare itself. The results displayed in Figure 5.4 are, however, encouraging and indicate that statistical studies may be able to estimate underlying nonlinear effects from categorical data.

It should be noted that for likelihood calculations, Monte Carlo estimates obtained by running the filter independently for each parameter set is probably not recommended. Some reduction in variability is obtained by using common random numbers, but in our situation this did not suffice to give smooth enough likelihood surfaces. A more promising alternative is the simulated likelihood ratio method (Billio, Monfort and Robert 1998), in which the filter (a smoothed version) is run only for one parameter set (the null set) and the likelihood ratio can be estimated at all other parameter sets by ratios of full likelihoods (including the underlying state process). This approach is also not really satisfactory because, although a smooth surface was obtained, the bias was too high for parameter sets insufficiently close to the null set.

5.4   Concluding remarks

In this paper we have been concerned with filtering for problems with low-dimensional state vectors. In such situations, *deterministic* filters may be preferable. Some results on the errors involved are given. It is shown that the errors in likelihood calculations propagate linearly in time. However, since the errors at each step can be made very small with a moderate number of particles, the linear increase is in many situations not a problem.

Constructing quadrature filters is a matter of selecting particles and their associated weights. In contrast to Monte Carlo filters, whereby the particles are automatically laid out in the high-probable regions, particles now have to be user-specified. Although many sophisticated designs are possible, we prefer to select the particles marginally on each variable. This results in many particles with weights close to zero, but this is outweighed by a much smaller computational burden and easier implementation. Further, for situations in which the information in the observations is sparse, parti-

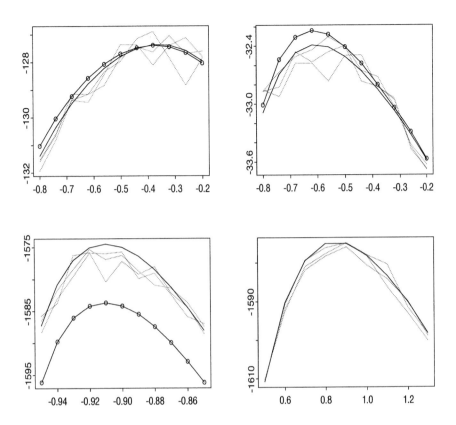

Figure 5.3. Estimated profile (log-)likelihood curves for the AR(2) process with Poisson (upper left), binary (upper right) and categorical (lower left and right) observations. Scenarios are given in Table 5.1. The x-axis displays the a_2 values used, except in the last plot, where the x-axis represents β-values. The quadrature filter estimate is given as a solid curve, and the estimate based on the Schnatter filter is given as a solid curve with dots (not given in the last plot). The three dashed dotted curves are different Monte Carlo estimates.

cles can successfully be specified directly from properties of the stationary distribution.

Deterministic filters are compared with Monte Carlo filters on some simulated data based on models representing animal populations. For these problems, likelihood-based inference is the primary concern, and large variabilities in these functions are problematic if the likelihood is to be maximised. In such cases, deterministic filters are superior because of their high accuracy.

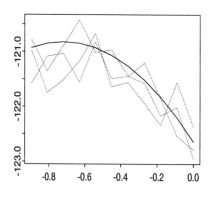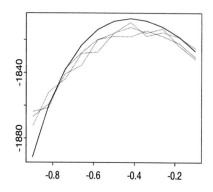

Figure 5.4. Estimated profile (log-)likelihood curves for the TAR(2) process with Poisson (left) and categorical (right) observations. Scenarios are given in Table 5.2. The x-axis displays the a_2 (b_2 in right plot) values used. The quadrature filter estimate is given as a solid curve. The three dashed dotted curves are different Monte Carlo estimates.

5.5   Appendix: Derivation of numerical errors

Equation (5.2.18) is verified by first noting that by (5.2.5) and (5.2.19)

$$\exp(\delta_t)\hat{p}(\mathbf{x}_{t+1}|\mathbf{y}_{1:t}) = \exp(\delta_t)\int p(\mathbf{x}_{t+1}|\mathbf{x}_t)\tilde{p}(\mathbf{x}_t|\mathbf{y}_{1:t})d\mathbf{x}_t + \eta(\mathbf{x}_{t+1}).$$

Hence,

$$\varepsilon(\mathbf{x}_{t+1}|\mathbf{y}_{1:t}) = \int p(\mathbf{x}_{t+1}|\mathbf{x}_t)[\exp(\delta_t)\tilde{p}(\mathbf{x}_t|\mathbf{y}_{1:t}) - p(\mathbf{x}_t|\mathbf{y}_{1:t})]d\mathbf{x}_t + \eta(\mathbf{x}_{t+1}),$$

or, by using (5.2.3) and (5.2.6),

$$\varepsilon(\mathbf{x}_{t+1}|\mathbf{y}_{1:t}) = \int p(\mathbf{x}_{t+1}|\mathbf{x}_t)p(\mathbf{y}_t|\mathbf{x}_t)[\exp(\delta_t)\widetilde{C}_t^{-1}\hat{p}(\mathbf{x}_t|\mathbf{y}_{1:t-1})$$
$$- C_t^{-1}p(\mathbf{x}_t|\mathbf{y}_{1:t-1})]d\mathbf{x}_t + \eta(\mathbf{x}_{t+1}),$$

which reduces to (5.2.18).

The proof of (5.2.25) starts by noting that, using (5.2.14), we may write

$$\varepsilon(\mathbf{x}_{t+1}|\mathbf{y}_{1:t}) = \exp(\delta_t)\sum_{i=1}^{N_t} p(\mathbf{x}_{t+1}|\mathbf{x}_t^{(i)})\hat{p}(\mathbf{x}_t^{(i)}|\mathbf{y}_{1:t}) - p(\mathbf{x}_{t+1}|\mathbf{y}_{1:t})$$

$$= \sum_{i=1}^{N_t} p(\mathbf{x}_{t+1}|\mathbf{x}_t^{(i)})[\exp(\delta_t)\hat{p}(\mathbf{x}_t^{(i)}|\mathbf{y}_{1:t}) - w_t^{(i)}p(\mathbf{x}_t^{(i)}|\mathbf{y}_{1:t})] +$$

$$\sum_{i=1}^{N_t} p(\mathbf{x}_{t+1}|\mathbf{x}_t^{(i)})p(\mathbf{x}_t^{(i)}|\mathbf{y}_{1:t}) - p(\mathbf{x}_{t+1}|\mathbf{y}_{1:t}).$$

Now, using (5.2.2) and (5.2.14), the first term is equal to

$$\sum_{i=1}^{N_t} p(\mathbf{x}_{t+1}|\mathbf{x}_t^{(i)})p(\mathbf{y}_t|\mathbf{x}_t^{(i)})w_t^{(i)}[\exp(\delta_t)\widehat{C}_t^{-1}\widehat{p}(\mathbf{x}_t^{(i)}|\mathbf{y}_{1:t-1})-C_t^{-1}p(\mathbf{x}_t^{(i)}|\mathbf{y}_{1:t-1})]$$

$$= C_t^{-1}\sum_{i=1}^{N_t} p(\mathbf{x}_{t+1}|\mathbf{x}_t^{(i)})p(\mathbf{y}_t|\mathbf{x}_t^{(i)})w_t^{(i)}[\exp(\delta_{t-1})\widehat{p}(\mathbf{x}_t^{(i)}|\mathbf{y}_{1:t-1})-p(\mathbf{x}_t^{(i)}|\mathbf{y}_{1:t-1})]$$

by noticing that $\exp(\delta_t) = \exp(\delta_{t-1})\widehat{C}_t C_t^{-1}$. The expression in the bracket is now equal to $\varepsilon(\mathbf{x}_t|\mathbf{y}_{1:t-1})$, showing that

$$\varepsilon(\mathbf{x}_{t+1}|\mathbf{y}_{1:t}) = C_t^{-1}\sum_{i=1}^{N_t} p(\mathbf{x}_{t+1}|\mathbf{x}_t^{(i)})p(\mathbf{y}_t|\mathbf{x}_t^{(i)})w_t^{(i)}\varepsilon(\mathbf{x}_t|\mathbf{y}_{1:t-1}) +$$

$$\sum_{i=1}^{N_t} p(\mathbf{x}_{t+1}|\mathbf{x}_t^{(i)})p(\mathbf{x}_t^{(i)}|\mathbf{y}_{1:t}) - p(\mathbf{x}_{t+1}|\mathbf{y}_{1:t}). \qquad (5.5.15)$$

By (5.2.20) an upper limit for the first term on the right side of (5.5.15) is

$$r_t C_t^{-1}\sum_{i=1}^{N_t} w_t^{(i)}p(\mathbf{x}_{t+1}|\mathbf{x}_t^{(i)})p(\mathbf{y}_t|\mathbf{x}_t^{(i)})p(\mathbf{x}_t^{(i)}|\mathbf{y}_{1:t-1}),$$

which can be simplified to

$$r_t\sum_{i=1}^{N_t} w_t^{(i)}p(\mathbf{x}_{t+1}|\mathbf{x}_t^{(i)})p(\mathbf{x}_t^{(i)}|\mathbf{y}_{1:t}),$$

using (5.2.2) again. By (5.2.23), this is less than or equal to $r_t(1 + \varepsilon)p(\mathbf{x}_{t+1}|\mathbf{y}_{1:t})$. By the same condition, the absolute value of the last two terms on the right hand side of (5.5.15) is less or equal to $\varepsilon p(\mathbf{x}_{t+1}^{(i)}|\mathbf{y}_t)$. Combining these upper limits gives

$$\varepsilon(\mathbf{x}_{t+1}|\mathbf{y}_{1:t}) \le [r_t(1+\varepsilon)+\varepsilon]p(\mathbf{x}_{t+1}|\mathbf{y}_{1:t}),$$

showing that

$$r_{t+1} \le r_t(1+\varepsilon)+\varepsilon,$$

from which (5.2.25) can be proven easily by recursion. Now,

$$|\varepsilon(\mathbf{x}_{t+1}|\mathbf{y}_{1:t})| = |\exp(\delta_t)\hat{p}(\mathbf{x}_{t+1}|\mathbf{y}_{1:t}) - p(\mathbf{x}_{t+1}|\mathbf{y}_{1:t})|$$

$$= |\exp(\delta_t)\sum_{i=1}^{N_t} p(\mathbf{x}_{t+1}|\mathbf{x}_t^{(i)})\hat{p}(\mathbf{x}_t^{(i)}|\mathbf{y}_{1:t}) - p(\mathbf{x}_{t+1}|\mathbf{y}_{1:t})|$$

$$\leq |\sum_{i=1}^{N_t} p(\mathbf{x}_{t+1}|\mathbf{x}_t^{(i)})[\exp(\delta_t)\hat{p}(\mathbf{x}_t^{(i)}|\mathbf{y}_{1:t}) - w_t^{(i)}p(\mathbf{x}_t^{(i)}|\mathbf{y}_{1:t})]|$$

$$+ |\sum_{i=1}^{N_t} p(\mathbf{x}_{t+1}|\mathbf{x}_t^{(i)})p(\mathbf{x}_t^{(i)}|\mathbf{y}_{1:t}) - p(\mathbf{x}_{t+1}|\mathbf{y}_{1:t})|$$

$$\leq |\sum_{i=1}^{N_t} p(\mathbf{x}_{t+1}|\mathbf{x}_t^{(i)})p(\mathbf{y}_t|\mathbf{x}_t^{(i)})w_t^{(i)}[\exp(\delta_t)\widehat{C}_t^{-1}\hat{p}(\mathbf{x}_t^{(i)}|\mathbf{y}_{1:t-1})$$

$$- C_t^{-1}p(\mathbf{x}_t^{(i)}|\mathbf{y}_{1:t-1})]| + \varepsilon p(\mathbf{x}_{t+1}|\mathbf{y}_{1:t})$$

$$= C_t^{-1}|\sum_{i=1}^{N_t} p(\mathbf{x}_{t+1}|\mathbf{x}_t^{(i)})p(\mathbf{y}_t|\mathbf{x}_t^{(i)})w_t^{(i)}[\exp(\delta_{t-1})\hat{p}(\mathbf{x}_t^{(i)}|\mathbf{y}_{1:t-1})$$

$$- p(\mathbf{x}_t^{(i)}|\mathbf{y}_{1:t-1})]| + \varepsilon p(\mathbf{x}_{t+1}|\mathbf{y}_{1:t}).$$

Hence,

$$|\varepsilon(\mathbf{x}_{t+1}|\mathbf{y}_{1:t})| \leq C_t^{-1}|\sum_{i=1}^{N_t} p(\mathbf{x}_{t+1}|\mathbf{x}_t^{(i)})p(\mathbf{y}_t|\mathbf{x}_t^{(i)})w_t^{(i)}\varepsilon(\mathbf{x}_t^{(i)}|\mathbf{y}_{1:t-1})|$$

$$+ \varepsilon p(\mathbf{x}_{t+1}|\mathbf{y}_{1:t})$$

$$\leq r_t C_t^{-1}\sum_{i=1}^{N_t} p(\mathbf{x}_{t+1}|\mathbf{x}_t^{(i)})p(\mathbf{y}_t|\mathbf{x}_t^{(i)})p(\mathbf{x}_t^{(i)}|\mathbf{y}_{1:t-1})w_t^{(i)} + \varepsilon p(\mathbf{x}_{t+1}|\mathbf{y}_{1:t})$$

$$= r_t|\sum_{i=1}^{N_t} w_t^{(i)}p(\mathbf{x}_{t+1}|\mathbf{x}_t^{(i)})p(\mathbf{x}_t^{(i)}|\mathbf{y}_{1:t}) + \varepsilon p(\mathbf{x}_{t+1}|\mathbf{y}_{1:t})$$

$$\leq r_t(1+\varepsilon)p(\mathbf{x}_{t+1}|\mathbf{y}_{1:t}) + \varepsilon p(\mathbf{x}_{t+1}|\mathbf{y}_{1:t}).$$

6

RESAMPLE–MOVE Filtering with Cross-Model Jumps

Carlo Berzuini
Walter Gilks

6.1   Introduction

In standard sequential imputation, repeated resampling stages progressively impoverish the set of particles, by decreasing the number of distinct values represented in that set. A possible remedy is Rao-Blackwellisation (Liu and Chen 1998). Another remedy, which we discuss in this chapter, is to adopt a hybrid particle filter, which combines importance sampling/resampling (Rubin 1988, Smith and Gelfand 1992) and Markov chain iterations. An example of this class of particle filters is the RESAMPLE–MOVE algorithm described in (Gilks and Berzuini 1999), in which the swarm of particles is adapted to an evolving target distribution by periodical resampling steps and through occasional Markov chain moves that lead each individual particle from its current position to a new point of the parameter space. These moves increase particle diversity. Markov chain moves had previously been introduced in particle filters (for example, (Berzuini, Best, Gilks and Larizza 1997, Liu and Chen 1998)), but rarely with the possibility of moving particles at *any* stage of the evolution process along *any* direction of the parameter space; this is, indeed, an important and innovative feature of RESAMPLE–MOVE. This allows, in particular, to prevent particle depletion along directions of the parameter space corresponding to static parameters, for example when the model contains unknown hyperparameters, a situation which is not addressed by the usual state filtering algorithms.

Hybrid particle filters of the kind mentioned above may also be appropriate to Bayesian problems involving model uncertainty, where the data are thought of as being generated by an unknown member of a given set of models. Situations of this kind arise in a variety of filtering applications, for example in on-line identification of a target of unknown type, in sequential forecasting in the presence of model uncertainty, and in statistical signal processing when the signal follows an unknown pattern. Interest may focus on model selection, i.e., which model generated the data, or in computing a predictive distribution of interest by averaging over the candidate models. In applications involving multiple models, the particles are scattered in a complex parameter space, which is the union of model–specific subspaces. In this context, a merit of the hybrid particle filters we are going to discuss is that they allow particles to jump from one model subspace to another. This may be useful to recover from occasional depletion of some subspaces, and moreover, in those applications involving nested models, to take advantage of relationships between parameters of different subspaces. A practical and methodological discussion of some of the relevant issues is also given in (Andrieu et al. 1999b). We illustrate the methods with the aid of a target tracking application adapted from (Gordon et al. 1993).

6.2   Problem statement

Suppose sequential observations $y_1, y_2, \ldots, y_t, \ldots$ become available at times $1, 2, \ldots, t, \ldots$. Here and in the following, we let the subscript t index discrete time $t = 1, 2, \ldots$. We assume the observations are generated from a parametric model (later we shall consider the more general situation with multiple models) and we let the probability distribution of $y_{1:t} = (y_1, \ldots, y_t)$ under such a model be denoted by

$$p(y_{1:t}|\theta_t),$$

where $\theta_t \in \Omega_t$ represents a vector of model unknowns and missing data. These unknowns and missing data we call the model *parameters*. We assume that θ_t can evolve by increasing its size every time a new observation arrives, as occurs for example in hidden Markov analysis of sequential data, where a new hidden component is appended to the model with each new data observation. If we let θ_t^+ denote the (possibly empty) set of parameters entering the model at t, we have

$$\theta_{t+1} = (\theta_t, \theta_t^+). \tag{6.2.1}$$

Within a Bayesian approach to the problem, θ_t has a *prior* distribution $p(d\theta_t) = p(d\theta_{t-1})\, p(d\theta_{t-1}^+|\theta_{t-1})$.

In this chapter we consider (from a Bayesian standpoint) problems wherein inferences are required on-line, as occurs for example in an intensive care unit, where we are concerned with on-line detection of patient abnormalities based on a continuous stream of data generated by the monitoring devices. In problems of this kind, at time t, the distribution of interest, or target distribution, is the Bayesian posterior distribution

$$\pi_t(d\theta_t) = p(d\theta_t|y_{1:t}),$$

representing our state of uncertainty just after obtaining the observation y_t. This distribution is in general known only up to a normalising constant, due to the integral $\int p(y_{1:t}|\theta_t)\, p(d\theta_t)$ being in general intractable. Here we use a measure-theoretic notation in which $\pi_t(d\theta_t)$ is the probability under π_t of a small measurable subset $d\theta_t$ of Ω_t.

Due to the accruing data information, the target π_t will generally evolve with time, generating a smooth sequence

$$\pi_0(d\theta_0), \pi_1(d\theta_1), \dots, \pi_t(d\theta_t), \dots \qquad (6.2.2)$$

in which π_t and π_{t-1} are linked to each other through the following Bayes' recursion:

$$\pi_t(d\theta_t) \propto \pi_{t-1}(d\theta_{t-1}) \cdot p(d\theta_{t-1}^+|\theta_{t-1}, y_{1:t-1}) \cdot p(y_t|y_{1:t-1}, \theta_t).$$

Our aim, roughly speaking, is to *track* the target sequence. More formally, for any function of interest g_t of the unknown parameters, defined on the support of π_t, we want to estimate the conditional posterior expectation

$$\mathbf{E}_{\pi_t} g_t(\theta_t) := \int g_t(\theta_t)\, \pi_t(d\theta_t), \qquad (6.2.3)$$

where $g_t(\theta_t)$ is the value of g_t in the small subset $d\theta_t$ at time t, and the integral is over such subsets. If Ω_t is discrete and π_t is a probability mass function over it, the above integral takes the form

$$\mathbf{E}_{\pi_t} g_t(\theta_t) := \sum g_t(\theta_t)\, \pi_t(\theta_t),$$

whereas if Ω_t is continuous with density $\pi_t(\theta_t)$, then (6.2.3) reduces to

$$\mathbf{E}_{\pi_t} g_t(\theta_t) := \int g_t(\theta_t)\, \pi_t(\theta_t) d\theta_t.$$

6.3   The RESAMPLE–MOVE algorithm

Particle filters track the target sequence (6.2.2) by producing at each integer time t a discrete representation of π_t through a random collection of *particles*:

$$\Theta_t := (\theta_{t,1}, \theta_{t,2}, \dots, \theta_{t,N_t}), \qquad (6.3.4)$$

where $\theta_{t,j}$, the jth particle at time t, is a realised configuration of the parameter vector θ_t. The set Θ_t of particles we call the tth *generation*. We consider particle filters in which the generations are updated recursively, with generation $(t-1)$ updated into generation t at integer time t. The filters are designed in such a way that the empirical distribution of the particles in generation t converges to π_t as the number of contained particles increases. This means that for any function of interest $g_t(\theta_t)$, the Monte Carlo estimate

$$G_t(\theta_t) = \frac{1}{N_t} \sum_{j=1}^{N_t} g_t(\theta_{t,j}) \tag{6.3.5}$$

converges to the posterior expectation $\mathbf{E}_{\pi_t} g_t(\theta_t)$ in (6.2.3) for $N_t \to \infty$. Consider the particle filter scheme described in the box below.

BASIC SCHEME

At $t = 0$: create generation 0 by sampling, independently for $j = 1, \ldots, N_0$:

$$\theta_{0,j} \sim \pi_0(d\theta_0),$$

where $\sim$ denotes is sampled from.

At $t = 1, 2, \ldots$: create generation t by performing the following 2 steps:

1. augmentation: for each $i = 1, \ldots, N_{t-1}$, sample $\theta_{t-1,i}^+ \sim f_{t,i}(d\theta_{t-1}^+)$, where θ_{t-1}^+ is defined in (6.2.1). Then define a new particle

$$\theta_{t-1,i}^* = (\theta_{t-1,i}, \theta_{t-1,i}^+),$$

and let the ensemble of new particles be denoted by

$$\Theta_{t-1}^* := (\theta_{t-1,1}^*, \ldots, \theta_{t-1,N_{t-1}}^*).$$

2. evolution: the particles of generation t are created by sampling, independently for $j = 1, \ldots, N_t$:

$$\theta_{t,j} \sim Q_t(d\theta_t | \Theta_{t-1}^*), \tag{6.3.6}$$

where the conditional distribution Q_t is discussed later in this section.

According to the algorithm described in the box, at integer time t the previous generation $t-1$ is updated into a new generation t through a pair of steps: *augmentation* and *evolution*. The former is required only when new parameters are introduced into the model at t, causing the θ_t space to have a larger size than the θ_{t-1} space. In these situations, we augment each ith particle of generation $t-1$ by appending the appropriate extra components, $\theta_{t-1,i}^+$, after these are sampled from a distribution $f_{t,i}$, for example, from the conditional prior

$$f_{t,i}(d\theta_{t-1,i}^+) = p(d\theta_{t-1,i}^+|\theta_{t-1,i}, y_{1:t-1}), \tag{6.3.7}$$

or from the conditional posterior

$$f_{t,i}(d\theta_{t-1,i}^+) = p(d\theta_{t-1,i}^+|\theta_{t-1,i}, y_{1:t}). \tag{6.3.8}$$

Once each particle of generation $(t-1)$ has been upgraded by augmentation, generation $(t-1)$ is evolved into generation t according to (6.3.6).

The RESAMPLE–MOVE algorithm discussed in (Gilks and Berzuini 1999) is obtained as a special case of the scheme described in the box by applying two restrictions. The first assumes that $\pi_{t-1} \cdot f_{t,i}$ is strictly positive for every point in the support of π_t, so that after augmenting the generic i^{th} particle of generation $(t-1)$ we can assign it the weight

$$w_{t-1,i} \quad \propto \quad \frac{\pi_t(d\theta_{t-1,i}^*)}{\pi_{t-1}(d\theta_{t-1,i}) \cdot f_{t,i}(d\theta_{t-1,i}^+)}, \qquad \text{for } t > 1,$$

$$w_{t-1,i} \quad = \quad 1, \qquad \text{for } t = 1.$$

The weights $\{w_{t-1,i}\}$ correct for the fact that the particles of generation $(t-1)$ will under-represent π_t in certain regions of the parameter space, and over-represent π_t in others. In the context of Bayesian learning, it can be shown that $w_{t-1,i}$ has the simple form $p(y_t|\theta_{t-1,i}^*, y_{1:t-1})$ when $f_{t,i}$ is chosen to be the conditional prior (6.3.7). Other choices, such as the conditional posterior (6.3.8) may cause $w_{t-1,i}$ to be analytically intractable.

In order to discuss the second restriction, we introduce the notation $q_t(dz|x)$ for a transition kernel in the Ω_t space. For every x, the kernel q_t specifies a probability distribution on Ω_t, and thus prescribes a way of moving from a given point x of Ω_t to a new point z of the same space. In the following we shall assume that q_t is invariant with respect to π_t, so that if the starting point x is distributed according to π_t, then the destination z will also be distributed according to π_t. The second restriction on the scheme described in the box is to take the distribution Q_t in (6.3.6) to have the mixture form

$$Q_t(d\theta_t|\Theta_{t-1}^*) = \sum_{i=1}^{N_{t-1}} \frac{w_{t-1,i}}{\sum_{i=1}^{N_{t-1}} w_{t-1,i}} \cdot q_t(d\theta_t|\theta_{t-1,i}^*), \tag{6.3.9}$$

which prescribes that the particles of generation t are created independently, conditional on Θ_{t-1}^*. According to (6.3.9), we could create each jth particle of generation t by randomly selecting one particle from generation $(t-1)$, and then moving the selected particle in accordance with the kernel q_t[1]. The selection should be an importance-weighted resampling, with replacement, such that the probability of a given particle of generation $(t-1)$

---

[1]That is, the position of the particle after the move, given its initial position, x, is distributed according to $q_t(.|x)$.

being selected should be proportional to its weight. Unfortunately, such a procedure is time consuming. Several authors (see, for example, (Carpenter et al. 1999a) and (Liu and Chen 1998)) propose alternative resampling schemes, with attention to the computational cost of the procedure, and its impact on the variance of the estimator. A useful approach is to implement (6.3.9) by a procedure consisting of a resample step followed by a move step. In the resample step, every i^{th} particle of generation $t-1$ branches into a number m_i of copies of itself, this number being high for particles with a large weight. Then, in the move step, each of the generated copies is moved according to the q_t kernel, independently of each other conditional on Θ_{t-1}^*. If we are willing to leave $N_t = \sum_{i=1}^{N_{t-1}} m_i$ random, we may follow (Crisan et al. 1999) and compute m_i, $i = 1, \ldots, N_{t-1}$, as follows

1: compute $k_i = \frac{N_{t-1} \, w_{t-1,i}}{\sum_{i=1}^{N_{t-1}} w_{t-1,i}}$,

2: set m_i equal to the integer part of $(k_i + 1)$ with probability $frac(k_i)$, otherwise set it equal to the integer part of k_i. Here $frac(k)$ stands for fractional part of k. In this procedure N_t has expected value N_{t-1}. Properties of this scheme are discussed in (Crisan et al. 1999). Unfortunately, the scheme above does not ensure that the number of particles is constant, and cannot therefore prevent extinction or explosion of the algorithm. This can be avoided by a multinomial branching scheme, also discussed in (Crisan et al. 1999).

To demonstrate RESAMPLE–MOVE, we have applied the algorithm to a simple problem involving a single unknown parameter θ with values in $\mathcal{R}^1$, where the target sequence is a sequence of univariate Gaussian distributions with unit variance and a mean value which evolves with time t according to $0.5 \cdot t$. The results are shown in Figure 6.1, wherein time and θ values are represented by the horizontal and vertical axes. Each tth generation of particles is represented in the figure as a set of circles with abscissa t and radius proportional to the particle weight, or importance. Every important particle of generation $t-1$ tends to give birth to a numerous family of child particles in generation t. In the figure, particles of the same family are connected to their common parent by a line, and their number and location are random variables depending on the weight and location of their common parent. The depicted process periodically redistributes the particles so that their density tends to be higher in those regions of the parameter space having a high probability under the current target distribution. Selection and moving cooperate to achieve adaptation while avoiding reduction of particle diversity. Unimportant particles in each generation tend to remain unselected and killed. A more thorough illustration of the algorithm is given in the final sections of this chapter.

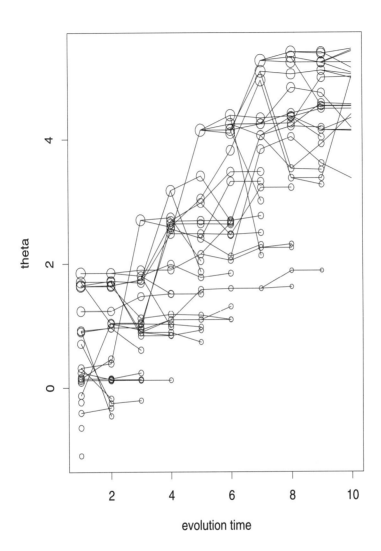

Figure 6.1. A graphic illustration of the RESAMPLE–MOVE algorithm in a simple univariate example. The vector of unknowns here is a scalar parameter θ with values in $\mathcal{R}^1$ represented in the figure by the vertical axis. The target distribution, $\pi_t(\theta)$, is a Gaussian density with constant unit variance, whose mean grows with time as ($0.5\times$ evolution time). Each tth generation of particles, providing a set of realizations of θ, is represented in the figure as a set of circles with abscissa t and radius proportional to the particle weight. Each parent particle of generation t gives birth to a (possibly empty) family of child particles in the next generation. Parent and child particles are connected by lines. The expected size of each family is proportional to the weight of the common parent.

6.4   Comments

We have assumed that the transition kernel $q_t(.|x)$, which we use to move the particles at time t, has invariant distribution π_t. Therefore, if a particle at time t, before being moved, is distributed according to π_t, it will be distributed according to π_t also after the move, formally

$$\pi_t(\mathrm{d}z) = \int q_t(\mathrm{d}z|x)\pi_t(\mathrm{d}x). \tag{6.4.10}$$

Methods for constructing transition kernels with the property (6.4.10) are familiar. Two popular methods are the Gibbs sampler (Geman and Geman 1984) and the Metropolis–Hastings scheme (Metropolis, Rosenbluth, Rosenbluth, Teller and Teller 1953, Hastings 1970), both discussed in (Tierney 1994). In particular, if we choose the Metropolis–Hastings method, at time t, if x is the current position of the particle we are going to move, a new candidate position z' for the particle is drawn from an essentially arbitrary distribution $m_t(.|x)$, called the proposal distribution. Then, with probability

$$\alpha(x, z') = \min\left(1, \frac{\pi_t(z') \cdot m_t(x|z')}{\pi_t(x) \cdot m_t(z'|x)}\right), \tag{6.4.11}$$

the proposed candidate is accepted and the particle moved to position $z = z'$. Otherwise, the candidate value is rejected and the particle left in position $z = x$. If the kernel mixes badly with respect to π_t (that is, if the particles do not move fluidly across the important region of the parameter space), particles of the same generation will tend to be highly correlated to each other through common ancestor particles in recent generations. A possible remedy might be to iterate the above proposal–acceptance procedure several times on each particle within the same move.

Notice that if m_t is symmetric, $m_t(z|x) = m_t(x|z)$, as would be the case when using a normal proposal, the acceptance probability α reduces to $\min(1, \pi_t(\mathrm{d}z')/\pi_t(\mathrm{d}x))$.

In a typical application with multi-dimensional Ω_t, it will be necessary to devise different types of move, for example moves with different proposal distribution. The various types of move are combined to form a sort of hybrid particle filter by allowing each particle at time t to randomly choose a subset of the available types and to carry out the corresponding moves, one after the other, within the tth stage of updating. Move choice will depend on preassigned move type probabilities, which may vary with t and across particles. By introducing move types operating along different subspaces of the parameter space, we can have certain sets of model parameters updated less frequently than others, or certain types of move invoked less frequently than others in certain periods of the evolution process. An illustration of the various possibilities is given in Section 6.7.

The computational cost of a specific type of move will depend in part on the cost of evaluating the Metropolis ratio, $\pi_t(dz')/\pi_t(dx)$, in (6.4.11). For certain types of move this involves a small and fixed number of components of θ_t, independent of t. For other types of move, the cost of evaluating the Metropolis ratio may quickly grow with t, but even in this case we may try to design the transition kernel q_t in (6.4.10) as a random combination of move types, in such a way that the global expected cost of the move step does not grow too quickly with t.

It may be instructive to compare the computational complexity of a given RESAMPLE – MOVE scheme with that of a corresponding standard MCMC implementation. One way to achieve this is to compare a quantity denoted by c_{MCMC}, representing the burden of computation of one MCMC move at time t with a quantity denoted by c_{RM}, representing the expected burden of computation of one particle move at time t within RESAMPLE–MOVE. In our illustrative study in Section 6.7, we have $c_{RM} < c_{MCMC} = \mathcal{O}(t)$. However, comparing algorithms exclusively in terms of complexity may be misleading, since a scheme with low complexity may in fact be useless because it mixes very badly (e.g. it gets stuck in one point of the parameter space).

The presented version of RESAMPLE–MOVE assumes that, at time $t = 0$, it is possible to sample the particles of generation 0 directly from the target distribution π_0. In most Bayesian applications, π_0 would be the prior distribution of the parameters, from which sampling would be easy for the large class of directed acyclic graphical models (DAGs). For non-DAG models, such as typically arise in the presence of spatial random effects, it may not be possible to sample π_0. One strategy, then, would be to set π_0 to a convenient approximation of the prior distribution, which nevertheless admits independent sampling, and then to set π_1 equal to the prior.

6.5 Central limit theorem

At current (integer) time t, for a generic function of interest $g_t(\theta_t)$, the estimator (6.3.5) yields an estimate G_t of $\mathbf{E}_{\pi_t} g_t(\theta_t)$, based on the most recent generation of particles. In (Gilks and Berzuini 1999) we present a central limit theorem for G_t, which guarantees its consistency and asymptotic normality as the number of particles grows to infinity. Under appropriate integrability conditions, the theorem states that

$$\frac{G_t - \mathbf{E}_{\pi_t} g(\theta_t)}{\sqrt{V_t(G_t)}} \;\Rightarrow\; N(0, 1),$$

as $N_1, \ldots, N_t \longrightarrow \infty$, in the sense that $N_1 \longrightarrow \infty$, then $N_2 \longrightarrow \infty$, and so on. The expression of $V_t(G_t)$, the variance of the estimator, involves intractable integrals (Gilks and Berzuini 1999), and can be approximated from the history of the particles up to t.

The mode of convergence of the above theorem is not directly relevant to practical contexts in which we might wish to consider behaviour for large N, where $N = N_1 = N_2 = \dots$. Such convergence results would require the use of a more complex interacting particle theory that used in (Gilks and Berzuini 1999). Also, whereas in some contexts we are interested in the convergence of the empirical measure of the particles *at a specific time t*, in others we might wish to prove convergence of the empirical set of *paths* to the posterior distribution of the whole trajectory $(\theta_1, \theta_2, \dots, \theta_t)$.

6.6   Dealing with model uncertainty

Sometimes the model of the data is subject to question. We have a countable collection of models for the y data, $\{\mathcal{M}_k; k = 1 \dots K\}$, and we are willing to assume that one of them is the true data-generating model. Which of the given models is the true one, we do not know. We call this model uncertainty. The models will often describe qualitatively different behaviours of the observed system. In a target tracking problem, for example, we might not know whether the target is an aeroplane or a missile. For that reason, we set up two distinct models of the observations corresponding to the two possible target types. Similar situations occur in the modelling of a patient under anaesthesia, where different types of behaviour may arise, and in the modelling of a financial market, where different types of market behaviour are possible. In these and other examples, acknowledging our uncertainty about the model enables us to use the incoming data to gradually learn about the unknown type, in accord with Bayes theorem, and to make our predictions more robust, to model mis-specification. We mention that we are assuming that the system behaves according to one of the models considered during the entire period of observation. In particular, we are not attempting here to model *changes* of behaviour of the system of interest over time.

We let the unknown variable k indicate the model, and we let the unknown vector $\theta_t^{(k)} \in \Omega_t^{(k)}$ represent the parameters in model $\mathcal{M}_k$ at time t. We treat k and $\theta_t^{(k)}$ as random variables drawn from a joint prior distribution

$$\mathrm{p}(k, \theta_t^{(k)}) = p(k) \cdot \mathrm{p}(\theta_t^{(k)} | k),$$

where $\mathrm{p}(k)$ represents the prior probability for Model $\mathcal{M}_k$. Our target distribution at time t is $\pi_t(k, \mathrm{d}\theta_t^{(k)})$, the joint posterior distribution for the model indicator and for the parameters of the indicated model, which is defined by

$$
\pi_t(k, d\theta_t^{(k)}) \propto \begin{cases} \mathrm{p}(k) \cdot \mathrm{p}(d\theta_t^{(k)}|k) & \text{if } t = 0, \\ \mathrm{p}(k) \cdot \mathrm{p}(d\theta_t^{(k)}|k) \cdot \mathrm{p}(y_{1:t}|k, \theta_t^{(k)}) & \text{if } t > 0, \end{cases} \tag{6.6.12}
$$

where $\mathrm{p}(y_{1:t}|k, \theta_t^{(k)})$ is the likelihood for data $y_{1:t}$ under model $\mathcal{M}_k$. The target distribution $\pi_t(k, \theta_t^{(k)})$ is defined on the union of the model–specific subspaces. For example, with two models, $K = 2$, the target distribution (6.6.12) is defined on the space $\{1\} \times \Omega_t^{(1)} \bigcup \{2\} \times \Omega_t^{(2)}$. Generally with K models, the target distribution (6.6.12) is defined on the space $\Gamma_t = \bigcup_{k=1}^{K} \{k\} \times \Omega_t^{(k)}$. The target (6.6.12) usually changes at each time t due to the incoming data, thereby generating the target sequence

$$
\pi_1(k, d\theta_1^{(k)}), \dots, \pi_t(k, d\theta_t^{(k)}), \dots . \tag{6.6.13}
$$

In order to track this sequence, we use a RESAMPLE – MOVE filter where particles of generation t are realisations of the pair $(k, \theta_t^{(k)})$ in the space Γ_t. If the filter works properly, the proportion of particles associated with model $\mathcal{M}_k$ in generation t will reflect the posterior distribution for that model at time t, $\pi_t(k)$, and the distribution of the values of the parameters $\theta_t^{(k)}$ among such particles will reflect the marginal posterior distribution of these parameters at time t, conditional on model $\mathcal{M}_k$.

The filtering scheme described in the previous sections will work in this situation, with some extra care. In particular, the augmentation step at time t will transform each generic particle (k, θ) currently associated with model $\mathcal{M}_k$ into a particle $(k, \theta^*) = (k, \theta, \theta^+)$, where θ^+ represents a realisation of the (possibly empty) set of parameters introduced in model $\mathcal{M}_k$ at time t. The θ^+ component is drawn from a distribution $f_t^{(k)}$, which depends on k. The augmented particle receives a weight

$$
w = \frac{\mathrm{p}(d\theta^+|k, \theta)\, \mathrm{p}(y_t|k, \theta^*, y_{1:t-1})}{f_t^{(k)}(d\theta^+)},
$$

where, as usual, the $\mathrm{p}(.)$ distributions are part of the model specification. If we choose $f_t^{(k)}(.)$ to be the conditional prior $\mathrm{p}(.|k, \theta)$, the above expression simplifies to

$$
w = \mathrm{p}(y_t|k, \theta^*, y_{1:t-1}). \tag{6.6.14}
$$

In the Move step, a particle at time t can either be left where it is, moved to a new point within its current subspace, or moved to a point in a different subspace using a cross–model jump. Cross–model jumps may be useful if the number of relevant models is high. During the recursive updating process, if some subspaces become occasionally depleted of particles, they may be repopulated on jump moves. Another motivation for cross–model jumps, in applications having some of the candidate models nested within

each other, is to take advantage of relationships between parameters of
different models to design sensible cross–model moves.

Formally, the Move step at time t requires that we specify a transition
kernel $q_t(k', d\theta'|k, \theta)$ that has invariant distribution (6.6.12), and takes a
particle currently in state (k, θ) to the subset $d\theta'$ of the subspace of model
$\mathcal{M}_{k'}$. A sufficient condition for the required invariance property is that the
kernel be reversible with respect to the distribution (6.6.12). Reversibility
can be achieved using the Metropolis–Hastings method, as follows. If the
particle we want to move is in state (k, θ), we draw a vector of random
numbers λ from a distribution that admits a density with respect to the
Lebesgue measure on $\mathcal{R}^{m_1}$, and set the proposed new particle state, (k', θ'),
to a deterministic function $g'(k, \theta, \lambda)$ such that

$$(k', \theta') = g'(k, \theta, \lambda),$$

which generally includes the possibility $k = k'$. This operation is generally
equivalent to drawing the proposed point from a probability distribution
$m'_t(k', d\theta'|k, \theta, \lambda)$, which admits a density $m'_t(k', \theta'|k, \theta, \lambda)$ with respect to
Lebesgue measure on $\mathcal{R}^{m_1}$. Similarly, we carry out the reverse move, from
(k', θ') to (k, θ), by drawing a vector of random numbers λ' from a distribution that admits a density with respect to the Lebesgue measure on
$\mathcal{R}^{m_2}$, and then setting the proposed state to a deterministic function g, i.e.,
$(k, \theta) = g(k', \theta', \lambda')$. This is equivalent to drawing from a probability dis-
tribution $m_t(k, d\theta|k', \theta', \lambda')$ which admits a density $m_t(k, \theta|k', \theta', \lambda')$ with
respect to Lebesgue measure on $\mathcal{R}^{m_2}$. In the Metropolis–Hastings method,
the candidate is accepted or rejected according to an acceptance probabil-
ity $\alpha = \min(1, \hat{\rho})$, and the question turns on the expression for $\hat{\rho}$ in order
to ensure reversibility. If we consider moving the particle from point (k, θ)
to the set $(k', d\theta')$, reversibility is ensured when

$$\hat{\rho} = \frac{\pi_t(k', d\theta'_t) \cdot m'_t(k, d\theta|k', \theta', \lambda')}{\pi_t(k, d\theta_t) \cdot m_t(k', d\theta'|k, \theta, \lambda)}. \tag{6.6.15}$$

In practice, however, because we move particles between points in space, not
sets, we need to consider the limit of $\hat{\rho}$ when the sets $d\theta$ and $d\theta'$ shrink to
points θ and θ'. Green (Green 1995) has studied the conditions under which
this limit exists. It turns out that the limit is guaranteed if the following
two conditions are satisfied. First, the transformation $(\theta', \lambda') \leftrightarrow (\theta, \lambda)$ must
be a bijection in the sense that

$$dim(\theta) + m_1 = dim(\theta') + m_2. \tag{6.6.16}$$

Second[2], the distribution $\pi_t(k, d\theta_t)$ must have a density $\pi_t(k, \theta)$ with respect to the Lebesgue measure $\mu(d\theta)$. Under the two conditions above, it

---

[2] Usually either m_1 or m_2 is set to zero, compatibly with the constraint.

can be shown that the limit of $\hat{\rho}$ exists and is given by

$$\rho = \frac{\pi_t(k',\theta') \cdot m'_t(k,\theta|k',\theta',\lambda')}{\pi_t(k,\theta) \cdot m_t(k',\theta'|k,\theta,\lambda)}. \tag{6.6.17}$$

In this case, what we have to do after a new candidate position (k',θ') is proposed for a particle currently in state (k,θ), is to accept or reject the proposal according to an acceptance probability $\alpha = \min(1,\rho)$. The ratio $\hat{\rho}$ admits a limit under less restrictive conditions than those we have just introduced. For a discussion of this topic, see Green (Green 1995).

Having run RESAMPLE–MOVE up to time t, the current ensemble of particles can be used to explore the posterior probability $\pi_t(k)$ that model k is the true data generating model, or the posterior distribution $\pi_t(\theta_t^{(k)}|k)$ of the parameters of model k, conditioned upon this model being true. In forecasting applications, particles can be used to approximate the predictive distribution for a specified, as yet unobserved, quantity z, by averaging over the set of models:

$$p_t(z|y_{1:t}) = \sum_k \pi_t(k) \int p_t(z|\theta_t^{(k)},k,y_{1:t})\, \pi_t(\theta_t^{(k)}|k)\, d\theta_t^{(k)}.$$

6.7   Illustrative application

We illustrate the method with the aid of the bearings-only tracking problem (Gordon et al. 1993) (Carpenter et al. 1999a) (Carpenter et al. 1999b) (Bergman 1999). We go through this example in rather great detail, hoping that this will help the reader get a sense of the many aspects involved in designing a RESAMPLE–MOVE filter for a specific application. The example is described in Figure 6.2: a ship is moving in the north–east plane through random smooth accelerations and decelerations. A stationary observer at the origin of the plane takes a noisy measurement z_t of the ship's angular position at each time t.

We adopt a simple discretisation of the problem, in which x_t and y_t denote the east and north co-ordinates of the ship at integer time t, with corresponding velocities $\dot{x}_t = x_{t+1} - x_t$ and $\dot{y}_t = y_{t+1} - y_t$. A possible model of the data-generating process, for $t > 0$, is

$$
\begin{aligned}
z_t &= arctan(y_t/x_t) + \mathrm{N}(0,\eta^2), \\
\dot{x}_t &\sim \mathrm{N}(\dot{x}_{t-1},\tau^2), \\
\dot{y}_t &\sim \mathrm{N}(\dot{y}_{t-1},\tau^2), \\
x_t &= x_{t-1} + \dot{x}_{t-1}, \\
y_t &= y_{t-1} + \dot{y}_{t-1},
\end{aligned}
\tag{6.7.18}
$$

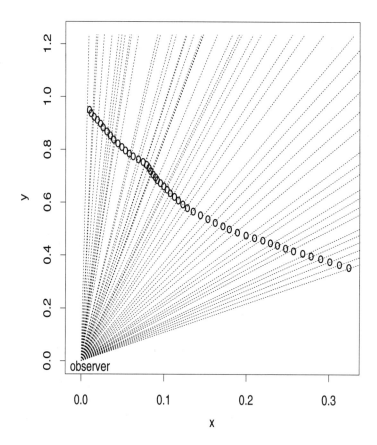

Figure 6.2. Our illustrative example involves a ship moving along a smooth trajectory in the $x-y$ plane where x and y represent the east and the north coordinates. The circles in the figure represent the true position of the ship at times $1, 2, 3, \ldots$. A stationary observer at the origin of the plane takes noisy measurements of the angular position of the ship at times $1, 2, 3, \ldots$, represented in the figure by the dotted lines.

where the notation $\sim$ stands for distributed as, and $N(a, b)$ denotes Gaussian distribution with mean a and variance b. In our experiment, the measurement error standard deviation, η, was set to 0.005. Notice that the ship's trajectory up to time t is entirely determined by the initial position, (x_0, y_0), and by the velocities $(\dot{x}_0, \dot{y}_0, \ldots, \dot{x}_t, \dot{y}_t)$. As we observe the ship through its motion, new data z_t accrue, along with new parameters $(\dot{x}_t, \dot{y}_t)$. At time t, the model contains the vector of unknowns

$$\theta_t = (\tau, x_0, y_0, \dot{x}_0, \dot{y}_0, \dot{x}_1, \dot{y}_1, \ldots, \dot{x}_t, \dot{y}_t),$$

of length $(5 + 2t)$, which describes the ship track up to time t, with $\theta_0 = (y_0, x_0, \dot{y}_0, \dot{x}_0)$ representing the *initial conditions*. From a Bayesian perspective, at time t the target is the posterior distribution of θ_t, which has the form

$$\pi_t(\theta_t) = p(\theta_t | z_{1:t}) \propto p(\theta_t) \cdot \prod_{i=1}^{t} p(z_i | \theta_i), \qquad (6.7.19)$$

and evolves in an expanding space Ω_t. The conditional likelihood $p(z_i|\theta_i)$ is fully determined by the model specification, and the prior distribution $p(\theta_t)$ is discussed later in this subsection. As t increases, our aim is to maintain a set of particles in Ω_t so as to estimate relevant aspects of π_t. In particular, the particles can be used at any given time t to approximate the posterior distribution for the current position of the ship (x_t, y_t), in the light of data $(z_1, \ldots, z_t)$ accumulated up to that point, or to predict the ship's position at a future time. The prior distribution component in (6.7.19) has the structure

$$p(\theta_t) = p(\theta_0) \cdot \exp\left(-.5\,\tau\,\dot{\mathbf{y}}_{0:t}'\,\mathbf{C}\,\dot{\mathbf{y}}_{0:t} - .5\,\tau\,\dot{\mathbf{x}}_{0:t}'\,\mathbf{C}\,\dot{\mathbf{x}}_{0:t}\right), \qquad (6.7.20)$$

where for a vector $\mathbf{v}$, $\mathbf{v}_{s:t}$ is a shorthand notation for $(v_1, \ldots, v_t)$. The prior on the initial conditions, $p(\theta_0)$, was chosen to represent a proper informative prior on the ship's initial conditions. The second term in (6.7.20) represents smoothness of the ship's accelerations, and $\mathbf{C}$ is the $(t+1) \times (t+1)$ tri-diagonal matrix

$$\mathbf{C} = \begin{vmatrix} 1 & -1 & 0 & 0 & \cdot & \cdot & \cdot & \cdot & \cdot & \cdot & \cdot & \cdot \\ -1 & 2 & -1 & 0 & \cdot & \cdot & \cdot & \cdot & \cdot & \cdot & \cdot & \cdot \\ 0 & -1 & 2 & -1 & \cdot & \cdot & \cdot & \cdot & \cdot & \cdot & \cdot & \cdot \\ \cdot & \cdot & \cdot & \cdot & \cdot & \cdot & \cdot & \cdot & \cdot & \cdot & \cdot & \cdot \\ \cdot & \cdot & \cdot & \cdot & \cdot & \cdot & \cdot & -1 & 2 & -1 & 0 \\ \cdot & \cdot & \cdot & \cdot & \cdot & 0 & 0 & 0 & -1 & 2 & -1 \\ \cdot & \cdot & \cdot & \cdot & \cdot & 0 & 0 & 0 & 0 & -1 & 1 \end{vmatrix}.$$

The matrix $\mathbf{C}$ is rank-deficient, which reflects the fact that the second term in (6.7.20) is invariant with respect to global level changes of coordinates and velocities. Thus, with a vague $p(\theta_0)$, the density (6.7.20) would be intrinsic.

6.7.1 *Applying RESAMPLE–MOVE*

At time t, upon the arrival of a new observed bearing z_t, the parameter space expands to incorporate the new element $(\dot{x}_t, \dot{y}_t)$, and we want to construct an updated, t^{th}, generation of particles, whose i^{th} member is a realisation of the ship's history up to that time, as represented by the vector

$$\theta_{t,i} = (\tau, x_0, y_0, \dot{x}_{0:t}, \dot{y}_{0:t}).$$

In the augmentation at time t, each particle $\theta_{t-1,i}$ of generation $t - 1$ is transformed into a new particle $\theta^*_{t-1,i}$ by appending to it a realisation of the new element $(\dot{x}_t, \dot{y}_t)$. We sample this element from its conditional prior, $N(\dot{x}_{t-1}, \tau^{-1}) \times N(\dot{y}_{t-1}, \tau^{-1})$, so that the weight $w_{t-1,i}$ of each ith particle has the simple form $p(z_t | \theta^*_{t-1,i})$. Because the prior ignores information about z_t, a large proportion of the particles, after the augmentation, will be in conflict with z_t, and consequently bound to be eliminated in the subsequent resampling. To avoid impoverishment of the set of particles, it is necessary that the particles be moved across the parameter space.

Before discussing details of the particle moving, we need to introduce some additional notation. For a generic matrix $\mathbf{C}$, we shall let $\mathbf{C}$[a:b,c:d] denote the sub-matrix of $\mathbf{C}$ comprising rows a through b and columns c through d, and we shall let $\mathbf{C}^{-1}$[a:b,c:d] denote the inverse of $\mathbf{C}$[a:b,c:d]. At time t, the conditional distribution for $\dot{\mathbf{x}}_{a:b}$ given the remaining part of the vector $\dot{\mathbf{x}}_{1:t}$ has the form

$$p(\dot{\mathbf{x}}_{a:b} | \dot{\mathbf{x}}_{1:(a-1)}, \dot{\mathbf{x}}_{(b+1):t}) = N(\mu_{\dot{x}_{a:b}}, \mathbf{C}^{-1}[a:b,a:b]), \qquad (6.7.21)$$

where the notation $\mu_{v_{a:b}}$ stands for

$$\mu_{v_{a:b}} := -\mathbf{C}^{-1}[a:b,a:b] \big(\mathbf{C}[a:b,1:(a-1)] \mathbf{v}_{1:(a-1)} + \mathbf{C}[a:b,(b+1):t] \mathbf{v}_{b+1:t} \big), (6.7.22)$$

under the convention that $\mathbf{v}_{a:b} = \mathbf{0}$ if either $a = 0$ or $b = 0$.

We use the following three types of move.

rescaling: all coordinate and velocity values representing the ship's trajectory, along both axes, are reduced or amplified by the same factor.

local perturbation: the particle trajectory is perturbed locally, by moving a chosen block of consecutive points of the trajectory to a new position, while leaving the remaining part of the trajectory unchanged,

τ**–move:** update the value of τ.

The first two types of move were implemented according to the Metropolis–Hastings scheme, whereas the third was implemented as a Gibbs move. As noted in (Carpenter et al. 1999b), the rescaling move leaves all the angles unchanged and hence does not affect the likelihood term in (6.7.19). Consider particle i representing a realisation h'_i of the ship's history, that is, $h'_i = (x'_{0:t}, y'_{0:t}, \dot{x}'_{0:t}, \dot{y}'_{0:t})$. By multiplying each component of this vector by a factor λ_{it}, we generate a candidate $h''_i = h'_i \cdot \lambda_{it}$. The factor λ_{it} is independently sampled from the uniform (dependence) proposal distribution

$U(r_1, 1/r_1)$, for a chosen constant r_1, $r_1 \in (0,1)$. The generated candidate, h_i'', goes through the usual Metropolis-Hastings accept-or-reject scheme. Note that the proposal density is equal to $r_1/(1 - r_1^2)$ everywhere in its support, and that if the probability of moving from h_i to h_i'' is non-zero, the probability of the reverse move is non-zero. It follows that this proposal distribution is symmetric. The rescaling move at time t involves the entire length of the ship's trajectory up to time t, and therefore its burden of computation is of order $\mathcal{O}(t)$. However, we can carry out the rescaling move progressively less frequently, with probability $\mathcal{O}(t^{-\delta})$, say, where $\delta > 0$.

The local perturbation move is applied first to the $\dot{x}$ vector, then to the $\dot{y}$ vector. For any given block $\dot{x}_{a:b}$ currently at value $\dot{\mathbf{x}}'_{a:b}$, a new candidate $\dot{\mathbf{x}}''_{a:b}$ can be drawn from the dependence proposal distribution $N(\dot{\mathbf{x}}'_{a:b}, r_{2,t} \cdot \mathbf{C}^{-1}[a:b,a:b])$. The spread of the distribution is controlled by $r_{2,t}$, and can be decreased over the evolution process as the size of the parameter space increases. To complete the move, we then generate a candidate $\dot{\mathbf{y}}''_{a:b}$ for the block $\dot{y}_{a:b}$ from a dependence proposal distribution that takes information about z_t into account. The candidate $(\dot{\mathbf{x}}''_{a:b}, \dot{\mathbf{y}}''_{a:b})$ is then either globally accepted or globally rejected. Notice that moving the block $(\dot{x}_{a:b}, \dot{y}_{a:b})$ does not involve any components of the ship's trajectory apart from those contained in this block, and that, moreover, we can make the block size arbitrarily high without incurring a high Metropolis–Hastings rejection rate. For this reason, we favoured this approach over that of using an independence proposal distribution such as (6.7.21). At each iteration of the algorithm, we could perturb (block by block) the whole past trajectory, but doing so would introduce a computational cost growing with time. Computational costs might be drastically reduced by perturbing only the most recent portion of each trajectory represented in the particle system. Unfortunately, this would allow the resampling steps to progressively reduce the number of distinct particle realisations of the early – and growing – part of the trajectory. The particle representation of a growing portion of trajectory might then progressively deteriorate, causing estimation of fixed parameters, specifically τ, to be based on more and more erroneous information, and possibly to degenerate through inadequate choice of the number of particles N_t. In view of this problem, it appears not to be a bad policy, at every t^{th} stage of the updating, to move a few recent blocks of trajectory, while giving every past block a chance to be moved. Because new data z_t will provide progressively less information on $(\dot{\mathbf{x}}_k, \dot{\mathbf{y}}_k)$, for each fixed k, we can arrange to update $(\dot{\mathbf{x}}_k, \dot{\mathbf{y}}_k)$ with diminishing frequency as t increases, such that the overall burden of computation for the local perturbation move at time t is of $\mathcal{O}(t^\beta)$ order for some $\beta < 1$. One may also allow the size of the blocks to gradually increase with t. However, we admit that the notion of allowing the system to progressively forget its past raises convergence issues, and we agree with (Andrieu et al. 1999b), who point out difficulties in establishing uniform convergence results for this type of algorithm. Along with these authors, we remark that the problem may be overcome by regarding the τ hyper-parameter as a time-varying random component of the model.

Concerning the τ-move, we take τ^{-1} to follow a-priori a Gamma distribution with mean d/c and variance $d/(c^2)$, denoted G(d, c). At time t, we can Gibbs-update this parameter, whose full conditional distribution is G(d', c'), with $d' = d+2t$ and $c' = c+\sum_{j=2}^{t}(\dot{x}_j - \dot{x}_{j-1})^2 + \sum_{j=2}^{t}(\dot{y}_j - \dot{y}_{j-1})^2$. Because the full conditional distribution involves all elements of the trajectory up to time t, the computational complexity of τ updating is $\mathcal{O}(t)$. However, because the marginal posterior distribution for τ will stabilise over time, we can exploit information on τ contained in the particles and update τ progressively less frequently, with probability $\mathcal{O}(t^{-\alpha})$, say, where $\alpha > 0$. Thus, the overall burden of computation of one evolution step at time t, summing the separate contributions of the three types of move, is $\mathcal{O}(t^{-\delta}) + \mathcal{O}(t^{\beta}) + \mathcal{O}(t^{-\alpha}) < \mathcal{O}(t)$. By comparison, in a typical MCMC algorithm, all the model parameters would need to be updated at every iteration, with an burden of computation for one iteration of MCMC at time t equal to $\mathcal{O}(t) + \mathcal{O}(t) + \mathcal{O}(t) = \mathcal{O}(t)$, summing the separate contributions for the rescaling move, the local perturbation move and the τ-updating. Of course, should t become large, the τ-move may become expensive, so a different model, one allowing τ to vary slowly with time, could be considered.

6.7.2   Simulation experiment

The equations (6.7.19), with initial conditions $x_0 = .01, y_0 = .95, \dot{x}_0 = .002, \dot{y}_0 = -0.013$, and with $\tau = 1e - 006$, were used to simulate a trajectory and a corresponding set of angular measurements, both represented in Figure 6.2. These data were sequentially incorporated into the RESAMPLE - MOVE filter of Section 6.7.1, with the following prior distribution on τ and on the ship's initial conditions τ:

$$
\begin{aligned}
p(x_0) &= \text{N}(0.0108, 0.04), \\
p(y_0) &= \text{N}(1.03, 0.4), \\
p(\dot{x}_0) &= \text{N}(0.002, 0.003), \\
p(\dot{y}_0) &= \text{N}(-0.013, 0.003), \\
p(\tau^{-1}) &= \text{G}(1, 0.00001)
\end{aligned}
\qquad (6.7.23)
$$

where G(a, b) stands for Gamma distribution with shape parameter a and rate parameter b, mean a/b and variance $a/(b^2)$. This prior is slightly displaced with respect to the true initial coordinates, x_0 and y_0, but not with respect to the true initial velocities, $\dot{x}_0$ and $\dot{y}_0$. Notice also that the prior mean for τ differs from the true value of this parameter. The number of particles was kept equal to 1000 over the entire filtering process.

On the same data, with the same number of particles, we ran a standard Sampling Importance Resampling (SIR) filter, which can be described as a special case of RESAMPLE–MOVE algorithm in which no evolution step takes place and no parameter value is changed once it has been sampled. Figure 6.3 compares the performance of RESAMPLE–MOVE and SIR on the same data. In that figure, the ship's true trajectory is represented as a solid curve, whose x- and y-coordinates represent east and north. The output of RESAMPLE–MOVE is represented by empty circles, and that of SIR by filled squares. For each tth measurement time, $t = 1, \ldots, 50$, and for each of the two filters, the plot contains a circle (or square) representing the particle approximation to the filtered mean of the bivariate $\pi_t(x_t, y_t)$ posterior distribution, as an estimate for the ship's position at t given the bearings available at t. Both filters start with an overestimate of the ship's initial position due to the conflict between data and prior. Compare the progressive departure of the SIR output from truth with the capability of the RESAMPLE–MOVE filter to gradually gain track of the ship.

6.7.3   Uncertainty about the type of target

So far, we have assumed we have the true model of the ship dynamics. In the following we relax this assumption by introducing two types of ship with different dynamic behaviour, and by treating the type of ship as an unknown variable of the problem. This kind of situation is discussed in Section 6.6.

As a (somewhat contrived) example of the situation, suppose there are two types of ship. Type 1 ships are assumed to follow the model (6.7.19), which we shall henceforth refer to as Model 1, $\mathcal{M}_1$. Conditional on this model, we assume our uncertainty about the ship's initial conditions to be represented by (6.7.23). Type 2 ships are assumed to move along a straight line with constant y-coordinate, according to what we shall refer to as Model 2, $\mathcal{M}_2$:

$$
\begin{aligned}
z_t &= arctan(y_0/x_t) + \mathrm{N}(0, \eta^2), \\
\dot{x}_t &\sim \mathrm{N}(\dot{x}_{t-1}, \tau^2), \\
x_t &= x_{t-1} + \dot{x}_{t-1}, \\
\mathrm{p}(x_0) &= \mathrm{N}(0.0108, 0.04) \\
\mathrm{p}(y_0) &= \mathrm{N}(1.03, 0.4) \\
\mathrm{p}(\dot{x}_0) &= \mathrm{N}(0.002, 0.003) \\
\mathrm{p}(\tau^{-1}) &= \mathrm{G}(1, 0.00001)
\end{aligned}
$$

with η set to 0.005. Notice that Type 2 ships keep a constant y coordinate over time, but are allowed smooth accelerations and decelerations along the x-direction of the same entity (controlled by τ) as ships of Type 1. Also, the measurement equation is assumed to be the same under the two models.

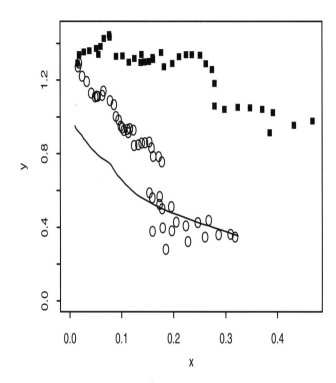

Figure 6.3. Comparing the performance of RESAMPLE–MOVE and a standard sequential importance resampling (SIR) filter. The true ship trajectory is represented in the figure as a solid curve. Empty circles and filled circles represent, respectively, RESAMPLE–MOVE and SIR tracking of the ship.

In a RESAMPLE–MOVE filter for this problem, a particle at time t can either live in the subspace of Model $\mathcal{M}_1$, given by $\{1\} \times \Omega_t^{(1)}$, or in the subspace of Model $\mathcal{M}_2$, given by $\{2\} \times \Omega_t^{(2)}$, where points in $\Omega_t^{(1)}$ are realisations of the vector $(\tau, y_0, x_0, \dot{y}_0, \dot{x}_0, \dot{y}_1, \dot{x}_1, \ldots, \dot{y}_t, \dot{x}_t)$, and points in $\Omega_t^{(2)}$ are realisations of $(\tau, y_0, x_0, \dot{x}_0, \dot{x}_1, \ldots, \dot{x}_t)$. The union of the two subspaces we denote by $\Gamma_t = \{1\} \times \Omega_t^{(1)} \bigcup \{2\} \times \Omega_t^{(2)}$. For particles associated with model $\mathcal{M}_1$, augmentation proceeds as described in the previous section, whereas for particles associated with model $\mathcal{M}_2$ it proceeds similarly except that a component $\dot{x}_t$, but not $\dot{y}_t$, is appended at time t.

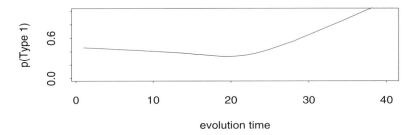

Figure 6.4. Evolution of the filtered mean for the posterior probability of the ship being of type 1.

Concerning the Move step, a particle at time t can either be left where it is, moved to a new point within its current subspace, or moved to a point in a different subspace. The move types described in Section 6.7.1 can be used to move the particle within its current subspace. Cross-model moves, from one subspace to another, can be useful in the present application because the two models considered are nested, and therefore the obvious relationships between parameters of the two models may be exploited to ensure a good rate of acceptance for this type of move. For example, to move a particle currently at point $(\tau, y_0, x_0, \dot{x}_0, \dot{x}_1, \ldots, \dot{x}_t)$ under Model $\mathcal{M}_2$ to a point in the subspace of model $\mathcal{M}_1$, we sample a sequence $(\dot{y}_0', \dot{y}_0', \dot{y}_1', \ldots, \dot{y}_t')$ and then determine the angle-preserving vector $(\dot{x}_0', \dot{x}_1', \ldots, \dot{x}_t')$, so that the proposed position in the subspace of model $\mathcal{M}_1$ is $(\tau, y_0, x_0, \dot{y}_0', \dot{x}_0', \dot{y}_1', \dot{x}_1', \ldots, \dot{y}_t', \dot{x}_t')$. The reverse move, from a point $(\tau, y_0, x_0, \dot{y}_0, \dot{x}_0, \dot{y}_1, \dot{x}_1, \ldots, \dot{y}_t, \dot{x}_t)$ in the subspace of model $\mathcal{M}_1$ to a point in the subspace of model $\mathcal{M}_2$ can be carried out simply by sampling $\dot{y}_0'$ and then proposing the new position $(\tau, y_0', x_0, \dot{x}_0, \dot{x}_1, \ldots, \dot{x}_t)$ in the subspace of model $\mathcal{M}_2$.

We applied the filter just described to the simulated data considered in Section 6.7.2, assigning an equal prior probability of 0.5 to each of the two models entertained. Three thousand particles were used for this experiment. We tested the ability of the filter to learn about the unknown type of ship from the incoming angular data. Figure 6.4 plots the filtered probability $\pi_t(k = 1)$ that the ship is of Type 1 (that is, that it moves according to Model 1), based on the bearings observed up to t, for $t = 1, \ldots, 50$. Initially, this probability is equal to the prior guess, 0.5. Then, as the angular measurements accumulate, the ship is gradually recognised as being of Type 1.

6.8   Conclusions

We have discussed the possibility of combining principles of sequential resampling and Markov chain iterations within a unified hybrid particle filtering scheme, and we have described an example of this approach, viz. our RESAMPLE–MOVE filter. This filter allows the particles to move along any direction of the parameter space at any stage of the evolution process. This improves adaptation to the evolving target, and prevents critical subspaces of the parameter space to become depleted of particles over successive updates. Consequently, we expect RESAMPLE–MOVE to outperform standard sequential imputation methods, especially in those dynamic applications involving static unknown hyper-parameters.

In this chapter we have also emphasised the ability of hybrid particle filters to deal with dynamic problems involving model uncertainty.

We would like to conclude with the following question. In cases when the data do not arrive sequentially, and there is consequently no need for on-line inference, can particle filters be preferable to standard MCMC techniques? For example, with reference to the application of Section 6.7, involving multiple models, suppose that the whole sequence of data $z_{1:T}$ is acquired at one point in time, T, and we are interested in identifying the type of target. Would a RESAMPLE – MOVE filter that works through the data sequentially still be preferable to standard MCMC equipped with reversible–jump technology? We believe so. One reason is, with a limited number of models, the particle filter does not necessarily rely on jump moves, which are notoriously difficult to design and possibly subject to a very high rejection rate. More generally, the particle filtering approach may be less dependent on the ability to design clever moves. Whereas a standard MCMC analysis estimates the quantities of interest on the basis of the entire path of one or a few non-interacting chains, thus relying on the ability of each of these chains to explore the entire posterior space, a particle filter estimates the quantities of interest from the current value of a large ensemble of inter-acting particles, and problems of mixing are attenuated by the possibility that the particles may interact with each other. We suggest, on the basis of these considerations, that hybrid particle filters may also be very useful in cases involving a single, static, target distribution.

7

Improvement Strategies for Monte Carlo Particle Filters

Simon Godsill
Tim Clapp

7.1 Introduction

The particle filtering field has seen an upsurge in interest over recent years, and accompanying this upsurge several enhancements to the basic techniques have been suggested in the literature. In this paper we collect a group of these developments that seem to be particularly important for time series applications and give a broad discussion of the methods, showing the relationships between them. We firstly present a general importance sampling framework for the filtering/smoothing problem and show how the standard techniques can be obtained from this general approach. In particular, we show that the auxiliary particle filtering methods of (Pitt and Shephard: this volume) fall into the same general class of algorithms as the standard bootstrap filter of (Gordon et al. 1993). We then develop the ideas further and describe the role of MCMC resampling as proposed by (Gilks and Berzuini: this volume) and (MacEachern, Clyde and Liu 1999). Finally, we present a generalisation of our own in which MCMC resampling ideas are used to traverse a sequence of 'bridging' densities which lie between the prediction density and the filtering density. In this way it is hoped to reduce the variability of the importance weights by attempting a series of smaller, more manageable moves at each time step.

Much of the particle filtering literature has been concerned with filtering for nonlinear models in tracking applications. There are many other important contexts in which sequential methods are needed, and here we explore a class of time-evolving models that can be used to explore the spectral evolution of a wide range of datasets: the time-varying auto-regression. Auto-regressions have been used to model a wide range of phenomena from earthquake data and EEG traces to financial time series. We have applied them extensively to the analysis and processing of speech and audio data (Godsill and Rayner 1998a, Godsill and Rayner 1998b, Godsill

1997), where it is typically important to model an additive background noise component in addition to the desired signal itself. Moreover, in such applications it is worth while to track the time-varying spectral structure which is present in the data, and this can be achieved by extending the standard models to have time-varying parameters; see (Rajan, Rayner and Godsill 1997, Kitagawa and Gersch 1996, West and Harrison 1997, West, Prado and Krystal 1999, Prado, West and Krystal 1999) for examples of their theory and application in a statistical context. In this paper we study a simple motivating TVAR model which captures the basic characteristics of many time-varying datasets. A Gaussian random walk is used to model the autoregressive parameter variation with time and signals are assumed observed in white Gaussian noise. More elaborate models incorporating stability constraints and more physically-based time evolution structures are discussed in (Godsill, Doucet and West 2000). Here we work with the most simple model because it is straightforward to construct efficient implementations of all of the above variants on particle filtering, and we describe in detail the resulting algorithms. Simulations are then carried out in order to compare the merits of the various approaches, both in terms of estimation performance and computational burden. In our initial discussions we will assume the standard Markovian state-space model

$$x_t \sim f(x_t|x_{t-1}) \qquad \text{State evolution density}$$
$$y_t \sim g(y_t|x_t) \qquad \text{Observation density}$$

where $\{x_t\}$ are unobserved states of the system and $\{y_t\}$ are observations made over some time interval $t \in \{1, 2, ..., T\}$. $f(.|.)$ and $g(.|.)$ are pre-specified state evolution and observation densities. The initial state has density $p(x_1)$.

7.2 General sequential importance sampling

Here we formulate the sequential importance sampling method in terms of updates to the smoothing density. Once we have samples drawn from the smoothing density, it is straightforward to discard those that are not required if filtering is the main objective. Suppose we have at time t weighted particles $\{x_{1:t}^{(i)}, w_t^{(i)}; i = 1, 2, ..., N\}$ drawn from the smoothing density $p(x_{1:t}|y_{1:t})$. We can represent this information using the particulate approximation to the density

$$p(x_{1:t}|y_{1:t}) \simeq \sum_{i=1}^{N} w_t^{(i)} \delta(x_{1:t} - x_{1:t}^{(i)}),$$

where $\sum_{i=1}^{N} w_t^{(i)} = 1$. In order to update the smoothing density from time $t-1$ to time t, we factorise it as follows:

$$p(x_{1:t}|y_{1:t}) = p(x_{1:t-1}|y_{1:t-1}) \times \frac{g(y_t|x_t)f(x_t|x_{t-1})}{p(y_t|y_{1:t-1})} \qquad (7.2.1)$$

where the denominator is simply a fixed normalising constant. The derivation presented here is slightly non-standard, involving an importance function over all the smoothed states. This allows us to consider a range of methods, including those with elements of smoothing as well as filtering, within a single framework. For more standard interpretations, see the review chapters of this book.

Ideally, we wish to update the particles at time $t-1$ in order to obtain joint draws from $p(x_{1:t}|y_{1:t})$. Using importance sampling, we can achieve this by generating states from an importance density $q(x_{1:t})$, which may implicitly depend upon the particles at time $t-1$ and observed data $y_{1:t}$, as required. The importance weight, from equation (7.2.1), is then given by

$$w_t \propto p(x_{1:t-1}|y_{1:t-1}) \times \frac{g(y_t|x_t)f(x_t|x_{t-1})}{q(x_{1:t})}$$

For standard filtering problems, it is customary to fix the values of the particles at time $t-1$ and generate the new states x_t conditional upon those particles. The importance function for each particle can in this case take the form $q(x_{1:t}) = q(x_t|x_{1:t-1}) \sum_{i=1}^{N} v_{t-1}^{(i)} \delta(x_{1:t-1} - x_{1:t-1}^{(i)})$, that is, we draw $x_{1:t-1}$ from a particulate representation with normalised weights $v_{t-1}^{(i)}$ and then generate x_t conditionally from a proposal $q(x_t|x_{1:t-1})$. Substituting this and the particulate approximation for $p(x_{1:t-1}|y_{1:t-1})$ into the weight expression leads to

$$w_t \propto \frac{w_{t-1}}{v_{t-1}} \times \frac{g(y_t|x_t)f(x_t|x_{t-1})}{q(x_t|x_{1:t-1})}$$

where $x_{1:t} = \{x_{1:t-1}, x_t\}$ is a particle drawn from the joint distribution $q(x_{1:t})$. The weights v_{t-1} serve to amplify or diminish the numbers of different particles in the population, thus allowing the importance sampler the potential to neglect particles with low probability and increase the numbers of particles with high probability. A standard approach chooses $v_t = w_t$, in which case particles are selected with probability proportional to their weights in the approximation to $p(x_{1:t-1}|y_{1:t-1})$. This case is equivalent to the bootstrap filter of (Gordon et al. 1993) in which particles $x_{1:t-1}$ are resampled according to their weights w_{t-1} before making a conditional draw for the new state $x_t \sim f(x_t|x_{t-1})$. However, particles that had large weights at time $t-1$ should not necessarily have large weights when propagated to t, in the light of an informative observation y_t. An improved choice of weighting function may then be $v_{t-1} \propto p(x_{1:t-1}|y_{1:t})$ because particles are amplified or diminished directly according to their posterior distribution conditional upon all the data, including the new point y_t. This is the principle behind the auxiliary particle filter (Pitt and Shephard: this volume), although in practice $p(x_{1:t-1}|y_{1:t})$ will be unavailable analytically and so v_{t-1} must be approximated in their scheme (see below for implementation details of the TVAR model).

A note on selection schemes. The general framework described above involves a random resampling step, in which particles are resampled randomly with replacement with probabilities $\{v_t^{(i)}\}$. This random resampling introduces unnecessary additional Monte Carlo variation and can be replaced with any partly deterministic scheme such that the expected number selected of each particle equals Nv_t. Such schemes can be designed to be much more computationally efficient than the fully random selection scheme, which simply selects particles independently from the multinomial distribution defined by the weights. There are several such schemes; see (Kitagawa 1996, Carpenter et al. 1999a, Liu and Chen 1998), but we do not discuss their relative merits here. Simply note that in all of our simulations the residual resampling method of (Liu and Chen 1998) is used. A further option is not to perform selection at every time step, accumulating weights through time and only performing selection when an appropriate measure of weight degeneracy, such as effective sample size (Liu and Chen 1995), reaches some threshold. In all our simulations, a selection step is carried out at every iteration, although we admit that further improvements could be made by less frequent selection operations.

In any of the above variants on the sequential importance sampling /resampling schemes, we have at time t an approximate set of draws from the joint smoothing density, $p(x_{1:t}|y_{1:t})$. In practice, for all the schemes, excepting perhaps the most general version that allows an importance function $q(x_{1:t})$ over the complete state sequence, we expect a high degree of degeneracy between the smoothed state sequences at times significantly less than t. This is because the importance function leaves the state sequences up to time $t-1$ unchanged in the move to t. This can lead to problems in pure filtering problems (where only $p(x_t|y_{1:t})$ is of interest), which become even more severe in cases when an element of smoothing is also required. This degeneracy is particularly troublesome is the case of models with fixed (non-dynamic) hyper-parameters. Here the smoothing density is augmented by fixed dimension parameters θ to $p(x_{1:t}, \theta|y_{1:t})$, which can be factorised as

$$p(x_{1:t}, \theta|y_{1:t}) = p(x_{1:t-1}, \theta|y_{1:t-1}) \times \frac{g(y_t|x_t, \theta)f(x_t|x_{t-1}, \theta)}{p(y_t|y_{1:t-1})}.$$

All the above ideas can be applied in principle to this case, augmenting the particles to include θ as well as $x_{1:t}$ at each time step: $\{x_{1:t}^{(i)}, \theta^{(i)}, w_t^{(i)}; i = 1, 2, ..., N\}$. The degeneracy problem becomes particularly severe here, however, since there is no means of generating new θ values as time proceeds when the particulate approximation is employed. We do not address this fixed parameter issue in our simulations, noting instead that it is an important problem that can be tackled using the Markov chain methods of the next sections or employing kernel density ideas (Liu and West: this volume).

In the next section we describe techniques allowing for regeneration of new state values through the use of Markov chains, and hence go some way towards solving the degeneracy issues associated with standard particle filtering methods.

7.3   Markov chain moves

Using the methods of the previous section, we have the means to generate approximate draws from $p(x_{1:t}|y_{1:t})$. As stated, however, there can be problems of degeneracy in the particle filter when the target distribution $p(x_{1:t}|y_{1:t})$ does not overlap significantly with the prediction distribution $p(x_{1:t}|y_{1:t-1})$. One way to overcome the deficiency is to incorporate Markov chain moves into the scheme (MacEachern et al. 1999)(Gilks and Berzuini: this volume). Since we wish particles to be drawn from $p(x_{1:t}|y_{1:t})$, it seems reasonable to design Markov chain transition kernels $T(x_{1:t}, dx'_{1:t})$, having $p(x_{1:t}|y_{1:t})$ as stationary distribution. There are many well known ways for achieving this (Gilks et al. 1996), including the Gibbs sampler and Metropolis-Hastings methods. We will explicitly develop MCMC samplers for the TVAR model in the simulation section. Intuitively, if the particles $\{x_{1:t}^{(i)}\}$ are truly drawn from $p(x_{1:t}|y_{1:t})$, then the Markov chain kernel applied to any of the particles will simply generate new state sequences which are also drawn from the desired distribution. Moreover, degeneracy will be less of a problem since particles can move independently of one another to different points in the state-space. More importantly, perhaps, if $\{x_{1:t}^{(i)}\}$ are not drawn accurately from $p(x_{1:t}|y_{1:t})$, because of the approximation of importance sampling with finite N, the Markov chain will nudge the particles closer to the target distribution. Thus, we have a means of improving both degeneracy amongst particles and the accuracy of the samples in the particle cloud. In practice for filtering problems with large datasets the transition kernels will not operate on the entirety of the state history, but rather on recent elements back to some fixed or varying time lag l.

These methods also help significantly with the fixed-parameter problem, since a Markov chain transition kernel can be applied for generating new values of the non-dynamic parameter θ. Typically, θ might be drawn from its conditional distribution $p(\theta|x_{1:t},y_{1:t})$ either exactly in a single move step if the distributions allow (Clapp and Godsill 1999), or iteratively using a Markov chain kernel. One troublesome aspect of this, however, is that the conditional distribution will in general depend upon the entire state history $x_{1:t}$, which means that computational burden and memory requirements will be a problem for large datasets. For certain cases, however, such as the conditionally Gaussian parameter model, this distribution may be summarised by simple sufficient statistics that are then efficiently updated at each time step.

7.3.1   The use of bridging densities with MCMC moves

The combination of importance sampling with Markov chain moves is powerful, and helps to reduce significantly the problems associated with sequential Monte Carlo. However, there are situations requiring a very large number of MCMC iterations in order to reach the target distribution. In particular this can occur when the likelihood for the new data point is centred far from the points sampled from the importance distribution $q(x_t|x_{1:t-1})$. The MCMC sampler may then have to converge from a set of atypical points lying in the tails of the target distribution. This will be an expensive procedure, since it must be repeated for all samples and there may be no reliable way of judging whether convergence has actually been achieved.

Here we explore a new version of the sequential Monte Carlo schemes outlined above in which a sequence of bridging densities is placed between the initial distribution at time $t-1$ and the target distribution at time t. It is hoped that by replacing a single large transition with a series of smaller transitions the Monte Carlo variation of the importance weights can be reduced. All the above ideas combining importance sampling with Markov chain moves can be used to good effect within the new scheme. The initial distribution in the sequence is denoted by $\pi_0 = q(x_{1:t})$ and the final distribution $\pi_{M+1} = p(x_{1:t}|y_{1:t})$. Intervening bridging distributions $\pi_1, \pi_2, ..., \pi_M$ are then defined according to some appropriate schedule such as

$$\pi_m(x_{1:t}) \propto q(x_{1:t})^{\alpha_m} p(x_{1:t}|y_{1:t})^{1-\alpha_m}, \qquad (7.3.2)$$

with $1 > \alpha_1 > \alpha_2 ... > \alpha_M > 0$. For example, if the importance distribution is of the form $q(x_{1:t}) = f(x_t|x_{t-1}) \sum_{i=1}^{N} w_{t-1}^{(i)} \delta(x_{1:t-1} - x_{1:t-1}^{(i)})$, which corresponds to the distribution used in the bootstrap filter of (Gordon et al. 1993), then we have

$$\pi_m(x_{1:t}) \propto p(x_{1:t-1}|y_{1:t-1}) f(x_t|x_{t-1}) g(y_t|x_t)^{1-\alpha_m}, \qquad (7.3.3)$$

and it is evident that the bridging densities gradually introduce the likelihood as $g(y_t|x_t)^{1-\alpha_m}$, starting with a flat likelihood ($\alpha_0 = 1$) and gradually sharpening it until $\alpha_{M+1} = 0$.

The task is now to move the particle cloud through this sequence of densities by any available means. A MCMC scheme for achieving this in a batch context for multi-modal target distributions, called annealed importance sampling, has been introduced by (Neal 1998). In our sequential context, an algorithm can be devised as follows. Firstly, generate states from the proposal distribution $q(x_{1:t})$ and denote these $\{x_{1:t}^{(i)(0)}, w_t^{(i)(0)}\}$ with $w_t^{(i)(0)} = w_{t-1}^{(i)}$. Then at step m, for $m = 1, ..., M + 1$, compute the incremental importance weight for the transition between $\pi_{m-1}(x_{1:t})$ and $\pi_m(x_{1:t})$, i.e.

$$w_t^{(i)(m)} \propto w_t^{(i)(m-1)} \frac{\pi_m(x_{1:t}^{(i)(m-1)})}{\pi_{m-1}(x_{1:t}^{(i)(m-1)})}.$$

We now have nominally a weighted sample from $\pi_m(x_{1:t})$. We can optionally perform a selection step here with probabilities $\{ w_{t+1}^{(i)(m)}\}$ if the weights are too degenerate and reset $\{ w_{t+1}^{(i)(m)}\}$ to uniform. Now, exactly as in the MCMC schemes above for a single target density, we can improve the particle stock by drawing $x_{1:t}^{(i)(m)}$ from $T_m(x_{1:t}^{(i)(m-1)}, dx'_{1:t})$, where $T_m(x_{1:t}^{(i)(m-1)}, dx'_{1:t})$ is a Markov chain transition kernel designed to have $\pi_m(x_{1:t})$ as invariant distribution. The importance weights are hence unchanged. Finally, after step $M + 1$, we have a weighted sample from the target distribution $\pi_{M+1}(x_{1:t}) = p(x_{1:t}|y_{1:t})$, as required, with weights $w_t^{(i)} = w_t^{(i)(M+1)}$. (Neal 1998) proves that such a procedure generates valid importance weights for the final samples $\{x_{1:t}^{(i)(M+1)}\}$ according to the target distribution $p(x_{1:t}|y_{1:t})$.

Notice that the special case with $M = 0$ (i.e. no bridging densities) corresponds to the MCMC resampling algorithms of (MacEachern et al. 1999) (Gilks and Berzuini: this volume). We have generalised Neal's procedure slightly here by allowing an optional resampling at each intermediate step. This has sometimes been found to be beneficial.

7.4 Simulation example: TVAR model in noise

The time-varying autoregressive (TVAR) model can be used to model parametrically a signal with time-varying frequency content. Here we work with the simplest TVAR structure and develop particle filtering approaches based upon the improvement strategies described above. It is assumed that the TVAR signal is observed in additive white Gaussian noise, which models any background noise present in the measurement environment. This aspect of the model is important in acoustic applications such as speech and audio (Godsill and Rayner 1998a)(Godsill and Rayner 1998b)(Godsill 1997) and adds an interesting challenge to the parameter estimation process. The TVAR signal process $\{z_t\}$ is generated as a linear weighted sum of previous signal values:

$$z_t = \sum_{i=1}^{P} a_{i,t} z_{t-i} + e_t = a_t^T z_{t-1:t-P} + e_t. \qquad (7.4.4)$$

Here $a_t = (a_{1,t}, a_{2,t}, ..., a_{P,t})^T$ is the Pth order AR coefficient vector at time t. The innovation sequence $\{e_t\}$ is assumed independent and Gaussian with time-varying variance $\sigma^2_{e_t}$. Hence we may write the conditional density for z_t as

$$f(z_t | z_{t-1:t-P}, a_t, \sigma^2_{e_t}) = \mathcal{N}\left(a_t^T z_{t-1:t-P}, \sigma^2_{e_t}\right).$$

The signal is assumed to be observed in independent Gaussian noise

$$y_t = z_t + v_t, \tag{7.4.5}$$

so that we may write the density for the observation y_t as $g(y_t | z_t, \sigma_{v_t}) = \mathcal{N}\left(z_t, \sigma^2_{v_t}\right)$, where once again $\sigma^2_{v_t}$ may be time-varying. For our simulations, a Gaussian random walk model is assumed for the log-standard deviations $\phi_{e_t} = \log(\sigma_{e_t})$ and $\phi_{v_t} = \log(\sigma_{v_t})$:

$$f(\phi_{e_t} | \phi_{e_{t-1}}, \sigma^2_{\phi_e}) = \mathcal{N}\left(\phi_{e_{t-1}}, \sigma^2_{\phi_e}\right), \tag{7.4.6}$$

$$f(\phi_{v_t} | \phi_{v_{t-1}}, \sigma^2_{\phi_v}) = \mathcal{N}\left(\phi_{v_{t-1}}, \sigma^2_{\phi_v}\right). \tag{7.4.7}$$

The model now requires specification of the time variation in a_t itself. One of the simplest choices of all is a first-order auto-regression directly on the coefficients

$$f(a_t | a_{t-1}, \sigma^2_a) = \mathcal{N}\left(\alpha a_{t-1}, \sigma^2_a I_P\right), \tag{7.4.8}$$

where α is a constant just less than 1.

More elaborate schemes of this sort are possible, such as a smoothed random walk involving AR coefficients from further in the past, or a non-diagonal covariance matrix, but none of these modifications make a significant difference to the computations required for the particle filter. In this paper we do not consider the stability of the TVAR model; see (Godsill et al. 2000) for more detailed consideration of this issue.

The state-space TVAR model is now fully specified. The unobserved state vector is $x_t = (z_{t:t-P+1}, a_t, \phi_{e_t}, \phi_{v_t})$. Hyper-parameters σ^2_a, $\sigma^2_{\phi_e}$ and $\sigma^2_{\phi_v}$ are assumed pre-specified and fixed in all the simulations. The initial state probability is specified as i.i.d. Gaussian. Notice that we shall refer to ϕ_{e_t} and ϕ_{v_t} interchangeably with σ_{e_t} and σ_{v_t} in the subsequent discussions, as is appropriate to the context. Also, we shall refer to x_t or its individual components $(z_{t:t-P+1}, a_t, \phi_{e_t}, \phi_{v_t})$ interchangeably.

7.4.1   Particle filter algorithms for TVAR models

In the models specified above, there is some rich analytic structure. For example, conditionally on the signal sequence and noise variance sequences, the AR parameter sequence $\{a_t\}$ is Gaussian and can be estimated using the Kalman filter recursions. Similarly, the signal sequence $\{z_t\}$ is Gaussian conditionally upon $\{a_t\}$ and the variance sequences and can also be estimated using the Kalman filter. This suggests a range of Rao-Blackwellised versions of the particle filter in which one or other of the sequences $\{a_t\}$ or $\{z_t\}$ is marginalised sequentially from the problem using the Kalman filter. These approaches lead to lower-dimensional state-spaces for the particle filter and can be expected to lead to greater robustness (Doucet, Godsill and Andrieu 2000). However, they come with a computational penalty, since Kalman filter recursions must be applied at every time step for every particle. We do not explore these approaches here, for reasons of computational complexity, and also because we lose some of the generality of the particle filter: a small change to the model structure by introducing some nonlinearity or non-Gaussianity (such as those in (Godsill et al. 2000)) would immediately destroy the analytic structure present and invalidate the approach. Some of these Rao-Blackwellised versions of the particle filter are investigated for a speech application in (Vermaak, Andrieu, Doucet and Godsill 1999).

Firstly, consider the choice of importance function for the various filtering approaches. The prior importance function, as used in the original bootstrap filter implementations (Gordon et al. 1993) requires a random draw from the state transition density, as in

$$\phi_{e_t} \sim \mathcal{N}\left(\phi_{e_{t-1}}, \sigma^2_{\phi_e}\right), \quad \phi_{v_t} \sim \mathcal{N}\left(\phi_{v_{t-1}}, \sigma^2_{\phi_v}\right), \quad a_t \sim \mathcal{N}\left(\alpha a_{t-1}, \sigma^2_a I_P\right), \quad (7.4.9)$$

$$z_t \sim \mathcal{N}\left(a_t^T z_{t-1:t-P}, \sigma^2_{e_t}\right), \quad (7.4.10)$$

where I_P denotes the P-dimensional identity matrix.

This is a very simple and efficient procedure and in the basic bootstrap filtering framework an importance weight can be computed as

$$w_t \propto g(y_t | z_t, \sigma_{v_t}) \quad (7.4.11)$$

In fact, we have found algorithms based on this procedure to perform very poorly, largely as a result of step (7.4.9), which can generate unstable realisations of the AR parameters (a sufficient condition for stability of a TVAR model is that the instantaneous poles, i.e. the roots of the polynomial $(1 - \sum_{i=1}^{P} a_{i,t} p^{-i})$, lie strictly within the unit circle – when poles move outside the unit circle, unstable oscillatory behaviour is observed in the generated signal). Hence, when the final step (7.4.10) is performed, large and oscillating values of the signal are generated and the filter can diverge entirely from the correct state sequence. This problem can be fixed by using a slightly more clever importance function taking account of the fact that z_t is conditionally Gaussian. We can then sample z_t from its conditional posterior distribution $p(z_t | x_{t-1}, y_t, a_t, \sigma_{e_t}, \sigma_{v_t}) \propto f(z_t | z_{t-1:t-P}, a_t, \sigma_{e_t}) g(y_t | z_t, a_t, \sigma_{v_t})$, which is Gaussian. The incorporation of the effect of y_t in this manner is generally found to be sufficient to stop the filter from diverging because of stability problems. Working through the algebra, step (7.4.10) is replaced by

$$z_t \sim \mathcal{N}\left(\widehat{z}_t, 1/\phi_t\right), \quad (7.4.12)$$

where

$$\widehat{z}_t = \theta_t/\phi_t, \quad \theta_t = y_t/\sigma_{v_t}^2 + a_t^T z_{t-1:t-P}/\sigma_{e_t}^2 \text{ and } \phi_t = 1/\sigma_{v_t}^2 + 1/\sigma_{e_t}^2. \quad (7.4.13)$$

The resulting importance function is then summarised as

$$q(x_t|x_{t-1}, y_t) = f(\phi_{v_t}|\phi_{v_{t-1}}, \sigma_{\phi_v}^2)$$
$$\times f(\phi_{e_t}|\phi_{e_{t-1}}, \sigma_{\phi_e}^2) f(a_t|a_{t-1}, \sigma_a^2) p(z_t|x_{t-1}, y_t, a_t, \sigma_{e_t}, \sigma_{v_t}) \quad (7.4.14)$$

and the importance weight is modified accordingly to

$$w_t \propto \frac{g(y_t|x_{t-1}) f(x_t|x_{t-1})}{q(x_t|x_{t-1}, y_t)}, \quad (7.4.15)$$

$$\propto p(y_t|z_{t-1:t-P}, a_t, \sigma_{e_t}^2, \sigma_{v_t}^2), \quad (7.4.16)$$

$$\propto \frac{(2\pi)^{1/2}}{\phi_t^{1/2}} N(y_t|\widehat{z}_t, \sigma_{v_t}^2) N\left(\widehat{z}_t|a_t^T z_{t-1:t-P}, \sigma_{e_t}^2\right), \quad (7.4.17)$$

where $N(a|\mu, \sigma^2)$ denotes the Gaussian density function with mean μ and variance σ^2, evaluated at a. This form of importance function involves only a modest computational overhead and is adopted in all of the ensuing simulations. It is somewhat more restrictive than the general form because it requires us to be able to marginalise z_t, but we note that this will be possible for any non-Gaussian noise distribution that can be reparameterised as a scale mixture of Gaussians, see (Godsill and Rayner 1998a)(Godsill and Rayner 1998b)(Godsill 1997) (Shephard 1994) for discussion of such reparameterisations in related modelling contexts. This form of importance function (7.4.14) forms the basic building block in all of our algorithms.

7.4.2 Bootstrap (SIR) filter

The bootstrap filter (Gordon et al. 1993) is constructed using sampling importance resampling (SIR) with the importance function defined in (7.4.14). The algorithm is summarised as follows[1].

---

[1] Notice that in this implementation the number of particles is fixed at N, whereas the full scheme of (Gordon et al. 1993) allows for prior boosting, with a larger number of particles generated at the sample stage.

1. Set $t = 0$. Draw $x_0^{(i)} \sim p(x_0)$, $i = 1, ..., N$

2. For $t = 1$ to T
   - $x_t^{(i)} \sim q(x_t|x_{t-1}^{(i)}, y_t)$, $i = 1, ..., N$; see (7.4.14)
   - $w_t^{(i)} \propto p(y_t|z_{t-1:t-P}^{(i)}, a_t^{(i)}, \sigma_{e_t}^{2(i)}, \sigma_{v_t}^{2(i)})$, $i = 1, ..., N$,

   $\sum_{i=1}^{N} w_t^{(i)} = 1$, see (7.4.17)
   - Select with replacement from $\{x_t^{(i)}\}$ with probability $\{w_t^{(i)}\}$

3. End

7.4.3   Auxiliary particle filter (APF)

As described in section 7.2, the auxiliary particle filter (Pitt and Shephard: this volume) attempts to reduce the variability of the importance
weights by resampling the particles at time $t-1$ with probability close to
$p(x_{t-1}|y_{1:t})$, i.e. allowing one step of look-ahead to improve the particle
stock at time t. In a basic SIR implementation of the APF, $p(x_{t-1}|y_{1:t})$
is approximated by making a prediction $\hat{x}_t$ of x_t from each particle x_{t-1},
which is then substituted in as a single point mass approximation to the
transition density $f(x_t|x_{t-1})$. The first stage of the process selects x_{t-1}
with replacement according to the approximation of $p(x_{t-1}|y_{1:t})$. The sec-
ond stage proceeds as in the basic bootstrap filter, generating new states
x_t randomly from the state transition density $f(x_t|x_{t-1})$. The importance
weights are straightforwardly computed (Pitt and Shephard: this volume)
as $w_t \propto g(y_t|x_t)/g(y_t|\hat{x}_t)$. In our implementation, we generate the new
states as before from (7.4.14) in order to introduce greater robustness. The
weights are then modified to

$$w_t = \frac{p(y_t|z_{t-1:t-P}, a_t, \sigma_{e_t}^2, \sigma_{v_t}^2)}{p(y_t|z_{t-1:t-P}, \hat{a}_t, \hat{\sigma}_{e_t}^2, \hat{\sigma}_{v_t}^2)}, \qquad (7.4.18)$$

where $\hat{a}_t, \hat{\sigma}_{e_t}^2, \hat{\sigma}_{v_t}^2$ are deterministic predictions of $a_t, \sigma_{e_t}^2, \sigma_{v_t}^2$ made from
the previous states. We use here simply the mode of the state transition
densities, i.e. $\hat{a}_t = a_{t-1}$, $\hat{\sigma}_{e_t}^2 = \sigma_{e_{t-1}}^2$, $\hat{\sigma}_{v_t}^2 = \sigma_{v_{t-1}}^2$, which corresponds to our
expectation that AR coefficients and noise variances will change slowly with
time. Notice that for this importance function the generated values of z_t
once again do not appear in the weight equation because we have eliminated

a possible source of extra Monte Carlo variation. The APF implementation can be summarised below.[2]

---

1. Set $t = 0$. Draw $x_0^{(i)} \sim p(x_0)$, set $w_0^{(i)} = 1/N$, $i = 1, ..., N$

2. For $t = 1$ to T

   - $\widehat{a}_t^{(i)} = \alpha a_{t-1}^{(i)}$, $\widehat{\sigma}_{e_t}^{2(i)} = \sigma_{e_{t-1}}^{2(i)}$, $\widehat{\sigma}_{v_t}^{2(i)} = \sigma_{v_{t-1}}^{2(i)}$, $i = 1, ..., N$

   - $w_t^{(i)} \propto w_{t-1}^{(i)} p(y_t | z_{t-1:t-P}^{(i)}, \widehat{a}_t^{(i)}, \widehat{\sigma}_{e_t}^{2(i)}, \widehat{\sigma}_{v_t}^{2(i)})$, $\sum_{i=1}^{N} w_t^{(i)} = 1$, $i = 1, ..., N$

   - Select with replacement from $\{x_{t-1}^{(i)}\}$ with probability $\{w_t^{(i)}\}$

   - $x_t^{(i)} \sim q(x_t | x_{t-1}^{(i)}, y_t)$, $i = 1, ..., N$, see (7.4.14)

   - $w_t^{(i)} \propto \dfrac{p(y_t | z_{t-1:t-P}^{(i)}, a_t^{(i)}, \sigma_{e_t}^{2(i)}, \sigma_{v_t}^{2(i)})}{p(y_t | z_{t-1:t-P}^{(i)}, \widehat{a}_t^{(i)}, \widehat{\sigma}_{e_t}^{2(i)}, \widehat{\sigma}_{v_t}^{2(i)})}$, $\sum_{i=1}^{N} w_t^{(i)} = 1$, see (7.4.17)

3. End

We can see that there is a modest computational overhead in the weight computations compared with the standard bootstrap filter, but for some problems this may be offset by gains in estimation performance.

7.4.4   MCMC resampling

In order to implement any of the MCMC-based schemes discussed above, we need to design an effective Markov chain transition kernel with $p(x_t | x_{t-1}, y_t)$ as its invariant distribution. For the model chosen here it is straightforward to implement a Gibbs sampler that draws random samples from the full conditionals for z_t, a_t, $\sigma_{e_t}^2$ and $\sigma_{v_t}^2$ in turn. Similar samplers have been found to work well in the batch-based case (Godsill and Rayner 1998a)(Godsill and Rayner 1998b) (Godsill 1997) and can be expected to move very rapidly towards the target distribution when implemented for a single state x_t. We already have the full conditional for z_t in (7.4.12). The full conditional for a_t is also Gaussian:

$$p(a_t | z_t, \sigma_{e_t}^2, \sigma_{v_t}^2, x_{t-1}, y_t) = \mathcal{N}\left(\Phi_t^{-1} \Theta_t, \Phi_t^{-1}\right); \qquad (7.4.19)$$

---

[2]Note that Pitt and Shephard, in their original implementation, suggested an additional resampling stage at the end of each time step. We omit this here because it would increase the Monte Carlo variability of the filter.

$$\Phi_t = I_P/\sigma_a^2 + Z_{t-1}/\sigma_{e_t}^2, \tag{7.4.20}$$

$$\Theta_t = \alpha a_{t-1}/\sigma_a^2 + z_{t-1:t-P} z_{t-1}/\sigma_{e_t}^2, \tag{7.4.21}$$

$$Z_{t-1} = z_{t-1:t-P} z_{t-1:t-P}^T. \tag{7.4.22}$$

Hence it is straightforward to implement the Gibbs sampler for these two components using standard Gaussian sampling techniques. Notice, however, that sampling for a_t is computationally far more demanding than any of the other steps we have so far considered when the model order P is high. The full conditionals for $\sigma_{e_t}^2$ and $\sigma_{v_t}^2$ are non-Gaussian, owing to the log-transformation in the model for their evolution. MCMC can be implemented for these steps using, for example, Metropolis-Hastings moves. In our implementations we do not sample these parameters at the MCMC stage because they can be expected to move very slowly with time (the correct invariant distribution is of course maintained when these parameters are not moved by the MCMC sampler). When implementing MCMC steps with bridging densities, we will need to sample from a modified target density. We rewrite $p(x_t|x_{t-1}, y_t)$ in terms of the importance function and a weighting term:

$$p(x_t|x_{t-1}, y_t) \propto q(x_t|x_{t-1}, y_t) p(y_t|z_{t-1:t-P}, a_t, \sigma_{e_t}^2, \sigma_{v_t}^2),$$

and then modify it to achieve the intermediate bridging densities:

$$\pi_m(x_t|x_{t-1}, y_t) \propto q(x_t|x_{t-1}, y_t) p(y_t|z_{t-1:t-P}, a_t, \sigma_{e_t}^2, \sigma_{v_t}^2)^{1-\alpha_m}.$$

In this way we can achieve a gradual transition from the original importance function ($\alpha_m = 1$) and the final target density ($\alpha_m = 0$). The full conditionals for Gibbs sampling are also modified for these intermediate densities. The full conditional for z_t remains unchanged. The full conditional for a_t is modified as follows:

$$\pi_m(a_t|z_t, \sigma_{e_t}^2, \sigma_{v_t}^2, x_{t-1}, y_t) = \mathcal{N}\left((\Phi_t^m)^{-1}\Theta_t^m, (\Phi_t^m)^{-1}\right); \tag{7.4.23}$$

$$\Phi_t^m = I_P/\sigma_a^2 + \frac{Z_{t-1}}{\sigma_{e_t}^2}\left((1-\alpha_m) + \frac{\alpha_m}{\phi_t \sigma_{e_t}^2}\right), \tag{7.4.24}$$

$$\Theta_t^m = \alpha a_{t-1}/\sigma_a^2 + z_{t-1:t-P} z_{t-1}/\sigma_{e_t}^2 - \alpha_m y_t z_{t-1:t-P}/(\sigma_{e_t}^2 + \sigma_{v_t}^2), \tag{7.4.25}$$

$$Z_{t-1} = z_{t-1:t-P} z_{t-1:t-P}^T. \tag{7.4.26}$$

The required density ratio for the sampler with bridging densities is

$$\pi_m(x_t|x_{t-1})/\pi_{m-1}(x_t|x_{t-1}) = p(y_t|z_{t-1:t-P}, a_t, \sigma_{e_t}^2, \sigma_{v_t}^2)^{\alpha_{m-1}-\alpha_m} \tag{7.4.27}$$

The MCMC algorithm with bridging densities can thus be summarised as follows.

1. Set $t = 0$. Draw $x_0^{(i)} \sim p(x_0)$, $w_t^{(i)} = 1/N$, $i = 1, ..., N$

2. For $t = 1$ to T

   - $x_t^{(i)(0)} \sim q(x_t | x_{t-1}^{(i)}, y_t)$, $i = 1, ..., N$, see (7.4.14)
   - $w_t^{(i)(0)} = w_{t-1}^{(i)}$, $i = 1, ..., N$
   - For $m = 1$ to $M + 1$

     (Optional 1): Select with replacement from the particle set $\{x_t^{(i)(m-1)}, w_t^{(i)(m-1)}\}$, where

     $$w_t^{(i)(m)} \quad \propto \quad w_t^{(i)(m-1)} \frac{\pi_m(x_t^{(i)(m-1)} | x_{t-1}^{(i)})}{\pi_{m-1}(x_t^{(i)(m-1)} | x_{t-1}^{(i)})}, \quad i =$$

     $1, ..., N$, and $\sum_{i=1}^{N} w_t^{(i)(m)} = 1$, see (7.4.27)

     $$a_t^{(i)(m)} \sim \pi_m(a_t | z_t^{(i)(m-1)}, \sigma_{e_t}^{2(i)}, \sigma_{v_t}^{2(i)}, x_{t-1}^{(i)}, y_t), \quad i =$$
     $1, ..., N$, see (7.4.23)
     $$z_t^{(i)(m)} \sim p(z_t | x_{t-1}^{(i)}, y_t, a_t^{(i)(m)}, \sigma_{e_t}^{2(i)}, \sigma_{v_t}^{2(i)}), i = 1, ..., N,$$
     see (7.4.12)
     End
   - $w_t^{(i)} \propto w_t^{(i)(M+1)}$, $i = 1, ..., N$
   - (Optional 2) Select with replacement from $\{x_t^{(i)}\}$ with probability $\{w_t^{(i)}\}$

   End

This general algorithm includes both the techniques of (MacEachern et al. 1999) and (Gilks and Berzuini: this volume) as special cases. To achieve the (MacEachern et al. 1999) method, set $M = 0$ and do not carry out the first (Optional 1) selection step; the second (Optional 2) selection step can be carried out when the effective sample size or some other suitable measure of degeneracy becomes too low. To achieve the (Gilks and Berzuini: this volume) method set $M = 0$ and perform the first (Optional 1) selection step at every time t; do not perform the second (Optional 2) selection step. Notice that neither of these special cases incorporates the bridging density ideas, since with $M = 0$ there is no gradual transition from prediction density to target densities via intermediate bridging steps.

7.4.5   Simulation results

We have tested all the above algorithms extensively with the TVAR model. It is certainly possible to devise scenarios where none of the techniques work reliably, or where an impractically huge number of particles are required.

One example of such a scenario results when the evolution variance σ_a^2 is set particularly low relative to the observation noise variance $\sigma_{v_t}^2$. In such cases the state trajectories very frequently diverge altogether from the true values, making a large number of particles essential. In other milder conditions, all of the methods work quite well with small particle cloud size.

In the simulations we aim to study several questions. The first is whether auxiliary particle filters, which are competitive in terms of computational load with the bootstrap filter, are a worth-while modification for this class of models. Secondly, we explore the performance of the various MCMC-based procedures, aiming to discover whether these approaches, which can be considerably more demanding in computer power than the bootstrap filter, can give benefits in terms of robustness and accuracy of estimation. Within this class of MCMC algorithms we wish to explore the relative merits of the (MacEachern et al. 1999) approach, which does not involve a selection step before the MCMC moves, and the resample-move method of (Gilks and Berzuini: this volume), which always involves a selection step prior to the MCMC step. Finally, we wish to evaluate whether the use of bridging densities can significantly aid estimation for hostile environments. The simulations below go some way to answering these questions for the TVAR model class, although it is difficult to draw any general conclusions for other models and data conditions not tested here.

Experiments are carried out with multiple realisations of synthetic data. In this way we can go some way towards evaluating the expected performance of each of the methods in terms of mean-squared error. We estimate conditional means from the Monte Carlo output, focusing in particular upon the AR coefficients themselves, as follows:

$$E[a_t|y_{1:t}] \simeq \widehat{\mu}_t = \sum_{i=1}^{N} w_t^{(i)} a_t^{(i)}.$$

Since the conditional mean is the minimum mean-squared error estimator for a parameter, we can evaluate the performance of each algorithm by averaging the squared error between $\widehat{\mu}_t$ and the true parameter value over many realisations of the data with the same hyper-parameters.

In the first set of simulations we compare the performance of the auxiliary particle filter (APF) with a basic bootstrap (SIR) filter. The simulations are carried out over 100 independent data realisations for a fourth order model ($P = 4$) with fixed hyper-parameters. These parameters are given by $\alpha = 0.996$, $\sigma_a = 0.03$ and $\sigma_{\phi_e} = 0.01$. Notice that in these simulations the observation noise variance is fixed to $\sigma_{v_t}^2 = 1$ and not estimated – it was found that estimation of this parameter sometimes led to divergence of the filter when the number of particles was low (i.e. the filter loses track entirely of the correct parameter trajectory and only regains it by chance). The remaining states are initialised randomly (both in generating the data realisations and in the subsequent filtering for $t = 0$) from an initial prior distribution $p(x_0)$ which is Gaussian and independent over all elements of x_0. Elements of a_0 and $z_{0:-P+1}$ each have variance 4 and mean 0, while ϕ_{e_0} has variance 1 and mean 1. A typical realisation of the AR coefficients and observed signal is shown in figure 7.1. In figure 7.2 some results are shown

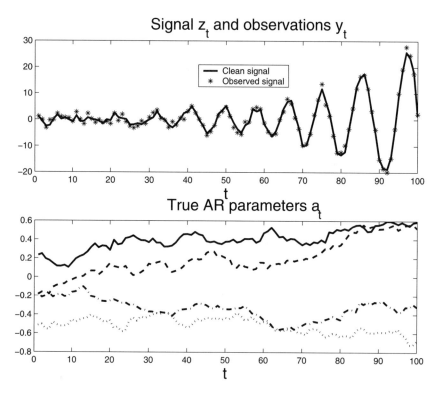

Figure 7.1. Typical signal and parameter trajectories.

from the APF with $N = 1000$. It can be seen that the filter is able to track the coefficient variation with time quite well, although the percentiles show a great deal of uncertainty about the precise values. A comparative result is now given for the SIR filter and the APF. Both filters are run on the same 100 realisations of the data, with 500 particles. The mean-squared error is computed at each time point for all 100 realisations and the result is plotted in figure 7.3. As expected, the estimation error is much larger at the start of the data, and decreases once the filter has seen enough data to be sure of the parameter trajectory. The APF shows a small improvement in mean-squared error at all time points. This small but systematic improvement has been observed in nearly all our trials, and so we conclude that the APF is a worth-while modification in this class of models. Furthermore, the APF has been found more robust to stability problems in

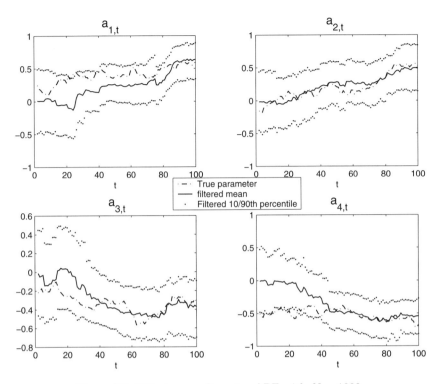

Figure 7.2. Filtering results using APF with $N = 1000$.

the signal estimates than the SIR, which can occasionally generate large errors when the AR model is unstable. Next we consider the use of MCMC sampling steps in the filter, both with and without bridging densities. It should be remembered that the extra computational requirements for these procedures are very considerable, so that any improvements would have to be significant to justify their use. In simulations with artificially generated data of the type discussed so far, we have not found a significant improvement in performance over the standard SIR or APF methods. However, the aim of the MCMC procedures is to robustify the filtering in challenging situations where the filtering density changes suddenly from one time point to the next. This can occur when outliers are present in the data, or step changes occur in the parameter trajectories. These situations very rarely arise when generating synthetic data from a known model, as above. However, they are much more likely to occur when working with real data, where un-modelled irregularities can occur in large datasets. It is in these situations that we can see the usefulness of the MCMC procedures. Quite a good illustration can be obtained by studying wrongly initialised filters. Here the initial state probability is chosen to be quite distant from the true initial state and with a low variance. Hence, the filter starts off along the

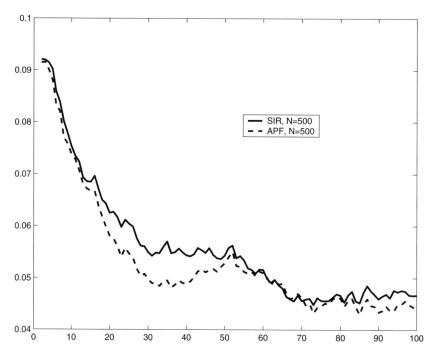

Figure 7.3. Comparison of MSE values between APF and SIR ($N = 500$).

wrong trajectory, and it is expected that the likelihood at each successive time point will be very far from the prediction density. In this simulation, the hyper-parameters are $\alpha = 0.996$, $\sigma_a = 0.01$ and $\sigma_{\phi_e} = 0.01$. The smaller value of σ_a means that the AR coefficients vary more slowly with time than before. The initial density for the coefficients has its mean offset by 0.1 from the initial density used to generate the data. Results of filtering for various schemes are shown in figure 7.4, plotting the first AR coefficient, $a_{1,t}$, which was typical of the performance for other parameters and other realisations of the data under the same conditions. The top two figures show that the APF is unable to track the parameter successfully, even when 10,000 particles are used. The bottom left graph shows the standard MCMC method without bridging densities. Six MCMC iterations are used at each time step. This method is able to track the parameter trajectories quite successfully after 30-40 time steps, although it never quite succeeds in locking on to the true trajectory. The bottom right shows the use of MCMC with bridging densities. The bridging schedule is fixed throughout, with $M = 5$ and $\{\alpha_1, ..., \alpha_5\} = \{0.9, 0.7, 0.5, 0.3, 0.1\}$. The computational requirement is thus almost identical to that of the standard MCMC approach. The state trajectories lock on to the true path slightly more rapidly than MCMC, and this is borne out by the overall mean-squared error figures of 0.0169 for the MCMC and 0.0085 for MCMC with bridging densities. In both cases, there was no selection step prior to carrying out the MCMC moves, i.e. the weights were accumulated right through to the end of the MCMC steps. Other simulations that included a selection step prior to the MCMC steps gave even worse mean-squared error figures for this example, although we are reluctant to draw general conclusions from such a small-scale trial.

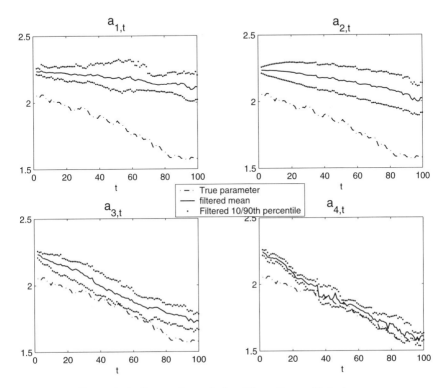

Figure 7.4. Comparison of algorithms for coefficient $a_{1,t}$.

7.5   Summary

In this chapter we have discussed various techniques for improving the basic particle filter, including a novel one employing bridging densities and MCMC moves. The application of these techniques has been illustrated by a detailed case study of the time-varying autoregressive model. From simulations we conclude that the auxiliary particle filter gives a slight but systematic improvement in performance. The case for MCMC methods is less clear-cut, given their significant addition to the computational burden. However, we have demonstrated challenging cases wherein the MCMC techniques, both with and without bridging densities, can lead to dramatically better performance than SIR or APF. In our experiments with bridging densities, we used a fixed schedule of annealing. However, there are unresolved issues here concerning how a schedule should be chosen, and whether it can be adapted at each time step to the demands of the situation; for example, the number of MCMC steps to be applied and whether to use different numbers of steps for each particle. More in-depth discussion and detail on these topics can be found in (Clapp 2000).

7.6   Acknowledgements

We wish to thank the referees for their constructive comments and suggestions, as well as those in the Signal Processing Group at Cambridge who have proof-read and commented on this work.

8

Approximating and Maximising the Likelihood for a General State-Space Model

Markus Hürzeler
Hans R. Künsch

8.1   Introduction

Assume that in a general state-space model the state transitions $p(\mathbf{x}_t|\mathbf{x}_{t-1})$ depend on a parameter τ, and the observations densities $p(\mathbf{y}_t|\mathbf{x}_t)$ on a parameter η. These can be for instance the noise variances in the state and the observation density, or some parameters related to the mean value of these densities. Both τ and η can be multi-dimensional. The combined parameter $(\tau', \eta')'$ will be denoted by θ. We discuss here the estimation of θ. For state-space models, the Bayesian viewpoint is often adopted and prior distributions for the unknown parameters are introduced. We discuss this approach briefly, but concentrate mainly on the frequentist approach where one has to compute and maximise the likelihood. Exact methods are usually not feasible, but the Monte Carlo methods allow us to approximate the likelihood function.

8.2   Bayesian methods

If one uses static MCMC methods to approximate the conditional distribution of the states given the data, then one can include updates on τ and η easily. Thus, one can make simultaneous inferences on the states and the parameters. The graph describing the dependence between all variables contains two additional nodes for τ and η, and – assuming that the priors for τ and η are independent – we have

$$p(\tau \mid \mathbf{x}_{0:T}, \mathbf{y}_{1:T}, \eta) = p(\tau \mid \mathbf{x}_{0:T}) \propto p(\tau)p(\mathbf{x}_0 \mid \tau) \prod_{t=1}^{T} p(\mathbf{x}_t \mid \mathbf{x}_{t-1}, \tau)$$

and

$$p(\eta \mid \mathbf{x}_{0:T}, \mathbf{y}_{1:T}, \tau) = p(\eta \mid \mathbf{x}_{0:T}, \mathbf{y}_{1:T}) \propto p(\eta) \prod_{t=1}^{T} p(\mathbf{y}_t \mid \mathbf{x}_t, \eta).$$

Exponential families for $p(\mathbf{x}_t \mid \mathbf{x}_{t-1}, \tau)$ and $p(\mathbf{y}_t \mid \mathbf{x}_t, \eta)$ with conjugate priors make the updating of τ and η particularly easy, but Metropolis updates can handle most other situations. This approach has been used in many published examples, e.g. Carlin et al. (1992) and Shephard (1994).

If one uses a recursive Monte Carlo filter as described in other chapters of this book, one can proceed by including the parameters among the states $\mathbf{x}_t$. The state transitions then become

$$\theta_t = \theta_{t-1}$$
$$\mathbf{x}_t \mid (\mathbf{x}_{t-1}, \theta_{t-1}) \sim p(\mathbf{x}_t \mid \mathbf{x}_{t-1}, \tau_{t-1}) d\mathbf{x}_t,$$

the initial distribution has the density

$$p(\mathbf{x}_0, \theta_0) = p(\tau)p(\eta)p(\mathbf{x}_0 \mid \tau),$$

and the observation densities for $\mathbf{y}_t$ given $(\mathbf{x}_t, \theta_t)$ are simply

$$p(\mathbf{y}_t \mid \mathbf{x}_t, \eta_t).$$

This has been discussed by Kitagawa (1998); see also Kitagawa (2001: this volume).

The main difficulty in this approach is that for all t the filter sample $(\theta_t^{(i)})$ is a sub-sample of the prior sample $(\theta_0^{(i)})$. But the prior sample will have only few values in the region where the posterior is concentrated so that we do not obtain precise information about the location, spread and shape of the posterior. It thus seems advantageous to add a small noise to the transitions for θ_t:

$$\theta_t = \theta_{t-1} + V_t,$$

presumably with a decreasing variance for V_t so as not to discount information from early observations too much. Considerations from recursive estimation suggest taking the variance of V_t of the order t^{-2} because by so doing the means of $(\theta_t^{(i)})$ and of $(\theta_{t-1}^{(i)})$ differ by a quantity of the order t^{-1}.

8.3 Pointwise Monte Carlo approximation of the likelihood

The easiest way to approximate the likelihood function at a fixed value of the parameter θ starts from

$$
\begin{aligned}
L(\theta \mid \mathbf{y}_{1:T}) &= p(\mathbf{y}_{1:T} \mid \theta) = \prod_{t=1}^{T} p(\mathbf{y}_t \mid \mathbf{y}_{1:t-1}, \theta) \\
&= \prod_{t=1}^{T} \int p(\mathbf{y}_t \mid \mathbf{x}_t, \eta) p(\mathbf{x}_t \mid \mathbf{y}_{1:t-1}, \theta) d\mathbf{x}_t \qquad (8.3.1) \\
&= \prod_{t=1}^{T} \mathbb{E}[p(\mathbf{y}_t \mid \mathbf{X}_t, \eta) \mid \mathbf{y}_{1:t-1}, \theta]. \qquad (8.3.2)
\end{aligned}
$$

In order to approximate this expectation, we need prediction samples $(\mathbf{x}_{\mathrm{pr},t}^{(j)})$ which are (approximately) distributed according to $p(\mathbf{x}_t \mid \mathbf{y}_{1:t-1}, \theta) d\mathbf{x}_t$. But these can be easily generated from filter samples $(\mathbf{x}_{t-1}^{(j)})$ which are (approximately) distributed according to $p(\mathbf{x}_{t-1} \mid \mathbf{y}_{1:t}, \theta) d\mathbf{x}_{t-1}$. We just take

$$
\mathbf{x}_{\mathrm{pr},t}^{(j)} \sim p(\mathbf{x}_t \mid \mathbf{x}_{t-1}^{(j)}, \theta) d\mathbf{x}_t.
$$

We thus can approximate

$$
\log L(\theta \mid \mathbf{y}_{1:T}) - T \log N
$$

by

$$
\sum_{t=1}^{T} \log \left(\sum_{j=1}^{N} p(\mathbf{y}_t \mid \mathbf{x}_{\mathrm{pr},t}^{(j)}, \theta) \right). \qquad (8.3.3)
$$

The disadvantage of this formula is that, if we need the likelihood at another value of θ, we have to generate new filter and prediction samples for this θ. Moreover, the likelihood function obtained in this way will be very noisy since the Monte Carlo errors for different values of θ will be independent. One therefore has to apply some smoothing before attempting to find the maximum.

8.3.1 Examples

AR(1) process with additive observation noise

We illustrate this pointwise Monte Carlo approximation method for the following AR(1) process with additive noise.

$$
\begin{aligned}
X_t &= \theta_0 \cdot X_{t-1} + V_t \\
Y_t &= X_t + W_t,
\end{aligned}
$$

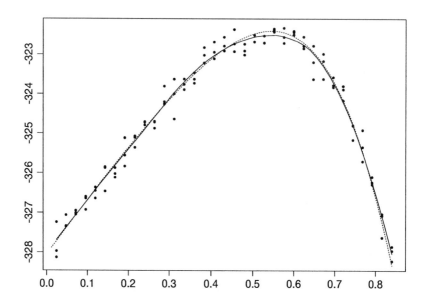

Figure 8.1. AR(1) process: Monte Carlo likelihood approximation for different θ-values of the AR(1) parameter. For comparison, a smoothed curve (solid line) and the true likelihood function (dotted) are drawn.

with $\theta_0 = 0.7$ and $V_t \sim \mathcal{N}(0, 0.7^2)$, $W_t \sim \mathcal{N}(0, 1)$ independent of each other. Notice that Y_t is an ARMA(1,1) process. We choose this example because we have the likelihood available in closed form.

In Figure 8.1 we have calculated the above approximation of the likelihood over the equally spaced grid from 0 to 0.85 with the sample size $N = 500$ for a fixed realisation $\mathbf{y}_{1:200}$ of the above process. The comparison of the fitted curve smoothed by the method "loess" (see Cleveland and Devlin (1988)) and the true likelihood function, which is known in this example, shows no essential difference.

In Figure 8.2 we see that by enlarging the sample size N the variance and the bias of the errors in this Monte Carlo likelihood tends to zero. Since the scale of y-axes is proportional to $\sqrt{N}$, we see that the variance of the errors tends to zero with order N. For each θ-value, we show the box plots of $n = 100$ independent replicates of the likelihood approximation. We used two different methods to generate the filter and prediction samples called direct and kernel method; see Hürzeler and Künsch (1998). We see that sampling according to the kernel filter algorithm introduces for $\theta = 0.6$ and $\theta = 0.8$ a small bias in the likelihood approximation.

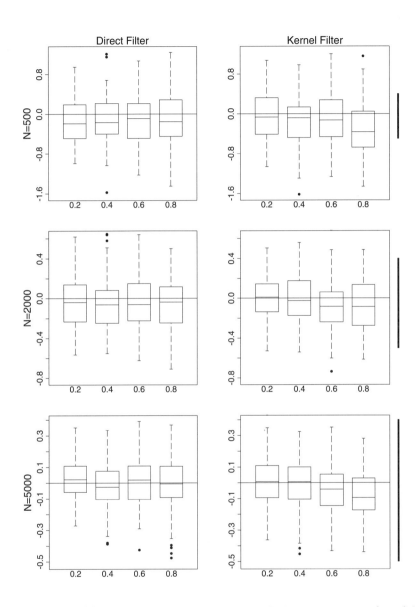

Figure 8.2. AR(1) process: The error of the Monte Carlo approximation (8.3.3) for log likelihood for different θ-values. Each box plot is based on 100 independent Monte Carlo samples. The bar at the right hand side allows comparison of the different scales in the y-axes. The bar length is proportional to $\sqrt{N}$.

Nonlinear model

As a second example we consider the nonlinear model

$$X_t = \alpha \cdot X_{t-1} + \beta \cdot \frac{X_{t-1}}{1 + X_{t-1}^2} + \gamma \cos(1.2t) + V_t \qquad (8.3.4)$$

$$Y_t = \frac{X_t^2}{20} + W_t, \qquad (8.3.5)$$

with $V_t \sim \mathcal{N}(0, \sigma_v^2)$ and $W_t \sim \mathcal{N}(0, \sigma_w^2)$ independent of each other. This model goes back to Andrade et al. (1978) and has been used as a kind of nonlinear benchmark model; see Kitagawa (1987). The data of length $T = 100$ were generated according to this model with $\alpha = 0.5$, $\beta = 25$, $\gamma = 8$, $\sigma_v^2 = 10$ and $\sigma_w^2 = 1$. We fix the three parameters α, β, γ of the state equation at their true value and take the two variances σ_v^2 of the state error and σ_w^2 of the observation error as unknown parameters. Figure 8.3 shows a smoothed polynomial surface. We used a 41×41 grid and at each grid point we calculated the approximate likelihood up to a constant with sample size $N = 500$. Again, all calculations were done for a fixed realisation $\mathbf{y}_{1:T}$. Here we also see that alternating maximisation for each parameter separately will be quite slow, since the main axes of the contour curves are not parallel to the parameter axes.

8.4   Approximation of the likelihood function based on filter samples

Using importance sampling, we can approximate the whole likelihood function based on a single set of prediction samples for a fixed parameter θ_0. Let us define filter and prediction importance weights

$$w_t(\mathbf{x}_t, \theta, \theta_0) = \frac{p_t(\mathbf{x}_t \mid \mathbf{y}_{1:t}, \theta)}{p_t(\mathbf{x}_t \mid \mathbf{y}_{1:t}, \theta_0)}$$

and

$$w_{\mathrm{pr},t}(\mathbf{x}_t, \theta, \theta_0) = \frac{p_t(\mathbf{x}_t \mid \mathbf{y}_{1:t-1}, \theta)}{p_t(\mathbf{x}_t \mid \mathbf{y}_{1:t-1}, \theta_0)}$$

respectively. Then we can write

$$p(\mathbf{y}_t \mid \mathbf{y}_{1:t-1}, \theta) = \mathbb{E}[p(\mathbf{y}_t \mid \mathbf{X}_t, \eta) \mid \mathbf{y}_{1:t-1}, \theta]$$
$$= \mathbb{E}[p(\mathbf{y}_t \mid \mathbf{X}_t, \eta) w_{\mathrm{pr},t}(\mathbf{X}_t, \theta, \theta_0) \mid \mathbf{y}_{1:t-1}, \theta_0]. \quad (8.4.6)$$

Multiplying these terms together gives the likelihood. Obviously, we can approximate the expectation on the right by sample means. However, the weights $w_{\mathrm{pr},t}$ are not available in closed form. The only thing we can do is

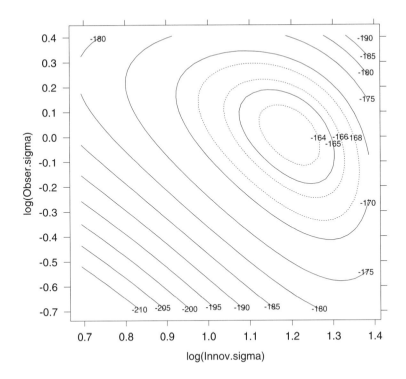

Figure 8.3. Nonlinear model: contour plot of the Monte Carlo likelihood approximation (up to a constant) is shown.

to write down recursions for the importance weights w_t and $w_{\mathrm{pr},t}$ as follows:

$$w_{\mathrm{pr},t}(\mathbf{x}_t, \theta, \theta_0) = \frac{\int p(\mathbf{x}_t \mid \mathbf{x}_{t-1}, \tau) p(\mathbf{x}_{t-1} \mid \mathbf{y}_{1:t-1}, \theta) d\mathbf{x}_{t-1}}{\int p(\mathbf{x}_t \mid \mathbf{x}_{t-1}, \tau_0) p(\mathbf{x}_{t-1} \mid \mathbf{y}_{1:t-1}, \theta_0) d\mathbf{x}_{t-1}}$$

$$= \frac{\int p(\mathbf{x}_t \mid \mathbf{x}_{t-1}, \tau) w_{t-1}(\mathbf{x}_{t-1}, \theta, \theta_0) p(\mathbf{x}_{t-1} \mid \mathbf{y}_{1:t-1}, \theta_0) d\mathbf{x}_{t-1}}{\int p(\mathbf{x}_t \mid \mathbf{x}_{t-1}, \tau_0) p(\mathbf{x}_{t-1} \mid \mathbf{y}_{1:t-1}, \theta_0) d\mathbf{x}_{t-1}},$$

$$(8.4.7)$$

and

$$w_t(\mathbf{x}_t, \theta, \theta_0) = \frac{p(\mathbf{y}_t \mid \mathbf{x}_t, \eta)}{p(\mathbf{y}_t \mid \mathbf{x}_t, \eta_0)} w_{\mathrm{pr},t}(\mathbf{x}_t, \theta, \theta_0) \frac{p(\mathbf{y}_t \mid \mathbf{y}_{1:t-1}, \theta_0)}{p(\mathbf{y}_t \mid \mathbf{y}_{1:t-1}, \theta)}, \qquad (8.4.8)$$

The integrals in these recursions are again approximated by sample averages over the filter and prediction samples. We start with

$$w_0(\mathbf{x}_0, \theta, \theta_0) = \frac{p(\mathbf{x}_0 \mid \tau)}{p(\mathbf{x}_0 \mid \tau_0)}.$$

and proceed recursively. Assuming that we already have $(w_{t-1}(\mathbf{x}_{t-1}^{(j)}, \theta, \theta_0))$, we compute first the weights $(w_{\mathrm{pr},t}(\mathbf{x}_{\mathrm{pr},t}^{(j)}, \theta, \theta_0))$ and $(w_{\mathrm{pr},t}(\mathbf{x}_t^{(j)}, \theta, \theta_0))$ with equation (8.4.7). In a second step we compute the densities $p(\mathbf{y}_t \mid \mathbf{y}_{t-1}, \theta)$ and $p(\mathbf{y}_t \mid \mathbf{y}_{t-1}, \theta_0)$ with equation (8.4.6). Finally, we compute the weights $(w_t(\mathbf{x}_t^{(j)}, \theta, \theta_0))$ with equation (8.4.8). This closes one circle of the recursion. Note that the algorithm is computationally intensive because in each recursion we have to sum over N terms for N different values. Thus, the number of operations is of the order $O(TN^2)$.

If the difference between θ and θ_0 becomes large, the approximations by arithmetic means become unreliable because a few weights dominate. The quality of the approximation can be measured by summation of the 10% largest weights $w_{t-1}(\mathbf{x}_{t-1}^{(j)}, \theta, \theta_0)$. If this sum exceeds 90%, then we generate new filter samples at the current θ, which becomes the new θ_0. In this way we have an iterative approximation/maximisation scheme which gives a sequence of estimates $(\hat{\theta}_i)$ approximating the maximum likelihood estimator.

8.5   Approximations based on smoother samples

Likelihood methods are easier as soon as we have available a smoother sample distributed according to the joint density of $\mathbf{X}_{0:T}$ given $\mathbf{y}_{1:T}$. We denote such a sample by $(\mathbf{x}_{\mathrm{sm},1:T}^{(j)})$. Unfortunately, such a smoother sample is more difficult to generate. Conditionally on $\mathbf{y}_{1:T}$, $\mathbf{X}_T, \mathbf{X}_{T-1}, \ldots$ is a Markov chain backward in time with transition densities

$$p(\mathbf{x}_t \mid \mathbf{x}_{t+1}, \mathbf{y}_{1:T}) = p(\mathbf{x}_t \mid \mathbf{x}_{t+1}, \mathbf{y}_{1:t}) \propto p(\mathbf{x}_t \mid \mathbf{y}_{1:t}) p(\mathbf{x}_{t+1} \mid \mathbf{x}_t)$$

(we take the parameter θ as fixed and suppress it from the notation). Thus, in order to generate a smoother sample, we start with the filter sample $\mathbf{x}_{\mathrm{sm},T}^{(j)} = \mathbf{x}_T^{(j)}$ at time T and then proceed backward in time by generating $\mathbf{x}_{\mathrm{sm},t}^{(j)}$ according to

$$\mathrm{const.} \cdot p(\mathbf{x}_t \mid \mathbf{y}_{1:t}) p(\mathbf{x}_{\mathrm{sm},t+1}^{(j)} \mid \mathbf{x}_t) d\mathbf{x}_t.$$

Because the filter density $p(\mathbf{x}_t \mid \mathbf{y}_{1:t})$ is not available in closed form, we have to use the filter samples $(\mathbf{x}_t^{(i)})$, which therefore need to be stored. It is usually an advantage not to replace $p(\mathbf{x}_t \mid \mathbf{y}_{1:t})$ directly by the empirical distribution concentrated on the sample $(\mathbf{x}_t^{(i)})$, but to go one step back and use

$$p(\mathbf{x}_t \mid \mathbf{y}_{1:t}) \approx \mathrm{const.} \cdot p(\mathbf{y}_t \mid \mathbf{x}_t) \sum_{i=1}^{N} p(\mathbf{x}_t \mid \mathbf{x}_{t-1}^{(i)});$$

see Hürzeler and Künsch (1998).

Once a new observation y_{T+1} becomes available, we have to regenerate in principle the whole smoother sample. However, in practice, only a small number of state variables at the end are affected by the information contained in the new observation so that it is usually sufficient to generate $(\mathbf{x}^{(j)}_{sm,T-k:T+1})$ for a fixed value k.

8.5.1   Approximation of the likelihood function

It is possible to approximate the difference $L(\theta)-L(\theta_0)$ based on a smoother sample distributed according to the joint density of $\mathbf{x}_{0:T}$ given $\mathbf{y}_{1:T}$ for a fixed value of θ_0. By definition

$$\log L(\theta) = \log(p(\mathbf{y}_{1:T} \mid \theta)) = \log\left(\int p(\mathbf{x}_{0:T}, \mathbf{y}_{1:T} \mid \theta) \prod_{t=0}^{T} d\mathbf{x}_t \right),$$

which implies

$$\log L(\theta) - \log L(\theta_0) = \log\left(\frac{p(\mathbf{y}_{1:T} \mid \theta)}{p(\mathbf{y}_{1:T} \mid \theta_0)} \right)$$

$$= \log\left(\mathbb{E}\left[\frac{p(\mathbf{X}_{0:T}, \mathbf{y}_{1:T} \mid \theta)}{p(\mathbf{X}_{0:T}, \mathbf{y}_{1:T} \mid \theta_0)} \mid \mathbf{y}_{1:T}, \theta_0 \right] \right). \quad (8.5.9)$$

The joint likelihood can be written easily and the expectation can be approximated by a sample mean. This leads to

$$\log L(\theta) - \log L(\theta_0) \approx \log\left(\frac{1}{N} \sum_{j=1}^{N} \frac{p(\mathbf{x}^{(j)}_{sm,0} \mid \tau)}{p(\mathbf{x}^{(j)}_{sm,0} \mid \tau_0)} \right.$$

$$\left. \prod_{t=1}^{T} \frac{p(\mathbf{x}^{(j)}_{sm,t} \mid \mathbf{x}^{(j)}_{sm,t-1}, \tau) p(\mathbf{y}_t \mid \mathbf{x}^{(j)}_{sm,t}, \eta)}{p(\mathbf{x}^{(j)}_{sm,t} \mid \mathbf{x}^{(j)}_{sm,t-1}, \tau_0) p(\mathbf{y}_t \mid \mathbf{x}^{(j)}_{sm,t}, \eta_0)} \right).$$

Evaluation of this sum needs operations of the order $O(TN)$, but as noted above, generating the smoother sample requires more time than generating the filter sample.

In practice, one has to again use an iterative approximation/ maximisation scheme as described for the filter samples above.

8.5.2   Stochastic EM-algorithm

A final possibility to find the maximum likelihood estimator is to use a stochastic version of the EM-algorithm (MCEM). The E-step of the EM-algorithm consists of computing

$$
\begin{aligned}
Q(\theta \mid \theta_0) &= \mathbb{E}[\log p(\mathbf{X}_{0:T}, \mathbf{y}_{1:T} \mid \theta) \mid \mathbf{y}_{1:T}, \theta_0] \\
&= \mathbb{E}[\log p(\mathbf{X}_0 \mid \tau) \mid \mathbf{y}_{1:T}, \theta_0] \\
&+ \sum_{t=1}^{T} \mathbb{E}[\log p(\mathbf{X}_t \mid \mathbf{X}_{t-1}, \tau) \mid \mathbf{y}_{1:T}, \theta_0] \\
&+ \sum_{t=1}^{T} \mathbb{E}[\log p(\mathbf{y}_t \mid \mathbf{X}_t, \eta) \mid \mathbf{y}_{1:T}, \theta_0],
\end{aligned}
$$

and we can approximate the expectations by arithmetic means with the smoother samples. The M-step consists in maximising $Q(\theta|\theta_0)$ with respect to θ in order to obtain the next θ_0; see Hürzeler (1998).

Stochastic EM algorithms have been discussed by many people, e.g. by Wei and Tanner (1990), Chan and Ledolter (1995).

The difficulty in implementing all these iterative approximation/ maximisation methods is in the allocation of computing resources. In the beginning, the smoother or filter sample can be rather small because no precise approximation of the likelihood or of $Q(\theta|\theta_0)$ is needed to obtain a reasonable better θ_0. Later on in the iteration, this changes, and from some point on the iterates start to fluctuate randomly unless the Monte Carlo sample size is increased.

Discarding the simulations at previous values of θ_0 seems wasteful. Methods have been proposed which utilise all simulations, see Geyer (1993) and Delyon et al. (1999).

8.6   Comparison of the methods

8.6.1   AR(1) process

We use again the AR(1) process with additive observation noise for illustration

$$
\begin{aligned}
X_t &= \theta \cdot X_{t-1} + V_t \\
Y_t &= X_t + W_t
\end{aligned}
$$

with $\theta_0 = 0.7$ and $V_t \sim \mathcal{N}(0, 0.7^2)$, $W_t \sim \mathcal{N}(0,1)$ independent of each other. We estimate θ for a fixed realisation of observations $\mathbf{y}_{1:200}$. The maximum likelihood estimation by the exact Kalman recursion is $\hat{\theta} = 0.548$. Remember that with state-space models non-stationary AR processes can also be treated. Thus, we do not require $|\theta| < 1$.

In Figure 8.4 we see that the filter approximation is quite near to the true likelihood function over a wide range of θ's, even for a small sample size N. On the other hand, the approximation based on smoothing samples is really only a local one.

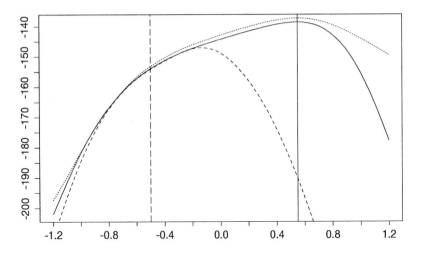

Figure 8.4. AR(1) process with additive noise: the true likelihood function $\log L(\theta)$ (solid line), the local filter approximation based on (8.4.6)-(8.4.8) (dotted line) and the local smoother approximation based on (8.5.9) (broken line) are drawn. The true MLE is $\widehat{\theta}_{ML} = 0.55$ (solid vertical line) and the samples are produced for $\theta_0 = -0.5$ (broken vertical line).

In Figures 8.5 and 8.6 we see that with a bad starting value, e.g., $\theta_0 = -0.5$, the smoothing approximation needs more iterations than the filter approximation until $\widehat{\theta}_k$ fluctuates around the maximum likelihood estimator. But for both starting values, convergence is rather fast.

We now compare the three estimation methods – *FILT* from Section 8.4, *SMOO* from Section 8.5.1 and *MCEM* from the Section 8.5.2 in more detail. For the setup we used an iterative approximation/ maximisation scheme $(\widehat{\theta}_i; i = 0, \ldots k)$ as described above. For the method *FILT* we used $k = 6$ iterations, for the other two methods we used $k = 21$. For all methods, the starting value of the iterations was $\widehat{\theta}_0 = 0.7$ and the final estimate was the median of the last 4 iterations . For the methods *SMOO* and *MCEM* we lose some accuracy since we only use the last four estimates of the sequences. But this is done to properly compare the results with the method *FILT*.

In Figure 8.7 the box plots of 25 repetitions of the above methods are shown. The samples were calculated with the direct algorithm and with the kernel method (see Hürzeler and Künsch (1998)). Comparing the direct and the kernel algorithm we see that the estimation methods based on kernel samples have a larger variation and bias. As expected, enlarging the sample size N reduces the variation of the estimates and the bias of method *FILT*.

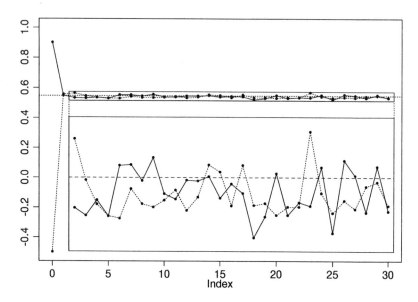

Figure 8.5. AR(1) process with additive noise: MLE sequences $(\widehat{\theta}_i)$ with starting value $\theta_0 = 0.9$ and -0.5 respectively for the local filter approximation method are shown ($N = 100$). The box has a zooming factor of 15.

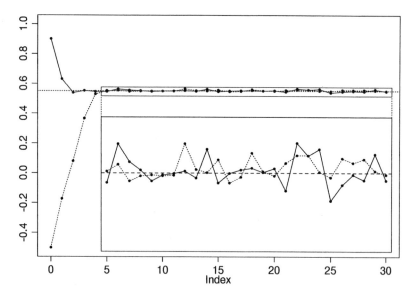

Figure 8.6. AR(1) process with additive noise: MLE sequences (θ_i) with starting value $\theta_0 = 0.9$ and -0.5 respectively for the local smoothing approximation method are shown ($N = 500$). The box has a zooming factor of 15.

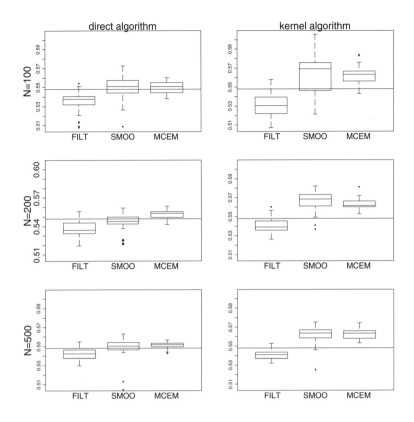

Figure 8.7. AR(1) process: box plots of the three estimation methods (*FILT*, *SMOO* and *MCEM*) for different sample sizes N. As a sampling method, both the direct and kernel algorithms are used.

8.6.2   Nonlinear example, 3 parameters

We consider again the nonlinear state-space model (8.3.4) – (8.3.5). Now we take both variances as known and estimate $\theta = (\alpha, \beta, \gamma)$. We have used the direct method with sample size $N = 200$ and as starting value $\theta = (0.9, 18, 10)$, which is quite distant from the maximum likelihood estimator. We see in Figure 8.8 that *MCEM* needs several iteration steps before it gets close to the maximum-likelihood estimator. The method *SMOO*, which is based on smoothing samples, needs only two iteration steps. The method *FILT* converges quickly as well but one notices that the variation is quite large.

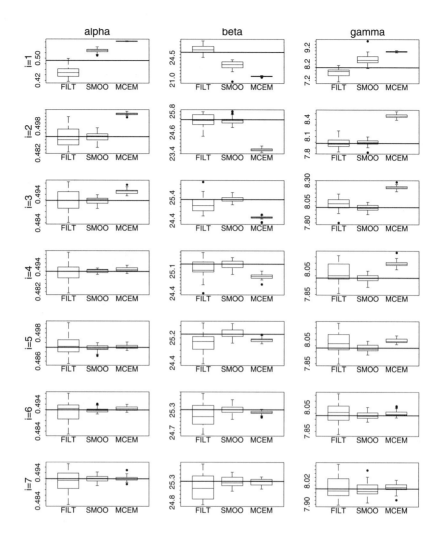

Figure 8.8. Nonlinear model, 3 parameters: Each boxplot shows 25 repetitions of the sequences $\widehat{\theta}_i$, $i = 1, 2, \ldots, 7$ (from top to bottom). From left to right we have the three parameters α, β and γ.

Computation times

In Section 8.4 we mentioned that the filter likelihood approximation is most computationally intensive. For our test series of the nonlinear model, we compared the run time needed to generate the required samples and do one maximisation step for the three methods, filter likelihood, smoothing likelihood and MCEM. In Table 8.1 the corresponding computing times are given in seconds. The calculations were done on a Sun Ultra-2 with 256MB

RAM and a SparcUltra processor (200MHz). The approximate order of the calculation time in dependence of N and T is confirmed.

	Filter	Smoother	MCEM
$T = 50, N = 50$	33	1.8	1.1
$T = 100, N = 100$	236	9.4	7.6
$T = 100, N = 200$	814	19.1	14.2
approx. order	$O(T \cdot N \cdot N)$	$O(T \cdot N)$	$O(T \cdot N)$

Table 8.1. Nonlinear example: computing times (in seconds) for the different parameter estimation methods, and their approximate dependencies on the time series length T and the sample size N are given.

8.6.3 Nonlinear model, 5 parameters

To show the performance of the parameter estimation methods in a higher dimensional case, we apply the MCEM algorithm to the same nonlinear model (8.3.4) – (8.3.5) with all five parameters $\theta = (\alpha, \beta, \gamma, \sigma_v^2, \sigma_w^2)$ unknown. As starting value we used $\theta_0 = (0.9, 18, 10, 10, 1)$. Figure 8.9 shows five of ten MCEM sequences for the hyper-parameters. We applied the direct method with sample size $N = 1000$. The approximative MLE $\widehat{\theta} = (0.49, 25.53, 7.92, 8.6, 1.03)$ is the median of the last eleven elements of all ten sequences, indicated in the figures by a horizontal line. Remember that the calculations were carried out for one fixed realisation $\mathbf{y}_{1:T}$ of the model. Here a grid search would be quite difficult because n nodes for each component of the parameter lead to a grid of size n^5.

8.6.4 Discussion

If we have a starting value θ_0 far away from the true parameter, then it is practically impossible to generate a smoothing sample; the data do not fit the wrong model. So we recommend in the first steps to use the filter likelihood approximation with a small sample size N. This only requires filter samples and leads to a good starting value for the smoothing or MCEM parameter estimation methods. Then we can also enlarge the sample N size without run time problems.

8.7 Recursive estimation

Recursive estimation of the parameters in a hidden Markov model has been discussed by Holst and Lindgren (1991) and by Rydén (1997). A recursive estimator has the form

$$\widehat{\theta}_{t+1} = \widehat{\theta}_t + \gamma_t h_{t+1}(\mathbf{y}_{1:t+1}, \widehat{\theta}_t).$$

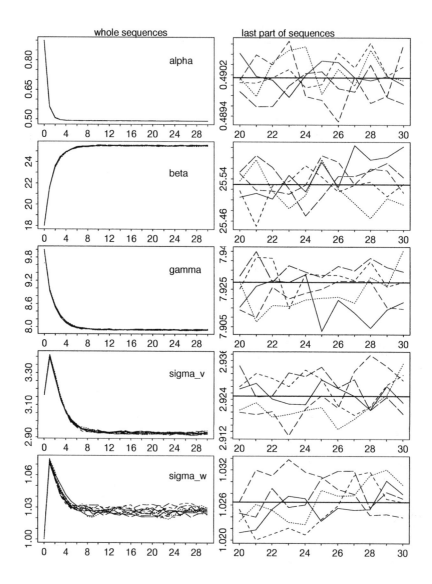

Figure 8.9. Nonlinear model comprising five parameters: five of ten MCEM sequences $\widehat{\theta}_i$ are shown. The samples of size $N = 1000$ were obtained by the direct filter and smoothing algorithm. The baseline in the right figures is equal to an approximate maximum likelihood estimate calculated from the last part of all ten sequences.

Here γ_t is typically of the form $(t+1)^{-1}H_t$ with H_t a matrix converging to a constant, and h_t is a so-called score function. It is crucial to choose this score function such that it is easily computable and leads to a consistent estimator. Holst and Lindgren (1991) essentially propose

$$h_{t+1} = \mathbb{E}\left[\frac{\partial}{\partial\theta}(\log p(\mathbf{X}_{t+1} \mid \mathbf{X}_t, \tau) + \log p(\mathbf{y}_{t+1} \mid \mathbf{X}_{t+1}, \eta) \mid \mathbf{y}_{1:t+1}, \theta\right].$$

Since it is in general not possible to compute this conditional expectation, it must be approximated by a sample mean. The drawback is that as a new observation arrives, one has to redo the recursive filtering from time zero because one now needs the filter samples with the parameter $\widehat{\theta}_t$, not with $\widehat{\theta}_{t-1}$. One is of course tempted to ignore this and use the filter samples under $\widehat{\theta}_{t-1}$ and do just one update with $\widehat{\theta}_t$. Unfortunately, we have neither theoretical nor empirical justification for such an approach.

Rydén (1997) proposes essentially

$$h_{t+1} = \frac{\partial}{\partial\theta}\log p(\mathbf{y}_{t-m+2:t+1} \mid \theta)$$

for a fixed m. Again, one will need Monte Carlo methods to evaluate the likelihood of $\mathbf{y}_{t-m+2:t+1}$, and one encounters the same problem as in the Holst and Lindgren (1991) proposal.

9

Monte Carlo Smoothing and Self-Organising State-Space Model

Genshiro Kitagawa
Seisho Sato

9.1 Introduction

A Monte Carlo method for nonlinear non-Gaussian filtering and smoothing and its application to self-organising state-space models are shown in this paper.

The linear-Gaussian state-space model, the Kalman filter and fixed-interval smoother are useful tools for time series analysis; they have applications in various fields of science, engineering and economics, and so on (Kalman 1960, Sage and Melsa 1971, Anderson and Moore 1979). However, with the expansion of applications, the necessity of nonlinear or non-Gaussian state-space modelling became apparent. Various extensions of Kalman filtering and approximations to non-Gaussian filters have been developed (Sage and Melsa 1971, Fahrmeir and Kaufmann 1991, Schnatter 1992). West, Harrison and Migon (1986) introduced a generalised dynamic linear model. Kitagawa (1987) directly generalised the state-space model to the case wherein either the system noise or the observation noise is non-Gaussian. According to these methods, recursive formulas for filtering and smoothing were derived and implemented using numerical computations. Many non-standard situations in time series analysis, such as abrupt model parameter changes, time series with outliers and with skewed distributions, can be handled well with this modelling and computational technique. On the other hand, a difficulty with this numerical method is that its intensive use of a computer, both in terms of memory and computational time; it is therefore difficult to apply to models with high dimensional states.

In this paper, we show a Monte Carlo method for state-space filtering and smoothing (Gordon et al. 1993, Kitagawa 1996). The algorithm is based on the approximation of successive prediction and filtering density

functions by many particles that can be considered as independent realisations from those distributions. The virtue of this algorithm is that it can
be applied to the very large class of nonlinear non-Gaussian higher dimensional state-space models, provided the dimensions of the system noise and
the observation noise are not too high. We show some numerical examples
to illustrate the method, as well as several associated problems.

To remedy the difficulty associated with the use of the Monte Carlo filter
and the non-Gaussian filter in parameter estimation, a self-organising state-
space model is also shown (Kitagawa 1998). In this approach, the state
vector is augmented by the unknown parameters of the model, and the
state and parameters are estimated simultaneously by the recursive filter
and smoother. This type of recursive estimation has been tried previously,
as reflected by the engineering literature (e.g. (Ljung 1979, Solo 1980)).
Such studies were based on the extended Kalman filter. It is very likely
that the main reason for the difficulty in recursive parameter estimation
was the lack of a practical nonlinear non-Gaussian smoothing algorithm.
In the method shown in this paper, more accurate approximations to the
marginal posterior densities of both state and parameters are obtained by
the Monte Carlo smoother.

9.2   General state-space model and state estimation

9.2.1   The model and the state estimation problem

Consider a nonlinear non-Gaussian state-space model for the time series
y_t,

$$\mathbf{x}_t = F_t(\mathbf{x}_{t-1}, \mathbf{v}_t) \tag{9.2.1}$$
$$\mathbf{y}_t = H_t(\mathbf{x}_t, \varepsilon_t), \tag{9.2.2}$$

where $\mathbf{x}_t$ is an unknown state vector, $\mathbf{v}_t$ and ε_t are the system noise and
the observation noise possibly with non-Gaussian densities $q_t(\mathbf{v})$ and $r_t(\varepsilon)$,
respectively. (9.2.1) and (9.2.2) are called the system model (or state model)
and the observation model. The initial state $\mathbf{x}_0$ is assumed to be distributed
according to the density $p_0(\mathbf{x})$. $F_t(\mathbf{x}, \mathbf{v})$ and $H_t(\mathbf{x}, \varepsilon)$ are possibly nonlinear
functions of the state and the noise inputs.

This model is an extension of the well-known state-space model

$$\mathbf{x}_t = F_t\mathbf{x}_{t-1} + G_t\mathbf{v}_t \tag{9.2.3}$$
$$\mathbf{y}_t = H_t\mathbf{x}_t + \varepsilon_t, \tag{9.2.4}$$

with Gaussian white noise sequences, $\mathbf{v}_t \sim N(0, V_t)$ and $\varepsilon_t \sim N(0, W_t)$.

The above nonlinear non-Gaussian state-space model specifies the con-
ditional density of the state given the previous state, $Q_t(\mathbf{x}_t|\mathbf{x}_{t-1})$, and that
of the observation given the current state, $R_t(\mathbf{y}_t|\mathbf{x}_t)$. This is the essential

feature of the state-space model, and it is sometimes convenient to express the model in the following general state-space model form based on these two conditional distributions:

$$\mathbf{x}_t \quad \sim \quad Q_t(\cdot \, |\mathbf{x}_{t-1}) \tag{9.2.5}$$

$$\mathbf{y}_t \quad \sim \quad R_t(\cdot \, |\mathbf{x}_t). \tag{9.2.6}$$

With this general state-space model, it is possible to handle discrete valued time series as well as discrete state models.

9.2.2   Non-Gaussian filter and smoother

One of the most important problems to overcome in state-space modelling is the estimation of the state vector $\mathbf{x}_t$ from the observations, and many problems in time series analysis, such as prediction, decomposition and parameter estimation can be solved by using this estimated state vector. The problem of state estimation can be formulated as the evaluation of the conditional density $p(\mathbf{x}_t|\mathbf{y}_{1:s})$, where $\mathbf{y}_{1:s} \triangleq \{\mathbf{y}_1, \dots, \mathbf{y}_s\}$. Corresponding to the three distinct cases, $t > s$, $t = s$ and $t < s$, the conditional distribution, $p(\mathbf{x}_t|\mathbf{y}_{1:s})$, is called the predictor, the filter and the smoother, respectively.

For the standard linear-Gaussian state-space model, each density can be expressed by a Gaussian density, and its mean vector and variance-covariance matrix can be obtained by the computationally efficient Kalman filter and smoothing algorithms (Kalman 1960, Anderson and Moore 1979, Sage and Melsa 1971).

For general state-space models, however, the conditional distributions become non-Gaussian and their distributions cannot be completely specified by the mean vectors and the variance-covariance matrices. Therefore, various types of approximations to, or assumptions about, the distributions have been used to obtain recursive formulas for state estimation, e.g. the extended Kalman filter (Anderson and Moore 1979), the Gaussian-sum filter (Alspach and Sorenson 1972) and the dynamic generalised linear model (West, Harrison and Migon 1985).

However, at least in principle, the following non-Gaussian filter and smoother (Kitagawa 1987) can yield an arbitrarily precise posterior density.

Non-Gaussian Filter

$$p(\mathbf{x}_t|\mathbf{y}_{1:t-1}) \quad = \quad \int p(\mathbf{x}_t|\mathbf{x}_{t-1})p(\mathbf{x}_{t-1}|\mathbf{y}_{1:t-1})d\mathbf{x}_{t-1}$$

$$p(\mathbf{x}_t|\mathbf{y}_{1:t}) \quad = \quad \frac{p(\mathbf{y}_t|\mathbf{x}_t)p(\mathbf{x}_t|\mathbf{y}_{1:t-1})}{p(\mathbf{y}_t|\mathbf{y}_{1:t-1})}, \tag{9.2.7}$$

where $p(\mathbf{y}_t|\mathbf{y}_{1:t-1})$ is the predictive distribution of $\mathbf{y}_t$ and is defined by

$$\int p(\mathbf{y}_t|\mathbf{x}_t)p(\mathbf{x}_t|\mathbf{y}_{1:t-1})d\mathbf{x}_t.$$

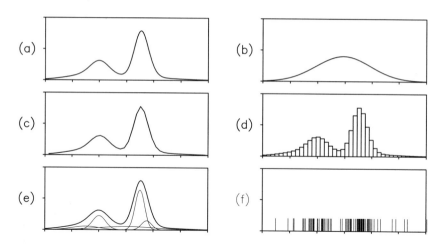

Figure 9.1. Approximations of non-Gaussian distribution. (a) True, (b) Gaussian approximation, (c) piecewise linear function, (d) step function, (e) Gaussian-mixture and (f) Monte Carlo.

Non-Gaussian Smoother

$$p(\mathbf{x}_t|\mathbf{y}_{1:T}) = p(\mathbf{x}_t|\mathbf{y}_{1:t}) \int \frac{p(\mathbf{x}_{t+1}|\mathbf{x}_t)p(\mathbf{x}_{t+1}|\mathbf{y}_{1:T})}{p(\mathbf{x}_{t+1}|\mathbf{y}_{1:t})} d\mathbf{x}_{t+1}. \qquad (9.2.8)$$

Various applications of this non-Gaussian filter and smoother are shown in (Kitagawa 1987, Kitagawa and Gersch 1996). However, the direct implementation of the formula requires computationally intense numerical integration that can be applied only to discrete-state models or lower-dimensional, say less than four-dimensional, continuous-state models.

9.3   Monte Carlo filter and smoother

9.3.1   Approximation of non-Gaussian distributions

The nonlinear or non-Gaussian filters/smoothers mentioned above are based on approximation by non-Gaussian state densities. Figure 9.1 (a) shows an example of a non-Gaussian state density. The extended Kalman filter yields only the state mean and its variance-covariance matrix, and hence it can be considered an approximation of the three state densities by a single Gaussian density (Figure 9.1 (b)). In the non-Gaussian filter and smoother, approximation is by a step function or a piecewise linear function, as shown in Figure 9.1 (c) and (d). In Gaussian-sum filter, approximation is by a mixture of several Gaussian densities (Figure 9.1 (e)).

In contrast to these approximations, in the Monte Carlo filtering (Gordon et al. 1993, Kitagawa 1996), we approximate each density function by many (say 10,000) particles, which can be considered to be realisations from that distribution. Specifically, each state distribution is expressed by using N particles as follows:

$$
\begin{aligned}
\{\mathbf{p}_t^{(1)}, \dots, \mathbf{p}_t^{(N)}\} &\sim p(\mathbf{x}_t|\mathbf{y}_{1:t-1}) &\text{Predictor} \\
\{\mathbf{f}_t^{(1)}, \dots, \mathbf{f}_t^{(N)}\} &\sim p(\mathbf{x}_t|\mathbf{y}_{1:t}) &\text{Filter} \\
\{\mathbf{s}_{t|T}^{(1)}, \dots, \mathbf{s}_{t|T}^{(N)}\} &\sim p(\mathbf{x}_t|\mathbf{y}_{1:T}) &\text{Smoother}
\end{aligned}
$$

We can approximate $p(\mathbf{x}_t|\mathbf{y}_{1:t-1})$ by the probability function

$$
\Pr(\mathbf{x}_t = \mathbf{p}_t^{(j)}|\mathbf{y}_{1:t-1}) = \frac{1}{N}, \qquad \text{for } j = 1, \cdots, N. \tag{9.3.9}
$$

This is equivalent to approximating the true distribution function by an empirical distribution function

$$
\frac{1}{N} \sum_{i=1}^{N} I(\mathbf{x}, \mathbf{p}_t^{(i)}), \tag{9.3.10}
$$

where $I(\mathbf{x}, \mathbf{a})$ is the indicator function defined by $I(\mathbf{x}, \mathbf{a}) = 1$ if $\mathbf{x} \geq \mathbf{a}$ and $I(\mathbf{x}, \mathbf{a}) = 0$ otherwise.

9.3.2 Monte Carlo filtering

Based on the Monte Carlo approximation, it can be shown that a set of realisations expressing the one step ahead predictor $p(\mathbf{x}_t|\mathbf{y}_{1:t-1})$ and the filter $p(\mathbf{x}_t|\mathbf{y}_{1:t})$ can be obtained recursively as follows (Kitagawa 1996).

Monte Carlo Filter

1. *Generate a random number* $\mathbf{f}_0^{(j)} \sim p_0(\mathbf{x})$ *for* $j = 1, \dots, N$.

2. *Repeat the following steps for* $t = 1, \dots, T$.

   2-1 *Generate a random number* $\mathbf{v}_t^{(j)} \sim q(\mathbf{v})$, *for* $j = 1, \dots, N$.
   2-2 *Compute* $\mathbf{p}_t^{(j)} = F(\mathbf{f}_{t-1}^{(j)}, \mathbf{v}_t^{(j)})$, *for* $j = 1, \dots, N$.
   2-3 *Compute* $w_t^{(j)} = p(\mathbf{y}_t|\mathbf{p}_t^{(j)})$ *for* $j = 1, \dots, N$.
   2-4 *Generate* $\mathbf{f}_t^{(j)}$, $j = 1, \dots, N$ *by the resampling of* $\mathbf{p}_t^{(1)}, \dots, \mathbf{p}_t^{(N)}$.

9.3.3 Derivation of the Monte Carlo filter

In the following two subsections, we briefly show separate derivations of the Monte Carlo based algorithm for recursive evaluation of the predictor and the filter.

One step ahead prediction

Let $\{\mathbf{f}_{t-1}^{(1)}, \ldots, \mathbf{f}_{t-1}^{(N)}\}$ and $\{\mathbf{v}_t^{(1)}, \cdots, \mathbf{v}_t^{(N)}\}$ be independent realisations of $p(\mathbf{x}_{t-1}|\mathbf{y}_{1:t-1})$ and the system noise $\mathbf{v}_t$, respectively. Specifically, for $j = 1, \ldots, N$

$$\mathbf{f}_{t-1}^{(j)} \sim p(\mathbf{x}_{t-1}|\mathbf{y}_{1:t-1}), \quad \mathbf{v}_t^{(j)} \sim q(\mathbf{v}). \tag{9.3.11}$$

Then the predictive distribution $p(\mathbf{x}_t|\mathbf{y}_{1:t-1})$ can be expressed by

$$
\begin{aligned}
p(\mathbf{x}_t|\mathbf{y}_{1:t-1}) &= \iint p(\mathbf{x}_t, \mathbf{x}_{t-1}, \mathbf{v}_t|\mathbf{y}_{1:t-1}) d\mathbf{v}_t d\mathbf{x}_{t-1} \\
&= \iint p(\mathbf{x}_t|\mathbf{x}_{t-1}, \mathbf{v}_t) p(\mathbf{v}_t) p(\mathbf{x}_{t-1}|\mathbf{y}_{1:t-1}) d\mathbf{v}_t d\mathbf{x}_{t-1} \qquad (9.3.12) \\
&= \iint \delta(\mathbf{x}_t - F(\mathbf{x}_{t-1}, \mathbf{v}_t)) p(\mathbf{v}_t) p(\mathbf{x}_{t-1}|\mathbf{y}_{1:t-1}) d\mathbf{v}_t d\mathbf{x}_{t-1},
\end{aligned}
$$

where $\delta(x)$ denotes the delta function. Therefore, if we define $\mathbf{p}_t^{(j)}$ by

$$\mathbf{p}_t^{(j)} = F(\mathbf{f}_{t-1}^{(j)}, \mathbf{v}_t^{(j)}), \tag{9.3.13}$$

$\{\mathbf{p}_t^{(1)}, \cdots, \mathbf{p}_t^{(N)}\}$ can be considered independent realisations from the distribution $p(\mathbf{x}_t|\mathbf{y}_{1:t-1})$.

Filtering

Given the observation $\mathbf{y}_t$ and the particle $\mathbf{p}_t^{(j)}$, we compute the importance weight $w_t^{(j)}$ of the particle $\mathbf{p}_t^{(j)}$ with respect to the observation $\mathbf{y}_t$ by

$$w_t^{(j)} = p(\mathbf{y}_t|\mathbf{p}_t^{(j)}).$$

The posterior probability of the particle is then obtained by

$$
\begin{aligned}
\Pr(\mathbf{x}_t = \mathbf{p}_t^{(i)}|\mathbf{y}_{1:t}) &= \Pr(\mathbf{x}_t = \mathbf{p}_t^{(i)}|\mathbf{y}_{1:t-1}, \mathbf{y}_t) \\
&= \frac{p(\mathbf{y}_t|\mathbf{p}_t^{(i)})\Pr(\mathbf{x}_t = \mathbf{p}_t^{(i)}|\mathbf{y}_{1:t-1})}{\sum_{j=1}^{N} p(\mathbf{y}_t|\mathbf{p}_t^{(j)})\Pr(\mathbf{x}_t = \mathbf{p}_t^{(j)}|\mathbf{y}_{1:t-1})} \\
&= \frac{w_t^{(i)}}{\sum_{j=1}^{N} w_t^{(j)}}. \tag{9.3.14}
\end{aligned}
$$

This means that the distribution function of $\Pr(\mathbf{x}_t = \mathbf{p}_t^{(i)}|\mathbf{y}_{1:t})$ can be expressed by a step function

$$\frac{1}{\sum_{j=1}^{N} w_t^{(j)}} \sum_{i=1}^{N} w_t^{(i)} I(\mathbf{x}, \mathbf{p}_t^{(i)}), \tag{9.3.15}$$

which has jumps at $\mathbf{p}_1, \cdots, \mathbf{p}_N$ with step sizes proportional to $w_t^{(1)}, \cdots,$ $w_t^{(N)}$.

For the next step of prediction, it is necessary to represent this distribution function by an empirical distribution of the form

$$\frac{1}{N} \sum_{i=1}^{N} I(\mathbf{x}, \mathbf{f}_t^{(i)}). \qquad (9.3.16)$$

This can be done by generating particles $\{\mathbf{f}_t^{(1)}, \cdots, \mathbf{f}_t^{(N)}\}$ by the resampling of $\{\mathbf{p}_t^{(1)}, \cdots, \mathbf{p}_t^{(N)}\}$ with probabilities

$$\mathrm{Pr}(\mathbf{f}_t^{(i)} = \mathbf{p}_t^{(j)} | \mathbf{y}_{1:t}) = \frac{w_t^{(j)}}{w_t^{(1)} + \cdots + w_t^{(N)}}. \qquad (9.3.17)$$

Figure 9.2 shows one cycle of the Monte Carlo filtering when, for purposes of illustration, we used only 100 particles ($N = 100$). A simple one-dimensional linear state-space model is assumed, and, for comparison, the exact distributions are computed by the non-Gaussian filter based on numerical integration. The plots in the left column show the exact densities, particles and histograms obtained from the particles. The plots in the right column show the exact distribution functions and the empirical distribution functions obtained from the particles. Plots (a) and (b) show the distributions of $\mathbf{f}_{n-1}^{(j)}$ and $\mathbf{v}_n^{(j)}$. Plot (c) shows the particles and the predictive distribution obtained by step (2-2). In plot (d), the vertical bars show the importance weights $w_t^{(j)}$ computed in step (2-3). Even with the same set of particles, by using this importance weight as the step size in the empirical distribution, the step function (9.3.15) becomes a good approximation to the exact distribution function (see plot (d)-right). Plot (e) shows the particle obtained by resampling in step (2-4). From the bars in the left plot, it can be seen that particles in the right half are drawn several times and many particles in the left half are not drawn at all. However, from the right plot, it can be seen that these particles can yield a good approximation to the exact distribution function.

9.3.4   Monte Carlo smoothing

The above algorithm for Monte Carlo filtering can be extended to smoothing by a simple modification. The details of the derivation of the algorithm are shown in (Kitagawa 1996, Clapp and Godsill 1999).

An algorithm for smoothing is obtained by replacing step (2-4) of the algorithm for filtering by

(2-4S) *Generate* $\{(\mathbf{s}_{1|t}^{(j)}, \cdots, \mathbf{s}_{t-1|t}^{(j)}, \mathbf{s}_{t|t}^{(j)})^T,\ j = 1, \ldots, N\}$ *by the resampling of* $\{(\mathbf{s}_{1|t-1}^{(j)}, \cdots, \mathbf{s}_{t-1|t-1}^{(j)}, \mathbf{p}_t^{(j)})^T,\ j = 1, \ldots, N\}$.

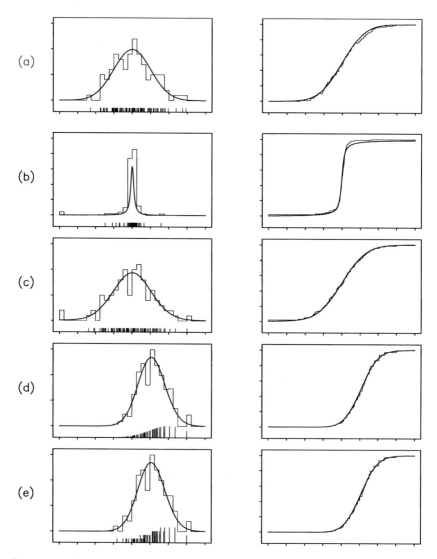

Figure 9.2. One cycle of Monte Carlo filtering. Exact density, histogram and particles (left plots) and exact distributions and empirical distributions (right plots). (a) Previous state, (b) system noise, (c) predictive distributions, (d) filter distributions and (e) filter distributions after resampling.

In this modification, the particles of the past state $s_{1|t-1}^{(j)}, \ldots s_{t-1|t-1}^{(j)}$, are preserved and $\{(s_{1|t-1}^{(j)}, \ldots, s_{t-1|t-1}^{(j)}, p_t^{(j)})^T, j = 1, \ldots, N\}$ is resampled with the same w weights as in the one obtained by step (2-4).

It is worth noting that this algorithm realises fixed interval smoothing for nonlinear non-Gaussian state-space models. However, in practice, since the number of realisations is finite, the repetition of the resampling (2-4S) will gradually diminish the number of different realisations in $\{s_{i|t}^{(1)}, \ldots, s_{i|t}^{(N)}\}$ and eventually loses the accuracy of the distribution.

To remedy this problem, the step (2-4S) needs to be replaced by (2-4L) *For fixed L, generate* $\{(s_{t-L|t}^{(j)}, \cdots, s_{t-1|t}^{(j)}, s_{t|t}^{(j)})^T, j = 1, \ldots, N\}$ *by the resampling of* $\{(s_{t-L|t-1}^{(j)}, \cdots, s_{t-1|t-1}^{(j)}, p_t^{(j)})^T, j = 1, \ldots, N\}$.

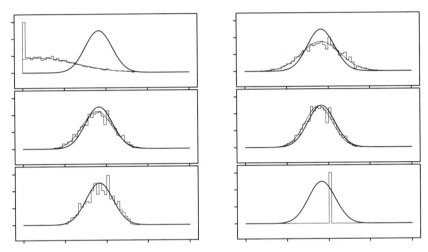

Figure 9.3. Fixed-interval and fixed-lag smoothers. (a) Filter, (b) $t = 5$, (c) $t = 10$, (d) $t = 20$, (e) $t = 100$, (f) $t = 400$,

Interestingly, this is equivalent to using the L-lag fixed-lag smoother (Anderson and Moore 1979). In Figure 9.3, the filter, the fixed-interval smoother, and the fixed-lag smoothers with $L = 5, 10, 20, 100$ and 400 for sample state-space model and the sample size $T = 400$ are shown. Each plot shows the fixed-lag smoothed distribution $p(x_1|y_{1:t})$ (thin curve), the exact fixed-interval smoother $p(x_1|y_{1:400})$ (bold curve) and the histogram obtained from the smoothed particles $s_{1|t}^{(j)}$, for $t = 1, 5, 10, 20, 100$ and 400. For $t = 1$, the histogram is a close approximation to the filter distribution $p(x_1|y_1)$, but is quite different from the fixed-interval smoother. As the lag-time t increases, the fixed-lag smoother $p(x_1|y_{1:t})$ becomes a closer approximation to the fixed-interval smoother $p(x_1|y_{1:400})$. However, as t increases, the approximation of the histogram to the fixed-lag smoother worsens, and at $t = 400$ collapses to one bin. The increase of lag, L, will improve the accuracy of the $p(x_t|y_{1:t+L})$ as an approximation to $p(x_t|y_{1:T})$, but it is very likely to decrease the accuracy of $\{s_{t|T}^{(1)}, \cdots, s_{t|T}^{(N)}\}$ as representatives of $p(x_t|y_{1:t+L})$. Since $p(x_t|y_{1:t+L})$ usually converges rather quickly to $p(x_t|y_{1:T})$, it is recommended not to take L too large.

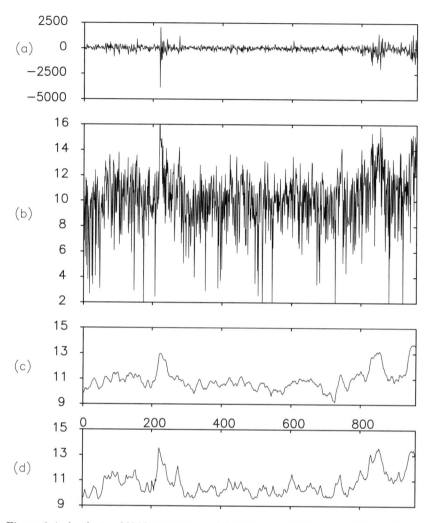

Figure 9.4. Analysis of Nikkei 225 data: (a) difference of the series, (b) nonlinear transformation of the data, (c) volatility estimated by the Gaussian model, (d) volatility estimated by the exact model.

9.3.5   Non-Gaussian smoothing for the stochastic volatility model

Figure 9.4 (a) shows the first difference of Nikkei 225 price index data from January 1987 to December 1990. Increases in amplitude at around $t = 200$ and after 800 are obvious. In financial engineering, the standard deviation of the return series is called volatility, which is modelled and estimated with great difficulty. In this subsection, we consider the difference of the original stock price data. However, the same method can be applied to the return series.

The changing amplitude of the time series can be expressed by a nonlinear observation model

$$y_t = \sigma_t \varepsilon_t, \tag{9.3.18}$$

where ε_t is a Gaussian white noise with mean 0 and the variance 1, and σ_t corresponds to the time-varying amplitude and is called the volatility of the series. For the estimation of the volatility, we assume a model for the change of the volatility, such as

$$\log \sigma_t^2 = \alpha + \beta \log \sigma_{t-1}^2 + v_t, \quad \varepsilon_t \sim N(0, \tau^2). \tag{9.3.19}$$

By taking the square of the observations, and then logarithms (Figure 9.4 (b)), we obtain a linear state-space model for $\log y_t^2$.

$$\log \sigma_t^2 = \alpha + \beta \log \sigma_{t-1}^2 + v_t \tag{9.3.20}$$
$$\log y_t^2 = \log \sigma_t^2 + \log w_t^2. \tag{9.3.21}$$

Here $\log \sigma_t^2$ is the state of the model, the distribution of the noise $\log w_t^2$ being given by

$$\frac{1}{\sqrt{2\pi}} \exp \left\{ \frac{w}{2} - \frac{e^w}{2} \right\}. \tag{9.3.22}$$

The well known method for estimating this volatility or σ_t^2 is to approximate the exact noise distribution by a Gaussian distribution, $N(\eta, \pi^2/2)$, with $\eta = -1.2704$, and apply the Kalman filter to estimate the volatility. Figure 9.4 (c) shows the estimate of $\log \sigma_n^2$ obtained by this Gaussian approximation.

By using the Monte Carlo filter and smoother, it is possible to estimate the state by using the exact noise density. Further, by using the nonlinear state-space model, the stochastic volatility model can be directly expressed as

$$\log \sigma_t^2 = \alpha + \beta \log \sigma_{t-1}^2 + v_t \tag{9.3.23}$$
$$y_t = \sigma_t \varepsilon_t. \tag{9.3.24}$$

In this model, nonlinearity appears only in the observation model.

Table 9.1. Comparison of approximated and exact methods

Model	log-lk	AIC	τ^2	α	β
Approximation	-2191.0	13545.3	0.0546	9.60	0.9686
Exact	-6634.0	13274.0	0.1035	10.83	0.9529

Table 9.1 shows the results of the estimation by the Gaussian approximation and the exact methods. The log-likelihood of the linear Gaussian

approximation is computed with respect to the transformed data, $\log y_t^2$, and cannot be compared with that of the exact method. However, the AIC values that take into account the Jacobian of the nonlinear data transformation are comparable. It can be seen that the AIC of the exact model is smaller than the Gaussian approximation by about 270, indicating that the fit of the exact model is significantly better than that of the Gaussian model.

Comparing the results by two methods, the estimate of β is fairly close to the exact value, whereas that of α is smaller than the exact value. However, by correcting the bias of the noise distribution, η, we have $\hat{\alpha} = 9.60 + 1.27 = 10.87$, and this is a reasonable approximation to the exact value. On the other hand, the estimate of the variance becomes about a half of the exact one. This suggests that, by the approximated method, the estimated volatility becomes too smooth in comparison with the results from the exact model. Figure 9.4 (d) show the estimated $\log \sigma_t^2$ by the exact model by the Monte Carlo smoother with $N = 100,000$ and $L = 20$.

9.3.6   Nonlinear Smoothing

We consider the application of the non-Gaussian filter and the smoother for the analysis of the data generated by the following model, which was originally considered by (Andrade Netto, Gimeno and Mendes 1978), and analysed by (Carlin, Polson and Stoffer 1992, Gordon et al. 1993, Hürzeler and Künsch 1998, Kitagawa 1991, Kitagawa and Gersch 1996)

$$x_t = \frac{1}{2}x_{t-1} + \frac{25x_{t-1}}{1 + x_{t-1}^2} + 8\cos(1.2n) + v_t$$

$$y_t = \frac{x_t^2}{20} + \varepsilon_t, \tag{9.3.25}$$

with $v_t \sim N(0, 1)$ and $\varepsilon_t \sim N(0, 10)$. The problem is to estimate the unobserved true signal x_t (Figure 9.5 (a)) only from the sequence of observations $y_{1:T}$ (Figure 9.5 (b)) assuming that the model (9.3.25) is known. It should be noted here that, since the value of the state x_t is squared in the observation equation in (9.3.25), it is difficult to correctly identify whether the state x_t is positive or negative. The extended Kalman filter does not work well for this problem. Figure 9.5 (c) shows the time changes of the median and 5% and 95% points of the marginal posterior distributions of the signal obtained by the Monte Carlo filter/smoother with $N = 10,000$ particles. Using plot (a) as a reference, it can be seen that the Monte Carlo filter yielded a very reasonable estimate of the signal.

The key to the success of the nonlinear filtering and smoothing is that the Monte Carlo filter can reasonably express the non-Gaussian state densities. Figure 9.6 (a) and (b) respectively show the marginal posterior state distributions $p(x_t|y_{1:t+j})$, $j = -1, 0, 1, 2$, for $t = 30$ and 43. For both

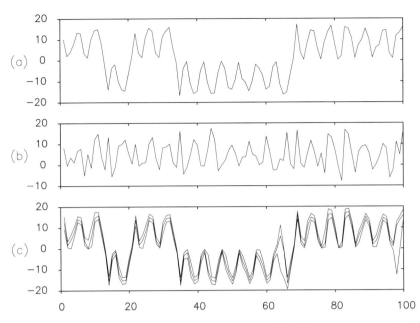

Figure 9.5. Nonlinear smoothing (Kitagawa (1996)): (a) unknown signal, (b) observed time series, (c) Monte Carlo estimate.

$t = 30$ and 43, the predictive distribution $p(x_t|y_{1:t-1})$ and the filter distribution $p(x_t|y_{1:t})$ are bi-modal. However, in the 2-lag smoother for $t = 30$, $p(x_{30}|y_{1:32})$, the left-half of the distribution disappeared and the distribution becomes uni-modal. The same phenomenon can be seen for $t = 43$. In this case, the right peak is higher than the left in the predictive and filter distributions. However, in the smoother distributions, the peak in the right half domain disappears. When this happens, the extended Kalman filter is very likely to yield an estimate with reverse sign.

Figure 9.7 (a) shows the predictive distributions of the state, $p(x_t|_{1:90})$ for $t = 91, \ldots , 100$. It can be seen that the shape of the distribution changes with time and it becomes bimodal for $t \geq 98$. On the other hand, (b) shows the predictive distributions of the observations, $p(y_t|\mathbf{y}_{1:90})$, which gradually become more diffuse as t increases.

9.4   Self-organising state-space models

9.4.1   Likelihood of the model and parameter estimation

The state-space model, (9.2.1) and (9.2.2), usually contains several unknown parameters, such as the variances of the noises and the coefficients of the functions F_t and H_t. The vector consisting of such unknown param-

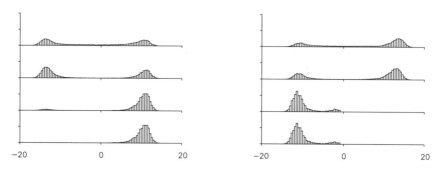

Figure 9.6. Fixed-point smoothing for $t = 30$ (left) and $t = 43$ (right). From top to bottom: predictive distributions, filter distributions, 1-lag smoothers and 2-lag smoothers.

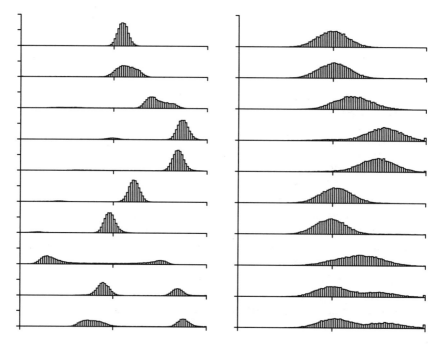

Figure 9.7. Predictive distributions of the state (left) and the observation (right) for $t = 91, \ldots, 100$.

eters is hereafter denoted by $\boldsymbol{\theta}$. For example, the predictive density can be denoted as $p(\mathbf{x}_t|\mathbf{y}_{1:t-1}, \boldsymbol{\theta})$. In general, the likelihood of the time series model specified by the parameter $\boldsymbol{\theta}$ is obtained by

$$L(\boldsymbol{\theta}) = p(\mathbf{y}_1, \cdots, \mathbf{y}_T|\boldsymbol{\theta}) = \prod_{t=1}^{T} p(\mathbf{y}_t|\mathbf{y}_{1:t-1}, \boldsymbol{\theta}), \qquad (9.4.26)$$

where $p(\mathbf{y}_t|\mathbf{y}_{1:t-1}, \boldsymbol{\theta})$ is the conditional density function of $\mathbf{y}_t$ given $\mathbf{y}_{1:t-1}$ and is obtained by

$$p(\mathbf{y}_t|\mathbf{y}_{1:t-1}, \boldsymbol{\theta}) = \int p(\mathbf{y}_t|\mathbf{x}_t, \boldsymbol{\theta})p(\mathbf{x}_t|\mathbf{y}_{1:t-1}, \boldsymbol{\theta})d\mathbf{x}_t. \qquad (9.4.27)$$

Using the Monte Carlo filter, this term can be approximated by

$$p(\mathbf{y}_t|\mathbf{y}_{1:t-1}, \boldsymbol{\theta}) \approx \frac{1}{N}\sum_{j=1}^{N} w_t^{(j)}. \qquad (9.4.28)$$

9.4.2 Self-organising state-space model

In principle, the maximum likelihood estimate of the parameter $\boldsymbol{\theta}$ is obtained by maximising the log-likelihood

$$\ell(\boldsymbol{\theta}) = \log L(\boldsymbol{\theta}) = \sum_{t=1}^{T} \log p(\mathbf{y}_t|\mathbf{y}_{1:t-1}, \boldsymbol{\theta}). \qquad (9.4.29)$$

However, in practice the sampling error due to the approximation (9.4.28) and long computing time sometimes renders the maximum likelihood method impractical. In this case, instead of estimating the parameter $\boldsymbol{\theta}$ by the maximum likelihood method, we consider a Bayesian estimation by augmenting the state vector $\mathbf{x}_t$ with unknown parameter $\boldsymbol{\theta}$ as

$$\mathbf{z}_t = \begin{bmatrix} \mathbf{x}_t \\ \boldsymbol{\theta}_t \end{bmatrix}, \qquad (9.4.30)$$

where $\boldsymbol{\theta}_t = \boldsymbol{\theta}$.

The state-space model for this augmented state vector $\mathbf{z}_t$ is easily obtained from the original state-space model and the obvious relation $\boldsymbol{\theta}_t = \boldsymbol{\theta}_{t-1}$, namely

$$\begin{aligned} \mathbf{z}_t &= F^*(\mathbf{z}_{t-1}, \mathbf{v}_t) \\ \mathbf{y}_t &= H^*(\mathbf{z}_t, \varepsilon_t), \end{aligned} \qquad (9.4.31)$$

where the nonlinear functions $F^*(\mathbf{z}, \mathbf{v})$ and $H^*(\mathbf{z}, \varepsilon)$ are defined by $F^*(\mathbf{z}, \mathbf{v}) = [F(\mathbf{x}, \mathbf{v}), \boldsymbol{\theta}]^T$, $H^*(\mathbf{z}, \varepsilon) = H(\mathbf{x}, \varepsilon)$.

Assume that we obtain the posterior distribution $p(\mathbf{z}_t|\mathbf{y}_{1:T})$ given the entire observations $\mathbf{y}_{1:T} = \{\mathbf{y}_1, \cdots, \mathbf{y}_T\}$. Since the original state vector $\mathbf{x}_t$ and the parameter vector $\boldsymbol{\theta}$ are included in the augmented state vector $\mathbf{z}_t$, it immediately yields the marginal posterior densities of both the parameter and of the original state. Various applications of this method are shown in (Kitagawa 1998).

This method of Bayesian simultaneous estimation of the parameter of the state-space model can be easily extended to a time-varying parameter situation where the parameter $\boldsymbol{\theta} = \boldsymbol{\theta}_t$ evolves with time t. Actually, the original formulation of the self-organising state-space does not work well when we use the Monte Carlo filter and smoother. This is because, since the parameters do not have their own system noises, the distribution gradually collapses as time proceeds. In that case, by allowing the parameter to change gradually, namely by assuming the model

$$\boldsymbol{\theta}_t = \boldsymbol{\theta}_{t-1} + \mathbf{v}_n, \tag{9.4.32}$$

we can obtain a reasonable estimate of the parameter. Liu and West show ways of improving the method of adding artificial noise to stationary parameters, see Liu and West (2001: this volume).

9.5   Examples

9.5.1   Self-organising smoothing for the stochastic volatility model

In Section 3.5, we considered the estimation of the stochastic volatility by the Monte Carlo smoother, assuming the model

$$
\begin{aligned}
y_t &= \sigma_t \varepsilon_t \\
\log \sigma_t^2 &= \alpha + \beta \log \sigma_{t-1}^2 + \log \varepsilon_t^2,
\end{aligned} \tag{9.5.33}
$$

or equivalently

$$
\begin{aligned}
y_t &= e^{\alpha_t/2} \sigma_t \varepsilon_t \\
\log \sigma_t^2 &= \beta \log \sigma_{t-1}^2 + v_t.
\end{aligned} \tag{9.5.34}
$$

The parameters of the models such as the regression coefficients α and β, and the variance of the noise inputs σ^2, were estimated by the maximum likelihood method. In this section, we consider the simultaneous estimation of these parameters $\boldsymbol{\theta}_t = (\alpha_t, \beta_t, \log \tau_t^2)^T$ and the stochastic volatility σ.

To construct the self-organising state-space model, we augment the original state $\log \sigma_t^2$ with the unknown parameter α_t, β_t and $\log \tau_t^2$,

$$
\mathbf{z}_t = \begin{bmatrix} \log \sigma_t^2 \\ \alpha_t \\ \beta_t \\ \log \tau_t^2 \end{bmatrix}. \tag{9.5.35}
$$

For time evolution of the parameters, we assume the models

$$
\begin{aligned}
\alpha_t &= \alpha_{t-1} + u_{t1} \\
\beta_t &= \beta_{t-1} + u_{t2} \\
\log \tau_t^2 &= \log \tau_{t-1}^2 + u_{t3},
\end{aligned} \tag{9.5.36}
$$

where $u_{t1} \sim N(0, \xi_1^2)$, $u_{t2} \sim N(0, \xi_2^2)$, $u_{t3} \sim N(0, \xi_3^2)$. The stochastic volatility model (9.5.34), together with the component models, (9.5.36) can be expressed in state-space model form:

$$
\begin{bmatrix} \log \sigma_t^2 \\ \alpha_t \\ \beta_t \\ \log \tau_t^2 \end{bmatrix} = \begin{bmatrix} \log \sigma_{t-1}^2 \\ \alpha_{t-1} \\ \beta_{t-1} \\ \log \tau_{t-1}^2 \end{bmatrix} + \begin{bmatrix} \tau_{t-1} v_t \\ u_{t1} \\ u_{t2} \\ u_{t3} \end{bmatrix} \tag{9.5.37}
$$

$$
y_t = e^{x_t} \varepsilon_t.
$$

Therefore, by estimating the state vector of this model, we can obtain the estimates of the stochastic volatility and the parameters of the stochastic volatility model simultaneously.

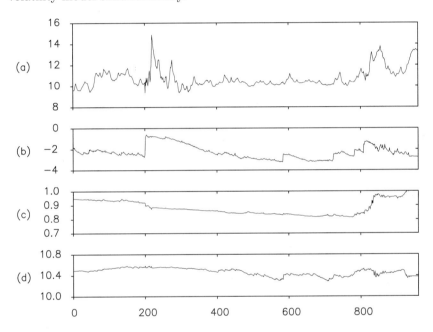

Figure 9.8. Results of the self-organising state-space model: (a) estimated volatility, (b) variance of the noise τ^2, (c) coefficient α, (d) coefficient β.

Figure 9.8 (a) shows the estimated $\log \sigma_t^2$ obtained by the Monte Carlo smoother with $N = 100,000$ and $L = 20$. The increase in the volatility at around $t = 200$ and after $t = 800$ is clearly detected. Plot (b) shows the time-varying variance of the noise of the stochastic volatility model. At the time corresponding to the increase in the volatility, the log-variance $\log \tau_t^2$ also increased abruptly at around $t = 200$. The increase of the volatility is not so sudden after $t = 800$, and $\log \tau_t^2$ did not increase as significantly. Plots (c) and (d) show the time-changes of the coefficients α_t and β_t, respectively. It can be seen that the coefficient α_t does not change as significantly, but β_t decreases gradually until $t = 800$, then abruptly increases to a value close to 1.

9.5.2 Time series with trend and stochastic volatility

Stock price and exchange rate data seem to share the characteristic that volatility increases after a significant change of trend. For modelling such series, we can generalise the above model to

$$y_t = T_t + \sigma_t \varepsilon_t, \tag{9.5.38}$$

where T_t is a trend, σ_t is the volatility and ε_t is a standard Gaussian white noise sequence. For simultaneous estimation of trend and variance, we introduce the smoothness prior models

$$\begin{aligned} T_t &= 2T_{t-1} - T_{t-2} + \varepsilon_t \\ \log \sigma_t^2 &= 2\log \sigma_{t-1}^2 - \log \sigma_{t-2}^2 + \delta_t, \end{aligned} \tag{9.5.39}$$

where ε_t and δ_t are white noise sequences specified below. Log-transformation is used to assure the positivity of the volatility σ_t. The models (9.5.38) and (9.5.39) can be expressed in the state-space representation (9.2.1) and (9.2.2) with the four-dimensional state vector $x_t = (T_t, T_{t-1}, \log \sigma_t^2, \log \sigma_{t-1}^2)^T$, two-dimensional system noise $v_t = (\varepsilon_t, \delta_t)^T$ and a nonlinear function $H(x_t, \varepsilon_t) = H((T_t, T_{t-1}, \log \sigma_t^2, \log \sigma_{t-1}^2)^T, \varepsilon_t) = T_t + \sigma_t \varepsilon_t$:

$$\begin{bmatrix} T_t \\ T_{t-1} \\ \log \sigma_t^2 \\ \log \sigma_{t-1}^2 \end{bmatrix} = \begin{bmatrix} 2 & -1 & 0 & 0 \\ 1 & 0 & 0 & 0 \\ 0 & 0 & 2 & -1 \\ 0 & 0 & 1 & 0 \end{bmatrix} \begin{bmatrix} T_{t-1} \\ T_{t-2} \\ \log \sigma_{t-1}^2 \\ \log \sigma_{t-2}^2 \end{bmatrix} + \begin{bmatrix} 1 & 0 \\ 0 & 0 \\ 0 & 1 \\ 0 & 0 \end{bmatrix} \begin{bmatrix} v_t \\ u_t \end{bmatrix} \tag{9.5.40}$$

Figures 9.9 (a) and (b) show the original Nikkei 225 data from January 1987 to December 1990, and the trend estimated by the Monte Carlo smoother with $N = 10,000$ and $L = 20$, respectively. To distinguish from the original series, $T_t - 5000$ is plotted. Black Monday occurred at around $t = 200$ and the bubble burst at around $t = 800$. Plot (c) shows the identified noise sequence, ε_t. The amplitude of the series become large after these events. At the corresponding times, increases in volatility after Black Monday at around $t = 200$, and the crash of the bubble at around $t = 800$, are clearly detected.

This model can be easily extended to include a stationary component

$$y_t = T_t + p_t + \sigma_t \varepsilon_t, \tag{9.5.41}$$

where the stationary component p_t is expressed by

$$p_t = \sum_{j=1}^{m} a_t p_{t-j} + u_t. \tag{9.5.42}$$

Moreover, it is possible to develop a model that takes into account the correlation between the change of the trend and the change of the volatility (Kitagawa and Sato 2000).

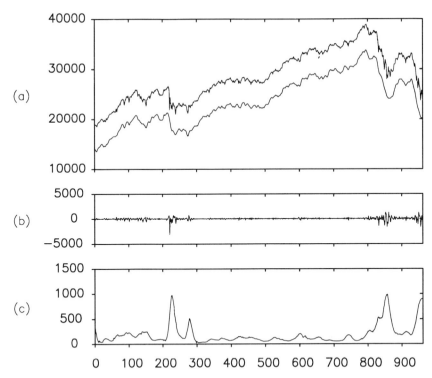

Figure 9.9. Estimation of the volatility of stock price data (Kitagawa (1996)): (a) original series and trend, (b) noise and (c) volatility.

9.6 Conclusion

The general state-space model is a useful tool for Bayesian time series modelling. Recursive filtering and smoothing formulae corresponding to the Kalman filter and smoother for linear Gaussian state-space models can be obtained for the general state-space model, and can be realised by implementations based on Monte Carlo approximation.

In the estimation of the parameters of the model, we may have difficulty using the Monte Carlo filter because the computed log-likelihood contains a sampling error. To remedy this, we also developed a self-organising state-space model wherein the state and unknown parameters are estimated simultaneously. Numerical examples show the usefulness of this method.

10

Combined Parameter and State Estimation in Simulation-Based Filtering

Jane Liu
Mike West

10.1 Introduction and historical perspective

Much of the recent and current interest in simulation-based methods of sequential Bayesian analysis of dynamic models has been focused on improved methods of filtering for time-varying state vectors. We now have quite effective algorithms for time-varying states, as represented throughout this volume. Variants of the auxiliary particle filtering algorithm (Pitt and Shephard 1999b), in particular, are of proven applied efficacy in quite elaborate models. However, the need for more general algorithms that deal simultaneously with both fixed model parameters and state variables is especially pressing. We simply do not have access to efficient and effective methods of treating this problem, especially in models with realistically large numbers of fixed model parameters. It is a very challenging problem.

A short historical commentary will be of interest as context to the developments presented in this chapter. In the statistics literature, simulation-based filtering can be seen as a natural outgrowth of converging interests in the late 1980s and early 1990s. For several decades, researchers involved in sequential analysis of dynamic models, both in statistics and various engineering fields, have been using discrete numerical approximations to sequentially updated posterior distributions in various mixture modelling frameworks. This literature has involved methods for both time-evolving states and fixed parameters, and is exemplified by the important class of adaptive multi-process models used in Bayesian forecasting since the early 1970s (Harrison and Stevens 1976, Smith and West 1983, West and Harrison 1997). During the 1980s, this naturally led to larger-scale analyses using discrete grids of parameter values, though the combinatorial explosion of grid sizes with increasing parameter dimension limited this line of development. Novel methods using efficient quadrature-based, adaptive numerical integration ideas were introduced later in the 1980s (Pole 1988, Pole, West and Harrison 1988, Pole and West 1990). These methods described how discrete grids – of both fixed model parameters and state variables – themselves change over time as data are processed, sequentially adapting the discrete posterior approximations by generating new "samples" as well as associated "weights". This work recognised the utility of the Markov evolution equations of dynamic models in connection with the generation of new grids of values for time-evolving state variables. It similarly recognised and addressed the practically critical issues of "diminishing weights" on unchanging grids of parameter values, and the associated need for some method of interpolation and smoothing to "regenerate" grids of values for fixed model parameters. In these respects, this adaptive deterministic approach naturally anticipated future developments of simulation-based approaches. Again, however, parameter dimension limited the broader applicability of such approaches.

The end of the 1980s saw a developing interest in simulation-based methods. Parallel developments in the early 1990s eventually led to publication of different but related approaches (West 1993a, West 1993b, Gordon et al. 1993). All the approaches involve methods of evolving and updating discrete sets of sampled state vectors, and the associated weights on such sampled values, as they evolve in time. It has become standard to refer to sampled values as "particles."

The above referenced works, and others, have again highlighted the utility of the convolution structure of Markov evolution equations for state variables in generating Monte Carlo samples of time-evolving states. Most approaches recognise, and attempt to address, the inherent problems of degrading approximation accuracy as particulate representations are updated over time – the issues of particle "attrition" in resampling methods and of "weight degeneracy" in reweighting methods. These issues are particularly acute in approaches that aim to deal with fixed model parameters as part of an extended state vector. This deficiency was addressed in the approaches of West (1993a,b). Openly recognising the need for some kind of interpolation/smoothing of "old" parameter particles to generate new values, this author used local smoothing based on modified kernel density estimation methods, the modifications being based on Bayesian reasoning and geared towards adjusting for the over-smoothing problems inherent in standard kernel methods. In later years, this approach has been enhanced. For example, it has been extended to include variable shapes of multivariate kernels to reflect changing patterns of dependencies among parameters in different regions of parameter space (Givens and Raftery 1996), as explicitly anticipated in West (1993a,b). A related approach may be referred to as the "artificial evolution" method for model parameters. This relates to the work of Gordon et al. (1993), who introduced the idea of adding additional random disturbances – or "roughening penalties" – to sampled state vectors in an attempt to deal with degeneracy. Extending this idea to fixed model parameters leads to a synthetic method of generating new sample points for parameters via the convolution implied by this "artificial evolution." This neat, ad-hoc idea is easy to implement, but suffers the obvious drawback that it "throws away" information about parameters in assuming them to be time-varying when they are, in fact, fixed. The same drawback is seen when using the idea in its original form for dynamic states.

In this chapter, we take as our starting point these approaches to dealing with fixed model parameters. We address ways to reconcile these approaches and how to embed a generalised algorithm within a sequential auxiliary particle filtering context. In Section 10.2, we discuss particle filtering for state variables, and describes the general framework for combined filtering on parameters and state variables. Section 10.3 focuses on the problems arising in simulation-based filtering for parameters, reviews the kernel and artificial evolution methods, identifies the inherent structural similarities in these methods, and introduces a modified and easily implemented approach that improves upon both by resolving the problem of information loss that each implies. Returning to the general framework in Section 10.4, we describe a general algorithm that extends the standard auxiliary particle filtering approach for state variables to include model parameters. We give a simple example for illustration, and then, in Section 10.5, report some experiences in applying the method to multivariate dynamic factor models for financial time series. In Section 10.6, we conclude with some summary comments, discussion and suggestions for new research directions.

10.2 General framework

10.2.1 Dynamic model and analysis perspective

Assume a Markovian dynamic model for sequentially observed data vectors y_t, $(t = 1, 2, \ldots)$, in which the state vector at time t is x_t and the fixed parameter vector is θ. The model is specified at each time t by the observation equation defining the observation density

$$p(y_t | x_t, \theta), \qquad (10.2.1)$$

and the Markovian evolution equation, or state equation, defining the transition density

$$p(x_t | x_{t-1}, \theta). \qquad (10.2.2)$$

Each y_t is conditionally independent of past states and observations given the current state x_t, and the parameter θ, and x_t is conditionally independent of past states and observations given x_{t-1} and θ. This covers a very broad class of models with practical application (West and Harrison 1997).

Sequential Monte Carlo methods aim to sequentially update Monte Carlo sample approximations to sequences of posterior distributions $p(\boldsymbol{x}_t, \boldsymbol{\theta}|D_t)$ where $D_t = \{D_{t-1}, \boldsymbol{y}_t\}$ is the information set at time t. Thus, at time t this posterior is represented by a discrete sample of points and weights (the latter possibly, though not necessarily, uniform weights). On observing the new observation $\boldsymbol{y}_{t+1}$ it is desired to produce a sample from the current posterior $p(\boldsymbol{x}_{t+1}, \boldsymbol{\theta}|D_{t+1})$. There have been many contributors to theoretical and applied aspects of research in this area in recent years (West 1993b, Gordon et al. 1993, Kitagawa 1998, Liu and Chen 1998, Berzuini et al. 1997, Pitt and Shephard 1999b) and the field has grown dramatically both within the statistical sciences and in related fields including, especially, various branches of engineering (Doucet 1998). The current volume represents a comprehensive catalogue of recent and current work.

The most effective methods all use the state equation to generate sample values of the current state vector $\boldsymbol{x}_{t+1}$ based on past sampled values of the state $\boldsymbol{x}_t$. This is critical to the utility and performance of discrete approximation methods, as the generation of new sets of states from what is usually a continuous state transition density allows the posterior approximations to move around in the state-space as the state evolves and new data are processed. Below, we focus exclusively on auxiliary particle filters as developed in Pitt and Shephard (1999b), variants of which are, in our opinion, the most effective methods currently available – though that may change as the field evolves.

10.2.2   Filtering for states

Consider a model with no fixed parameters, or in which $\boldsymbol{\theta}$ is assumed known, so that the focus is entirely on filtering for the state vector. Standing at time t, suppose we have a sample of current states $\{\boldsymbol{x}_t^{(1)}, ..., \boldsymbol{x}_t^{(N)}\}$ and associated weights $\{\omega_t^{(1)}, ..., \omega_t^{(N)}\}$ that together represent a Monte Carlo importance sample approximation to the posterior $p(\boldsymbol{x}_t|D_t)$. This includes, of course, the special case of equal weights in which we have a direct posterior sample. Time evolves to $t+1$, we observe $\boldsymbol{y}_{t+1}$, and want to generate a sample from the posterior $p(\boldsymbol{x}_{t+1}|D_{t+1})$. Theoretically

$$p(\boldsymbol{x}_{t+1}|D_{t+1}) \propto p(\boldsymbol{y}_{t+1}|\boldsymbol{x}_{t+1})p(\boldsymbol{x}_{t+1}|D_t), \qquad (10.2.3)$$

where $p(\boldsymbol{x}_{t+1}|D_t)$ is the prior density of $\boldsymbol{x}_{t+1}$ and $p(\boldsymbol{y}_{t+1}|\boldsymbol{x}_{t+1})$ is the likelihood function. The second term here – the prior for the state at time $t+1$ – is implied by the state equation as $\int p(\boldsymbol{x}_{t+1}|\boldsymbol{x}_t)p(\boldsymbol{x}_t|D_t)d\boldsymbol{x}_t$. Under the Monte Carlo approximation to $p(\boldsymbol{x}_t|D_t)$, this integral is replaced by a weighted summation over the sample points $\boldsymbol{x}_t^{(k)}$, so that the required update in equation (10.2.3) becomes

$$p(\boldsymbol{x}_{t+1}|\boldsymbol{D}_{t+1}) \propto p(\boldsymbol{y}_{t+1}|\boldsymbol{x}_{t+1}) \sum_{k=1}^{N} w_t^{(k)} p(\boldsymbol{x}_{t+1}|\boldsymbol{x}_t^{(k)}). \qquad (10.2.4)$$

To generate Monte Carlo approximations to this density, an old and natural idea is to sample from $p(\boldsymbol{x}_{t+1}|\boldsymbol{x}_t^{(k)})$ for $k = 1, ..., N$, evaluate the corresponding values of the weighted likelihood function $w_t^{(k)} p(\boldsymbol{y}_{t+1}|\boldsymbol{x}_{t+1})$ at each draw, and then use the normalised weights as the new weights of the samples. This basic particle filter is an importance sampling method closely related to those of West(1993a,b). A key problem is that the sampled points come from the current prior of $\boldsymbol{x}_{t+1}$, and the resulting weights may be very small on many points in cases of meaningful separation of the prior and the likelihood function based on $\boldsymbol{y}_{t+1}$. West(1993a,b) developed an effective method of adaptive importance sampling to address this. The idea of auxiliary particle filtering (see Pitt and Shephard, 1999b, and the chapter by these authors in this volume) is similar in spirit, but has real computational advantages. It works as follows. Incorporate the likelihood function $p(\boldsymbol{y}_{t+1}|\boldsymbol{x}_{t+1})$ under the summation in equation (10.2.4) to give

$$p(\boldsymbol{x}_{t+1}|\boldsymbol{D}_{t+1}) \propto \sum_{k=1}^{N} w_t^{(k)} p(\boldsymbol{x}_{t+1}|\boldsymbol{x}_t^{(k)}) p(\boldsymbol{y}_{t+1}|\boldsymbol{x}_{t+1})$$

and generate samples of the current state, as follows. For each $k = 1, \dots, N$, select an estimate $\boldsymbol{\mu}_{t+1}^{(k)}$ of $\boldsymbol{x}_{t+1}$, such as the mean or mode of $p(\boldsymbol{x}_{t+1}|\boldsymbol{x}_t^{(k)})$. Evaluate the weights

$$g_{t+1}^{(k)} \propto w_t^{(k)} p(\boldsymbol{y}_{t+1}|\boldsymbol{\mu}_{t+1}^{(k)}).$$

A large value of $g_{t+1}^{(k)}$ indicates that $\boldsymbol{x}_t^{(k)}$, when "evolving" to time $t+1$, is likely to be more consistent with the datum $\boldsymbol{y}_{t+1}$ than not. Then indicators j are sampled with probabilities proportional to $g_{t+1}^{(j)}$, and values $\boldsymbol{x}_{t+1}^{(j)}$ of the current state are drawn from $p(\boldsymbol{x}_{t+1}|\boldsymbol{x}_t^{(j)})$ based on these "auxiliary" indicators. These sampled states are essentially importance samples from the time $t+1$ posterior, and have associated weights

$$w_{t+1}^{(j)} = \frac{p(\boldsymbol{y}_{t+1}|\boldsymbol{x}_{t+1}^{(j)})}{p(\boldsymbol{y}_{t+1}|\boldsymbol{\mu}_{t+1}^{(j)})}. \qquad (10.2.5)$$

Posterior inferences at time $t+1$ can be based directly on these sampled values and weights, or we can resample according to the importance weights $w_{t+1}^{(j)}$ to obtain an equally weighted set of states representing a direct Monte Carlo approximation to the required posterior in equation (10.2.3).

10.2.3 Filtering for states and parameters

In the general model with fixed parameters $\boldsymbol{\theta}$, extend the sample-based framework as follows. Standing at time t, we now have a combined sample

$$\{\boldsymbol{x}_t^{(j)}, \boldsymbol{\theta}_t^{(j)} : j = 1, \ldots, N\}$$

and associated weights

$$\{\omega_t^{(j)} : j = 1, \ldots, N\}$$

representing an importance sample approximation to the time t posterior $p(\boldsymbol{x}_t, \boldsymbol{\theta}|D_t)$ for both parameter and state. Notice that the t suffix on the $\boldsymbol{\theta}$ samples here indicate that they are from the time t posterior, *not* that $\boldsymbol{\theta}$ is time-varying. Time evolves to $t + 1$, we observe $\boldsymbol{y}_{t+1}$, and now want to generate a sample from $p(\boldsymbol{x}_{t+1}, \boldsymbol{\theta}|D_{t+1})$. Bayes' theorem gives this as

$$\begin{aligned} p(\boldsymbol{x}_{t+1}, \boldsymbol{\theta}|D_{t+1}) &\propto p(\boldsymbol{y}_{t+1}|\boldsymbol{x}_{t+1}, \boldsymbol{\theta})p(\boldsymbol{x}_{t+1}, \boldsymbol{\theta}|D_t) \\ &\propto p(\boldsymbol{y}_{t+1}|\boldsymbol{x}_{t+1}, \boldsymbol{\theta})p(\boldsymbol{x}_{t+1}|\boldsymbol{\theta}, D_t)p(\boldsymbol{\theta}|D_t), \quad (10.2.6) \end{aligned}$$

where the form chosen in the last equation makes explicit the notion that the theoretical density function $p(\boldsymbol{\theta}|D_t)$ is an important ingredient in the update.

If $\boldsymbol{\theta}$ were known, equation (10.2.6) simplifies: $p(\boldsymbol{\theta}|D_t)$ is degenerate and we drop the known parameter from the conditioning statements. This leads to equation (10.2.3), and the auxiliary particle method applies for filtering on the state vector. Otherwise, it is now explicit that we have to deal with the problem of not knowing the form of the theoretical density function $p(\boldsymbol{\theta}|D_t)$ in order to obtain combined filtering on the parameter and the state. The next section reviews the two main historical approaches.

10.3   The treatment of model parameters

10.3.1 Artificial evolution of parameters

In dealing with time-varying states, Gordon et al.(1993) suggested an approach to reducing sample degeneracy/attrition by adding small random disturbances (referred to as "roughening penalties") to state particles between time steps, in addition to any existing evolution noise contributions. In the literature since then, this idea has been extrapolated to fixed model parameters. One version of the idea adds small random perturbations to all the parameter particles under the posterior at each time point before evolving to the next. This specific method has an interpretation as arising from an extended model in which the model parameters are viewed as if they were, in fact, time-evolving – an "artificial evolution." That is, consider a different model in which $\boldsymbol{\theta}$ is replaced by $\boldsymbol{\theta}_t$ at time t, and simply include $\boldsymbol{\theta}_t$

in an augmented state vector. Then add an independent, zero-mean normal increment to the parameter at each time. That is,

$$\boldsymbol{\theta}_{t+1} = \boldsymbol{\theta}_t + \boldsymbol{\zeta}_{t+1}$$
$$\boldsymbol{\zeta}_{t+1} \sim N(\mathbf{0}, \boldsymbol{W}_{\mathbf{t+1}}) \tag{10.3.7}$$

for some specified variance matrix $\boldsymbol{W}_{t+1}$ and where $\boldsymbol{\theta}_t$ and $\boldsymbol{\zeta}_{t+1}$ are conditionally independent given D_t. With the model recast with the corresponding augmented state vector, the standard filtering methods for states alone, such as the auxiliary particle filter, now apply. The key motivating idea is that the artificial evolution provides the mechanism for generating new parameter values at each time step in the simulation analysis, so helping to address sample attrition in reweighting methods that stay with the same sets of parameter points between time steps.

Among the various issues and drawbacks of this approach, the key one is simply that fixed model parameters are, well, fixed! Pretending that they are in fact time-varying implies an artificial "loss of information" between time points, resulting in posteriors that are, eventually, far too diffuse relative to the theoretical posteriors for the actual fixed parameters. To date, there has been no resolution to this issue: if one adopts a model in which all parameters are subject to independent random shocks at each time point, the precision of resulting inferences is inevitably limited.

However, an inherent interpretation in terms of kernel smoothing of particles leads to a modification of this artificial evolution method in which the problem of information loss is avoided. First, we discuss the basic form of kernel smoothing as introduced and developed in West (1993).

10.3.2   Kernel smoothing of parameters

Understanding the imperative to develop some method of smoothing for approximation of the required density $p(\boldsymbol{\theta}|D_t)$ in equation (10.2.6), West(1993b) developed kernel smoothing methods that provided the basis for rather effective adaptive importance sampling techniques. This represented extension to sequential analysis of basic mixture modelling ideas in West (1993a).

Standing at time t, suppose we have current posterior parameter samples $\boldsymbol{\theta}_t^{(j)}$ and weights $\omega_t^{(j)}$, $(j = 1, ..., N)$, providing a discrete Monte Carlo approximation to $p(\boldsymbol{\theta}|D_t)$. Again remember that the t suffix on $\boldsymbol{\theta}$ here indicates that the samples are from the time t posterior; $\boldsymbol{\theta}$ is not time-varying. Write $\overline{\boldsymbol{\theta}}_t$ and $\boldsymbol{V}_t$ for the Monte Carlo posterior mean and variance matrix of $p(\boldsymbol{\theta}|D_t)$, computed from the Monte Carlo sample $\boldsymbol{\theta}_t^{(j)}$ with weights $\omega_t^{(j)}$. The smooth kernel density form of West (1993a,b) is given by

$$p(\boldsymbol{\theta}|D_t) \approx \sum_{j=1}^{N} \omega_t^{(j)} N(\boldsymbol{\theta}|\boldsymbol{m}_t^{(j)}, h^2 \boldsymbol{V}_t), \tag{10.3.8}$$

with the following defining components. First, $N(\cdot|\boldsymbol{m}, \boldsymbol{S})$ is a multivariate normal density mean $\boldsymbol{m}$ and variance matrix $\boldsymbol{S}$, so that the above density is a mixture of $N(\boldsymbol{\theta}|\boldsymbol{m}_t^{(j)}, h^2\boldsymbol{V}_t)$ distributions weighted by the sample weights $\omega_t^{(j)}$. Kernel rotation and scaling uses $\boldsymbol{V}_t$, the Monte Carlo posterior variance, and the overall scale of kernels is a function of the smoothing parameter $h > 0$. Standard density estimation methods suggest that h be chosen as a slowly decreasing function of N, so that kernel components are naturally more concentrated about their locations $\boldsymbol{m}_t^{(j)}$ for larger N. West(1993a,b) suggests taking slightly smaller values than the conventional kernel methods as a general rule. As we discuss below, our new work has led to a quite different perspective on this issue.

The kernel locations $\boldsymbol{m}_t^{(j)}$ are specified using a shrinkage rule introduced by West(1993a,b). Standard kernel methods would suggest $\boldsymbol{m}_t^{(j)} = \boldsymbol{\theta}_t^{(j)}$ so that kernels are located about existing sample values. However, this results in a kernel density function that is *over-dispersed* relative to the posterior sample, in the sense that the variance of the resulting mixture of normals is $(1 + h^2)\boldsymbol{V}_t$, always larger than $\boldsymbol{V}_t$. This is a most significant flaw in the sequential simulation; an over-dispersed approximation to $p(\boldsymbol{\theta}|D_t)$ will lead to an over-dispersed approximation to $p(\boldsymbol{\theta}|D_{t+1})$, and the consequent "loss of information" will accumulate as the operation is repeated at future times. To correct this, West(1993a,b) introduced the idea of shrinkage of kernel locations. Take

$$\boldsymbol{m}_t^{(j)} = a\boldsymbol{\theta}_t^{(j)} + (1-a)\overline{\boldsymbol{\theta}}_t \qquad (10.3.9)$$

where $a = \sqrt{1 - h^2}$. With these kernel locations, the resulting normal mixture retains the mean $\overline{\boldsymbol{\theta}}_t$ and now has the correct variance $\boldsymbol{V}_t$, hence the over-dispersion is trivially corrected.

10.3.3   Reinterpreting artificial parameter evolutions

The undesirable "loss of information" implicit in equation (10.3.7) can be easily quantified. The Monte Carlo approximation $\{\boldsymbol{\theta}_t^{(j)}, \omega_t^{(j)}\}$ to $p(\boldsymbol{\theta}|D_t)$ has mean $\overline{\boldsymbol{\theta}}_t$ and variance matrix $\boldsymbol{V}_t$. Hence, in the evolution in equation (10.3.7) with the innovation $\boldsymbol{\zeta}_{t+1}$ independent of $\boldsymbol{\theta}_t$ as proposed, the implied prior $p(\boldsymbol{\theta}_{t+1}|D_t)$ has the correct mean $\overline{\boldsymbol{\theta}}_t$ but variance matrix $\boldsymbol{V}_t + \boldsymbol{W}_{t+1}$. The loss of information is explicitly represented by the component $\boldsymbol{W}_{t+1}$. Now, there is a close tie-in between this method and the kernel smoothing approach. To see this clearly, note that the Monte Carlo approximation to $p(\boldsymbol{\theta}_{t+1}|D_t)$ implied by equation (10.3.7), is also a kernel form, namely

$$p(\boldsymbol{\theta}_{t+1}|D_t) \approx \sum_{j=1}^{N} \omega_t^{(j)} N(\boldsymbol{\theta}_{t+1}|\boldsymbol{\theta}_t^{(j)}, \boldsymbol{W}_{t+1}) \qquad (10.3.10)$$

and this, as we have seen, is over-dispersed relative to the required or "target" variance $\boldsymbol{V}_t$.

It turns out that we can correct for this over-dispersion as follows. The key is to note that our kernel method is effective because of the use of location shrinkage. This shrinkage pushes samples $\boldsymbol{\theta}_t^{(j)}$ values towards their mean $\overline{\boldsymbol{\theta}}_t$ before adding a small degree of "noise" implied by the normal kernel. This suggests that the artificial evolution method should be modified by introducing correlations between $\boldsymbol{\theta}_t$ and the random shock $\boldsymbol{\zeta}_{t+1}$. Assuming a non-zero covariance matrix, note that the artificial evolution equation (10.3.7) implies

$$V(\boldsymbol{\theta}_{t+1}|D_t) = V(\boldsymbol{\theta}_t|D_t) + \boldsymbol{W}_{t+1} + 2C(\boldsymbol{\theta}_t, \boldsymbol{\zeta}_{t+1}|D_t).$$

To correct to "no information lost" implies that we set

$$V(\boldsymbol{\theta}_{t+1}|D_t) = V(\boldsymbol{\theta}_t|D_t) = \boldsymbol{V}_t,$$

which then implies

$$C(\boldsymbol{\theta}_t, \boldsymbol{\zeta}_{t+1}|D_t) = -\boldsymbol{W}_{t+1}/2.$$

Hence, there must be a structure of negative correlations to remove the unwanted information loss effect. In the case of approximate joint normality of $(\boldsymbol{\theta}_t, \boldsymbol{\zeta}_{t+1}|D_t)$, this would then imply the conditional normal evolution in which

$$p(\boldsymbol{\theta}_{t+1}|\boldsymbol{\theta}_t) = N(\boldsymbol{\theta}_{t+1}|\boldsymbol{A}_{t+1}\boldsymbol{\theta}_t + (\boldsymbol{I} - \boldsymbol{A}_{t+1})\overline{\boldsymbol{\theta}}_t, (\boldsymbol{I} - \boldsymbol{A}_{t+1}^2)\boldsymbol{V}_t), \quad (10.3.11)$$

where

$$\boldsymbol{A}_{t+1} = \boldsymbol{I} - \boldsymbol{W}_{t+1}\boldsymbol{V}_t^{-1}/2.$$

The resulting Monte Carlo approximation to $p(\boldsymbol{\theta}_{t+1}|D_t)$ is then a generalised kernel form with complicated shrinkage patterns induced by the shrinkage matrix $\boldsymbol{A}_{t+1}$. We restrict ourselves here to the very special case in which the evolution variance matrix $\boldsymbol{W}_{t+1}$ is specified using a standard discount factor technique. Specifically, take

$$\boldsymbol{W}_{t+1} = \boldsymbol{V}_t(\frac{1}{\delta} - 1),$$

where δ is a discount factor in $(0, 1]$, typically around $0.95 - 0.99$. In this case, $\boldsymbol{A}_{t+1} = a\boldsymbol{I}$ with $a = (3\delta - 1)/2\delta$ and the conditional evolution density above reduces

$$p(\boldsymbol{\theta}_{t+1}|\boldsymbol{\theta}_t) \sim N(\boldsymbol{\theta}_{t+1}|a\boldsymbol{\theta}_t + (1 - a)\overline{\boldsymbol{\theta}}_t, h^2\boldsymbol{V}_t), \quad (10.3.12)$$

where

$$h^2 = 1 - a^2,$$

so that

$$h^2 = 1 - ((3\delta - 1)/2\delta)^2,$$

and we note that $a = \sqrt{1 - h^2}$. The resulting Monte Carlo approximation to $p(\boldsymbol{\theta}_{t+1}|D_t)$ is then precisely of the kernel form of equation (10.3.8), but now with a controlling smoothing parameter h specified directly via the discount factor.

We therefore have a version of the method of Gordon et al.(1993) applied to parameters that connects directly with kernel smoothing with shrinkage. This justifies the idea of an artificial evolution for fixed model parameters in a modification that removes the problem of information loss over time.

Notice also that the modified artificial evolution model of equation (10.3.12) may be adopted directly, without reference to the motivating discussion involving normal posteriors. This is clear from the following general result. Suppose $p(\boldsymbol{\theta}_t|D_t)$ has a finite mean $\bar{\boldsymbol{\theta}}_t$ and variance matrix $\boldsymbol{V}_t$, *whatever the global form of the distribution may be.* Suppose in addition that $\boldsymbol{\theta}_{t+1}$ is generated by the evolution model specified by equation (10.3.12). It is then easily seen that the mean and variance matrix of the implied marginal distribution $p(\boldsymbol{\theta}_{t+1}|D_t)$ are also $\bar{\boldsymbol{\theta}}_t$ and $\boldsymbol{V}_t$. Hence the connection of kernel smoothing with shrinkage, and the adjustment to fix the problem of information loss over time in artificial evolution approaches, is quite general. Note finally that the framework provides a direct method of specifying the scale of kernels via the single discount factor δ, as h (and a) are then simply determined. Generally, a higher discount factor – around 0.99 – will be relevant.

10.4   A general algorithm

Return now to the general filtering problem, that of sampling the posterior in equation (10.2.6). We have available the Monte Carlo sample $(\boldsymbol{x}_t^{(j)}, \boldsymbol{\theta}_t^{(j)})$ and weights $\omega_t^{(j)}$ ($j = 1, \ldots, N$), representing the joint posterior $p(\boldsymbol{x}_t, \boldsymbol{\theta}|D_t)$. Again, the suffix t on the parameter samples indicates the time t posterior, not time-variation. We adopt the kernel form of equation (10.3.8) as the marginal density for the parameter, following the earlier discussion. With the equivalent interpretation of this as arising from an artificial evolution with correlation structure, as just discussed, we can now apply an extended version of the auxiliary particle filter algorithm, incorporating the parameter with the state. The resulting general algorithm runs as follows.

1. For each $j = 1, ..., N$, identify the prior point estimates of (x_t, θ) given by $(\mu_{t+1}^{(j)}, m_t^{(j)})$ where

$$\mu_{t+1}^{(j)} = E(x_{t+1}|x_t^{(j)}, \theta_t^{(j)})$$

may be computed from the state evolution density and $m_t^{(j)} = a\theta_t^{(j)} + (1-a)\overline{\theta}_t$ is the jth kernel location from equation (10.3.9).

2. Sample an auxiliary integer variable from the set $\{1, ..., N\}$ with probabilities proportional to

$$g_{t+1}^{(j)} \propto w_t^{(j)} p(y_{t+1}|\mu_{t+1}^{(j)}, m_t^{(j)});$$

call the sampled index k.

3. Sample a new parameter vector $\theta_{t+1}^{(k)}$ from the kth normal component of the kernel density, namely

$$\theta_{t+1}^{(k)} \sim N(\cdot|m_t^{(k)}, h^2 V_t).$$

4. Sample a value of the current state vector $x_{t+1}^{(k)}$ from the system equation

$$p(\cdot|x_t^{(k)}, \theta_{t+1}^{(k)}).$$

5. Evaluate the corresponding weight

$$w_{t+1}^{(k)} \propto \frac{p(y_{t+1}|x_{t+1}^{(k)}, \theta_{t+1}^{(k)})}{p(y_{t+1}|\mu_{t+1}^{(k)}, m_t^{(k)})}.$$

6. Repeat step (2)-(5) a large number of times to produce a final posterior approximation $(x_{t+1}^{(k)}, \theta_{t+1}^{(k)})$ with weights $w_{t+1}^{(k)}$, as required.

Notice that the Monte Carlo sample size N can be different at each time point, if required. Also, we might be interested in over-sampling a rather larger set of values and then resampling according to the weights above in order to produce an equally weighted final sample. Further, it is generally appropriate to operate with parameters that are real-valued, when using the normal kernel method. Hence we routinely deal with log variances rather than variances, logit transforms of parameters restricted to a finite range – such as the autoregressive parameter in the following AR(1) example and the later dynamic factor model – and so forth.

Example

As a simple example in which the sequential updating is available in closed form for comparison, consider the AR(1) model in which $y_t = x_t$, a scalar, with $x_{t+1} \sim N(x_t\phi, 1)$. Here $\boldsymbol{\theta} = \phi$, a single parameter, and there is no unobserved state variable. Hence the focus is exclusively on the efficacy of learning the parameter (steps 1 and 4 of the general algorithm above are vacuous).

A realisation of length 897 was generated from this AR(1) model at $\phi = 0.8$. The sequential analysis was then performed over times $t = 1, \ldots, 897$, and the posterior approximation at $t = 897$ compared with the exact (normal) posterior for ϕ. The simulation-based analysis used $N = 5000$ sample points throughout. Figure 10.1 graphs the time trajectories of the posterior sample quantiles (2.5%, 25%, 50%, 75%, 97.5%) together with the exact posterior quantiles. Agreement is remarkable across the entire time period, with very little evidence of "build up" of approximation error. Table 10.1 provides a numerical comparison of exact and approximate quantiles for $p(\phi|D_{897})$ which further illustrates the accuracy. By comparison, a direct use of the artificial evolution method leads to a gradual decay of approximation accuracy due to the loss of information.

	0.025	0.25	0.50	0.75	0.975
exact:	0.7755	0.8005	0.8136	0.8267	0.8517
approx:	0.7758	0.7993	0.8119	0.8242	0.8482

Table 10.1. Posterior quantiles from the posteriors for the AR(1) parameter ϕ at $t = 897$.

10.5   Factor stochastic volatility modelling

In studies of dynamic latent factor models with multi-variate stochastic volatility components, recent Bayesian work has developed both MCMC methods and aspects of sequential analysis using versions of auxiliary particle filtering for states (Aguilar and West 2000, Pitt and Shephard 1999a). In these models, the state variables are latent volatilities of both common factor processes and of residual/idiosyncratic random terms specific to observed time series. As applied to exchange rate modelling, forecasting and portfolio analysis, Aguilar and West (1998) use MCMC methods to fit these complicated dynamic models to historical data, and then carry out sequential particle filtering over a long stretch of additional data providing a context for sequential forecasting and portfolio construction. In that example, these authors fix a full set of constant model parameters at estimated values taken as the means of posterior distributions based on the MCMC

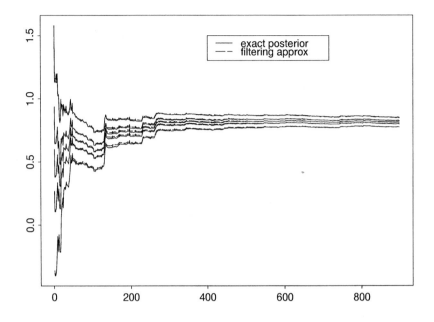

Figure 10.1. Time trajectories of posterior quantiles (2.5%, 25%, 50%, 75%, 97.5%) of the posteriors for the AR(1) parameter ϕ.

analysis of the initial (and very long) data stretch. The results are very positive from the financial time series modelling viewpoint. For many practical purposes, an extension of that approach involving periodic re-analysis of some recent historical data using full MCMC methods, followed by sequential analysis using auxiliary particle filtering solely on the time-varying states with model parameters fixed at most recently estimated values, is quite satisfactory. However, from the viewpoint of the use of sequential simulation technology in more interesting, and complicated models, this setting provides a very nice and somewhat challenging test-bed, especially when considering multi-variate time series on moderate dimensions. Hence our interest in exploring the general sequential algorithm of the previous section in this context.

We adopt the context and notation of Aguilar and West(1998), noting the very similar developments in Pitt and Shephard(1999a), and, that our models are based on those of the earlier authors (Kim, Shephard and Chib 1998). Begin with a q-variate time series of observations, in $\boldsymbol{y}_t$, ($t = 1, 2, \ldots$). In our example, this is a vector of observed daily exchange rates of a set of $q = 6$ national currencies relative to the US dollar. The dynamic latent factor model with multi-variate stochastic volatility components is defined and structured as follows.

At each time t, we have

$$\boldsymbol{y}_t = \boldsymbol{\alpha}_t + \boldsymbol{X}\boldsymbol{f}_t + \boldsymbol{\epsilon}_t, \qquad (10.5.13)$$

with the following ingredients.

- $\boldsymbol{y}_t$ is the q-vector of observation and $\boldsymbol{\alpha}_t$ is a q-vector representing a local level of the series.

- $\boldsymbol{X}$ is a $q \times k$ matrix called the *factor loadings matrix*.

- $\boldsymbol{f}_t$ is a k–vector which represents the vector of *latent factors* at time t; the $\boldsymbol{f}_t$ are assumed to be conditionally independent over time and distributed as $N(\boldsymbol{f}_t|\boldsymbol{0}, \boldsymbol{H}_t)$ where $\boldsymbol{H}_t = \mathrm{diag}(h_{t1}, \dots, h_{tk})$ is the diagonal matrix of instantaneous factor variances.

- $\boldsymbol{\epsilon}_t \sim N(\boldsymbol{\epsilon}_t|\boldsymbol{0}, \boldsymbol{\Psi})$ are idiosyncratic noise terms, assumed to be conditionally independent over time and with a diagonal variance matrix $\boldsymbol{\Psi} = \mathrm{diag}(\psi_1, \dots, \psi_k)$. The elements of $\boldsymbol{\Psi}$ are called the *idiosyncratic noise variances* of the series. We note that Aguilar and West (1998) use an extension of this model that, as in Pitt and Shephard (1999a), has time-varying idiosyncratic noise variances, but we do not consider here.

- $\boldsymbol{\epsilon}_t$ and $\boldsymbol{f}_s$ are mutually independent for all t, s.

Following earlier authors (Geweke and Zhou 1996), Aguilar and West (1998) adopt a factor loading matrix of the form

$$\boldsymbol{X} = \begin{pmatrix} 1 & 0 & 0 & \cdots & 0 \\ x_{2,1} & 1 & 0 & \cdots & 0 \\ \vdots & \vdots & \vdots & \cdots & 0 \\ x_{k,1} & x_{k,2} & x_{k,3} & \cdots & 1 \\ x_{k+1,1} & x_{k+1,2} & x_{k+1,3} & \cdots & x_{k+1,k} \\ \vdots & \vdots & \vdots & \cdots & \vdots \\ x_{q,1} & x_{q,2} & x_{q,3} & \cdots & x_{q,k} \end{pmatrix}. \qquad (10.5.14)$$

The reduced number of parameters in $\boldsymbol{X}$ ensures mathematical identification of the model and the lower-triangular form provides a nominal identification of the factors: the first series is driven by the first factor alone, the second series is driven by the first two factors, and so forth.

Stochastic volatility structures are defined for the sequences of conditional variances of the factors. For each $i = 1, \dots, k$, define $\lambda_{ti} = \log(h_{ti})$, and write $\boldsymbol{\lambda}_t = (\lambda_{t1}, \dots, \lambda_{tk})$. The set of log factor variances $\{\boldsymbol{\lambda}_t\}$ is modelled as a vector auto-regression of order one, VAR(1), to capture correlations in fluctuations in volatility levels. Specifically,

$$\boldsymbol{\lambda}_t = \boldsymbol{\mu} + \boldsymbol{\Phi}(\boldsymbol{\lambda}_{t-1} - \boldsymbol{\mu}) + \boldsymbol{\gamma}_t, \qquad (10.5.15)$$

with the following ingredients: $\boldsymbol{\mu} = (\mu_1, \dots, \mu_k)'$ is the underlying stationary volatility level, $\boldsymbol{\Phi} = \text{diag}(\phi_1, \dots, \phi_k)$ is a diagonal matrix with individual AR parameters ϕ_i for factor volatility process λ_{ti}, and the innovations vectors $\boldsymbol{\gamma}_t$ are conditionally independent and normal,

$$\boldsymbol{\gamma}_t \sim N(\boldsymbol{\gamma}_t | \mathbf{0}, \boldsymbol{U}) \qquad (10.5.16)$$

for some innovations variance matrix $\boldsymbol{U}$. This model differs from that of Pitt and Shephard (1999a) in several respects, an important one being that we allow non-zero off-diagonal entries in $\boldsymbol{U}$ to estimate dependencies in changes in volatility patterns across the factors. This turns out to be empirically supported and practically relevant in short-term exchange rate modelling. We mention that Aguilar and West (1998) also develop stochastic volatility model components for the variances $\boldsymbol{\Psi}$ of the idiosyncratic errors, but we do not explore them here.

We analyse the one-day-ahead returns on exchange rates over a period of several years in the 1990s, as in Aguilar and West (1998). Taking s_{ti} as the spot rate in US dollars for currency i on day t, the returns are simply $y_{ti} = s_{ti}/s_{t-1,i} - 1$ for currency $i = 1, \dots, q = 6$. The currencies are, in order, the German mark (DEM), Japanese yen (JPY), Canadian dollar (CAD), French franc (FRF), British pound (GBP) and Spanish peseta (ESP). Here we explore analysis of the returns over the period 12/1/92 to 8/9/96, for a total of 964 observations. We adopt the model as structured above, and take an assumed fixed return level $\boldsymbol{\alpha} = \boldsymbol{\alpha}$.

We first carried out an intense Bayesian analysis of the first 914 observations using the MCMC simulation approach of Aguilar and West (1998). At $t = 914$, we then have a full sample from the actual posterior, based on data up to that time, for all past latent factors, their volatilities, and all fixed model parameters. In terms of proceeding sequentially, we identify the relevant state variables and parameters as follows. First, notice that we can reduce the model equation (10.5.13) by integrating out the latent factors to give the conditional observation distribution

$$\boldsymbol{y}_t \sim N(\boldsymbol{y}_t | \mathbf{0}, \boldsymbol{X} \boldsymbol{H}_t \boldsymbol{X}' + \boldsymbol{\Psi}). \qquad (10.5.17)$$

Now, introduce the definitions of the state variable

$$\boldsymbol{x}_t \equiv \boldsymbol{H}_t$$

at time t, and the fixed model parameters

$$\boldsymbol{\theta} = \{\boldsymbol{X}, \boldsymbol{\Psi}, \boldsymbol{\mu}, \boldsymbol{\Phi}, \boldsymbol{U}\}.$$

In our example, the state variable $\boldsymbol{x}_t$ is three$-$dimensional, and the parameter $\boldsymbol{\theta}$ is 36$-$dimensional. As we discuss below, the sequential component of the study reported here treats $\boldsymbol{U}$ as fixed at a value based on the MCMC analysis of the first 914 observations, so that $\boldsymbol{\theta}$ reduces to 30 free model parameters, and the posterior at each time point is in 33 dimensions. For reference, the estimate of $\boldsymbol{U}$ is

$$E(\boldsymbol{U}|D_{914}) = \begin{pmatrix} 0.0171 & 0.0027 & 0.0009 \\ 0.0027 & 0.0194 & 0.0013 \\ 0.0009 & 0.0013 & 0.0174 \end{pmatrix},$$

based on the initial MCMC analysis over $t = 1, \ldots, 914$. The posterior standard deviations of elements of $\boldsymbol{U}$ at $t = 914$ are all of the order of 0.001-0.003, so there is a fair degree of uncertainty about $\boldsymbol{U}$ being ignored in the sequential analysis and comparison.

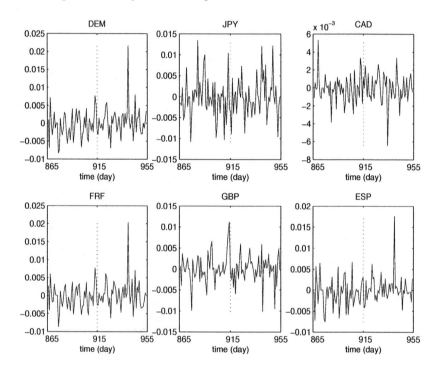

Figure 10.2. Exchange rate time series.

To connect with the general dynamic model framework, note that we now have the observation equation (10.2.1) defined by the model equation (10.5.17), and the evolution equation (10.2.2) given implicitly by the stochastic volatility model equations (10.5.15) and (10.5.16). We can now apply the general sequential filtering algorithm, and do so starting at time $t = 914$ with the full posterior $p(\boldsymbol{x}_{914}, \boldsymbol{\theta}|D_{914})$ available as a large Monte Carlo sample based on the MCMC analysis of all the data up to that time. It is relevant to note that the context here – with an informed prior

$p(\boldsymbol{x}_{914}, \boldsymbol{\theta}|D_{914})$ based on past data, is precisely that facing practical analysts in many fields in which further analysis, at least over short stretches of data, is required to be sequential. As noted above, we make one change: for reasons discussed below we fix the VAR(1) innovations variance matrix $\boldsymbol{U}$ at the estimate $E(\boldsymbol{U}|D_{914})$, the parameter $\boldsymbol{\theta}$ being reduced by removal of $\boldsymbol{U}$. Hence the particle filtering applies to the three state variables at each time point and the 30 model parameters. We then proceed to analyse further data, sequentially, over $t = 915, 916, \ldots$. Figure 10.2 displays a stretch of the data running from $t = 864$ to $t = 964$ with $t = 914$ marked. Our sequential filtering methods produces Monte Carlo approximations to each $p(\boldsymbol{x}_t, \boldsymbol{\theta}|D_t)$ over $t = 915, 916, \ldots$. Throughout the analysis, the Monte Carlo sample size is fixed at $N = 9000$ at each step. The kernel shrinkage and shapes are defined by means of the discount factor $\delta = 0.99$, implying $a = 0.995$ and $h = 0.1$. A final technical point to note is that we operate with the kernel method on parameters transformed so that normal kernels are appropriate; thus each of the μ_j and ϕ_j parameters is transformed to the logit scale, and the variance parameters ψ_j are logged (this follows West 1993a,b).

Our experiences in this study mirror those using the straight auxiliary particle filtering method when the parameters are assumed to be fixed (Aguilar and West 2000). That is, filtering on the volatilities is a more-or-less standard problem, and the state variable is in only three dimensions so performance is expected to be excellent. The questions of accuracy and performance in the extended context with a larger number of parameters become much more interesting, however, due to the difficulties inherent in dealing with discrete samples in higher dimensional parameter spaces. Inevitably, the accuracy of approximation is degraded relative to simple filtering on two or three time-evolving states. One way to define "performance" here is via comparison of the sequentially computed Monte Carlo approximations with posteriors with those based on a full MCMC analysis that refits the entire data set up to specified time points of interest. Our discussion here focuses specifically on this aspect in connection with inferences on the fixed model parameters. For a chosen set of times during the period of 50 observations between $t = 914$ and 964, we re-ran the full MCMC analysis of the factor model based on all the data up to that time, and explored comparisons with the sequential filtering-based approximations in which we begin filtering at $t = 914$. At any time t, the posterior from the MCMC analysis plays the role of the "true" posterior, or at least the "gold standard" by which to assess the performance of the filtering algorithm. Some relevant summaries appear in Figures 10.3 to 10.12. The first set, Figures 10.3 to 10.7, display summaries of the univariate marginal posterior distributions at $t = 924$. We refer to this as the 10-step analysis, because the sequential filter is run for just 10 steps from the starting position at $t = 914$. For each of the 30 fixed parameters, we display quantile plots comparing quantiles of the approximate posteriors from the MCMC and sequential analyses. The graphs indicate (with crosses) the posterior quantiles at 1, 5, 25, 50, 75, 95 and 99% of the posteriors, graphing the filtering-based quantiles versus those from the MCMC. The $y = x$ line is also drawn. From these graphs, it is evident that posterior margins are in excellent agreement (we could have added approximate intervals to these plots, based on methods of Bayesian density estimation, to represent uncertainty about the estimated quantile functions; for the large sample sizes of 9000 here, such intervals are extremely narrow except in the extreme tails, and just obscure the plots.) The poster margins computed by the auxiliary particle filtering (APF) and MCMC analyses are the same for all practical purposes. Only in the very extreme upper tail of two of the VAR model parameters – μ_1 and the logit of ϕ_1 – are there any deviations at all, and here the APF posterior is very slightly heavier tailed than that from the MCMC, but the differences are hardly worth a mention.

The remaining graphs, Figures 10.8 to 10.12, display similar quantile plots comparing the APF and MCMC posteriors $t = 964$. This provides a similar comparison but now for a 50-step analysis, the sequential filter running for 50 time points from the starting position at $t = 914$. Again we see a high degree of concordance in general, although the longer filtering period has introduced some discrepancies between the marginal posteriors, especially in the extreme tails of several of the margins. Some of the bigger apparent differences appear in the parameters ϕ and μ of the VAR volatility model component, indicated in Figures 10.11 and 10.12. It is also noteworthy that this period of 50 observations includes a point at around $t = 940$ where the series exhibits a real outlier, peaking markedly in the DEM, FRF, ESP and CAN series. Such events challenge sequential methods of any kind, and may play a role here in inducing small additional inaccuracies in the APF approximations by skewing the distribution of posterior weights at that time point. We do have ranges of relevant methods for model monitoring and adaptation to handle such events (West 1986, West and Harrison 1986, 1989 and 1997) though such methods are not applied in this study.

One additional aspect of the analysis worth noting is that the distributions of the sets of sequentially updated weights $\omega_t^{(j)}$ remain very well-behaved across the 50 update points. The shape is smooth and unimodal near the norm of $1/N = 1/9000$, with few weights deviating really far at all. Even at the outlier point the maximum weight is only is 0.004, fewer than 200 of the 9000 weights are less than 0.1/9000, and only 16 exceed 10/9000. All in all, we can view the analysis as indicating the utility of the filtering approach even over rather longer time intervals.

As earlier mentioned, the sequential simulation analysis fixes the volatility innovations variance matrix U at its prior mean $E(U|D_{914})$. This is because we have no easy way of incorporating a structured set of parameters such as U in the kernel framework – normal distributions do not apply

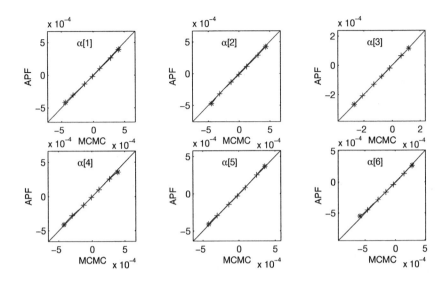

Figure 10.3. Q-Q plots of posterior samples of the α_j parameters in the 10-step analysis.

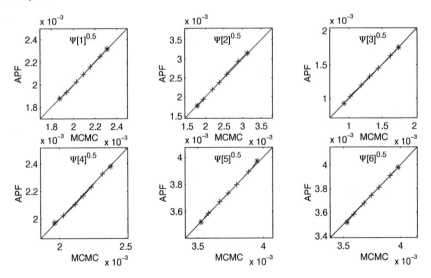

Figure 10.4. Q-Q plots of posterior samples of the $\sqrt{\psi_j}$ parameters in the 10-step analysis.

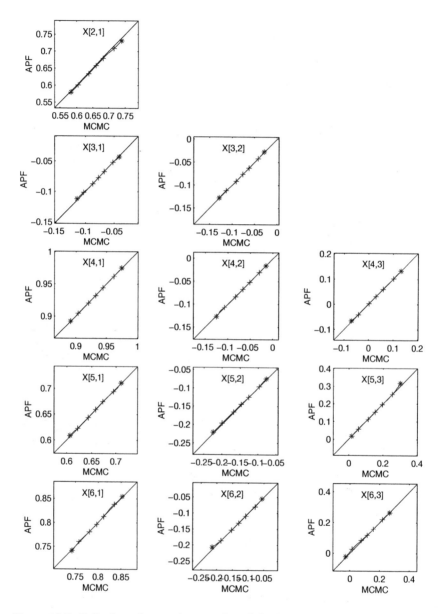

Figure 10.5. Q-Q plots of posterior samples of the X_{ij} parameters in the 10-step analysis.

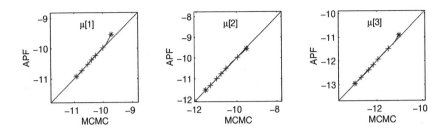

Figure 10.6. Q-Q plots of posterior samples of the μ_j parameters in the 10-step analysis.

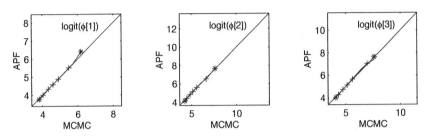

Figure 10.7. Q-Q plots of posterior samples of the logits of the ϕ_j parameters in the 10-step analysis.

to symmetric positive definite matrices of parameters. It may be that fixing this set of parameters induces some inaccuracies in the filtering analysis as compared with the MCMC analysis. With this in mind, we should expect to see some differences between the APF posterior and the MCMC, and these differences can be expected to be most marked in the margins for the volatility model parameters $\boldsymbol{\mu}$ and, most particularly, $\boldsymbol{\phi}$. The greatest differences in the 50-step analysis do indeed relate to $\boldsymbol{\phi}$, suggesting that some of the differences may indeed be due to the lack of proper accounting for the uncertainties about $\boldsymbol{U}$. Looking ahead, it is of interest to anticipate developments of kernel methods that might allow for such structuring, perhaps using normal kernels with elaborate reparametrisations, or perhaps with a combination of normal and non-normal kernels – though for the moment we have no way of doing this.

10.6   Discussion and future directions

In the moderate dimensional model above, the analysis certainly indicates the feasibility of sequential simulation-based filtering using the extended auxiliary particle filtering algorithm that incorporates several parameters in addition to state variables. Performance relative to the (almost) equiv-

218   Liu and West

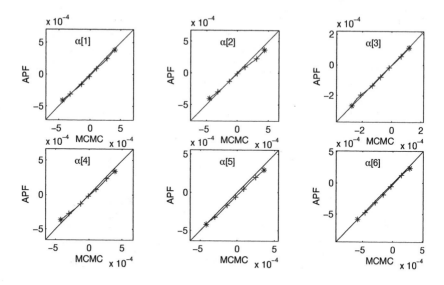

Figure 10.8. Q-Q plots of posterior samples of the α_j parameters in the 50-step analysis.

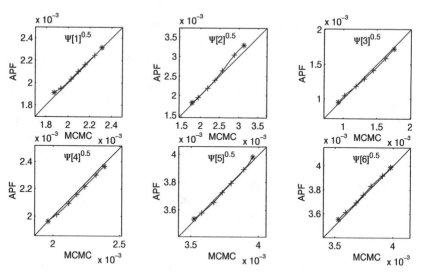

Figure 10.9. Q-Q plots of posterior samples of the $\sqrt{\psi_j}$ parameters in the 50-step analysis.

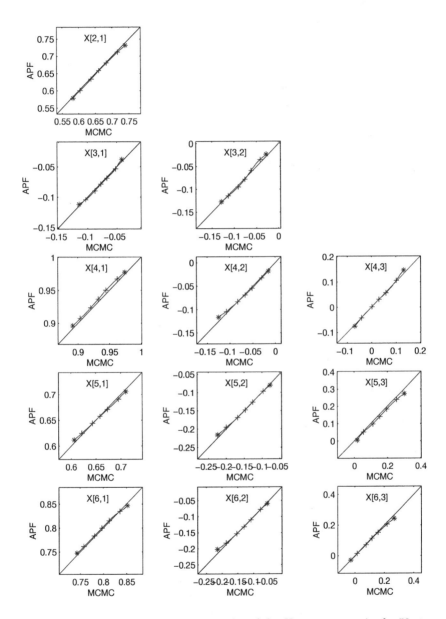

Figure 10.10. Q-Q plots of posterior samples of the X_{ij} parameters in the 50-step analysis.

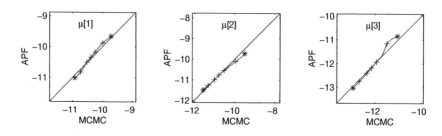

Figure 10.11. Q-Q plots of posterior samples of the μ_j parameters in the 50-step analysis.

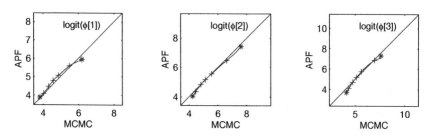

Figure 10.12. Q-Q plots of posterior samples of the logits of the ϕ_j parameters in the 50-step analysis.

alent MCMC analysis is excellent; for most practical purposes, the results are in good agreement with the MCMC results even in the 50-step analysis where some minor differences in tail behaviour are noted. We have indicated some possible reasons for these differences unrelated to the specific algorithm or the sequential context. If we ignore those issues and assume that all differences arise due to the inaccuracies inherent in sequential particle filtering, it is clear that there should be room for improvement. Before discussing some ideas and suggestions for improvements, we want to stress the relevance of context and goals. Sequential filtering inherently induces approximation errors that may tend to build up over time. In applied work, such as when using dynamic factor models in financial analysis, this must be accounted for and corrected. In existing applications of factor models with collaborators in the banking industry, the sequential filtering methods are used over only short time scales – the five days of a working week with daily time series, such as the exchange rate returns here presented. This is coupled with periodic updating based on a full MCMC analysis of a longer historical stretch of data (i.e., MCMC at the weekend based on the previous several months of data). The horizon of 10 days in the example is therefore very relevant, whereas the 50 day horizon is very long and perhaps unrealistic. In this context, the differences between the filtering-based and MCMC-based posterior quantile functions at lower time steps are quite negligible relative to those at the longer time step. This experience and perspective is consistent with our long-held view that sequential simulation-based filtering methods must always be combined with some form of periodic re-calibration based on off-line analysis performed with much more computational time available than the filtering methods are designed to accommodate.

Some final comments relate to possible extensions of the filtering algorithm that may improve posterior approximations. Questions of accuracy and adequacy arise in connection with the approximation of (typical) posteriors that exhibit varying patterns of dependencies among parameters in different regions of the parameter space, and also varying patterns of tail-weight. Discrete, sample-based approximations inevitably suffer problems of generating points far enough into the tails of fatter-tailed posteriors, especially in higher dimensions. It is sometimes helpful to use fatter-tailed kernels, such as T kernels (West 1993a, West 1993b), but this does not often help much and does not address the real need for more sensitive analytic approximation of *local* structure in the posterior; the kernel mixture of equation (10.3.8) is *global* in that the mixture components are based on the same "global" shrinkage center $\overline{\theta}_t$ and all have the same scales and shapes as determined by $h^2 V_t$. Very large numbers of such kernels are needed to truly adequately approximate posteriors that may evidence tails of differing weight in different dimensions (and fatter than normal tails), highly nonlinear relationships among the parameters and hence varying patterns of "local" scale and shape as we move around in θ space. We need to complement this suggestion with modifications that allow "differential shrinkage centers." West (1993a,b) discussed some of these issues, with suggestions about kernel methods with kernel-specific variance matrices in components in particular, and this idea was developed and implemented in certain non-sequential contexts in Givens and Raftery (1996). Development for implementation in sequential contexts remains an important research challenge.

A simple example helps to highlight these issues and underscores some suggestions for algorithmic extensions that follow. Consider a bimodal prior $p(\theta|D_t)$ in which one mode has the shape of a unit normal distribution and the other that of a normal but with a much larger variance. In using a kernel approximation based on a prior sample, we would expect to do very much better using a bimodal mixture in which sample points "near" one mode are shrunk towards that mode, and with kernel scalings that are higher for points "near" the second mode. The existing global kernel method uses global shrinkage to match the first two moments, but loses accuracy in less regular situations, as this example indicates.

A specific research direction that reflects these considerations completes our discussion. To begin, consider the existing framework and recall that *any* density function $p(\theta_t|D_t)$ may be arbitrarily well approximated by a mixture of normal distributions. Suppose, therefore, for theoretical discussion, that the density has exactly such a form, namely

$$p(\boldsymbol{\theta}_t|D_t) = \sum_{r=1}^{R} q_r N(\boldsymbol{\theta}_t|\boldsymbol{b}_r, \boldsymbol{B}_r) \tag{10.6.18}$$

for some parameters R and $\{q_r, \boldsymbol{b}_r, \boldsymbol{B}_r : r = 1, \ldots, R\}$ (these will all depend on t, though this is not made explicit in the notation, for clarity). In this case, the mean $\bar{\boldsymbol{\theta}}_t$ and variance matrix $\boldsymbol{V}_t$ are given by $\bar{\boldsymbol{\theta}}_t = \sum_{r=1}^{R} q_r \boldsymbol{b}_r$ and $\boldsymbol{V}_t = \sum_{r=1}^{R} q_r\{\boldsymbol{B}_r + (\boldsymbol{b}_r - \bar{\boldsymbol{\theta}}_t)(\boldsymbol{b}_r - \bar{\boldsymbol{\theta}}_t)'\}$. Suppose also, again, that $\boldsymbol{\theta}_{t+1}$ is generated by the evolution model specified by equation (10.3.12). It then easily follows that the implied marginal density $p(\boldsymbol{\theta}_{t+1}|D_t)$ is of the form

$$p(\boldsymbol{\theta}_{t+1}|D_t) = \sum_{r=1}^{R} q_k N(\boldsymbol{\theta}_{t+1}|a\boldsymbol{b}_r + (1-a)\bar{\boldsymbol{\theta}}_t, a^2 \boldsymbol{B}_r + (1-a^2)\boldsymbol{V}_t).$$

$$\tag{10.6.19}$$

Now, in general, this is not the same as the density of $(\boldsymbol{\theta}_t|D_t)$, though the mean and variance matrix match as mentioned above. In practice, a will be quite close to 1 so that the two distributions will be close, but not precisely the same in general. The exception is the case of a normal $p(\boldsymbol{\theta}_t|D_t)$, i.e., the case $R = 1$, when both $p(\boldsymbol{\theta}_t|D_t)$ and $p(\boldsymbol{\theta}_{t+1}|D_t)$ are $N(\cdot|\bar{\boldsymbol{\theta}}_t, \boldsymbol{V}_t)$. Otherwise, in the case of quite non-normal priors, the component means $\boldsymbol{b}_r$ will be quite separated, the local variance matrices $\boldsymbol{B}_r$ quite different in scale and structure. Hence, the location and scale/shape shrinkage effects in the components of the resulting mixture (10.6.19) tend to obscure the differences by the implied shrinking/averaging.

This discussion, referring back to the important use of normality of $\boldsymbol{\theta}_t$ in the theoretical tie-up between artificial evolution methods and kernel methods in Section 10.3.3, suggests the following development. Suppose that the distribution $p(\boldsymbol{\theta}_t|D_t)$ is indeed exactly of the form of equation (10.6.19). To focus on "local" structure in this distribution, introduce the component indicator variable r_t such that $r_t = r$ with probability q_r ($r = 1, \ldots, R$). Then $(\boldsymbol{\theta}_t|r_t = r, D_t) \sim N(\cdot|\boldsymbol{b}_r, \boldsymbol{B}_r)$. At this point we can apply the same line of reasoning about an artificial evolution to smooth a set of $\boldsymbol{\theta}_t$ samples, but now explicitly including the indicator r_t provides a focus on the *local* structure. This suggests the modification of the key evolution equation (10.3.12) to the local form

$$p(\boldsymbol{\theta}_{t+1}|\boldsymbol{\theta}_t, r_t = r, D_t) \sim N(\boldsymbol{\theta}_{t+1}|a\boldsymbol{\theta}_t + (1-a)\boldsymbol{b}_r, h^2 \boldsymbol{B}_r), \tag{10.6.20}$$

where a, h are as earlier defined. This conditional distribution is such that the implied marginal $p(\boldsymbol{\theta}_{t+1}|D_t)$ has precisely the same mixture form as $p(\boldsymbol{\theta}_t|D_t)$, so that the local structure is respected. This theoretical discussion therefore indicates and opens up a direction for development that, if implemented, can be expected to generate more accurate and efficient methods of smoothing posterior samples. To exploit this mixture theory will require, among other things, computationally and statistically efficient methods of identifying the parameters R and $\{q_r, \boldsymbol{b}_r, \boldsymbol{B}_r : r = 1, \ldots, R\}$ of the mixture in equation (10.6.18) based on an existing Monte Carlo sample (and weights) from that distribution. Some form of hierarchical clustering of sample points (and weights), such as utilised in West (1993a,b), will be needed, though a key emphasis lies with computational efficiency, so new clustering methods will be needed. Such developments, while challenging, will directly contribute in this context to usefully extend and improve the existing algorithms for sequential filtering on both parameters and states in higher dimensional dynamic models.

Acknowledgements

The authors are grateful for comments and fruitful discussions with Omar Aguilar, Simon Godsill, Neil Gordon, Pepe Quintana and an anonymous referee. This work was performed under partial support of NSF grant DMS-9704432 (USA) while the first author was a PhD student at Duke University. We should also like to acknowledge the support of CDC Investment Management Corporation.

11

A Theoretical Framework for Sequential Importance Sampling with Resampling

Jun S. Liu
Rong Chen
Tanya Logvinenko

11.1   Introduction

Monte Carlo filters (MCF) can be loosely defined as a set of methods that use Monte Carlo simulation to solve *on-line* estimation and prediction problems in a dynamic system. Compared with traditional filtering methods, simple, flexible — yet powerful — MCF techniques provide effective means to overcome computational difficulties in dealing with nonlinear dynamic models. One key element of MCF techniques is the recursive use of the importance sampling principle, which leads to the more precise name *sequential importance sampling* (SIS) for the techniques that are to be the focus of this article.

The earliest SIS method dates from the 1950s, (Hammersley and Morton 1954, Rosenbluth and Rosenbluth 1955) when scientists were interested in the computer simulation of a long-chain polymer. The polymer model they analyzed was the *self-avoid (random) walk* (SAW) on a d-dimensional lattice space (Kremer and Binder 1988). Figure 11.1 shows a realisation of SAW of length $N = 149$ (with 150 nodes/links), on a 2-D lattice. Each node on the chain represents the position of a monomer.

For a given spatial configuration of this polymer, say, $\boldsymbol{x}_N = (x_0, \dots, x_N)$, where x_j denotes the position of the jth monomer, one can compute the *potential energy* $U(\boldsymbol{x}_N)$ of this polymer according to natural laws in chemistry and physics. According to statistical physics, the probability of seeing such a configuration (in nature) follows the Boltzmann distribution

$$\pi_N(\boldsymbol{x}_N) = \frac{1}{Z_N}\exp\{-U(\boldsymbol{x}_N)/kT\},$$

A Self-Avoiding Walk of Length N=150

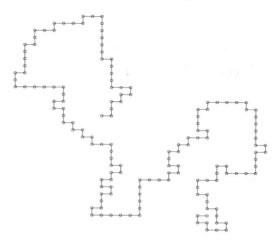

Figure 11.1. A SAW of length 150 on a 2-D lattice space. Without loss of generality, we started the chain at $(0,0)$ and placed its first step at $(0,1)$.

where Z_N is the normalising constant (also called the partition function). Temperature T and the Boltzmann constant k are assumed known. One is often interested in estimating Z_N or some *averages*, e.g., the mean squared extension $E_\pi \|x_N - x_0\|^2$, for this polymer system.

The simple energy function used by the Rosenbluths in their polymer studies is $U(\boldsymbol{x}_N) = 0$, meaning all allowable (self-avoiding) configurations receive equal probability. Generating a long SAW from this uniform distribution is not as simple as it first appears. The naive approach that generates simple random walks and accepts only those that do not self-cross becomes ineffective very rapidly (it will take about 50 million independently generated simple random walks in order to get a self-avoiding one of length 149). Another viable approach is to use the Metropolis algorithm (Metropolis et al. 1953), but the result was not very satisfactory because of the highly local nature of the Metropolis moves in a SAW (Kremer and Binder 1988). The most effective Markov chain based Monte Carlo method is the "slithering snake" method (Kremer and Binder 1988), which treats the SAW chain as a snake moving on the lattice space. However, this approach can not be generalised to handle non-constant energy function $U(\boldsymbol{x}_N)$. The *growth Monte Carlo* method of (Hammersley and Morton 1954) and (Rosenbluth and Rosenbluth 1955) is rather simple and effective. Instead of treating the whole configuration $\boldsymbol{x}_N$ directly, they propose to sequentially build up this configuration by adding one monomer a time, just like growing a molecule. To compensate for the bias induced by the sequential growth, an "importance weight" is computed and attached to the resulting polymer.

Although it has been nearly half a century, the growth Monte Carlo, together with its variations and improvements (Kremer and Binder 1988, Grassberger 1997, Wall and Erpenbeck 1959), remains the most attractive and versatile method for polymer simulations. As we will explain in the next section, the key component of this method is *sequential importance sampling* that can be very generally and simply formulated and applied to different problems. Indeed, the MCF techniques recently emerged from the fields of statistics and engineering follow exactly the same principle. The applications of these techniques include computer vision (Isard and Blake 1996), deconvolution of digital signals (Liu and Chen 1995), financial data modelling (Pitt and Shephard 1999b), genetic linkage analysis (Irwing, Cox and Kong 1994), radar signal analysis (Gordon et al. 1993), medical diagnosis (Berzuini et al. 1997), and the standard Bayesian computations (Kong, Liu and Wong 1994, Liu 1996b, MacEachern et al. 1999).

This chapter is organised as follows. Section 11.2 introduces the basic concept of a *weighted sample* and sequential importance sampling. Section 11.3 describes some major improvements of the SIS including resampling, rejection control and marginalisation. Section 11.4 explains the Monte Carlo filters for state-space models; Section 11.5 shows a few applications of the techniques described in Sections 11.3 and 11.4, and Section 11.6 concludes the article with a discussion. Detailed mathematical derivations can be found in the Appendix.

11.2   Sequential importance sampling principle

11.2.1   Properly weighted sample

Suppose we want to evaluate $\theta = E_\pi h(x)$ for some arbitrary function h under the target distribution $\pi(x)$ (given up to a normalising constant).

Definition 11.2.1. *A set of weighted random samples $\{(x^{(j)}, w^{(j)})\}_{j=1}^m$ is called proper with respect to π if for any square integrable function $h(\cdot)$,*

$$E[h(x^{(j)})w^{(j)}] = cE_\pi h(x), \quad for \ j = 1, \ldots, m,$$

where c is a normalising constant common to all the m samples.

With this set of weighted samples, we can estimate θ as

$$\hat{\theta} = \frac{1}{W} \sum_{j=1}^m w^{(j)} h(x^{(j)}), \tag{11.2.1}$$

where $W = \sum_{j=1}^m w^{(j)}$. For instance, if the $x^{(j)}$ are drawn from π directly, the set of $\{(x^{(j)}, 1)\}$ is proper. In the context of importance sampling (Marshall 1956), one draws $x^{(j)}$ from a trial distribution $q(x)$ and gives it a weight

$$w^{(j)} = \frac{\pi(x^{(j)})}{q(x^{(j)})}.$$

Then $\{(x^{(j)}, w^{(j)})\}$ is proper with respect to π. One reason for using the renormalised estimator (11.2.1) in an importance sampling framework is that one does not need to know the normalising constant for π (in Definition 11.2.1, constant c can be unknown).

Monte Carlo's view of the world is that any probability distribution π, regardless of its complexity, can always be *represented* by a discrete (Monte Carlo) sample from it. By "representation" we mean that any computation of expectations using π can be replaced to an acceptable degree of accuracy by using the empirical distribution resulting from the discrete sample. This viewpoint is also central to the *multiple imputation* (Rubin 1987a). When dealing with importance sampling, we see that π can be represented, at least conceptually, by any set of properly weighted samples. An important issue in applications, however, is how to find a convenient and efficient discrete representation (i.e., the representation set that is easily generated whose estimation is accurate).

11.2.2   Sequential build-up

Now we go back to the problem of estimating $E_\pi h(x)$ for a target distribution π. For example, in the SAW case x is the configuration (the set of positions of all "beads") of the polymer of length N; π is the uniform distribution on all SAWs of length N; and $h(x)$ is the mean squared extension of the SAW. If x has a "natural decomposition" — e.g. $x = (x_0, \ldots, x_N)$, as in the SAW case (N is suppressed from x) — we can imagine building up x sequentially by adding one monomer a time. According to the "telescope" law of probability

$$\pi(x) = \pi(x_0)\pi(x_1 \mid x_0) \cdots \pi(x_N \mid x_0, \ldots, x_{N-1}),$$

we should draw x_0 according to its marginal distribution under π, and then add x_1 conditional on x_0 according to $\pi(x_0, x_1)$ (with other components, $x_2, \ldots, x_N$, integrated out), etc. This task is not feasible in most applications, although exceptions exist (Diaconis and Shahshahani 1987).

Another strategy for sequential build-up is to draw x_0 from some other distribution q_0 which is reasonably close to π, and then sample x_1 conditional on x_0 from $q_1(x_1|x_0)$, and so on. We eventually obtain a sample of $x = (x_0, \ldots, x_N)$ according to the sampling distribution

$$q(x) = q_0(x_0)q_1(x_1 \mid x_0) \cdots q_N(x_N \mid x_0, \ldots, x_{N-1}).$$

To make this x proper with respect to π, we need to weight it by $w(x) = \pi(x)/q(x)$. This idea offers one more insight beyond the standard importance sampling (Marshall 1956): it is often fruitful to construct the trial distribution sequentially. In many applications, we do have a hint as to how to choose those q's. This "hint" can be more precisely formulated as an evolving *probabilistic dynamic system* (Liu and Chen 1998):

Definition 11.2.2. *A probabilistic dynamic system is a sequence of probability distributions defined on spaces with increasing dimensions: $\pi_t(\boldsymbol{x}_t)$ for $t = 0, 1, \ldots, N$, where $\boldsymbol{x}_t = (x_0, \ldots, x_t)$.*

Suppose this system is so defined that $\pi_t(\boldsymbol{x}_t)$ is the "best" distribution (closest to π in a certain sense) that we can come up with at time t and that the end distribution π_N is identical to the target distribution π; then we can let $q_t(x_t|\boldsymbol{x}_{t-1})$ be the same as or at least close to $\pi_t(x_t|\boldsymbol{x}_{t-1})$. In this scenario, we are able to update the importance weight recursively as

$$w_t = w_{t-1} \frac{\pi_t(\boldsymbol{x}_t)}{\pi_{t-1}(\boldsymbol{x}_{t-1})q_t(x_t \mid \boldsymbol{x}_{t-1})} = w_{t-1} \frac{\pi_t(\boldsymbol{x}_{t-1})}{\pi_{t-1}(\boldsymbol{x}_{t-1})} \frac{\pi_t(x_t \mid \boldsymbol{x}_{t-1})}{q_t(x_t \mid \boldsymbol{x}_{t-1})}. \quad (11.2.2)$$

In practice, we only need to know (11.2.2) up to a normalising constant, which is sufficient (and often more efficient) to estimate the quantity of interest (say, θ) by using (11.2.1). In order for the *sequential importance sampling* (SIS) method to work efficiently, we require that π_t evolve towards the target distribution π smoothly as t increases.

In simulating SAWs, the growth Monte Carlo (Hammersley and Morton 1954, Rosenbluth and Rosenbluth 1955) chooses π_t as the uniform distribution on all the SAWs of length t and uses $q_t(x_t \mid \boldsymbol{x}_{t-1}) = \pi_t(x_t \mid \boldsymbol{x}_{t-1})$. In other words, suppose at step $t-1$ the ending position x_{t-1} has k_{t-1} unoccupied neighbors (i.e., those neighbors that have not been visited by $x_0, \ldots, x_{t-2}$); then $q_t(x_t \mid \boldsymbol{x}_{t-1})$ places the tth monomer (i.e. x_t) uniformly at one of the k_{t-1} neighbors. Distribution $\pi_t(\boldsymbol{x}_{t-1})$ differs from $\pi_{t-1}(\boldsymbol{x}_{t-1})$ by giving more probability to those $\boldsymbol{x}_{t-1}$ with more "open ends" (i.e., bigger k_{t-1}). The weight of the SAW so generated can be computed recursively as

$$w_t = w_{t-1}k_{t-1}. \quad (11.2.3)$$

A byproduct of the SIS is that the sample mean of the unnormalised weights is an estimate of the normalising constant Z_N (Rosenbluth and Rosenbluth 1955): we can easily see in the SAW simulation that the unnormalised weight (11.2.3) satisfies the identity $Ew_N = Z_N$. A more general result can be found in (Kong et al. 1994).

11.3   Operations for enhancing SIS

Early researchers (Wall and Erpenbeck 1959) noticed that the simple application of the sequential build-up strategy is not good enough to simulate very long SAWs. Specifically, the SAWs simulated by the growth Monte Carlo run into "cages" (i.e., the number of unoccupied neighbors of x_t is zero at a time $t < N$) more and more frequently as N increases. (Wall and Erpenbeck 1959) introduced an improvement strategy in which one re-uses those successfully simulated partial SAWs instead of restarting from scratch. Recently, (Grassberger 1997) added a new trick to "prune" away those partial SAWs that are associated with very small weights (i.e., are "doomed"). This set of strategies is largely equivalent to the resampling approach employed either implicitly or explicitly by statisticians (Gordon et al. 1993, Liu and Chen 1995). Other possibilities, including marginalisa-tion and rejection control for improving the performance of SIS, have also been proposed (Doucet 1998, Liu and Chen 1998, Liu, Chen and Wong 1998).

11.3.1   Reweighting, resampling and reallocation

Suppose at time t we have a set of random samples $\mathcal{S}_t = \{(\boldsymbol{x}_t^{(j)}, w_t^{(j)})\}_{j=1}^m$ properly weighted with respect to π. By treating $\mathcal{S}_t$ as a discrete rep-resentation of π, we can generate another discrete representation as follows:

- For $j' = 1, \ldots, \tilde{m}$,
    - let $\tilde{\boldsymbol{x}}_t^{(j')}$ be $\boldsymbol{x}_t^{(j)}$ independently with probability propor-tional to $a^{(j)}$;
    - let the new weight of $\tilde{\boldsymbol{x}}_t^{(j')}$ be $\tilde{w}_t^{(j')} = w_t^{(j)}/a^{(j)}$; and
- return the new representation $\tilde{\mathcal{S}}_t = \{(\tilde{\boldsymbol{x}}_t^{(j')}, \tilde{w}_t^{(j')})\}_{j'=1}^{\tilde{m}}$.

The new set $\tilde{\mathcal{S}}_t$ thus formed is also (approximately) proper with respect to π (Rubin 1987b). It is not obvious, however, why resampling is useful. In fact, it does not help at all in a static importance sampling scheme. A few heuristics (Liu and Chen 1995) are as follows: (a) resampling can prune away those hopelessly unrepresentative samples (by giving them a small $a^{(j)}$) and (b) resampling can produce multiple copies of those good samples (by giving them a big $a^{(j)}$) to help generate better future samples in the SIS setting. Consequently, in a probabilistic dynamic system with SIS, resampling helps one *steer* towards the right direction. In light of these arguments, one should choose $a^{(j)}$ as a monotone function of $w_t^{(j)}$.

If we let $a^{(j)} = w_t^{(j)}$, then the foregoing scheme is exactly the same as the one described earlier (Gordon et al. 1993, Liu and Chen 1995). But having additional flexibility in choosing the sampling weights $a^{(j)}$ is rather intriguing and can potentially be very useful. For example, the $a^{(j)}$

can be chosen to reflect certain "future trend" (Meirovitch 1985, Pitt and Shephard 1999b) or be chosen to balance the need of diversity (i.e., having multiple distinct samples) with the need of focus (i.e., giving more presence to those samples with big weights). A generic choice is

$$a^{(j)} = \sqrt{w_t^{(j)}}, \tag{11.3.4}$$

as suggested by Professor W.H. Wong. More generally, we can let $a^{(j)}$ be $[w_t^{(j)}]^\alpha$, where α can vary according to the coefficient of variation of the w_t.

Another important point regarding the generation of $\tilde{S}_t$ is that the *extra variation* due to resampling is unnecessary and unwanted (Liu and Chen 1995). Instead of resampling, a more efficient approach is the partially deterministic *reallocation*. For example, the following scheme can be implemented for generating $\tilde{S}_t$ from S_t (we assume $\sum_{j=1}^m a^{(j)} = m$):

- For $j = 1, \cdots, m$,
    1. For $a^{(j)} \geq 1$,
        - retain $k_j = \lfloor a^{(j)} \rfloor$ copies of the sample $x_t^{(j)}$;
        - assign weight $\tilde{w}_t^{(j')} = w_t^{(j)}/k_j$ for each copy/
    2. For $a^{(j)} < 1$,
        - remove the sample with probability $1 - a^{(j)}$;
        - assign weight $\tilde{w}_t^{(j')} = w_t^{(j)}/a^{(j)}$ to the survived sample.
- Return set $\tilde{S}_t$ consisting of the new set of x_t's and $\tilde{w}_t$'s produced in the foregoing procedures.

This scheme tends to slightly decrease the total sample size. Alternatively, one can choose $k_j = \lfloor a^{(j)} \rfloor + 1$, which will tend to slightly increase the sample size. In order to maintain a fixed sample size, a residual-resampling strategy (Liu and Chen 1995) can be applied to make up the lost samples. A similar strategy has also been used for genetic algorithms (Baker 1987). Another resampling scheme that guarantees the fixed sample size (Doucet et al. 1998) draws k_j from certain distribution and then reweights the jth candidate as $w_j/\{k_j \Pr(k_j > 0)\}$.

11.3.2 Rejection control and partial rejection control

Another useful technique for rejuvenating a sequential importance sampler is the rejection control (RC) method (Liu et al. 1998), which can be understood as a combination of the rejection method (von Neumann 1951) and importance sampling. In the RC, one monitors the *coefficient of variation* of the importance weights for $x_t^{(j)}$, defined as

$$cv_t^2 = \frac{\mathrm{var}(w_t^{(j)})}{E^2(w_t^{(j)})}.$$

This cv^2 can be used to derive a heuristic criterion, i.e., the *effective sample size*,

$$\mathrm{ESS}_t = \frac{m}{1 + cv_t^2},$$

which is heuristically understood as the equivalent number of i.i.d. samples at time t (Liu, 1996). Once ESS_t drops below a threshold, say $\mathrm{ESS}_t \leq \alpha_0 m$ $(0 < \alpha_0 < 1)$, we call this time a *dynamic* check-point. The check-point sequence prescribed in advance (e.g. every 10 steps) will be called *static* check-points.

When encountering a new check-point t_k (i.e. the kth check-point), we compute a *control threshold* c_k that may be a quantity given in advance or the median or quantile of the $w_{t_k}^{(j)}$. Then we check every sample to decide whether to accept it according to probability $\min\{1, w_{t_k}^{(j)}/c_k\}$. In other words, those samples with weights greater than or equal to c_k are automatically accepted, whereas those with weights $< c_k$ are accepted with a probability. All accepted samples are given a new weight $w_{t_k}^{(j*)} = \max\{c_k, w_{t_k}^{(j)}\}$. All rejected ones are restarted from $t = 0$ and re-checked at all previous check-points. It has been shown (Liu et al. 1998) that the RC operation is *proper* in the sense of Monte Carlo estimation and it always increases the ESS. A problem with the RC is that its computation cost increases rapidly as t increases, although the threshold c_k can be adjusted by the user to compromise between computation cost and the ESS.

To overcome computational difficulties associated with the RC, we can implement the *partial rejection control* (PRC), which combines the RC with resampling and can be understood as a *delayed resampling*. More precisely, if t is the kth check-point (record as $t_k = t$), we do the following:

Partial Rejection Control

A. Compute the *control threshold* c_k. It can be either the median or a quantile (say, upper quartile) of the weights $w_t^{(1)}, \ldots, w_t^{(m)}$.

B. For $j = 1, \ldots, m$: accept the jth sample with probability

$$p_j = \min\left\{1, \frac{w_t^{(j)}}{c_k}\right\}.$$

If accepted, its weight is updated to $\max\{w_t^{(j)}, c_k\}$.

C. For any rejected sample at time t_k, we go back to the previous check-point t_{k-1} to draw a previous *partial sample* $\boldsymbol{x}_{t_{k-1}}^{(j)}$. This partial sample is drawn from set $\mathcal{S}_{t_{k-1}}$ with probability proportional to its weight at time t_{k-1} (i.e., $w_{t_{k-1}}^{(j)}$) and given an initial weight of $W_{t_{k-1}}/m$, where $W_{t_{k-1}} = \sum_{j=1}^{m} w_{t_{k-1}}^{(j)}$. Apply the SIS recursion to this partial sample until the current time t_k; conduct the same rejection control steps A and B as described above.

D. Repeat Step C until acceptance.

Remark 1: The PRC differs from the RC in two major ways: (a) when rejection occurs, the PRC goes back to the previous check point, whereas the RC goes all the way back to time 0; (b) the PRC uses resampling (from the pools of samples at the previous checkpoint) whereas the RC does not.

Remark 2: One of the benefits of RC is that it simultaneously generates almost-independent samples (some slight dependence can be caused by using an estimated threshold value c_k) and controls the coefficient of variation in the weight distribution. The PRC does not produce independent samples because it uses resampling to meet computational need. Both the RC and PRC can be seen as methods to bring future information into the generation of the current state variables. By allowing an opportunity to go back (in time) to regenerate samples, PRC is especially advantageous in dealing with sudden changes in the dynamic system (such as outliers in the observation). In contrast, a simple resampling may destroy most of the "good" samples if an outlier is present, for the reason that it is not able to take advantage of future information.

Remark 3: The PRC scheme is related to the rejection method proposed in (Hürzeler and Künsch 1998). Two major differences are: (a) the PRC does not require any knowledge on the precise "envelope" constant needed by the rejection method; and (b) the PRC invokes the rejection step only at a checkpoint and incorporates future information into the rejection decision. In contrast, the rejection step in (Hürzeler and Künsch 1998) has to be carried out at each filtering step and the rejection decision is based only on the current information.

11.3.3 *Marginalisation*

By *marginalisation* we mean that, whenever possible, one should analytically integrate out as many components from the system as possible, although this may sometimes introduce global dependency among the remaining variables. For example, by conditioning on a latent discrete vector I_t that indicates mixture component, one can easily integrate out the state variable x_t in a conditional Gaussian state-space model, but the remaining indicator vector I_t no longer has the Markovian structure (Chen and Liu 2000, Doucet 1998, Liu and Chen 1998). In the context of Markov chain Monte Carlo simulations, this operation has also been shown to improve a Gibbs sampling algorithm (Carter and Kohn 1996, Liu, Wong and Kong 1994).

A general formulation of *marginalisation* is as follows. Suppose each component x_s of x_t can also be decomposed "horizontally" as $x_s = (\xi_s, \nu_s)$, then we write $x_t = (\boldsymbol{\xi}_t, \boldsymbol{\nu}_t)$. If we can obtain the marginal distributions

$$\pi_t(\boldsymbol{\xi}_t) = \int \pi_t(x_t) d\boldsymbol{\nu}_t$$

for the dynamic system, and have the corresponding *marginalised* sampling distribution

$$q_t(\xi_t \mid \boldsymbol{\xi}_{t-1}) = \pi_t(\xi_t \mid \boldsymbol{\xi}_{t-1})$$

for sequential sampling, then the marginalised SIS performs better than the original one (MacEachern et al. 1999). This strategy is particularly useful for handling the linear state-space model with a mixture Gaussian error distribution and target tracking. Specifically, one can let ξ_t be the latent indicator for the Gaussian component from which the error term derives. Thus, conditional on $\boldsymbol{\xi}_t$, the process is linear and Gaussian, and can be dealt with by the Kalman filter (Kalman 1960).

11.4   Monte Carlo filter for state-space models

Monte Carlo methods for nonlinear non-Gaussian filtering for the state-space models have received considerable attention recently. A key observation in this chapter is the *mathematical equivalence* between the structure of these MCF methods and the general SIS framework described in Section 11.2.

11.4.1   The general state-space model

Consider the *generalised state-space model*

$$x_t \sim f_t(\cdot \mid \boldsymbol{x}_{t-1}),$$
$$y_t \sim g_t(\cdot \mid x_t),$$

where $\boldsymbol{x}_{t-1} = (x_0, \ldots, x_{t-1})$ is a *latent process*. When this process is a Markov chain, the model reduces to the ordinary state-space model (or the hidden Markov model). Let $\boldsymbol{y}_{t-1} = (y_1, \ldots, y_{t-1})$ be the observations available at time $t-1$. Suppose at time $t-1$ we have the posterior distribution of $\boldsymbol{x}_{t-1}$ as $p(\boldsymbol{x}_{t-1} \mid \boldsymbol{y}_{t-1})$, then the predictive distribution for x_t is

$$p(x_t \mid \boldsymbol{y}_{t-1}) = \int f_t(x_t \mid \boldsymbol{x}_{t-1}) p(\boldsymbol{x}_{t-1} \mid \boldsymbol{y}_{t-1}) d\boldsymbol{x}_{t-1}.$$

If we have a further observation y_t, the new posterior at time t becomes

$$p(\boldsymbol{x}_t \mid \boldsymbol{y}_t) \propto g_t(y_t \mid x_t) f_t(x_t \mid \boldsymbol{x}_{t-1}) p(\boldsymbol{x}_{t-1} \mid \boldsymbol{y}_{t-1}).$$

Of interest to researchers is the problem of estimating "true signal characteristics," say $h(x_t)$, *on-line*. Clearly, the optimal solution (under the MSE criterion) of this estimation problem is the *Bayes estimator*

$$\hat{h}_t = E[h(x_t) \mid \boldsymbol{y}_t] = \int h(x_t) p(\boldsymbol{x}_t \mid \boldsymbol{y}_t) d\boldsymbol{x}_t.$$

When the system is linear and Gaussian, this Bayes solution can be obtained recursively through the Kalman filter; when x_t only takes on a few discrete values, the solution can also be obtained by a recursive summation algorithm (known as the "peeling algorithm" in genetic linkage analysis). Otherwise, the Bayes estimator can only be obtained through costly numerical approximations, these often being impractical for many real problems.

By denoting $\pi_t(\boldsymbol{x}_t) = p(\boldsymbol{x}_t \mid \boldsymbol{y}_t)$, we see that the *posterior probability system* induced by the state-space model is just like the one we introduced in Section 11.2. If we ignore special features of the state-space model or of long-chain polymers, the system of interest here is identical in form to the polymer model. Therefore, the general SIS strategy, together with its improvements, can be applied. More interestingly, we do not have to take the dynamic system $\{\pi_t\}$ as the one directly suggested by the problem. Instead, we may choose to form a dynamic system with those $\pi_t(\boldsymbol{x}_t)$ reflecting more on the future trend. For example, we can carry out delayed, or some other forms, of estimations of x_t in order to bring in future information.

Under the SIS framework, we proceed from a discrete representation $S_{t-1} = \{(\boldsymbol{x}_{t-1}^{(j)}, w_{t-1}^{(j)})\}_{j=1}^m$ of the posterior distribution $p(\boldsymbol{x}_{t-1} \mid \boldsymbol{y}_{t-1})$ to the discrete representation S_t of $p(\boldsymbol{x}_t \mid \boldsymbol{y}_t)$ as follows. Draw a sample $x_t^{(j)}$ from a $q_t(x_t \mid \boldsymbol{x}_{t-1}^{(j)})$ to form $\boldsymbol{x}_t^{(j)} = (\boldsymbol{x}_{t-1}^{(j)}, x_t^{(j)})$; attach to it a weight computed recursively according to (11.2.2):

$$w_t^{(j)} = w_{t-1}^{(j)} \times \frac{f_t(x_t^{(j)} \mid \boldsymbol{x}_{t-1}^{(j)}) \, g_t(y_t \mid x_t^{(j)})}{q_t(x_t^{(j)} \mid \boldsymbol{x}_{t-1}^{(j)})}.$$

Then $\mathcal{S}_t = \{(\boldsymbol{x}_t^{(j)}, w_t^{(j)})\}_{j=1}^m$ is proper with respect to π_t. If the weights are too skewed (as measured by a low ESS), we can undertake resampling, or rejection control, or both, as described in Section 11.2. This defines the recursive procedure of SIS with resampling for the state-space model.

11.4.2 Conditional dynamic linear model and the mixture Kalman filter

As a special state-space model, the conditional dynamic linear model (CD LM) can be defined as:

$$\text{State equation:} \qquad x_t = G_{t,i_1} x_{t-1} + \epsilon_{t,i_1}, \quad \text{if } I_{t,1} = i_1, \quad (11.4.5)$$

$$\text{Observation equation:} \qquad y_t = H_{t,i_2} x_t + e_{t,i_2}, \quad \text{if } I_{t,2} = i_2; \quad (11.4.6)$$

where $\epsilon_{t,j_1} \sim \mathcal{N}(\boldsymbol{0}, A_{i_1})$, $e_{t,i_2} \sim \mathcal{N}(\boldsymbol{0}, B_{i_2})$, and the indicator vector $I_t = (I_{t,1}, I_{t,2})$ is a discrete latent variable with a prior distribution $\pi_t(I_t)$. It is straightforward to generalise this prior to allow for Markov dependence among the I_t. The CDLM is a direct generalisation of the dynamic linear model (DLM) (West and Harrison 1997) and is more flexible in dealing with outliers, sudden jumps, clutters, and other nonlinear features.

The SIS can be easily applied to directly treat the x_t's (which form a latent Markov chain). Many previously available Monte Carlo filters for this model have been designed in such a way (Avitzour 1995, Gordon et al. 1993, Kitagawa 1996). A more sophisticated algorithm, however, can be derived by making use of the conditional Gaussian structure: that is, when conditioning on the values of $I_1, \ldots, I_t$, the CDLM defined by (11.4.5) and (11.4.6) becomes a DLM, and all the x_s, $s = 1, \ldots, t$, can be integrated out recursively by using a standard Kalman filter. When the SIS is applied to treat the marginal dynamic system for the indicators, we obtain the *mixture Kalman filter* (MKF), which is always more accurate than those Monte Carlo filters dealing with $\boldsymbol{x}_t$ directly (Chen and Liu 2000). Detailed formulas are given in the appendix.

Remark 4: Several approaches on designing efficient Markov chain Monte Carlo algorithms for this type of model have been proposed (Carlin et al. 1992, Carter and Kohn 1996). Two key ideas employed in the design are the "collapsing" and "grouping" (Liu et al. 1994). In particular, (Carter and Kohn 1996) used a discrete variable to indicate the mixing component for each observation and integrated out the state variable x_t in their MCMC iterations. Here we advocate a similar strategy, *marginalisation*, for conducting sequential importance sampling.

Remark 5: Compared with those Monte Carlo filters applied to x_t, which uses a discrete sample to approximate the posterior distribution of x_t, MKF uses a Gaussian mixture distribution instead. Under CDLM the "true" target distribution is indeed a mixture of Gaussians, although the number of the mixture components increases exponentially with t. The advantage of the MKF can be seen as follows. Suppose the target distribution π is a mixture Gaussian with k components, with known μ_i, σ_i^2 and mixing probabilities p_i. The particle filter obtains m samples $x^{(j)}$ by first drawing an $I = i$ with probability p_i and then drawing $x^{(j)}$ from $N(\mu_i, \sigma_i^2)$. In contrast, the MKF directly draws m samples of the indicator I. Using a Rao-Blackwellisation argument, we can easily see that the latter is more efficient (MacEachern et al. 1999). In simulations, we have seen results in which an MKF with $m = 50$ provides a better estimation accuracy than a particle filter with $m > 1,000$. The downside of the MKF is that its implementation usually involves more analytical manipulations and more careful programming.

11.5   Some examples

11.5.1   A simple illustration

For illustration, we consider the following simple model:

$$
\begin{aligned}
x_t &= ax_{t-1} + \epsilon_t, \quad \text{with } \epsilon_t \sim N(0, \sigma_i^2) \text{ for } I_t = i; \\
y_t &= x_t + \eta_t \quad \text{with } \eta_t \sim N(0, \sigma_\eta^2).
\end{aligned}
$$

We assume that $P(I_t = i \mid I_{t-1}, x_{t-1}) = P(I_t = i) = p_i$, which can be easily extended to deal with the Markovian case. For this example, we can carry out the computation easily (see details in the Appendix).

We first simulated a system with $a{=}.9$, $\sigma_1{=}.5$, $\sigma_2{=}1.5$, $p_1{=}.7$, and $\sigma_\eta{=}.3$. The MKF estimated the state variable x_t (on-line) very accurately, even though several sudden jumps occurred in the system. With $m{=}50$ and without using PRC, we produced a result better than that in Kitagawa (1996), who routinely used $m{=}10,000$ "particles." We also applied a PRC with $\alpha = 0.8$ (i.e., the PRC is invoked whenever ESS_t drops below $0.8M$) together with the MKF. The MKF-PRC algorithm outperformed the plain MKF in terms of total mean-squared errors (MSE) (i.e., $\sum_{t=1}^{T} \|\hat{x}_t - x_t\|^2$, where $\hat{x}_t \approx E(x_t|y_t)$) of on-line estimation. Figure 11.2 (a) presents a plot of the total MSE of the MKF estimates (small circles) versus those of the MKF-PRC estimates (small dots) in 100 repeated experiments. As we expected, such a difference disappeared as m increased to 1,000. For a total of 100 repeated experiments, the plain MKF took only 36 seconds of CPU time on a Sun Ultra 2 workstation, and the MKF-PRC, 90 seconds.

To confirm these observations, we simulated another dataset from the system with a different setting: $a{=}.9$, $\sigma_1{=}.2$, $\sigma_2{=}1.2$, $p_1 \equiv P(J = 1){=}.9$,

(a)   (b)

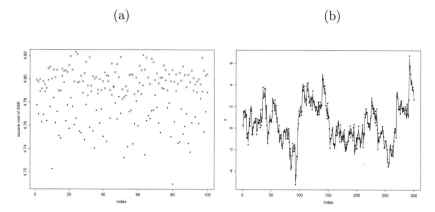

Figure 11.2. (a) The mean-squared-error of the estimates of x_t for using the MKF (circles) and the MKF-PRC (dots) in 100 repeated experiments. (b) The true x_t versus on-line estimation from MKF-PRC: dots represent observations (y_t); line, true values (x_t); dotted lines, estimates (i.e., $\hat{E}(x_t \mid \boldsymbol{y}_t)$).

and $\sigma_\eta=.8$. In this system, the noise level in the observation equation increased significantly. We found a similar improvement (more significant) of the MKF-PRC over the MKF (detail omitted). Figure 11.3 shows the comparison of the histograms of the logarithm of the importance weights under two schemes. As expected, the MKF-PRC had less variable importance weights.

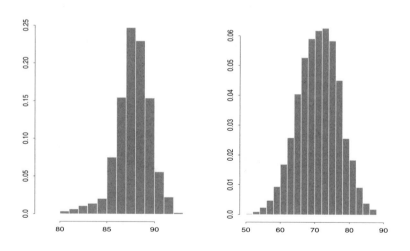

Figure 11.3. Histogram of the log-importance weights for MKF-PRC (left) and the simple MKF (right), respectively.

11.5.2 Target tracking with MKF

Target tracking on a 2-D plane can be modelled as follows: let $s_t = (s_{t,1}, s_{t,2})^T$ be the position vector, and let $v_t = (v_{t,1}, v_{t,2})^T$ be the velocity vector. They are supposed to evolve in the following way:

$$
\begin{pmatrix} s_{t,1} \\ s_{t,2} \\ v_{t,1} \\ v_{t,2} \end{pmatrix} = \begin{pmatrix} 1 & 0 & 1 & 0 \\ 0 & 1 & 0 & 1 \\ 0 & 0 & 1 & 0 \\ 0 & 0 & 0 & 1 \end{pmatrix} \begin{pmatrix} s_{t-1,1} \\ s_{t-1,2} \\ v_{t-1,1} \\ v_{t-1,2} \end{pmatrix} + \begin{pmatrix} \frac{1}{2} & 0 \\ 0 & \frac{1}{2} \\ 1 & 0 \\ 0 & 1 \end{pmatrix} \begin{pmatrix} \epsilon_{t,1} \\ \epsilon_{t,2} \end{pmatrix}
$$

$$
\begin{pmatrix} z_{t,1} \\ z_{t,2} \end{pmatrix} = \begin{pmatrix} s_{t,1} \\ s_{t,2} \end{pmatrix} + \begin{pmatrix} e_{t,1} \\ e_{t,2} \end{pmatrix}.
$$

The state equation innovation $\epsilon_t = (\epsilon_{t,1}, \epsilon_{t,2})^T$ is distributed as $\mathcal{N}(\mathbf{0}, \sigma_a^2 \mathbf{I})$. and the observation noise $e_t = (e_{t,1}, e_{t,2})^T$ follows $\mathcal{N}(\mathbf{0}, \sigma_b^2 \mathbf{I})$. Thus, the state equation for this tracking system is assumed, instead, to follow

$$
s_t = s_{t-1} + v_{t-1} + \frac{1}{2}\epsilon_t,
$$
$$
v_t = v_{t-1} + \epsilon_t,
$$

with $\epsilon_t \sim \mathcal{N}(\mathbf{0}, \sigma_a^2 \mathbf{I})$. More generally, by writing $x_t = (s_{t,1}, s_{t,2}, v_{t,1}, v_{t,2})^T$, we use the relationship $x_t = G x_{t-1} + \epsilon_t$, $\epsilon_t \sim \mathcal{N}(\mathbf{0}, \sigma_a^2 A)$, for tracking, where

$$
G = \begin{pmatrix} 1 & 0 & 1 & 0 \\ 0 & 1 & 0 & 1 \\ 0 & 0 & 1 & 0 \\ 0 & 0 & 0 & 1 \end{pmatrix}, \quad \text{and } A = \begin{pmatrix} \frac{1}{4} & 0 & \frac{1}{2} & 0 \\ 0 & \frac{1}{4} & 0 & \frac{1}{2} \\ \frac{1}{2} & 0 & 1 & 0 \\ 0 & \frac{1}{2} & 0 & 1 \end{pmatrix}.
$$

At a discrete time t, we observe a clutter of points, $y_t = \{y_{t,1}, \dots, y_{t,k_t}\}$ in a 2-D detection region with area Δ, in which the number of false signals follow a spatial Poisson process with rate λ. Generally speaking, the larger the detection region, the easier the tracking task but the more intensive the computation. A latent indicator variable I_t can be introduced, where $I_t = i > 0$ indicates that $y_{t,i}$ corresponds to the true signal z_t (a noisy measurement of s_t), and $I_t = 0$ means that z_t is not included in y_t. We can write down the conditional distribution of y_t as

$$
p(y_t \mid x_t, I_t = i) \propto \begin{cases} k_t (2\pi\sigma_b^2)^{-1} \exp\{-\frac{\|y_{t,i} - O x_t\|^2}{2\sigma_b^2}\}, & \text{if } i > 0, \\ \lambda, & \text{if } i = 0, \end{cases}
$$

$$(11.5.7)$$

where the observation matrix is

$$
O = \begin{pmatrix} 1 & 0 & 0 & 0 \\ 0 & 1 & 0 & 0 \end{pmatrix}.
$$

Because we only make observations in a finite detection region centered at a predictive location $(\hat{s}_{t,1}, \hat{s}_{t,2})$ (i.e., a guess of the target's move), say R_t, the "prior" probability $P(I_t = 0)$ has to be assessed as

$$\int P(z_t \notin R_t \mid x_t)p(x_t \mid x_{t-1})dx_t,$$

which is computable in principle. For the sake of simplicity, we ignore this technicality and assume that $P(I_t = 0) = 1 - p_d$ (i.e. the set y_t includes the true signal with probability p_d) and that $P(I_t = i \mid I_t > 0) = 1/k_t$.

Clearly, this tracking model has the form of a CDLM and the procedure outlined in Section 11.4.2 can be applied to marginalise the state variable. Thus, the resulting SIS only needs to operate on the indicator variable I_t (see the Appendix for detailed formulas) and takes the form of a MKF.

We simulated this tracking system with the observation noise level σ_a=.1, state innovation level σ_b=.5. The clutter of false signals is generated by a Poisson point process with rate λ=0.08 and the detection region at time t is a 10×10 square centered at the 1-step prediction of $s_t = (s_{t,1}, s_{t,2})$. The total tracking time was set as $T = 150$. One hundred independent replications of the above simulation were carried out and the MKF formulation described in the Appendix was applied to track the target. Figure 11.4 shows one of the simulated data (the line corresponds to the trace of the target being tracked).

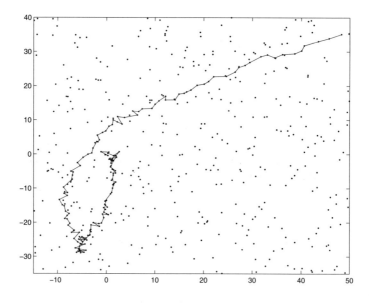

Figure 11.4. A simulation of the target tracking system. The dots connected by the line represent true positions of the target at each time step; the dots elsewhere represent confusing objects.

We applied four different SIS algorithms to each of the 100 simulated system. For ease of notations, we name the plain SIS without resampling as MKF; the one with systematic resampling every two steps as MKF-R2; the one with resampling every five steps as MKF-R5; and the MKF with partial rejection control a MKF-PRC. The MKF-PRC was operated with a sequence of dynamic check points corresponding to $\alpha_0 = 0.8$.

Our main interest in this simulation study was to compare the performances, in terms of both computing efficiency and tracking accuracy, of the four SIS strategies. For the first three methods, we let the number of "super-particles" (so named because each particle corresponds to one Kalman filter) m be 50 and for MKF-PRC, $m = 30$. Since our algorithms were implemented in MATLAB for illustration purposes, substantial improvement in terms of computational efficiency is expected if these methods were programmed in more basic languages such as Fortran or C. In Table 11.1, we record the total number of lost targets (defined as those for which $\max \|\hat{s}_t - s_t\| > C$) in 100 replications and the total amount of computing time (on SGI Challenge IRIX workstation) for each method. One of our tracking results (the estimates of target positions) is shown in Figure 11.5.

Methods	# Particles (m)	# Lost Targets	Computing Time
MKF	50	69 out of 100	39 min
MKF-R5	50	22 out of 100	74 min
MKF-R2	50	15 out of 100	87 min
MKF-PRC	30	10 out of 100	65 min

Table 11.1. A comparison between different sequential importance sampling strategies for the target tracking problem.

The MKF-PRC and MKF-R outperformed the plain MKF not only in terms of number of "lost" targets but also in the mean-squared errors (MSE) for those being correctly tracked (i.e., $\sum_{t=1}^{T} \|\hat{s}_t - s_t\|^2$, where $\hat{s}_t \approx E(s_t|\mathbf{y}_t)$) of on-line estimation. Figure 11.6 presents total MSEs of the three MKF estimates, i.e., the plain MKF, MKF-R2, and MKF-PRC, in 100 repeated experiments (m=50 for the plain MKF, and m=30 for the MKF-PRC). It is seen that MKF-R and MSE-PRC performed better than the plain MKF even among those tracked targets.

11.6 Discussion

In this article, we provide a theoretical framework for sequential importance sampling with resampling. We emphasise the general applicability of SIS and its enhancements, which ranges from molecular simulation to signal

(a)   (b)

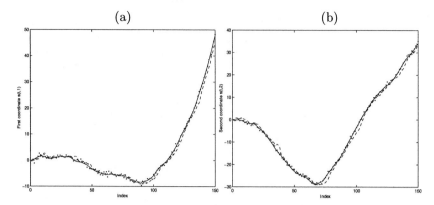

Figure 11.5. The true s_t versus its estimates from MKF-PRC. Dots represent true signals (z_t) (not necessarily observed); line, true value of s_t; dashed lines, estimates (i.e. $\hat{E}(s_t \mid \boldsymbol{y}_t)$). The first coordinate is plotted in (a) and the second in (b).

processing. We have also given examples showing how the two strategies, namely, partial rejection control and marginalisation, can be used to improve an SIS algorithm for on-line estimation with conditional dynamic linear models. Although the *marginalisation* operation may not be applicable to all nonlinear dynamic models, it usually results in dramatic improvements over the plain SIS or particle filter whenever applicable. Various resampling/reallocation schemes and the *partial rejection control* method can be broadly applied to all dynamic systems under the SIS setting. Such schemes assist in the design of effective Monte Carlo filters, and are also useful for general Monte Carlo computation (but not necessarily for on-line estimation problems). Our work demonstrates that the SIS is a powerful platform, enabling researchers to design various efficient sequential Monte Carlo algorithms for a large class of optimisation, estimation, and prediction problems.

11.7   Acknowledgements

The authors thank Dr. Faming Liang for assistance in computation, and Dr. Arnaud Doucet for valuable suggestions.

Appendix

The MKF recursion formula in Section 11.4.2

Suppose the initial value x_0, the coefficients G_{t,i_1} and H_{t,i_2}, and the variances σ_i^2 and δ_i^2 are given. After observing y_1 at time $t = 1$, we have a

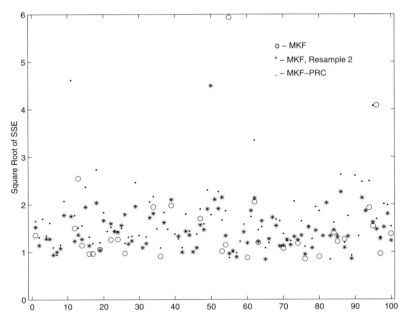

Figure 11.6. Comparison between the MSEs of the MKF, MKF-R2, and MKF-PRC for the estimation of s_t in 100 repeated simulations.

posterior distribution

$$p(I_1 = (i_1, i_2), x_1 \mid y_1, x_0) \propto p(y_1 \mid x_1, I_{1,1} = i_1)p(x_1 \mid x_0, I_{1,2} = i_2),$$

which gives us a marginal distribution

$$p(I_1 = (i_1, i_2) \mid y_1, x_0) \propto \int p(y_1 \mid x_1, I_{1,1} = i_1)p(x_1 \mid x_0, I_{1,2} = i_2)dx_1.$$

$$(11.7.8)$$

Distribution (11.7.8) can be computed precisely because it only involves manipulating Gaussian density functions. Thus, we can draw m i.i.d. or correlated (if a Markov chain sampler is used) samples, $I_1^{(1)}, \ldots, I_1^{(m)}$ from (11.7.8). For each $I_1^{(j)}$, we record the predictive mean $\mu_1^{(j)}$ and variance $V_1^{(j)}$ corresponding to the distribution $p(x_1 \mid y_1, x_0, I_1^{(m)})$, which is clearly Gaussian. Consequently, we have obtained a collection of m *Kalman filters*, $KF_1^{(1)}, \ldots, KF_1^{(m)}$, where $KF_1^{(j)} = (\mu_1^{(j)}, V_1^{(j)}, I_1^{(j)})$, at time $t = 1$. The weights $w_1^{(j)}$ associated with $KF_1^{(j)}$ are equal to one.

Suppose we have a collection of m Kalman filters, $KF_{t-1}^{(1)}, \ldots, KF_{t-1}^{(m)}$ at time $t - 1$. Each $KF_{t-1}^{(j)}$ corresponds to $(\mu_{t-1}^{(j)}, V_{t-1}^{(j)}, \boldsymbol{I}_{t-1}^{(j)})$, where $\boldsymbol{I}_{t-1}^{(j)} = (I_1^{(j)}, \ldots, I_{t-1}^{(j)})$ is the imputed latent indicator vector up to time $t - 1$. The

$\mu_{t-1}^{(j)}$ and $V_{t-1}^{(j)}$ are the mean vector and covariance matrix for the jth filter. Because the CDLM is reduced to a DLM when conditioning on $I_{t-1}^{(j)}$, vector $(\mu_{t-1}^{(j)}, V_{t-1}^{(j)})$ is simply the sufficient statistics at time $t-1$, an end-product of a standard Kalman filter. Each filter is associated with a weight $w_{t-1}^{(j)}$. The following algorithm gives rules on how to update these Kalman filters and their associated weights at time t.

The MKF recursion: for $j = 1, \dots, m,$

1. Obtain the predictive density for the new observation:

$$u_t^{(j)} = \sum_i \pi_t(I_t) \int p(y_t \mid x_t, I_t, KF_{t-1}^{(j)}) p(x_t \mid KF_{t-1}^{(j)}, I_t) dx_t,$$

and update the weight for the jth filter as

$$w_t^{(j)} = w_{t-1}^{(j)} \times u_t^{(j)}, \quad j = 1, \dots, m.$$

2. Impute the new indicator $I_t^{(j)}$ from its *a-posteriori* distribution

$$p(I_t \mid y_t, KF_{t-1}^{(j)}) \propto \pi_t(I_t) \int p(y_t \mid x_t, I_t, KF_{t-1}^{(j)}) p(x_t \mid KF_{t-1}^{(j)}, I_t) dx_t;$$

3. Update $KF_{t-1}^{(j)}$ to $KF_t^{(j)}$ by letting $I_t^{(j)} = (I_{t-1}^{(j)}, I_t^{(j)})$ and computing

$$\begin{aligned} \mu_t^{(j)} &= E(x_t \mid KF_{t-1}^{(j)}, y_t, I_t^{(j)}) \\ V_t^{(j)} &= \mathrm{var}(x_t \mid KF_{t-1}^{(j)}, y_t, I_t^{(j)}). \end{aligned}$$

At this time, an on-line estimate of the state-variable x_t is

$$\hat{x}_t = \frac{1}{W_t} \sum_{m=1}^M w_t^{(m)} \mu_t^{(m)},$$

where $W_t = w_t^{(1)} + \cdots + w_t^{(M)}$. Alternatively, we can also conduct a lag-k delayed estimate as

$$\hat{x}_{t-k} = \frac{1}{W_t} \sum_{m=1}^M w_t^{(m)} \mu_{t-k}^{(m)}.$$

(Liu and Chen 1998) shows that this procedure is proper.

Detailed computations involved in a MKF approach are very closely related to those in a standard KF method. Take any filter at time $t-1$, say $KF_{t-1}^{(j)} = (\mu_{t-1}^{(j)}, V_{t-1}^{(m)}, I_{t-1}^{(j)})$. For convenience, we omit the superscript (j) in the following derivation. At time t, we define

$$\mu_{+i_1} = G_{t,i_1} \mu_t, \quad V_{+i_1} = G_{t,i_1} A_{i_1} G_{t,i_1}^T + V_{t-1}.$$

Then by standard calculation, we have

$$p(x_t \mid I_{t,1} = i_1, KF_{t-1}) = \mathcal{N}(x_t \mid \boldsymbol{\mu}_{+i_1}, V_{+i_1})$$
$$p(y_t, x_t, I_t = (i_1, i_2) \mid KF_{t-1}) = \mathcal{N}(y_t \mid H_{t,i_2} x_t, B_{i_2}) \times \mathcal{N}(x_t \mid \boldsymbol{\mu}_{+i_1}, V_{+i_1}) \times \pi_t(i_1, i_2),$$

where $\mathcal{N}(x \mid \boldsymbol{\mu}, V)$ denotes a multivariate-Gaussian density. Hence, we can work out an explicit form of the required posterior distribution

$$p(x_t, I_t = (i_1, i_2) \mid y_t, KF_{t-1}) \propto C_t(i_1, i_2) \mathcal{N}(x_t \mid \boldsymbol{\mu}_t(i_1, i_2), V_t(i_1, i_2)),$$

where

$$\boldsymbol{\mu}_t(i_1, i_2) = V_t(i_1, i_2)(H_{t,i_2}^T B_{i_2}^{-1} y_t + V_{+i_1}^{-1} \boldsymbol{\mu}_{+i_1}),$$
$$V_t(i_1, i_2) = (H_{t,i_2}^T B_{i_2}^{-1} H_{t,i_2} + V_{+i_1}^{-1})^{-1},$$
$$C_t(i_1, i_2) = \pi_t(i_1, i_2) \exp\left\{ \frac{y_t^T B_{i_2}^{-1} y_t + \boldsymbol{\mu}_{+i_1}^T V_{+i_1}^{-1} \boldsymbol{\mu}_{+i_1} - (\boldsymbol{\mu}_t^T V_t^{-1} \boldsymbol{\mu}_t)|_{(i_1, i_2)}}{2} \right\}.$$

Thus, to update the filter, we first sample $I_t = (i_1, i_2)$ with probability proportional to $C_t(i_1, i_2)$; then we update $\boldsymbol{\mu}_t^{(m)}$ as $\boldsymbol{\mu}_t(i_1, i_2)$ and $V_t^{(m)}$ as $V_t(i_1, i_2)$. The weight for this updated filter is $w_t = w_{t-1} \times \sum_{i_1, i_2} C_t(i_1, i_2)$.

Computational details for the example in Section 11.5.1

Take any filter $KF_{t-1} = (\mu_{t-1}, V_{t-1}, \boldsymbol{I}_{t-1})$ at time $t-1$. We define

$$\mu_{+i} = a_i \mu_{t-1}; \quad V_{+i} = a_i^2 V_{t-1} + \sigma_i^2.$$

As a result, before we observe y_t, the distribution of x_t, conditional on $J_t = i$, is $N(\mu_{+i}, V_{+i})$. Thus, after seeing y_t, the posterior distribution of (J_t, x_t) is

$$
\begin{aligned}
p(J_t = i, X_t \mid KF_{t-1}, y_t) &\propto \pi_i \times N(x_t \mid \mu_{+i}, V_{+i}) \times N(y_t \mid x_t, \sigma_\eta^2) \\
&= C_t(i) \times N(x_t \mid \mu_t(i), V_t(i)),
\end{aligned}
$$

where

$$
\begin{aligned}
\mu_t(i) &= \left(\frac{\mu_{+i}}{V_{+i}} + \frac{y_t}{\sigma_\eta^2} \right) \left(\frac{1}{V_{+i}} + \frac{1}{\sigma_\eta^2} \right)^{-1}, \\
V_t(i) &= \left(\frac{1}{V_{+i}} + \frac{1}{\sigma_\eta^2} \right)^{-1}, \\
C_t(i) &= \frac{\pi_i}{\sqrt{2\pi}\sqrt{V_{+i} + \sigma_\eta^2}} \exp\left\{ -\frac{(y_t - \mu_{+i})^2}{2(V_{+i} + \sigma_\eta^2)} \right\}.
\end{aligned}
$$

Thus, when y_t is observed, the filter KF_{t-1}'s weight is adjusted to $w_t = w_{t-1} \times u_t$, where $u_t = \sum_i C_t(i)$. The indicator $I_t = i$ is drawn with a probability proportional to $C_t(i)$, which, together with the new mean and variance $\mu_t(i)$ and $V_t(i)$, forms the updated filter KF_t.

Computational detail for target tracking

Before observing y_t, however, the distribution of x_t can be written as

$$p(x_t \mid KF_{t-1}^{(m)}) = \mathcal{N}(x_t \mid G\boldsymbol{\mu}_{t-1}^{(m)}, \; \sigma_a^2 A + GV_{t-1}^{(m)}G^T) \equiv \mathcal{N}(x_t \mid \boldsymbol{\mu}_+, V_+).$$
$$(11.7.9)$$

For simplicity, we omit the superscript (m) with an understanding that all the following computations are conditioned on the mth filter $KF^{(m)}$.

Multiplying (11.7.9) with (11.5.7), we obtain the joint posterior of (x_t, I_t):

$$p(x_t, I_t = i \mid KF_{t-1}^{(m)}, y_t)$$

$$\propto \begin{cases} \dfrac{p_d}{2\pi\sigma_b^2} \exp\left\{-\dfrac{\|y_{t,i} - Ox_t\|^2}{2\sigma_b^2} - \dfrac{(x_t - \boldsymbol{\mu}_+)^T V_+^{-1}(x_t - \boldsymbol{\mu}_+)}{2}\right\} & \text{if } i > 0 \\[4mm] \lambda(1 - p_d)\exp\left\{-\dfrac{(x_t - \boldsymbol{\mu}_+)^T V_+^{-1}(x_t - \boldsymbol{\mu}_+)}{2}\right\} & \text{if } i = 0. \end{cases}$$
$$(11.7.10)$$

By further simplifying (11.7.10), we obtain

$$p(x_t, I_t = i \mid KF_{t-1}, y_t) \propto C_t(i)\mathcal{N}(x_t \mid \boldsymbol{\mu}^*(i), V^*(i)),$$

where $C_t(0) = \lambda(1 - p_d) \times (2\pi)^2 |V_+|^{1/2}$, $\boldsymbol{\mu}^*(0) = \boldsymbol{\mu}_+$, and $V^*(0) = V_+$; whereas for $i > 0$,

$$C_t(i) = (2\pi)^2 |V^*(i)|^{1/2} \times \frac{p_d}{2\pi\sigma_b^2}$$

$$\times \exp\left[-\frac{1}{2}\left\{\frac{\|y_{t,i}\|^2}{\sigma_b^2} + \boldsymbol{\mu}_+^T V_+^{-1}\boldsymbol{\mu}_+ - \boldsymbol{\mu}^*(i)^T V^*(i)^{-1}\boldsymbol{\mu}^*(i)\right\}\right]$$

$$\boldsymbol{\mu}^*(i) = V^*(i)\left(\frac{O^T y_{t,i}}{\sigma_b^2} + V_+^{-1}\boldsymbol{\mu}_+\right)$$

$$V^*(i) = \left(V_+^{-1} + \frac{O^T O}{\sigma_b^2}\right)^{-1}.$$

Therefore, the incremental weight of each $KF^{(m)}$ in the MKF is computed as $\sum_{i=0}^{k_t} C_t(i)$, and the imputation step can be achieved by drawing $p(I_t^{(m)} = i) \propto C_t(i)$. To complete the MKF, the mth filter is updated to $KF_t^{(m)} = (\boldsymbol{I}_t^{(m)}, \boldsymbol{\mu}_t^{(m)}, V_t^{(m)})$ with $\boldsymbol{I}_t^{(m)} = (\boldsymbol{I}_{t-1}^{(m)}, I_t^{(m)})$, where $\boldsymbol{\mu}_t^{(m)} = \boldsymbol{\mu}^*(i)$ and $V_t^{(m)} = V^*(i)$ for $I_t^{(m)} = i$.

12

Improving Regularised Particle Filters

Christian Musso
Nadia Oudjane
Francois Le Gland

12.1 Introduction

The optimal filter computes the posterior probability distribution of the state in a dynamical system, given noisy measurements, by iterative application of prediction steps according to the dynamics of the state, and correction steps taking the measurements into account. A new class of approximate nonlinear filter has been recently proposed, the idea being to produce a sample of independent random variables, called a particle system, (approximately) distributed according to this posterior probability distribution. The method is very easy to implement, even in high-dimensional problems, since it is sufficient in principle to simulate independent sample paths of the hidden dynamical system.

The earliest contribution in this field was (Gordon et al. 1993), which proposed to use sampling / importance resampling (SIR) techniques of (Rubin 1988) in the correction step. If the resampling step were skipped, then the method would reduce to the weighted particle method, a very inefficient method in which the measurements are only used to update the weights (likelihood) associated with the non-interacting particle system. The positive effect of the resampling step is to automatically concentrate particles in regions of interest of the state-space. However, it was already observed in (Gordon et al. 1993) that some difficulties could occur with the particle or bootstrap filter, if the hidden dynamical system was noise-free, or has very small noise, and also (for different reason however) if the observation noise had very small variance. Heuristics were proposed there to remedy these difficulties.

We present a more systematic approach in Section 12.2, one based on regularisation of the empirical distribution associated with the particle system, using the kernel method,; see (Silverman 1986). This results in two different approximations, called pre-regularised particle filter (pre-RPF) and post-regularised particle filter (post-RPF), depending on whether the regularisation step is taken before or after the correction step. Notice that an efficient implementation of the resampling step in the pre-RPF is the kernel filter (KF), which has been proposed in (Hürzeler and Künsch 1998). To handle efficiently the (difficult) case in which the observation noise has very small variance, we present an improved approximation in Section 12.3, based on a progressive correction (PC) principle, in which the correction step is split into sub-correction steps associated with a decreasing sequence of (fictitious) variance matrices for the observation noise. In Section 12.4, we present an additional improvement of the KF, called the local rejection regularised particle filter (L2RPF) that reduces the computing cost involved in the resampling step. In Section 12.5, simulation results are presented for several 2D–tracking problems, with bearings–only measurements, and with range and bearing measurements.

To be more specific, we consider the following model, with two sequences $(\mathbf{x}_t)_{t \geq 0}$ and $(\mathbf{y}_t)_{t \geq 0}$, called state and observation, and taking values in $\mathbb{R}^{n_x}$ and $\mathbb{R}^{n_y}$ respectively.

- In full generality, the state sequence $(\mathbf{x}_t)_{t \geq 0}$ is defined as an in-homogeneous Markov chain, with transition probability kernel Q_t, i.e.

$$\mathbb{P}(\mathbf{x}_t \in d\mathbf{x} \mid \mathbf{x}_{0:t-1}) = \mathbb{P}(\mathbf{x}_t \in d\mathbf{x} \mid \mathbf{x}_{t-1}) = Q_t(\mathbf{x}_{t-1}, d\mathbf{x}) ,$$

  for all $t \geq 1$, and with initial probability distribution p_0. For instance, $(\mathbf{x}_t)_{t \geq 0}$ could be defined by the equation

$$\mathbf{x}_t = F_t(\mathbf{x}_{t-1}, \mathbf{w}_t) ,$$

  where $(\mathbf{w}_t)_{t \geq 0}$ is a sequence of independent random variables, not necessarily Gaussian, independent of the initial state $\mathbf{x}_0$.

- The observation sequence $(\mathbf{y}_t)_{t \geq 0}$ is related to the state sequence $(\mathbf{x}_t)_{t \geq 0}$ by

$$\mathbf{y}_t = H_t(\mathbf{x}_t) + \mathbf{v}_t ,$$

  for all $t \geq 0$, where $(\mathbf{v}_t)_{t \geq 0}$ is a sequence of independent random variables, not necessarily Gaussian, independent of the state sequence $(\mathbf{x}_t)_{t \geq 0}$. It is assumed that $\mathbf{v}_t$ has a probability density, i.e. $\mathbf{v}_t \sim g_t(\mathbf{v}) \, d\mathbf{v}$, for all $t \geq 0$. This special form of the observation equation allows us to define the likelihood function (see below) which is necessary for the methods proposed here.

The problem of nonlinear filtering is to compute at each time t the conditional probability distribution $\pi_{t|t}$ of the state $\mathbf{x}_t$ given a realisation of the observation sequence $\mathbf{y}_{0:t} = (\mathbf{y}_0, \cdots, \mathbf{y}_t)$ up to time t. Even if the optimal filter is in general difficult to compute, its algorithm can be easily described in two steps. Introducing the conditional probability distribution $\pi_{t|t-1}$ of the state $\mathbf{x}_t$ given a realisation of the observation sequence $\mathbf{y}_{0:t-1} = (\mathbf{y}_0, \cdots, \mathbf{y}_{t-1})$ up to time $(t-1)$, the transition from $\pi_{t-1|t-1}$ to $\pi_{t|t}$ can be described by

$$\pi_{t-1|t-1} \xrightarrow[\text{Prediction}]{(1)} \pi_{t|t-1} = Q_t^* \, \pi_{t-1|t-1} \xrightarrow[\text{Correction}]{(2)} \pi_{t|t} = \Psi_t \cdot \pi_{t|t-1} \, .$$
$$(12.1.1)$$

1. Prediction: this step consists of the application of the transition probability kernel Q_t to $\pi_{t-1|t-1}$, i.e.

$$\pi_{t|t-1}(\mathrm{d}\mathbf{x}') = Q_t^* \, \pi_{t-1|t-1}(\mathrm{d}\mathbf{x}') = \int \pi_{t-1|t-1}(\mathrm{d}\mathbf{x}) \, Q_t(\mathbf{x}, \mathrm{d}\mathbf{x}') \, .$$
$$(12.1.2)$$

2. Correction: this step consists of Bayes rule. Since $\mathbf{y}_t$ is independent of the past observations $\mathbf{y}_{0:t-1}$ given $\mathbf{x}_t$, then introducing the likelihood function $\Psi_t(\mathbf{x}) = g_t(\mathbf{y}_t - H_t(\mathbf{x}))$, the correction step can be written as the projective product

$$\pi_{t|t}(\mathrm{d}\mathbf{x}) = \frac{\Psi_t(\mathbf{x}) \, \pi_{t|t-1}(\mathrm{d}\mathbf{x})}{\langle \pi_{t|t-1}, \Psi_t \rangle} = (\Psi_t \cdot \pi_{t|t-1})(\mathrm{d}\mathbf{x}) \, . \qquad (12.1.3)$$

Throughout this article, $S^N(\pi)$ denotes the N-empirical distribution from π, i.e.

$$S^N(\pi) = \frac{1}{N} \sum_{i=1}^{N} \delta_{\mathbf{x}^{(i)}} \quad \text{with } (\mathbf{x}^{(1)}, \cdots, \mathbf{x}^{(N)}) \text{ i.i.d. } \sim \pi \, .$$

Notice that the correction step applied to $S^N(\pi)$ yields

$$\Psi \cdot S^N(\pi) = \sum_{i=1}^{N} \omega^{(i)} \delta_{\mathbf{x}^{(i)}} \quad \text{with } (\mathbf{x}^{(1)}, \cdots, \mathbf{x}^{(N)}) \text{ i.i.d. } \sim \pi$$

$$\text{and } \omega^{(i)} = \Psi(\mathbf{x}^{(i)}) / \sum_{j=1}^{N} \Psi(\mathbf{x}^{(j)}) \, .$$

12.2   Particle filters

Particle methods are essentially based on Monte Carlo methods. The Monte Carlo principle allows approximation of a probability measure when a sample $(\mathbf{x}^{(1)}, \cdots, \mathbf{x}^{(N)})$ from that probability is given. This is a direct consequence of the law of large numbers, which states the weak convergence of $S^N(\pi)$ to π with rate $1/\sqrt{N}$.

12.2.1 The (classical) interacting particle filter (IPF)

The classical particle filter has appeared under several names in the literature, such as interacting particle filter (IPF) in (Del Moral 1996), sampling / importance resampling (SIR) in (Gordon et al. 1993), or branching particle filter (BPF) in (Crisan and Lyons 1997), where the method is developed for the continuous-time case. In all these cases, the transition from $\pi_{t-1|t-1}^N$ to $\pi_{t|t}^N$ is described by the following two steps.

$$\pi_{t-1|t-1}^N \xrightarrow[\text{Sampled}]{(1)} \pi_{t|t-1}^N = S^N(Q_t^* \, \pi_{t-1|t-1}^N) \xrightarrow[\text{Correction}]{(2)} \pi_{t|t}^N = \Psi_t \cdot \pi_{t|t-1}^N.$$

Prediction

$$(12.2.4)$$

In practice, those two steps consist of:

1. Sampled prediction
   - Resampling: generate the particles $(\mathbf{x}_{t-1|t-1}^{(1)}, \cdots, \mathbf{x}_{t-1|t-1}^{(N)})$ i.i.d. $\sim \pi_{t-1|t-1}^N$.
   - Evolution: independently for all i, generate the particles $\mathbf{x}_{t|t-1}^{(i)} \sim Q_t(\mathbf{x}_{t-1|t-1}^{(i)}, \cdot)$.

   Then set $\pi_{t|t-1}^N = S^N(Q_t^* \, \pi_{t-1|t-1}^N) = \dfrac{1}{N} \sum_{i=1}^N \delta_{\mathbf{x}_{t|t-1}^{(i)}}$.

2. Correction: for all i, compute
   - $\omega_t^{(i)} = \Psi(\mathbf{x}_{t|t-1}^{(i)})/\sum_{j=1}^N \Psi(\mathbf{x}_{t|t-1}^{(j)})$.

   Then set $\pi_{t|t}^N = \sum_{i=1}^N \omega_t^{(i)} \delta_{\mathbf{x}_{t|t-1}^{(i)}}$.

Remark 1. Here, resampling is equivalent to simulating the random vector $(N^{(1)}, \cdots, N^{(N)})$ representing the number of occurrences of each particle in the new system. This random vector follows a multinomial distribution with parameters $(N, N, (\omega_{t-1|t-1}^{(i)})_{1\leq i\leq N})$ and can be rapidly simulated, in order $O(N)$ steps, in the following way.

1. Generate directly ordered uniform variables $u^{(1)} \leq \cdots \leq u^{(N)}$, see the Malmquist theorem in (Devroye 1986, p. 212).

2. For all i, set $p_i = \sum_{j=1}^{i} \omega_{t-1|t-1}^{(j)}$, and return the number $N^{(i)}$ of $u^{(j)}$'s lying within the interval $[p_i, p_{i+1})$.

In the first step of the sampled prediction, i.e. in the resampling step, we generate N i.i.d. random variables according to the weighted discrete probability distribution $\pi_{t|t}^N = \sum_{i=1}^{N} \omega_t^{(i)} \delta_{\mathbf{x}_{t|t-1}^{(i)}}$. The more likely particles are selected, so that the particle system concentrates in regions of interest of the state-space. This produces a new particle system according to which several particles may have the same location. If the dynamical noise is small or nonexistent, the variety of the particle system decreases at each time step because of the accumulation of repetitions in the sample, and the particle system ultimately concentrates on a single point of the state-space. This phenomenon is called *particle degeneracy*, and causes the divergence of the filter.

Nevertheless, the IPF is proved to converge to the optimal filter, in the weak sense, with rate $1/\sqrt{N}$, but the error is not uniformly bounded in time (except under some strong mixing assumptions on the transition kernel Q_t see (Del Moral and Guionnet 1998b)) which explains why some divergent behaviors are still observed; see (Del Moral and Guionnet 1998a) for an analysis of the asymptotic behavior.

We propose to add a step in the preceding algorithm, to ensure the diversity of the particle system as time progresses. The resulting filters, called regularised particle filters (RPF), are presented below.

12.2.2 Regularised particle filters (RPF)

The main idea consists in changing the discrete approximation $\pi_{t|t}^N$ to a continuous approximation such that the resampling step is changed into simulations from an absolutely continuous distribution, hence producing a new particle system with N different particle locations. In doing this, we implicitly assume that the optimal filter $\pi_{t|t}$ has a smooth density, which is the case in most applications. From the theoretical point of view, this additional assumption allows us to produce strong approximations of the optimal filter, in L^1 or L^2 sense. In practice, this provides approximate filters which are much more stable in time than the IPF.

Regularisation of an empirical measure

Let the regularisation kernel K be a symmetric probability density function on $\mathbb{R}^{n_x}$, such that

$$K \geq 0 , \quad \int K(\mathbf{x})\,\mathrm{d}\mathbf{x} = 1 , \quad \int \mathbf{x}\,K(\mathbf{x})\,\mathrm{d}\mathbf{x} = 0 , \quad \int \|\mathbf{x}\|^2\,K(\mathbf{x})\,\mathrm{d}\mathbf{x} < \infty ,$$

and for any bandwidth $h > 0$, define the rescaled kernel

$$K_h(\mathbf{x}) = \frac{1}{h^{n_x}}\,K(\frac{\mathbf{x}}{h})$$

for any $\mathbf{x} \in \mathbb{R}^{n_x}$. For any probability distribution ν on $\mathbb{R}^{n_x}$, the regularisation of ν is the absolutely continuous probability distribution $K_h * \nu$ with density

$$\frac{\mathrm{d}(K_h * \nu)}{\mathrm{d}\mathbf{x}}(\mathbf{x}) = \int K_h(\mathbf{x} - \mathbf{u})\,\nu(\mathrm{d}\mathbf{u}) ,$$

where $*$ denotes the convolution operator. If $\nu = \Psi \cdot S^N(\pi) == \sum_{i=1}^{N} \omega^{(i)}\,\delta_{\mathbf{x}^{(i)}}$ is a discrete probability distribution on $\mathbb{R}^{n_x}$, where $(\mathbf{x}^{(1)}, \cdots, \mathbf{x}^{(N)})$ is a sample from π, then

$$\frac{\mathrm{d}(K_h * \nu)}{\mathrm{d}\mathbf{x}}(\mathbf{x}) = \frac{1}{h^{n_x}}\sum_{i=1}^{N} \omega^{(i)}\,K(\frac{1}{h}\,(\mathbf{x} - \mathbf{x}^{(i)})) = \sum_{i=1}^{N} \omega^{(i)}\,K_h(\mathbf{x} - \mathbf{x}^{(i)})$$

see Figure 12.1.

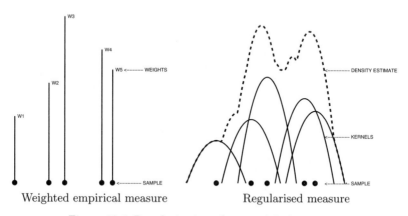

Weighted empirical measure   Regularised measure

Figure 12.1. Regularisation of an empirical measure

The kernel and bandwidth are chosen so as to minimise the mean integrated error $\mathbb{E}\|K_h * \nu - \Psi \cdot \pi\|_1$, or the mean integrated square error $\mathbb{E}\|K_h * \nu - \Psi \cdot \pi\|_2^2$ between the posterior distribution $\Psi \cdot \pi$ and the corresponding regularised weighted empirical measure. In the special case of a classical equally weighted sample, $\omega^{(i)} = 1/N$ for $i = 1, \cdots, N$, the density estimation theory, as described in (Silverman 1986, Devroye 1987), provides the optimal choice for the kernel,

$$K_{\mathrm{opt}}(\mathbf{x}) = \begin{cases} \dfrac{n_x + 2}{2c_{n_x}} (1 - \|\mathbf{x}\|^2) & \text{if } \|\mathbf{x}\| < 1 \\ 0 & \text{otherwise,} \end{cases} \qquad (12.2.5)$$

and when the underlying density is Gaussian with unit covariance matrix, the optimal choice for the bandwidth,

$$h_{\mathrm{opt}} = A(K)\, N^{-\frac{1}{n_x+4}} \quad \text{with} \quad A(K) = [8\, c_{n_x}^{-1}\, (n_x + 4)\, (2\sqrt{\pi})^{n_x}]^{\frac{1}{n_x+4}}\,, \qquad (12.2.6)$$

where c_{n_x} is the volume of the unit sphere of $\mathbb{R}^{n_x}$. K_{opt} defined above is called the Epanechnikov kernel. To reduce the computing cost of generating from the regularised measure we can replace the Epanechnikov kernel by the Gaussian kernel, the optimal bandwidth associated (when the underlying density is Gaussian with unit covariance matrix) is then

$$h_{\mathrm{opt}} = A(K)\, N^{-\frac{1}{n_x+4}} \quad \text{with} \quad A(K) = (4/(n_x + 2))^{\frac{1}{n_x+4}}\,. \qquad (12.2.7)$$

In the general case of an arbitrary underlying density π, we make two approximations when assuming that the density is Gaussian with covariance matrix S equal to the empirical covariance matrix of the sample $(\mathbf{x}^{(1)}, \cdots, \mathbf{x}^{(N)})$ from π. We then apply a linear transformation to achieve unit covariance (whitening). That is, $\mathbf{x}^{(i)}$ is changed into $A^{-1}\mathbf{x}^{(i)}$, where $A A^T = S$. The covariance matrix of the new particle system is then the unit matrix, and the bandwidth (12.2.6) or (12.2.7) can be used directly. This reduces to use the following rescaled regularisation kernel,

$$\frac{(\det A)^{-1}}{h^{n_x}}\, K(\frac{1}{h} A^{-1}\mathbf{x})\,. \qquad (12.2.8)$$

To handle the case of multi-modal densities, we chose $h = h_{\mathrm{opt}}/2$; see Figure 3.3 in (Silverman 1986) .

Remark 2. Generating from the Epanechnikov kernel (12.2.5) consists of generating $\sqrt{\beta}\, T$, where β follows a beta distribution with parameters $(n_x/2, 2)$, and T is uniformly distributed over the unit sphere of $\mathbb{R}^{n_x}$, see (Devroye and Györfi 1985, pp. 236–237).

Two types of regularised particle filters are proposed depending on whether the regularisation step is taken before or after the correction step.

The post–regularised particle filter

The post–regularised particle filter (post–RPF) has been proposed in (Musso and Oudjane 1998, Oudjane and Musso 1999) and compared with the IPF when applied to classic tracking problems such as the bearings–only problem or the range and bearing problem with multiple dynamic models. A theoretical analysis is developed in (Le Gland, Musso and Oudjane 1998).

We denote by $\tilde{\pi}^N_{t|t}$ the post–regularised approximation of $\pi_{t|t}$. The transition from $\tilde{\pi}^N_{t-1|t-1}$ to $\tilde{\pi}^N_{t|t}$ consists of these three steps

$$\tilde{\pi}^N_{t-1|t-1} \xrightarrow[\substack{\text{Sampled} \\ \text{Prediction}}]{(1)} \tilde{\pi}^N_{t|t-1} \xrightarrow[\text{Correction}]{(2)} \tilde{\nu}^N_t = \Psi_t \cdot \tilde{\pi}^N_{t|t-1}$$

$$\xrightarrow[\text{Regularisation}]{(3)} \tilde{\pi}^N_{t|t} = K_h * \tilde{\nu}^N_t \ .$$

(12.2.9)

Because of the regularisation step (3), the first stage of the sampled prediction (1) is changed. Instead of simply resampling the particle system, the following algorithm is implemented, independently for all i

---

1. Generate $I \in \{1, \cdots, N\}$, with $\mathbb{P}(I = j) = \omega^{(j)}_{t|t}$.

2. Generate $\varepsilon \sim K$, Epanechnikov (12.2.5) or Gaussian kernel.

3. Compute $\mathbf{x}^{(i)}_{t|t} = \mathbf{x}^{(I)}_{t|t} + h\,\Gamma_t\,\varepsilon$, with $h = h_{\text{opt}}$ given by (12.2.6) or (12.2.7). If whitening is used, $\Gamma_t = A_t$ the square root of the empirical covariance matrix, see (12.2.8), otherwise $\Gamma_t = I$.

In terms of complexity, this algorithm is similar to the IPF because it only requires N additional generations from the kernel K at each time step. Moreover, the post-RPF markedly improves upon the IPF performance, especially in cases of small dynamic noise, as would be expected.

The pre-regularised particle filter

A theoretical analysis of the pre-regularised particle filter (pre-RPF) is developed in (Le Gland et al. 1998). Otherwise, an improved version of the pre–RPF, the kernel filter (KF), is proposed in (Hürzeler and Künsch 1998). This latter version is adapted in Section 12.4 to reduce the computing cost.

We denote by $\bar{\pi}_{t|t}^{N}$ the pre-regularised approximation of $\pi_{t|t}$. The transition from $\bar{\pi}_{t-1|t-1}^{N}$ to $\bar{\pi}_{t|t}^{N}$ consists of the following three steps:

$$\bar{\pi}_{t-1|t-1}^{N} \xrightarrow[\substack{\text{Sampled} \\ \text{Prediction}}]{(1)} \bar{\pi}_{t|t-1}^{N} \xrightarrow[\text{Regularisation}]{(2)} \bar{\nu}_{t|t-1}^{N} = K_h * \bar{\pi}_{t|t-1}^{N}$$

$$\xrightarrow[\text{Correction}]{(3)} \bar{\pi}_{t|t}^{N} = \Psi_t \cdot \bar{\nu}_{t|t-1}^{N} .$$

$$(12.2.10)$$

The cost of the sampled prediction rises because it requires the implementation of the following rejection algorithm (see Section 12.3.1) independently for all i.

---

1. Generate I uniformly in $\{1, \cdots, N\}$.

2. Generate $\varepsilon \sim K$, Epanechnikov (12.2.5) or Gaussian kernel, and U uniformly on $[0, 1]$.

3. Put $\mathbf{X} = \mathbf{x}_{t|t-1}^{(I)} + h\,\Gamma_t\,\varepsilon$, with $h = h_{\text{opt}}$ given by (12.2.6) or (12.2.7). If whitening is used, $\Gamma_t = A_t$ the square root of the empirical covariance matrix, see (12.2.8), otherwise $\Gamma_t = I$.

4. If $\Psi_t(\mathbf{X}) > U \sup_{\mathbf{x} \in \mathbb{R}^{n_x}} \Psi_t(\mathbf{x})$, then return $\mathbf{x}_{t|t}^{(i)} = \mathbf{X}$, otherwise go to step 1.

Both RPF's are proved to converge to the optimal filter in the weak sense, with rate $(h^2 + 1/\sqrt{N})$. The term h^2 corresponds to the error owing to regularisation. When $h = 0$, one recovers the rate of convergence of the IPF. In the strong L^1 sense, the error estimate is proportional to $(h^2 + 1/\sqrt{N}\,h^{n_x})$ for the post–RPF and pre–RPF, but, as for the IPF, those error estimates are in general not uniformly bounded in time.

12.3 Progressive correction

In this section we propose methods to approximate the correction step in controlling the local cost, between $(t-1)$ and t, in terms of error and computing time.

Lemma 12.3.1. *Let π and π' be two probability distributions on $\mathbb{R}^{n_x}$, let Q be a transition probability kernel, and let Ψ be a positive bounded function* *in $\mathbb{R}^{n_x}$. Then the following inequalities hold:*

$$\|Q^* \pi - Q^* \pi'\|_{\mathrm{TV}} \leq \|\pi - \pi'\|_{\mathrm{TV}}, \qquad (12.3.11)$$

$$\|\Psi \cdot \pi - \Psi \cdot \pi'\|_{\mathrm{TV}} \leq \delta \|\pi - \pi'\|_{\mathrm{TV}}, \qquad (12.3.12)$$

with $\delta = \sup_{\mathbf{x} \in \mathbb{R}^{n_x}} \Psi(\mathbf{x})/\langle \pi, \Psi \rangle \geq 1$.

Inequality (12.3.12) is sharp, see (Hürzeler 1998, pp. 99–100), and shows that the correction step can induce tremendous errors when δ is high (which can occur when the measurement noise is small or when the state variable is unusual because then the denominator $\langle \pi, \Psi \rangle$ is small), whereas the prediction step is always a contraction, as seen from inequality (12.3.11). This is why we shall focus on the correction step. Henceforth, we are interested in methods to approximate the posterior probability distribution $\Psi \cdot \pi$. Remember that in the context of filtering the probability distribution π to be updated is the prediction filter (12.1.1). Afterwards, δ will be called the cost coefficient of the correction.

12.3.1   Focus on the correction step

In particle filtering, the main difficulties arise at the correction step. There are two kinds of problem encountered, depending on the method used to approximate the correction.

Weighting correction:

$\pi \to \Psi \cdot (S^N(\pi)) \to K_h * (\Psi \cdot (S^N(\pi)))$

As diagram (12.2.9) shows, this is the method used in the post–RPF. The correction step consists only in weighting the points of a sample from π. The likelihood of any other point of the state-space is not used. When measurements are accurate, which results in high δ, the likelihood function concentrates in a small region of the state-space that can be too small to contain points of the particle system. If that happens, we commonly observe the divergence of the filter during simulations, as suggested by (12.3.12). However, the computing time is cheap, and independent of δ.

Rejection correction:

$\pi \to K_h * (S^N(\pi)) \to \Psi \cdot (K_h * (S^N(\pi)))$

As diagram (12.2.10) shows, this is the method used in the pre–RPF. Here, the correction is applied directly to the regularised probability distribution. Hence, each point of the support of the probability density is updated and the problem induced by the discrete correction is solved. Unfortunately, this method requires to generate from π a random number of variables with negative binomial (or Pascal) distribution of parameter $(N, 1/\delta)$, and mean equal to $N\delta$ (as the sum of N independent variables geometrically distributed with parameter $1/\delta$). This implies that, when δ is high, we always observe a great increase of the computing time. However, during simulations we have never observed the divergence of the filter, even when δ is high.

Whatever the method chosen to approximate the correction, the case of high δ is difficult to deal with. It implies either an increase in the error or an increase in the computing time. The aim of progressive correction is to offset these outcomes.

12.3.2 Principle of progressive correction

In principle, progressive correction is based on the following elementary observation. If π is a probability distribution on $\mathbb{R}^{n_x}$, and if Ψ, Ψ_1 and Ψ_2 are three bounded non-negative functions defined on $\mathbb{R}^{n_x}$, such that $\Psi = \Psi_1 \, \Psi_2$, then it is equivalent to correct π by the function Ψ, or to correct π by Ψ_1 and then correct the resulting probability distribution by Ψ_2. In other words, $\Psi \cdot \pi = \Psi_2 \cdot (\Psi_1 \cdot \pi)$.

Progressive correction consists of splitting the correction step into several sub-correction steps for which the coefficient δ is well controlled. To do so, we need to choose a decomposition $(\Psi_1, \cdots, \Psi_n)$ of the likelihood function Ψ. This problem will be discussed in Section 12.3.3. For the moment, assume that this decomposition is given. Let $\nu_k = \Psi_k \cdot \nu_{k-1}$ denote the distribution resulting from the first k sub-corrections, $1 \le k \le n$, with $\nu_0 = \pi$. Let δ_k denote the cost coefficient corresponding to the kth sub-correction, i.e. $\delta_k = \sup\limits_{\mathbf{x} \in \mathbb{R}^{n_x}} \Psi_k(\mathbf{x})/\langle \nu_{k-1}, \Psi_k \rangle$. It is easy to check that the total cost coefficient is equal to the product of the sub-cost coefficients, i.e. $\delta = \delta_1 \cdots \delta_n$ as soon as all fictitious likelihood functions Ψ_k reach their maximums in the same point of the state-space. There are two kinds of progressive correction methods depending on the method used to implement each sub-correction step.

Progressive weighting correction

Let $\tilde{\nu}_k$ denote the approximation of ν_k by successive weighting corrections

$$\tilde{\nu}_k = K_h * (\Psi_k \cdot (S^N(\tilde{\nu}_{k-1}))) \, , \quad 1 \le k \le n \, , \quad \text{with } \tilde{\nu}_0 = \pi.$$

Just as δ_k, we define $\tilde{\delta}_k = \sup\limits_{\mathbf{x} \in \mathbb{R}^{n_x}} \Psi_k(\mathbf{x})/\langle \tilde{\nu}_{k-1}, \Psi_k \rangle$. Let $\mathcal{S}_k$ denote the σ-field generated by the random variables simulated in the transition from $\tilde{\nu}_0$ to $\tilde{\nu}_k$. Using inequality (12.3.12) and a generalisation of density estimation error estimates yields

Proposition 12.3.2. *If ν_{k-1} is absolutely continuous, with density given by $\dfrac{d\nu_{k-1}}{d\mathbf{x}} \in W^{2,1}$, then the local mean error induced by progressive*

weighting correction satisfies

$$\mathbb{E}[\,\|\tilde{\nu}_k - \nu_k\|_1 \mid \mathcal{S}_k\,] \leq \tilde{\alpha}\,h^2 + \tilde{\delta}_k\,[\,\frac{\tilde{\beta}}{\sqrt{N}\,h^{n_x}} + \|\tilde{\nu}_{k-1} - \nu_{k-1}\|_1\,]\,,\quad (12.3.13)$$

with $\tilde{\alpha} = \alpha(K, \Psi_k \cdot \nu_{k-1})$, and $\tilde{\beta} = \beta(K, \tilde{\nu}_{k-1})$.

Inequality (12.3.13) shows that to minimise the local error, at each time step, we have to choose Ψ_k such that $\tilde{\delta}_k$ is well controlled.

Progressive rejection correction

Let $\bar{\nu}_k$ denote the approximation of ν_k by successive rejection corrections

$$\bar{\nu}_k = \Psi_k \cdot (K_h * (S^N(\bar{\nu}_{k-1})))\,,\quad 1 \leq k \leq n\,,\quad \text{with } \bar{\nu}_0 = \pi.$$

By analogy with $\tilde{\delta}_k$, we define $\bar{\delta}_k = \sup_{\mathbf{x}\in\mathbb{R}^{n_x}} \Psi_k(\mathbf{x})/\langle\bar{\nu}_{k-1}, \Psi_k\rangle$. Using inequality (12.3.12) and classical density estimation error estimates yields

Proposition 12.3.3. *If ν_{k-1} is absolutely continuous, with density given by $\dfrac{d\nu_{k-1}}{d\mathbf{x}} \in W^{2,1}$, then the local mean error induced by progressive rejection correction satisfies*

$$\mathbb{E}[\,\|\bar{\nu}_k - \nu_k\|_1 \mid \mathcal{S}_k\,] \leq \bar{\delta}_k\,[\bar{\alpha}\,h^2 + \frac{\bar{\beta}}{\sqrt{N}\,h^{n_x}} + \|\bar{\nu}_{k-1} - \nu_{k-1}\|_1\,]\,,\quad (12.3.14)$$

with $\bar{\alpha} = \alpha(K, \nu_{k-1})$ and $\bar{\beta} = \beta(K, \bar{\nu}_{k-1})$.

In addition, the number of random variables required by progressive rejection correction (with the decomposition $(\Psi_1, \cdots, \Psi_n)$ introduced above) is a random number with mean $N\,(\bar{\delta}_1 + \cdots + \bar{\delta}_n)$. Recall that the mean number of simulations required by the direct rejection correction method is $N\,\delta = N\,(\delta_1 \cdots \delta_n)$. We can see here that progressive correction allows us to change the product $(\delta_1 \cdots \delta_n)$ into the sum $(\bar{\delta}_1 + \cdots + \bar{\delta}_n)$, which implies in general a lower computing cost.

12.3.3   Adaptive choice of the decomposition

An adaptive method is described to chose the decomposition $(\Psi_1, \cdots, \Psi_n)$. For simplicity, we consider only the case of an additive Gaussian measurement noise with covariance matrix R_t.

Assume that the decomposition $(\Psi_1, \cdots, \Psi_{k-1})$ is already given up to step $k - 1 < n$, and that the first $(k - 1)$ sub-corrections have already been computed. We are then, given $\bar{\nu}_{k-1}$, in the case of progressive rejection correction, and given $\tilde{\nu}_{k-1}$, in the case of progressive weighting correction, respectively. The aim is to find Ψ_k such that $\bar{\delta}_k \leq \delta_{\max}$, and $\tilde{\delta}_k \leq \delta_{\max}$ respectively, where $\delta_{\max} > 1$ is a control parameter given by the user and determining the maximum cost coefficient of each sub-correction step. Recall that

$$\frac{d\tilde{\nu}_k}{d\mathbf{x}}(\mathbf{x}) = \frac{d(K_h * (\Psi_k \cdot (S^N(\tilde{\nu}_{k-1}))))}{d\mathbf{x}}(\mathbf{x}) = \sum_{i=1}^{N} \tilde{\omega}_k^{(i)} K_h(\mathbf{x} - \tilde{\mathbf{x}}_{k-1}^{(i)}) ,$$

$$\frac{d\bar{\nu}_k}{d\mathbf{x}}(\mathbf{x}) = \frac{d(\Psi_k \cdot (K_h * (S^N(\bar{\nu}_{k-1}))))}{d\mathbf{x}}(\mathbf{x}) \propto \sum_{i=1}^{N} \Psi_k(\mathbf{x}) K_h(\mathbf{x} - \bar{\mathbf{x}}_{k-1}^{(i)}) ,$$

where $(\tilde{\mathbf{x}}_{k-1}^{(1)}, \cdots, \tilde{\mathbf{x}}_{k-1}^{(N)})$ i.i.d. $\sim \tilde{\nu}_{k-1}$ and $(\bar{\mathbf{x}}_{k-1}^{(1)}, \cdots, \bar{\mathbf{x}}_{k-1}^{(N)})$ i.i.d. $\sim \bar{\nu}_{k-1}$ respectively. Notice that we are interested in Ψ_k or any other proportional function. The choice of Ψ_k is made among functions of the family

$$\Psi_k(\mathbf{x}) = \exp\{-\frac{1}{2\lambda_k}(\mathbf{y}_t - H_t(\mathbf{x}))^T R_t^{-1}(\mathbf{y}_t - H_t(\mathbf{x}))\} , \qquad (12.3.15)$$

where $\mathbf{y}_t$ is the current measurement, and $\lambda_k > 0$ is a parameter to be determined. We make the following Monte Carlo approximations:

$$\tilde{\delta}_k = \frac{1}{\langle \tilde{\nu}_{k-1}, \Psi_k \rangle} \approx \frac{1}{\sum_{i=1}^{N} \Psi(\tilde{\mathbf{x}}_{k-1}^{(i)})} , \qquad \bar{\delta}_k = \frac{1}{\langle \bar{\nu}_{k-1}, \Psi_k \rangle} \approx \frac{1}{\frac{1}{N}\sum_{i=1}^{N} \Psi_k(\bar{\mathbf{x}}_{k-1}^{(i)})} .$$

Henceforth, the case of rejection correction and weighting correction are identical, so we shall focus on only one of them. The choice of λ_k is then determined by the equation

$$\min_{1 \le i \le N} \Psi_k(\bar{\mathbf{x}}_{k-1}^{(i)}) = 1/\delta_{\max} . \qquad (12.3.16)$$

It appears here that $\delta_{\max}$ is necessarily larger than 1, since Ψ_k is smaller than 1. Condition (12.3.16) is chosen because it automatically implies $\bar{\delta}_k \le \delta_{\max}$ and is sufficient to determine λ_k. Indeed, it follows from (12.3.16) that

$$\lambda_k = \frac{\max_{1 \le i \le N}(\mathbf{y}_t - H_t(\bar{\mathbf{x}}_{k-1}^{(i)}))^T R_t^{-1}(\mathbf{y}_t - H_t(\bar{\mathbf{x}}_{k-1}^{(i)}))}{2 \log(\delta_{\max})} .$$

Notice that the sequence $(\lambda_1, \cdots, \lambda_{k-1}, \cdots)$ is decreasing. Indeed, the particles concentrate at each sub-correction step closer to the real state, which implies that $\max_{1 \le i \le N}(\mathbf{y}_t - H_t(\bar{\mathbf{x}}_{k-1}^{(i)}))^T R_t^{-1}(\mathbf{y}_t - H_t(\bar{\mathbf{x}}_{k-1}^{(i)}))$ decreases. The size n of the decomposition is a-priori unknown and determined during the implementation of the algorithm. The algorithm runs as long as $\lambda_k \ge 1$. If $\lambda_k < 1$, then n is set to k, and $\Psi_n = \Psi/(\Psi_1 \cdots \Psi_{n-1})$.

Given $\delta_{\max}$ and a maximum number $n_{\max}$ of sub-correction steps, computing Ψ_k at each sub-correction step is done in the following two steps.

1. Compute $\lambda_k = \dfrac{\displaystyle\max_{1 \le i \le N} (H_t(\bar{\mathbf{x}}_{k-1}^{(i)}) - \mathbf{y}_t)^T R_t^{-1} (H_t(\bar{\mathbf{x}}_{k-1}^{(i)}) - \mathbf{y}_t)}{2 \log(\delta_{\max})}.$

2. If $\lambda_k \ge 1$, and $k \le n_{\max}$, then compute Ψ_k using (12.3.15), otherwise set $n = k$, and $\Psi_n = \Psi/(\Psi_1 \cdots \Psi_{k-1})$.

Remark 3. More generally, i.e. if the measurement noise is not necessarily Gaussian, it is still possible to implement the progressive correction principle as follows. Indeed, let

$$\Psi_k(\mathbf{x}) = [\Psi(\mathbf{x})]^{1/\lambda_k} = [g_t(\mathbf{y}_t - H_t(\mathbf{x}))]^{1/\lambda_k} ,$$

where $\mathbf{y}_t$ is the current measurement, and $\lambda_k > 0$ is a parameter to be determined. The choice of λ_k is determined by (12.3.16) again, which yields

$$\lambda_k = \frac{\displaystyle\max_{1 \le i \le N} [-\log g_t(\mathbf{y}_t - H_t(\bar{\mathbf{x}}_{k-1}^{(i)}))]}{\log(\delta_{\max})} .$$

Otherwise, the algorithm is the same as in the Gaussian case.

12.4 The local rejection regularised particle filter (L2RPF)

The L2RPF presented below uses regularisation, with the Epanechnikov kernel defined in (12.2.5) which is optimal in the L^2 sense, and the rescaling procedure (12.2.8). This filter is based on the kernel filter (KF) (Hürzeler and Künsch 1998), a local rejection method that produces samples according to (12.4.22) below and that is faster than the classical rejection method; see Section 12.2.2. Observing that the Epanechnikov kernel has compact support, the maximum $c_t^{(i)}$ of the likelihood is taken around each particle $\mathbf{x}_{t|t-1}^{(i)}$, and the computational cost gain is $\sum_{i=1}^{N} c_t^{(i)}/(N \max_{1 \le i \le N} c_t^{(i)})$. However, this cost remains high, especially when the particles are dispersed or when the variance of the measurement noise is small.

12.4.1 Description of the filter

We introduce the computing control parameter $\alpha_t \in [0, 1]$. At each time step, this adaptive parameter is computed so as to take into account the

computing capability by means of the evaluation of the acceptance probability of the rejection loop AL below. It allows us to alternate weighted
sample methods such as post–RPF, which generates samples according
to (12.4.23) below, and local rejection methods such as KF, which generates samples according to (12.4.22) below. The proposed L2RPF allows a
precise correction step in a given computing time. Given a prediction sam-
ple $(\mathbf{x}_{t|t-1}^{(1)}, \cdots, \mathbf{x}_{t|t-1}^{(N)})$ with covariance matrix $S_t = A_t A_t^T$, and a scalar
α_t, we use the following algorithm AL to generate a corrected sample
independently for all i.

---

1. Generate $I \in \{1, \cdots, N\}$, with $\mathbb{P}(I = j) \propto c_t^{(j)}(\alpha_t)$.

2. Generate $\varepsilon \sim K$, Epanechnikov kernel (12.2.5), and U uniformly
   on $[0, 1]$.

3. Put $\mathbf{X} = \mathbf{x}_{t|t-1}^{(I)} + h\, A_t\, \varepsilon$, with $h = h_{\text{opt}}$ given by (12.2.6).

4. If $\Psi_t(\mathbf{X}) \geq \alpha_t\, c_t^{(I)}(\alpha_t)\, U$, then return $\mathbf{x}_{t|t}^{(i)} = \mathbf{X}$, otherwise go to
   step 1.

The coefficients $c_t^{(i)}(\alpha_t)$ are computed below, and satisfy

$$c_t^{(i)}(\alpha_t) \geq \sup_{\mathbf{x} \in \Sigma_i(\alpha_t)} \Psi_t(\mathbf{x}) , \qquad (12.4.17)$$

where the supremum is taken on an ellipsoid centered at the particle $\mathbf{x}_{t|t-1}^{(i)}$

$$\Sigma_i(\alpha_t) = \{\mathbf{x} \in \mathbb{R}^{n_x} : (\mathbf{x} - \mathbf{x}_{t|t-1}^{(i)})^T S_t^{-1} (\mathbf{x} - \mathbf{x}_{t|t-1}^{(i)}) \leq \alpha_t^2\, h^2\} . \quad (12.4.18)$$

Proposition 12.4.1. *The L2RPF algorithm produces a sample according
to an absolutely continuous probability distribution* $\hat{\pi}_{t|t}^{\alpha_t}$, *with density*

$$\frac{d\hat{\pi}_{t|t}^{\alpha_t}}{d\mathbf{x}}(\mathbf{x}) \propto \sum_{i=1}^{N} c_t^{(i)}(\alpha_t) \min(1, \frac{\Psi_t(\mathbf{x})}{\alpha_t\, c_t^{(i)}(\alpha_t)}) K_h(A_t^{-1}(\mathbf{x} - \mathbf{x}_{t|t-1}^{(i)})). \quad (12.4.19)$$

Indeed, the probability distribution of the random variable $\mathbf{X}$ generated by AL is characterised as follows, for any test function ϕ defined on $\mathbb{R}^{n_x}$.

$$\mathbb{E}_{I,U,\varepsilon}[\phi(\mathbf{X})] = \mathbb{E}[\phi(\mathbf{x}_{t|t-1}^{(I)} + h\,A_t\,\varepsilon)\,\mathbf{1}_{(\Psi_t(\mathbf{x}_{t|t-1}^{(I)} + h\,A_t\,\varepsilon) \geq \alpha_t\,c_t^{(i)}(\alpha_t)\,U)}]$$

$$\propto \sum_{i=1}^{N} c_t^{(i)}(\alpha_t) \int \phi(\mathbf{x}_{t|t-1}^{(i)} + h\,A_t\,\mathbf{z})\,\min(1, \frac{\Psi_t(\mathbf{x}_{t|t-1}^{(i)} + h\,A_t\,\mathbf{z})}{\alpha_t\,c_t^{(i)}(\alpha_t)})\,K(\mathbf{z})\mathrm{d}\mathbf{z}$$

$$\propto \sum_{i=1}^{N} c_t^{(i)}(\alpha_t) \int \phi(\mathbf{x})\,\min(1, \frac{\Psi_t(\mathbf{x})}{\alpha_t\,c_t^{(i)}(\alpha_t)})\,K(\frac{1}{h}\,A_t^{-1}(\mathbf{x} - \mathbf{x}_{t|t-1}^{(i)}))\,\mathrm{d}\mathbf{x}$$

$$\propto \sum_{i=1}^{N} c_t^{(i)}(\alpha_t) \int \phi(\mathbf{x})\,\min(1, \frac{\Psi_t(\mathbf{x})}{\alpha_t\,c_t^{(i)}(\alpha_t)})\,K_h(A_t^{-1}(\mathbf{x} - \mathbf{x}_{t|t-1}^{(i)}))\,\mathrm{d}\mathbf{x}\;.$$

The next proposition computes the acceptance probability of the algorithm, that is, the probability that a sample goes out of AL.

Proposition 12.4.2. *The acceptance probability P_a of L2RPF is*

$$P_a(\alpha_t) = c \sum_{i=1}^{N} c_t^{(i)}(\alpha_t) \int \min(1, \frac{\Psi(\mathbf{x}_{t|t-1}^{(i)} + hA_t\mathbf{z})}{\alpha_t\,c_t^{(i)}(\alpha_t)})\,K(\mathbf{z})\,\mathrm{d}\mathbf{z} \quad (12.4.20)$$

$$\approx c \sum_{i=1}^{N} c_t^{(i)}(\alpha_t)\,\min(1, \frac{\Psi(\mathbf{x}_{t|t-1}^{(i)})}{\alpha_t\,c_t^{(i)}(\alpha_t)}), \quad (12.4.21)$$

with $c = 1/\sum_{i=1}^{N} c_t^{(i)}(\alpha_t).$

Equation (12.4.20) is derived as was (12.4.19), and approximation (12.4.21) is obtained using an expansion of $\Psi_t(\mathbf{x}_{t|t-1}^{(i)} + h\,A_t\,\mathbf{z})$ around $h = 0$. This approximation is in general precise; see simulations in Figure 12.4. Notice that with the choice made in Section 12.4.2 below for the coefficients $(c_t^{(i)}(\alpha_t))_{1\leq i\leq N}$, the probability of acceptance $P_a(\alpha_t)$ decreases when α_t increases. If $\alpha_t = 1$, then $\Sigma_i(\alpha_t)$ coincides with the support of $\mathbf{x} \mapsto K_h(A_t^{-1}(\mathbf{x} - \mathbf{x}_{t|t-1}^{(i)}))$; hence,

$$\hat{\pi}_{t|t}^{\alpha_t=1}(\mathbf{x}) \propto \sum_{i=1}^{N} \Psi_t(\mathbf{x})\,K_h(A_t^{-1}(\mathbf{x} - \mathbf{x}_{t|t-1}^{(i)}))\,, \quad (12.4.22)$$

which is exactly the KF density. But in this case P_a is minimal, i.e. the computing cost is maximal. On the other hand, if $\alpha_t = 0$, then $\Sigma_i(\alpha_t)$ reduces to the current particle $\mathbf{x}_{t|t-1}^{(i)}$, and taking equality in (12.4.17) yields

$c_t^{(i)}(\alpha_t) \propto w_t^{(i)}$; hence,

$$\hat{\pi}_{t|t}^{\alpha_t=0}(\mathbf{x}) \propto \sum_{i=1}^{N} w_t^{(i)} K_h(A_t^{-1}(\mathbf{x} - \mathbf{x}_{t|t-1}^{(i)})) , \qquad (12.4.23)$$

which is exactly the post–RPF density, with whitening. In this case $P_a = 1$, i.e. the computing cost is low. At each time step, the choice of α_t is done in the following way. Given a coarse discretisation of $[0, 1]$, we take the maximal value of α_t such that $P_a(\alpha_t) \geq P_a^{\min}$, where $P_a^{\min}$ is given by the computing capability. The higher α_t is, the better the correction. When α_t is chosen, we put $N_e = N/P_a(\alpha_t)$ the number of test samples that enter in the loop AL. For the problems presented in section 12.5, α_t is close to 0 for the first measurements, then increases to 1 when the particles concentrate on more likely regions of the state-space; see Figure 12.4. The L2RPF generalises both the post–RPF and the KF. The practical advantage of the L2RPF compared with the post–RPF is that it allows us to reduce significantly the number of particles. For example, the performances of the tracking problems are identical with $N = 1000$ in place of $N = 5000$. The additional cost is small. It was also observed that the variance of the particle system is smaller with the L2RPF (owing to the better correction step).

We now present a fast method to compute $c_t^{(i)}(\alpha_t)$.

12.4.2   Computing the coefficient $c_t^{(i)}(\alpha_t)$

By Lagrangian methods, we see that the coordinates $(x_1, \cdots, x_j, \cdots, x_{n_x})$ of any point in the ellipsoid $\Sigma_i(\alpha_t)$ centered on $\mathbf{x}_{t|t-1}^{(i)}$, see (12.4.18), verifies for all $j = 1, \cdots, n_x$,

$$x_j^{(i),\min} = x_j^{(i)} - \alpha_t h \sqrt{S_{jj}} \leq x_j \leq x_j^{(i)} + \alpha_t h \sqrt{S_{jj}} = x_j^{(i),\max} \quad (12.4.24)$$

where S_{jj} is the jth diagonal term of S. Indeed, the optimisation of x_j over $\Sigma_i(\alpha_t)$, with the constraint, $\mathbf{x}^T S^{-1} \mathbf{x} < \alpha^2 h^2$, gives the system

$$\frac{\partial L}{\partial \mathbf{x}} = \Gamma_j + 2\lambda S^{-1}\mathbf{x} = 0, \quad \mathbf{x}^T S^{-1}\mathbf{x} = \alpha^2 h^2 , \qquad (12.4.25)$$

where λ is the Lagrange multiplier, and $\Gamma_j^T = (0, \cdots, 1, \cdots, 0)$ with 1 in the jth position.

The solution of (12.4.25) is $\mathbf{x} = \pm \alpha_t h S_{\cdot j}/\sqrt{S_{jj}}$, where $S_{\cdot j}$ is the jth column of S. Let $C_i(\alpha_t)$ be the hypercube, $\Sigma_i(\alpha_t) \subset C_i(\alpha_t)$,

$$C_i(\alpha_t) = \{\mathbf{x}| x_j^{(i),\min} \leq x_j \leq x_j^{(i),\max}, 1 \leq j \leq n_x\} . \qquad (12.4.26)$$

$c_t^{(i)}(\alpha_t)$ will be chosen as the maximum of Ψ on $C_i(\alpha_t)$. We can also take the hypercube containing the ellipsoid $\Sigma_i(\alpha_t)$, i.e. the smallest hypercube with faces orthogonal to the eigenvectors of S_t. Assume that the measurement function H_k, $k = 1, \cdots, n_y$ is monotone in each coordinate in the neighbourhood of the current particle. For example, if we measure an angle $H_k(\mathbf{x}) = \arctan(x_1/x_2)$, H_k increases when x_1 increases and H_k decreases when x_2 increases (if x_1, $x_2 > 0$). The extreme values of H_k on the hypercube $C_i(\alpha_t)$ are for $k = 1, \cdots, n_y$,

$$\forall \mathbf{x} \in C_i(\alpha_t), \quad H_k^{\min} = H_k(\mathbf{x}_m^{\mathrm{extr}}) \leq H_k(\mathbf{x}) \leq H_k(\mathbf{x}_M^{\mathrm{extr}}) = H_k^{\max}, \quad (12.4.27)$$

where $\mathbf{x}_{(\cdot)}^{\mathrm{extr}}$ is one of the vertices of $C_i(\alpha_t)$ with coordinates $x_j^{(i),\min}$ or $x_j^{(i),\max}$ (12.4.24).

Assume that the components $(v_1, \cdots, v_{n_y})$ of the measurement noise v are independent, (otherwise it suffices to whiten the measurement vector) with densities $(g_1, \cdots, g_{n_y})$ decreasing around the origin. It can be seen that the maximum of the likelihood on $C_i(\alpha_t)$ verifies

$$\sup_{x \in \Sigma_i(\alpha_t)} \Psi(\mathbf{x}) \leq \sup_{x \in C_i(\alpha_t)} \prod_{k=1}^{n_y} g_k(\mathbf{y}_k - H_k(\mathbf{x})) \quad (12.4.28)$$

$$= \prod_{k=1}^{n_y} g_k(\mathbf{y}_k - H_k^{\mathrm{extr}}(\mathbf{x})) = c_t^{(i)}(\alpha_t), \quad (12.4.29)$$

where $H_k^{\mathrm{extr}} = H_k^{\min}$ if $\mathbf{y}_k \leq H_k^{\min}$, $H_k^{\mathrm{extr}} = \mathbf{y}_k$ if $H_k^{\min} \leq \mathbf{y}_k \leq H_k^{\max}$ and $H_k^{\mathrm{extr}} = H_k^{\max}$ if $\mathbf{y}_k \geq H_k^{\max}$.

12.5   Applications to tracking problems

We present three two-dimensional-tracking problems to which L2RPF is applied. Classical post–RPF has also been applied, for which the results are similar to the L2RPF in terms of estimator standard deviation (std), as defined below. But it was observed that the variances of the particle clouds are larger than when L2RPF was used, owing to the correction error. To estimate the error committed by approximate filters on each coordinate j of the state, we do the following Monte Carlo approximations with M runs.

$$\{\mathbb{E}[\hat{\mathbf{x}}_{j,t} - \mathbb{E}(\hat{\mathbf{x}}_{j,t})]^2\}^{1/2} \approx \{\frac{1}{M} \sum_{k=1}^{M} [\hat{\mathbf{x}}_{j,t|t}^{(k)} - \frac{1}{M} \sum_{k=1}^{M} \hat{\mathbf{x}}_{j,t|t}^{(k)}]^2 \}^{1/2} \quad [\text{std}]$$

$$\mathbb{E}(\hat{\mathbf{x}}_{j,t}) - \mathbb{E}(\mathbf{x}_{j,t}) \approx \frac{1}{M} \sum_{k=1}^{M} [\hat{\mathbf{x}}_{j,t|t}^{(k)} - \mathbf{x}_{j,t}], \quad [\text{bias}]$$

where $\mathbf{x}_{j,t}$ is the jth coordinate of the real state $\mathbf{x}_t$ (x–y position and velocity) of the target, and $\hat{\mathbf{x}}_{j,t|t}^{(k)}$ is the jth coordinate of the filter estimate for the kth Monte Carlo run. The number of Monte Carlo runs is $M = 50$.

The filters initialisation, $\hat{\mathbf{x}}_{0|-1}$ is a Gaussian variable centered on the true state $\mathbf{x}_0$, with covariance matrix $P_{0|0}$. The number of particles used for the L2RPF is $N = 1000$. The acceptance probability is $P_{\mathrm{a}}^{\min} = 0.2$, and the associated computing cost is about 20 times greater than if the extended Kalman filter (EKF) is employed. In the problems considered below, there is no dynamic noise, and we can easily compute the Cramer–Rao lower bound (CRLB). This gives the minimal std of any unbiased estimator. For another application of the CRLB in particle filtering, see Bergman (2001: this volume).

12.5.1 Range and bearing

The target follows a uniform straight motion (USM). Noisy measurements of the distance between the origin and the target, and the angle between the horizontal line and the line of sight are available. That is, we have
$$H_t(\mathbf{x}) = \begin{pmatrix} \sqrt{x_1^2 + x_3^2} \\ \arctan(x_3/x_1) \end{pmatrix}, \text{ where } (x_1, x_3) \text{ denote the horizontal and}$$
vertical positions, and (x_2, x_4) the horizontal and vertical velocities.

Progressive correction

We compare the performance of the post–RPF (with $N = 5000$ particles), and the post–RPF with progressive correction (PC) (with $N = 1000$ particles) with the CRLB, as a function of the range measurement noise std. The observation time is $T = 500s$ with inter-observation time $\Delta = 5s$, and the initial state is $\mathbf{x}_0 = (50\text{km}, 0\text{m/s}, 50\text{km}, 5\text{m/s})$ with $P_{0|0} = [\mathrm{diag}(0.5\text{km}, 50\text{m/s}, 0.5\text{km}, 50\text{m/s})]^2$. For each value of the range measurement noise std on abscissa, we report on ordinate the path error of the horizontal position and velocity estimates, averaged between time $20s$ and $500s$. The std of angle measurement noise is 1 degree. The maximum cost coefficient is $\delta_{\max} = 10$, and the maximum number of sub-correction steps is $n_{\max} = 25$.

In Figure 12.2 we report the std of the position and velocity estimates. We can see that PC highly improves the performance of the post–RPF, since it gives some efficient results (close to CRLB) uniformly with the measurement noise std. In Figure 12.3 (left) we report the ratio of the computing time of the post–RPF with and without PC, w.r.t. the computing time of the EKF, for each value of the measurement noise std. We can observe that when PC is used, the computing time decreases when the std of the measurement noise increases, from 24 times greater than for the EKF to less than 12 times. In the same way, we can observe in Figure 12.3 (right) that $n_{\max} = 25$ is only reached for the first measurements, for problems with small measurement noise (std=0.1m, 3.3m) where δ is known to be high. Hence, with PC the computing time is automatically adjusted to the problem's complexity.

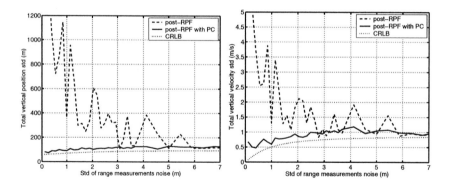

Figure 12.2. Vertical position and velocity std.

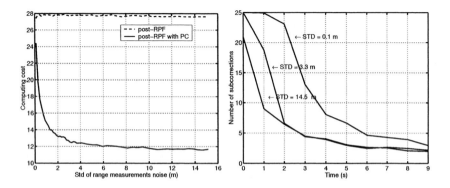

Figure 12.3. Evolution of the computing cost.

L2RPF

Below, the L2RPF and the EKF are compared with the CRLB. Here, the initial state is $\mathbf{x}_0=$(5km, -20m/s, 5km, -20m/s) with $P_{0|0}=$[diag(0.5km, 50m/s, 0.5km, 50m/s)]^2, and the observation time is $T=$200s with $\Delta=$1s. The std of range and angle measurements noise is equal to 1m (case of accurate measurements) and 1 degree, respectively.

We can observe in Figure 12.4 that the control parameter is close to 0 for the first measurements (post–RPF) and increases to 1 (KF). Notice that the theoretical P_a is close to the empirical, around the value 0.65. We observe in Figure 12.5 that the bias is close to 0, and in Figure 12.6 that unlike the EKF, the L2RPF converges rapidly to the CRLB.

12.5.2   Bearings–only

The target has a USM initialised on $\mathbf{x}_0=$(10km, -10m/s, 10km, 10m/s), with $P_{0|0}=$[diag(5km, 30m/s, 5km, 30m/s)]^2. The observer is initially at

Time (s)	10	50	120	160	200
Pa					
Emp.	0.333	0.634	0.555	0.709	0.673
Theo.	0.210	0.631	0.554	0.708	0.673

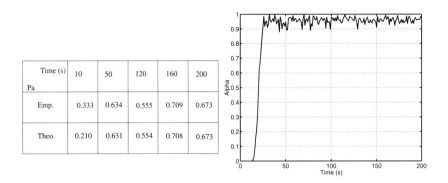

Figure 12.4. Acceptance probability P_a and control parameter α.

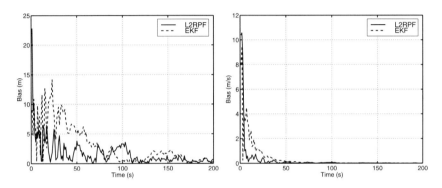

Figure 12.5. Horizontal position and velocity bias.

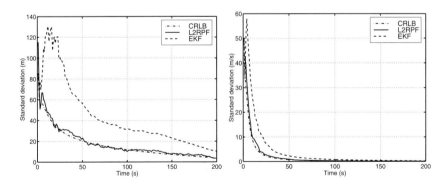

Figure 12.6. Horizontal position and velocity std.

the origin with USM (50m/s, 0m/s), then maneuvers at time $t = 100s$ to follow another USM (-50m/s, 50m/s) until $T = 200s$. The observer measures every $\Delta=1$s a noisy angle with std 0.5 degree.

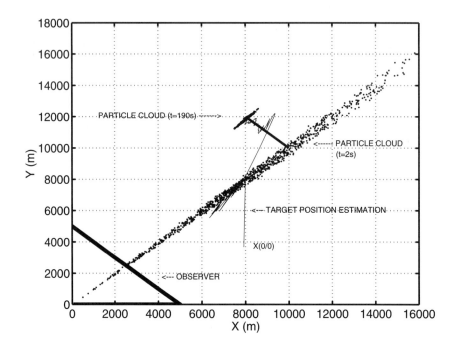

Figure 12.7. Real and estimated paths.

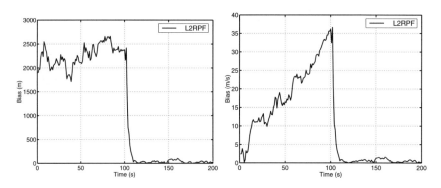

Figure 12.8. Horizontal position and velocity bias.

Figure 12.7 shows the configuration and the trajectory estimate (center of the cloud) of the L2RPF. As can be seen in Figure 12.9, the std of the horizontal position and velocity of the target is close to the CRLB. The corresponding bias is null after the observer maneuvers, see Figure 12.8.

Recall that this model is not identifiable until the observer maneuvers, see (Nardone and Aidala 1981), and indeed we have observed the divergence of the EKF in 5 runs out of 50, hence results for the EKF are not reported here.

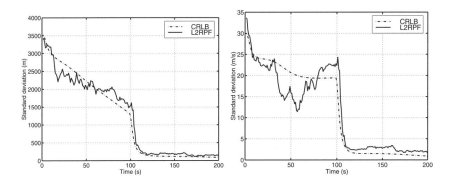

Figure 12.9. Horizontal position and velocity std.

12.5.3 Multiple model particle filter (MMPF)

By means of the formalism of interacting multiple model (IMM) (Blom and Bar-Shalom 1988), the dynamic model of the target is estimated among some given models. This is a case of multi–modality. We suppose that $(\theta_t)_{t\geq 0}$ is a Markov chain with finite state-space S, and transition probability matrix $p = (p(k,\ell))_{k,\ell \in S}$, which determines the dynamic model between time t and $(t+1)$. Therefore, we can apply the theory of particle filters with the new augmented state $(\mathbf{x}_t, \theta_t)$. Assuming that θ_t is independent of $\mathbf{x}_{t-1}$, given θ_{t-1}, and introducing the transition probability kernel

$$\mathbb{P}(\mathbf{x}_t \in d\mathbf{x}' \mid \theta_t = \ell, \mathbf{x}_{t-1} = \mathbf{x}) = Q_t(\ell, \mathbf{x}, d\mathbf{x}') ,$$

the prediction step is given by

$$\mathbb{P}(\mathbf{x}_t \in d\mathbf{x}', \theta_t = \ell \mid \mathbf{y}_{0:t-1})$$

$$= \sum_{k \in S} p(k, \ell) \int Q_t(\ell, \mathbf{x}, d\mathbf{x}') \, \mathbb{P}(\mathbf{x}_{t-1} \in d\mathbf{x}, \theta_{t-1} = k \mid \mathbf{y}_{0:t-1}) . \tag{12.5.30}$$

If we have a sample $(\mathbf{x}_{t-1}^{(1)}, \theta_{t-1}^{(1)}, \cdots, \mathbf{x}_{t-1}^{(N)}, \theta_{t-1}^{(N)})$, the following algorithm produces a predicted sample according to (12.5.30): independently for all i

1. Generate $\theta_t^{(i)} \in S$, with $\mathbb{P}[\theta_t^{(i)} = \ell] = p(\theta_{t-1}^{(i)}, \ell)$.

2. Generate $\mathbf{x}_{t|t-1}^{(i)} \sim Q_t(\theta_t^{(i)}, \mathbf{x}_{t-1}^{(i)}, \cdot)$.

The correction step is done with L2RPF. In our simulation, the target can have 2 dynamic models: USM and turn with constant angular velocity $\Omega = 0.006$ rad/s. Figure 12.10 shows the configuration. The observer located at the origin measures bearing (std=1 degree) and range (std=20m) every 10s. The target is in USM during 600s, then in turn during 800s, and finally in USM until T=2000s. The initial state is set as follows: $\mathbf{x}_0$=(50km,0m/s,50km,5m/s) with $P_{0|0}$=[diag(450m, 63m/s, 42m,

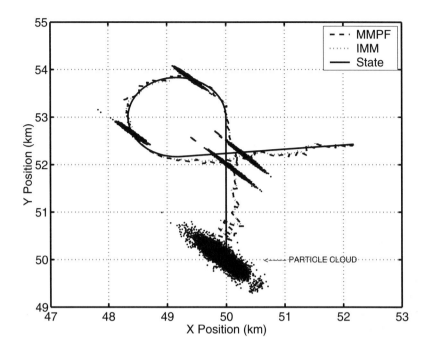

Figure 12.10. Real and estimated paths.

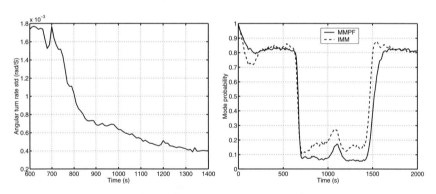

Figure 12.11. Angular turn rate and mode probability estimates.

60m/s)]^2. The initial USM mode probability is $\mathbb{P}(\theta_0 = 1) = 0.99$, and the transition probability matrix is $\begin{pmatrix} 0.98 & 0.02 \\ 0.02 & 0.98 \end{pmatrix}$. Classic IMM filter and MMPF are compared. MMPF estimates the angular turn rate (dimension of the state is 5). The IMM knows this rate (otherwise, for this context, IMM is not stable). Nevertheless, the behavior of the two filters is similar.

We can see in Figure 12.11 that the probability of the USM mode and the angular turn rate are well estimated by MMPF.

13

Auxiliary Variable Based Particle Filters

Michael K. Pitt
Neil Shephard

13.1 Introduction

We model a time series $\{y_t, t = 1, ..., n\}$ using a state-space framework with the $\{y_t | \alpha_t\}$ being independent and with the state $\{\alpha_t\}$ assumed to be Markovian. The task will be to use simulation to estimate $f(\alpha_t | \mathcal{F}_t)$, $t = 1, ..., n$, where $\mathcal{F}_t$ is contemporaneously available information. We assume a known measurement density $f(y_t | \alpha_t)$ and the ability to simulate from the transition density $f(\alpha_{t+1} | \alpha_t)$. Sometimes we will also assume that we can evaluate $f(\alpha_{t+1} | \alpha_t)$.

Filtering can be thought of as the repeated application of the iteration

$$f(\alpha_{t+1} | \mathcal{F}_{t+1}) \propto f(y_{t+1} | \alpha_{t+1}) \int f(\alpha_{t+1} | \alpha_t) dF(\alpha_t | \mathcal{F}_t). \qquad (13.1.1)$$

This implies the data can be processed in a single sweep, updating our knowledge about the states as we receive more information. This is straightforward if $\alpha_{t+1} | \alpha_t$ has a finite set of known discrete points of support, because (13.1.1) can be computed exactly. When the support is continuous and the integrals cannot be analytically solved, then numerical methods have to be used.

There have been many attempts to provide algorithms which approximate the filtering densities. Important recent work includes (Kitagawa 1987), (West 1993b), (Gerlach, Carter and Kohn 1999) and those papers reviewed in (West and Harrison 1997, Ch. 13 and 15). Here we use simulation to carry out filtering following an extensive recent literature. Our approach is to extend the particle filter using an auxiliary variable, an idea that first appeared in (Pitt and Shephard 1999b). The literature on particle filtering is reviewed extensively in previous chapters of this book, so we shall not review it here.

The outline of the paper is as follows. In Section 13.2 we analyse the statistical basis of particle filters and focus on its weaknesses. In Section 13.3 we review the main focus of the chapter, which is an auxiliary particle filter method. Section 13.4 discusses fixed-lag filtering, while Section 13.5 uses stratified sampling to improve the performance of the algorithm.

13.2   Particle filters

13.2.1   The definition of particle filters

Particle filters are the class of simulation filters that recursively approximate the filtering random variable $\alpha_t|\mathcal{F}_t = (y_1, ..., y_t)'$ by a set of particles $\alpha_t^1, ..., \alpha_t^M$, with associated discrete probability masses $\pi_t^1, ..., \pi_t^M$. Hence, a continuous variable is approximated by a discrete one with random support. These discrete points are thought of as samples from $f(\alpha_t|\mathcal{F}_t)$. In our work we shall always set the $\{\pi_t^M\}$ to be equal to $\frac{1}{M}$, where M is taken to be very large. Then we require, that as $M \to \infty$, the particles can be used to increasingly well approximate the density of $\alpha_t|\mathcal{F}_t$.

Particle filters treat the discrete support generated by the particles as the true filtering density, chaining this argument to produce a new density

$$\widehat{f}(\alpha_{t+1}|\mathcal{F}_{t+1}) \propto f(y_{t+1}|\alpha_{t+1}) \sum_{j=1}^{M} f(\alpha_{t+1}|\alpha_t^j), \qquad (13.2.2)$$

the empirical filtering density as an approximation to the true filtering density (13.1.1). Generically, particle filters sample from this density to produce new particles $\alpha_{t+1}^1, ..., \alpha_{t+1}^M$. This procedure can then be iterated through the data. We will call a particle filter fully adapted if it produces independent and identically distributed samples from (13.2.2).

13.2.2   Sampling the empirical prediction density

One way of sampling from the empirical prediction density is to think of

$$\frac{1}{M} \sum_{j=1}^{M} f(\alpha_{t+1}|\alpha_t^j)$$

as a 'prior' density $\widehat{f}(\alpha_{t+1}|\mathcal{F}_t)$ which is combined with the likelihood $f(y_{t+1}|\alpha_{t+1})$ to produce a posterior. We can sample from $\widehat{f}(\alpha_{t+1}|Y_t)$ by choosing α_t^j with probability $\frac{1}{M}$, and then drawing from $f(\alpha_{t+1}|\alpha_t^j)$. If we can also evaluate $f(y_{t+1}|\alpha_{t+1})$ up to proportionality, this leaves us with three sampling methods to draw from $f(\alpha_{t+1}|\mathcal{F}_{t+1})$.

1. Sampling/importance resampling.

2. Acceptance sampling.

3. Markov chain Monte Carlo (MCMC).

In the rest of this section we write the prior as $f(\alpha)$ and the likelihood as $f(y|\alpha)$, abstracting from subscripts and conditioning arguments, in order to briefly describe these methods in this context.

Sampling/importance resampling (SIR)

This method (Rubin 1987b, Smith and Gelfand 1992) draws $\alpha^1, ..., \alpha^R$ from $f(\alpha)$ and then associates with each of these draws the weights π_j where

$$w_j = f(y|\alpha^j), \qquad \pi_j = \frac{w_j}{\sum_{i=1}^{R} w_i}, \qquad j = 1, ..., R.$$

The weighted sample will converge, as $R \to \infty$, to a non-random sample from the desired posterior $f(\alpha|y)$ as $R^{-1} \sum_{i=1}^{R} w_i \xrightarrow{p} f(y)$. The non-random sample can be converted into a random sample of size M by resampling the $\alpha^1, ..., \alpha^R$ using weights $\pi_1, ..., \pi_R$. This requires $R \to \infty$ and $R >> M$. The use of this method has been suggested in the particle filter framework by (Gordon et al. 1993), (Kitagawa 1996), (Berzuini et al. 1997) and (Isard and Blake 1996).

To understand the efficiency of the SIR method, it is useful to think of SIR as an approximation to the importance sampler of the moment

$$E_{f\pi}\{h(\alpha)\} = \int h(\alpha)\pi(\alpha)dF(\alpha),$$

by

$$\frac{1}{R}\sum_{j=1}^{R} h(\alpha^j)\pi(\alpha^j),$$

where $\alpha^j \sim f(\alpha)$ and $\pi(\alpha) = f(y|\alpha)/f(y)$. (Liu 1996a) suggested the variance of this estimator is approximately (for slowly varying $h(\alpha)$) proportional to $E_f\{\pi(\alpha)^2\}/R$. Hence the SIR method will become imprecise as the π_j become more variable, which can happen if the likelihood is highly peaked compared with the prior.

Adaption

The above SIR algorithm samples from $f(\alpha|y)$ by making *blind* proposals $\alpha^1, ..., \alpha^R$ from the prior, ignoring that we know the value of y. This is the main feature of the initial particle filter proposed by (Gordon et al. 1993). We say that a particle filter is adapted if we make proposals that take into account the value of y.

An adapted SIR based particle filter has the following general structure.

1. Draw from $\alpha^1, ..., \alpha^R \sim g(\alpha|y)$.

2. Evaluate $w_j = f(y|\alpha^j)f(\alpha^j)/g(\alpha^j|y)$, $j = 1, ..., R$.

3. Resample amongst the $\{\alpha^j\}$ using weights proportional to $\{w_j\}$ to produce a sample of size M.

Although this looks attractive, for a particle filter $f(\alpha)$ represents the density $f(\alpha_{t+1}|\mathcal{F}_t) = \sum_{j=1}^{M} f(\alpha_{t+1}|\alpha_t^j)$, which implies we have at least to evaluate $M \times R$ densities in order to generate M samples from $f(\alpha|y)$. Given that M and R are typically very large, adaption is not generally feasible for SIR based particle filters.

Rejection and MCMC sampling

Exactly the same remarks hold for rejection sampling. A blind rejection sampling based particle filter will simulate from $f(\alpha)$ and accept with probability $\pi(\alpha) = f(y|\alpha)/f(y|\alpha_{\max})$, where $\alpha_{\max} = \arg\max_\alpha f(y|\alpha)$. This has been proposed by (Hürzeler and Künsch 1998) and used on a univariate log-normal stochastic volatility model by (Kim et al. 1998). Again, the rejection worsens if the $var_f \{\pi(\alpha)\}$ is high and adaption is difficult, as it will typically involve evaluating $f(\alpha)$, making it computationally infeasible.

Another approach to SIR is the use of a blind MCMC method (see (Gilks et al. 1996) for a review). In this context the MCMC accepts a move from a current state α^i to $\alpha^{i+1} \sim f(\alpha)$ with probability

$$\min\left\{1, \frac{f(y|\alpha^{i+1})}{f(y|\alpha^i)}\right\},$$

otherwise it sets $\alpha^{i+1} = \alpha^i$. Once again, if the likelihood is highly peaked there may be a large amount of rejection, which will mean the Markov chain will have a great deal of dependence. This suggests adapting, when possible, the MCMC method to draw from $g(\alpha|y)$, and then accept these draws with probability

$$\min\left\{1, \frac{f(y|\alpha^{i+1})f(\alpha^{i+1})}{f(y|\alpha^i)f(\alpha^i)} \frac{g(\alpha^i|y)}{g(\alpha^{i+1}|y)}\right\}.$$

The problem with this, as with the previous adapted SIR, is that evaluating $f(\alpha)$ is very expensive.

13.2.3   Weaknesses of particle filters

The particle filter based on SIR has two basic weaknesses. The first is well known: when there is an outlier, the weights π_j will be very unevenly distributed, and so requiring an extremely large value of R for the draws to be close to samples from the empirical filtering density. This is of particular concern if the measurement density $f(y_{t+1}|\alpha_{t+1})$ is highly sensitive to α_{t+1}. Notice this is not a matter of having too small a value of M. Instead, the difficulty is, given that degree of accuracy, how to efficiently sample from (13.2.2). We show how to do this in the next section.

The second weakness holds for particle filters in general. As $R \to \infty$, the weighted samples can be used to arbitrarily well approximate (13.2.2). However, the tails of $\frac{1}{M} \sum_{j=1}^{M} f(\alpha_{t+1}|\alpha_t^j)$ usually poorly approximate the true tails of $\alpha_{t+1}|\mathcal{F}_t$ because of the use of the mixture approximation. As a result, (13.2.2) can only inadequately approximate the true $f(\alpha_{t+1}|\mathcal{F}_{t+1})$ when there is an outlier. Hence, the second difficulty to be overcome: how to improve the empirical prediction density's behaviour in the tails? We discuss this in Section 13.4.

13.3   Auxiliary variable

13.3.1   The basics

A fundamental problem with conventional particle filters is that their mixture structure means that it is difficult to adapt the SIR, rejection or MCMC sampling methods without significantly slowing the running of the filter. (Pitt and Shephard 1999b) have argued that many of these problems are reduced when we perform particle filtering in a higher dimension. In this section, we review that argument.

Our task will be to sample from the joint density $f(\alpha_{t+1}, k|\mathcal{F}_{t+1})$, where k is an index, on the mixture in (13.2.2). Let us define

$$f(\alpha_{t+1}, k|\mathcal{F}_{t+1}) \propto f(y_{t+1}|\alpha_{t+1})f(\alpha_{t+1}|\alpha_t^k), \qquad k = 1, ..., M. \quad (13.3.3)$$

If we draw from this joint density and then discard the index, we produce a sample from the empirical filtering density (13.2.2) as required. We call k an auxiliary variable because it is present simply to aid the task of the simulation. Generic particle filters of this type will be labelled auxiliary particle filters.

We can now sample from $f(\alpha_{t+1}, k|\mathcal{F}_{t+1})$ using SIR, rejection sampling or MCMC. The SIR idea will be to make R proposals $\alpha_{t+1}^j, k^j \sim g(\alpha_{t+1}, k|\mathcal{F}_{t+1})$ and then construct resampling weights

$$w_j = \frac{f(y_{t+1}|\alpha_{t+1}^j)f(\alpha_{t+1}^j|\alpha_t^{k^j})}{g(\alpha_{t+1}^j, k^j|\mathcal{F}_{t+1})}, \qquad \pi_j = \frac{w_j}{\sum_{i=1}^{R} w_i}, \qquad j = 1, ..., R.$$

We have complete control over the design of $g(.)$, which can depend on y_{t+1} and α_t^k, in order to make the weights more even. Thus, this method is adaptable and extremely flexible. In the next subsection we shall give a convenient generic suggestion for the choice of $g(.)$.

Rejection sampling for auxiliary particle filtering could also be used in this context. An example appears in Section 13.3.3. We can also make proposals for an MCMC variate of the auxiliary particle filter from $\alpha_{t+1}^{(i+1)}, k^{(i+1)} \sim g(\alpha_{t+1}, k|\mathcal{F}_{t+1})$, where $g(\alpha_{t+1}, k|\mathcal{F}_{t+1})$ is some arbitrary density; these moves are then accepted with probability

$$\min\left\{1, \frac{f(y_{t+1}|\alpha_{t+1}^{(i+1)})f(\alpha_{t+1}^{(i+1)}|\alpha_t^{k^{(i+1)}})}{f(y_{t+1}|\alpha_{t+1}^{(i)})f(\alpha_{t+1}^{(i)}|\alpha_t^{k^{(i)}})} \frac{g(\alpha_{t+1}^{(i)}, k^{(i)}|\mathcal{F}_{t+1})}{g(\alpha_{t+1}^{(i+1)}, k^{(i+1)}|\mathcal{F}_{t+1})}\right\}$$

$$= \min\left(1, \frac{w_{i+1}}{w_i}\right).$$

A special case of this argument has appeared in (Berzuini et al. 1997), who put $g(\alpha_{t+1}, k|\mathcal{F}_{t+1}) \propto f(\alpha_{t+1}|\alpha_t^k)$, which means the method is again blind.

13.3.2   A generic SIR based auxiliary proposal

The method

Here we will give a generic $g(.)$ that can be broadly applied. We will base our discussion on the SIR algorithm, although we could equally have used an MCMC method. We approximate (13.3.3) by

$$g(\alpha_{t+1}, k|\mathcal{F}_{t+1}) \propto f(y_{t+1}|\mu_{t+1}^k)f(\alpha_{t+1}|\alpha_t^k), \qquad k = 1, ..., M,$$

where μ_{t+1}^k is the mean, the mode, a draw, or some other likely value associated with the density of $\alpha_{t+1}|\alpha_t^k$. The form of the approximating density is designed so that

$$g(k|\mathcal{F}_{t+1}) \propto \int f(y_{t+1}|\mu_{t+1}^k)dF(\alpha_{t+1}|\alpha_t^k) = f(y_{t+1}|\mu_{t+1}^k).$$

Thus, we can sample from $g(\alpha_{t+1}, k|\mathcal{F}_{t+1})$ by simulating the index with probability $\lambda_k \propto g(k|\mathcal{F}_{t+1})$, then sampling from the transition density given the component $f(\alpha_{t+1}|\alpha_t^k)$. We call the λ_k the first stage weights.

The implication is that we will simulate from particles associated with large predictive likelihoods. Having sampled the joint density of $g(\alpha_{t+1}, k|\mathcal{F}_{t+1})$ R times, we carry out a reweighting, putting on the draw (α_{t+1}^j, k^j) the weights proportional to the so-called second stage weights

$$w_j = \frac{f(y_{t+1}|\alpha_{t+1}^j)}{f(y_{t+1}|\mu_{t+1}^{k^j})}, \qquad \pi_j = \frac{w_j}{\sum_{i=1}^R w_i}, \qquad j = 1, ..., R.$$

The hope is that these second stage weights are much less variable than those for the original SIR method. We might resample from this discrete distribution to produce a sample of size M.

By making proposals having high conditional likelihoods, we reduce the costs of sampling many times from particles which have very low likelihoods and so will not be resampled at the second stage of the process. Statistical efficiency of the sampling procedure is thereby improved, meaning we can reduce the value of R substantially.

To measure the statistical efficiency of these procedures, we argued earlier that we could look at minimising $E\left\{\pi(\alpha)^2\right\}$. Here we compare a standard SIR with a SIR based on our auxiliary variable. Then, for a standard SIR based particle filter, for large R,

$$E\left\{\pi(\alpha)^2\right\} = \frac{\frac{1}{M}\sum_{k=1}^{M}\int f(y_{t+1}|\alpha_{t+1})^2 dF(\alpha_{t+1}|\alpha_t^k)}{\left\{\frac{1}{M}\sum_{k=1}^{M}\int f(y_{t+1}|\alpha_{t+1})dF(\alpha_{t+1}|\alpha_t^k)\right\}^2} = \frac{M\sum_{k=1}^{M}\lambda_k^2 f_k}{\left(\sum_{k=1}^{M}\lambda_k f_k^*\right)^2},$$

where

$$f_k = \int\left\{\frac{f(y_{t+1}|\alpha_{t+1})}{f(y_{t+1}|\mu_{t+1}^k)}\right\}^2 dF(\alpha_{t+1}|\alpha_t^k)$$

and

$$f_k^* = \int\left\{\frac{f(y_{t+1}|\alpha_{t+1})}{f(y_{t+1}|\mu_{t+1}^k)}\right\} dF(\alpha_{t+1}|\alpha_t^k).$$

The same calculation for a SIR based auxiliary variable particle filter gives

$$E\left\{\pi_\alpha(\alpha)^2\right\} = \frac{\sum_{k=1}^{M}\lambda_k f_k}{\left(\sum_{k=1}^{M}\lambda_k f_k^*\right)^2},$$

which shows an efficiency gain if

$$\sum_{k=1}^{M}\lambda_k f_k < M\sum_{k=1}^{M}\lambda_k^2 f_k.$$

If f_k does not vary over k, then the auxiliary variable particle filter will be more efficient as $\sum_{k=1}^{M}\lambda_k\frac{1}{M} = \frac{1}{M} \leq \sum_{k=1}^{M}\lambda_k^2$. More likely is that f_k will depend on k, but only mildly as $f(\alpha_{t+1}|\alpha_t^k)$ will be typically quite tightly peaked (much more tightly peaked than $f(\alpha_{t+1}|Y_t)$) compared with the conditional likelihood.

Example: a time series of angles

The model

In this section we compare the performance of the particle and auxiliary particle filter methods for an angular time series model, the bearings-only model. We consider the simple scenario described by (Gordon et al. 1993). The observer is considered stationary at the origin of the $x - z$ plane and the ship is assumed to gradually accelerate or decelerate randomly over time. We use the following discretisation of this system, where

$\alpha_t = (x_t, vx_t, z_t, vz_t)'$,

$$\alpha_{t+1} = \begin{pmatrix} 1 & 1 & 0 & 0 \\ 0 & 1 & 0 & 0 \\ 0 & 0 & 1 & 1 \\ 0 & 0 & 0 & 1 \end{pmatrix} \alpha_t + \sigma_\eta \begin{pmatrix} \frac{1}{2} & 0 \\ 1 & 0 \\ 0 & \frac{1}{2} \\ 0 & 1 \end{pmatrix} u_t, \quad u_t \sim \mathsf{NID}(0, \mathsf{I}).$$

$$(13.3.4)$$

In obvious notation x_t, z_t represent the ship's horizontal and vertical position at time t and vx_t, vz_t represent the corresponding velocities. The state evolution is thus a VAR(1) of the form $\alpha_{t+1} = T\alpha_t + Hu_t$. The model indicates that the source of state evolution error is due to the accelerations being white noise. The initial state describes the ship's starting positions and velocities $\alpha_1 \sim \mathsf{NID}(a_1, P_1)$. This prior together with the state evolution of (13.3.4) describes the overall prior for the states.

Our model will be based on a mean direction $\mu_t = \tan^{-1}(z_t/x_t)$. The measured angle will be assumed to be wrapped Cauchy with density (see for example (Fisher 1993, p. 46))

$$f(y_t|\mu_t) = \frac{1}{2\pi} \frac{1 - \rho^2}{1 + \rho^2 - 2\rho\cos(y_t - \mu_t)}, \quad 0 \le y_t < 2\pi, \quad 0 \le \rho \le 1.$$

$$(13.3.5)$$

ρ is termed the mean resultant length.

The simulated scenario

In order to assess the relative efficiency of the particle filter and the basic auxiliary method discussed in section 13.3.2, we have closely followed the setup described by (Gordon et al. 1993). They consider $\sigma_\eta = 0.001$ and $\sigma_\varepsilon = 0.005$, where $z_t|\mu_t \sim NID(\mu_t, \sigma_\varepsilon^2)$. We choose $\rho = 1 - \sigma_\varepsilon^2$ (yielding the same circular dispersion) for our wrapped Cauchy density. The actual initial starting vector of this is taken to be $\alpha_1 = (-0.05, 0.001, 0.2, -0.055)'$. By contrast with (Gordon et al. 1993), we wish to have an extremely accurate and tight prior for the initial state. This is because we want the variance of quantities arising from the filtered posterior density to be small enabling reasonably conclusive evidence to be formulated about the relative efficiency of the auxiliary method to the standard method. We therefore take $a_1 = \alpha_1$ and have a diagonal initial variance P_1 with the elements $0.01 \times (0.5^2, 0.005^2, 0.3^2, 0.01^2)$ on the diagonal.

Figure 13.1 illustrates a realisation of the model for the above scenario with $T = 10$. The ship is moving in a South-Easterly direction over time. The trajectories given by the posterior filtered means from the particle SIR method and the auxiliary SIR method ($M = 300$, $R = 500$ in both cases) are both fairly close to the true path despite the small amount of simulation used.

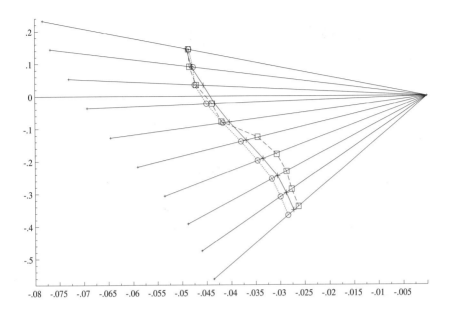

Figure 13.1. Plot of the angular measurements from origin, the true trajectory (solid line, crosses), the particle filtered mean trajectory (broken line, boxes) and the auxiliary particle mean trajectory (dotted line, circles). Ship moving South-East. $T = 10$, $M = 300$, $R = 500$.

Monte Carlo comparison

The two methods are now compared using a Monte Carlo study of the above scenario with $T = 10$. The "true" filtered mean is calculated for each replication by using the auxiliary method with $M = 100,000$ and $R = 120,000$. Within each replication, the mean squared error for the particle method for each component of the state over time is evaluated by running the method, with a different random number seed, S times and recording the average of the resulting squared difference between the particle filter's estimated mean and the "true" filtered mean. Hence for replication i, state component j, at time t we calculate

$$MSE_{i,j,t}^{P} = \frac{1}{S}\sum_{s=1}^{S}(\overline{\alpha}_{t,j,s}^{i} - \widetilde{\alpha}_{t,j}^{i})^{2},$$

where $\overline{\alpha}_{t,j,s}^{i}$ is the particle mean for replication i, state component j, at time t, for simulation s and $\widetilde{\alpha}_{t,j}^{i}$ is the "true" filtered mean replication i, state component j, at time t. The log mean squared error for component j

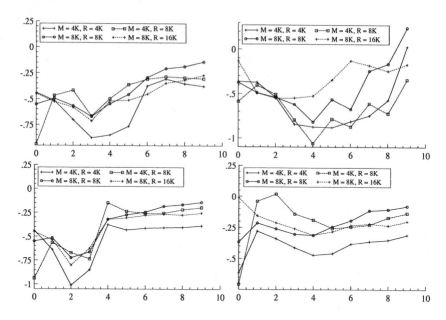

Figure 13.2. Plot of the relative mean square error performance (on the log-scale) of the particle filter and the auxiliary based particle filter for the bearings only tracking problem. Numbers below zero indicate a superior performance by the auxiliary particle filter. In these graphs $M = 4,000$ or $8,000$ while $R = M$ or $R = 2M$. Throughout, SIR is used as the sampling mechanism. Top left: $\alpha_{t1} = x_t$, bottom left: $\alpha_{t3} = z_t$, while top right: $\alpha_{t2} = vx_t$ and bottom right: $\alpha_{t4} = vz_t$.

at time t is obtained as

$$LMSE_{j,t}^P = \log \frac{1}{REP} \sum_{i=1}^{REP} MSE_{i,j,t}^P. \qquad (13.3.6)$$

The same operation is performed for the auxiliary method to deliver the corresponding quantity $LMSE_{j,t}^{AM}$. For this study, we set use $REP = 40$ and $S = 20$. We allow $M = 4,000$ or $8,000$, and for each of these values we set $R = M$ or $2M$. Figure 13.2 shows the relative performance of the two methods for each component of the state vector over time. For each component j, the quantity $LMSE_{j,t}^{AM} - LMSE_{j,t}^P$ is plotted against time. Values close to 0 indicate that the two methods are broadly equivalent in performance, whilst negative values indicate that the auxiliary method performs better than that associated with the standard particle filter.

The graphs give the expected result with the auxiliary particle filter typically being more precise, but with the difference between the two methods falling as R increases.

13.3.3   Examples of adaption

Basics

Although the generic scheme above can usually reduce the variability of the second-stage weights, sometimes there are convenient alternative adaption schemes that exploit the specific structure of the time series model, allowing us to achieve even more efficient samplers. If we can achieve exactly equal weights, then we say that we have *fully adapted* the procedure to the model, because we can then produce i.i.d. samples from (13.2.2). This situation is particularly interesting in that we are then close to the assumptions made by (Kong et al. 1994) for their sequential importance sampler.

Nonlinear Gaussian measurement model

In the Gaussian measurement case, the absorption of the measurement density into the transition equation is particularly convenient. Consider a nonlinear transition density with $\alpha_{t+1}|\alpha_t \sim N\{\mu(\alpha_t), \sigma^2(\alpha_t)\}$ and $y_{t+1}|\alpha_{t+1} \sim N(\alpha_{t+1}, 1)$. Then,

$$f(\alpha_{t+1}, k|\mathcal{F}_{t+1}) \propto f(y_{t+1}|\alpha_{t+1})f(\alpha_{t+1}|\alpha_t^k) = g_k(y_{t+1})f(\alpha_{t+1}|\alpha_t^k, y_{t+1}),$$

where $\sigma_k^{-*2} = 1 + \sigma^{-2}(\alpha_t^k)$ and

$$f(\alpha_{t+1}|\alpha_t^k, y_{t+1}) = N(\mu_k^*, \sigma_k^{*2}), \qquad \mu_k^* = \sigma_k^{*2}\left\{\frac{\mu(\alpha_t^k)}{\sigma^2(\alpha_t^k)} + y_{t+1}\right\}.$$

This implies that the first stage weights are

$$g_k(y_{t+1}) \propto \frac{\sigma_k^*}{\sigma(\alpha_t^k)}\exp\left\{\frac{\mu_k^{*2}}{2\sigma_k^{*2}} - \frac{\mu(\alpha_t^k)^2}{2\sigma^2(\alpha_t^k)}\right\}.$$

The Gaussian measurement density implies the second stage weights are all equal.

An example of this is a Gaussian ARCH model (see, for example, (Bollerslev, Engle and Nelson 1994)) observed with independent Gaussian error. So we have

$$y_t|\alpha_t \sim N(\alpha_t, \sigma^2), \qquad \alpha_{t+1}|\alpha_t \sim N(0, \beta_0 + \beta_1\alpha_t^2).$$

This model is fully adaptable. It has received a great deal of attention in the econometric literature because it has some attractive multivariate generalisations: see the work by (Diebold and Nerlove 1989), (Harvey, Ruiz and Sentana 1992) and (King, Sentana and Wadhwani 1994). As far as we know, no likelihood methods exist in the literature for the analysis of this type of model (and its various generalisations), although a number of very good approximations have been suggested.

Log-concave measurement densities

Suppose again that $f(\alpha_{t+1}|\alpha_t^k)$ is Gaussian, but the measurement density is log-concave as a function of α_{t+1}. We might then extend the above argument by Taylor expanding $\log f(y_{t+1}|\alpha_{t+1})$ to a second order term, again around μ_{t+1}^k, to give the approximation

$$\log g(y_{t+1}|\alpha_{t+1}, \mu_{t+1}^k) = \log f(y_{t+1}|\mu_{t+1}^k) + (\alpha_{t+1} - \mu_{t+1}^k)' \frac{\partial \log f(y_{t+1}|\mu_{t+1}^k)}{\partial \alpha_{t+1}}$$

$$+ \frac{1}{2}(\alpha_{t+1} - \mu_{t+1}^k)' \frac{\partial^2 \log f(y_{t+1}|\mu_{t+1}^k)}{\partial \alpha_{t+1} \partial \alpha_{t+1}'} (\alpha_{t+1} - \mu_{t+1}^k),$$

then

$$g(\alpha_{t+1}, k|\mathcal{F}_{t+1}) \propto g(y_{t+1}|\alpha_{t+1}; \mu_{t+1}^k) f(\alpha_{t+1}|\alpha_t^k).$$

Rearranging, we can express this as

$$g(\alpha_{t+1}, k|\mathcal{F}_{t+1}) \propto g(y_{t+1}|\mu_{t+1}^k) g(\alpha_{t+1}|\alpha_t^k, y_{t+1}; \mu_{t+1}^k),$$

which means we could simulate the index with probability proportional to $g(y_{t+1}|\mu_{t+1}^k)$ and then draw from $g(\alpha_{t+1}|\alpha_t^k, y_{t+1}, \mu_{t+1}^k)$. The resulting reweighted sample's second stage weights are proportional to the (it is hoped) fairly even weights for

$$w_j = \frac{f(y_{t+1}|\alpha_{t+1}^j) f(\alpha_{t+1}|\alpha_t^{k^j})}{g(y_{t+1}|\mu_{t+1}^{k^j}) g(\alpha_{t+1}^j|\alpha_t^{k^j}, y_{t+1}, \mu_{t+1}^{k^j})}$$

$$= \frac{f(y_{t+1}|\alpha_{t+1}^j)}{g(y_{t+1}|\alpha_{t+1}^j; \mu_{t+1}^{k^j})}, \qquad \pi_j = \frac{w_j}{\sum_{i=1}^R w_i}, \qquad j = 1, ..., R.$$

Thus, we can exploit the special structure of the model, if available, to improve upon the auxiliary particle filter.

Stochastic volatility and rejection sampling

The same argument carries over when we use a first order Taylor expansion to construct $g(y_{t+1}|\alpha_{t+1}, \mu_{t+1}^k)$, but in this case we know that $g(y_{t+1}|\alpha_{t+1}, \mu_{t+1}^k) \geq f(y_{t+1}|\alpha_{t+1})$ for any value of μ_{t+1}^k due to the assumed log-concavity of the measurement density. Thus,

$$\begin{aligned} f(\alpha_{t+1}, k|Y_{t+1}) &\propto f(y_{t+1}|\alpha_{t+1}) f(\alpha_{t+1}|\alpha_t^k) \\ &\leq g(y_{t+1}|\alpha_{t+1}; \mu_{t+1}^k) f(\alpha_{t+1}|\alpha_t^k) \\ &= g(y_{t+1}|\mu_{t+1}^k) g(\alpha_{t+1}|\alpha_t^k, y_{t+1}; \mu_{t+1}^k) \\ &\propto g(\alpha_{t+1}, k|Y_{t+1}). \end{aligned}$$

Thus we can perform rejection sampling from $f(\alpha_{t+1}, k|\mathcal{F}_{t+1})$ by simply sampling k with probability proportional to $g(y_{t+1}|\mu_{t+1}^k)$, and then drawing α_{t+1} from $g(\alpha_{t+1}|\alpha_t^k, y_{t+1}; \mu_{t+1}^k)$. This pair is then accepted with probability

$$f(y_{t+1}|\alpha_{t+1})/g(y_{t+1}|\alpha_{t+1}; \mu_{t+1}^k).$$

This argument applies to the stochastic volatility (SV) model

$$y_t = \epsilon_t \beta \exp(\alpha_t/2), \quad \alpha_{t+1} = \phi \alpha_t + \eta_t, \qquad (13.3.7)$$

where ϵ_t and η_t are independent Gaussian processes with variances of 1 and σ^2. This model produces a Martingale difference sequence which is not i.i.d. because of the changing scale of the process. In particular, the $\{y_t^2\}$ are serially dependent. In terms of the parameters of the model, β has the interpretation as the modal volatility, ϕ the persistence in the volatility shocks and σ_η^2 is the volatility of the volatility. This model has attracted much recent attention in the econometrics literature as a way of generalising the Black-Scholes option pricing formula to allow volatility clustering in asset returns; see, for instance, (Hull and White 1987). MCMC methods have been used on this model by, for instance, (Jacquier, Polson and Rossi 1994), (Shephard and Pitt 1997) and (Kim et al. 1998).

For this model $\log f(y_{t+1}|\alpha_{t+1})$ is concave in α_{t+1} so that, for $\mu_{t+1}^k = \phi \alpha_t^k$,

$$\log g(y_{t+1}|\alpha_{t+1}; \mu_{t+1}^k) = const - \frac{1}{2}\alpha_{t+1} - \frac{y_t^2}{2\beta^2}\exp(-\mu_{t+1}^k)\{1 - (\alpha_{t+1} - \mu_{t+1}^k)\}.$$

The implication is that

$$g(\alpha_{t+1}|\alpha_t^k, y_{t+1}; \mu_{t+1}^k) = N\left[\mu_{t+1}^k + \frac{\sigma^2}{2}\left\{\frac{y_t^2}{\beta^2}\exp\left(-\mu_{t+1}^k\right) - 1\right\}, \sigma^2\right]$$

$$= N(\mu_{t+1}^{*k}, \sigma^2).$$

Similarly,

$$g(y_{t+1}|\mu_{t+1}^k) = \exp\left\{\frac{1}{2\sigma^2}\left(\mu_{t+1}^{*k2} - \mu_{t+1}^{k2}\right)\right\}\exp\left\{-\frac{y_t^2}{2\beta^2}\exp\left(-\mu_{t+1}^k\right)\left(1 + \mu_{t+1}^k\right)\right\}.$$

Finally, the log-probability of acceptance is

$$-\frac{y_t^2}{2\beta^2}\left[\exp(-\alpha_{t+1}) - \exp(-\mu_{t+1}^k)\left\{1 - (\alpha_{t+1} - \mu_{t+1}^k)\right\}\right].$$

Notice that as σ^2 falls to zero so the acceptance probability goes to one.

Finally, the same argument holds when we use a SIR algorithm instead of rejection sampling. The proposals are made in exactly the same way, but now, instead of computing log-probabilities of accepting, these become log-second stage weights.

Simulation experiment

The basic SV model was defined in Section 13.3.3. We construct the compound daily returns on the US dollar against the UK pound from the first

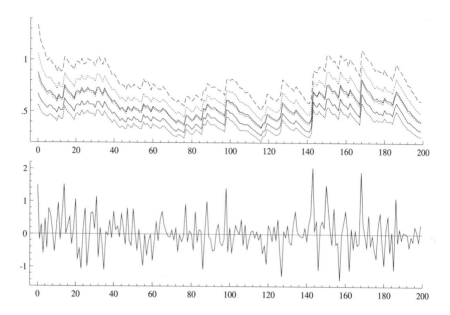

Figure 13.3. Bottom graph shows the daily returns on the US dollar against sterling from the first day of trading in 1997 for 200 trading days. We display in the top graph the posterior filtered mean (heavy line) of $\beta\exp(\alpha_t/2)|Y_t$, together with the 5, 20, 50, 80, 95 percentage points of the distribution. Notice the median is always below the mean. $M = 5,000$, $R = 6,000$.

day of trading in 1997 and for the next 200 days of active trading. This data is discussed in more detail in (Pitt and Shephard 1999c), where the parameters of the model were estimated using Bayesian methods. Throughout we take $\phi = 0.9702$, $\sigma_\eta = 0.178$ and $\beta = 0.5992$, the posterior means of the model for a long time series of returns up until the end of 1996.

Figure 13.3 graphs these daily returns against time. The figure also displays the estimated quantiles of the filtering density, $f\{\beta\exp(\alpha_t/2)|\mathcal{F}_t\}$ computed using an auxiliary particle filter. Throughout the series, we set $M = 5,000$, $R = 6,000$. We have also displayed the posterior mean of the filtering random variable. This is always slightly above the posterior median because $\alpha_t|Y_t$ is close to being symmetric.

The picture shows that the filtered volatility jumps up more quickly than it tends to go down. This reflects the fact that the volatility is modelled on the log scale.

To compare the efficiency of the simple particle filter, our basic auxiliary particle filter and the (rejection based) fully adapted particle filter discussed in Section 13.3.3., we again conducted a simulation measuring mean square error for each value of t, using the above model and again having $n = 50$. The data were simulated using the model parameters discussed above. The

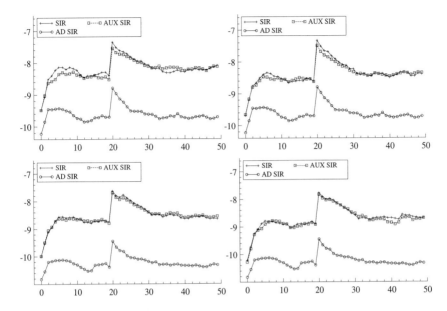

Figure 13.4. Plot of the mean square error performance (on the log-scale) of the particle filter to the auxiliary based particle filter and an adapted particle filter. The lower the number, the more efficient the method. Top graphs have $M = 2,000$, the bottom have $M = 4,000$. The left graphs have $R = M$, while the right ones have $R = 2M$

results are reported (using a log scale) in Figure 13.4. To make the problem slightly more realistic and challenging we set $\varepsilon_{21} = 2.5$ for each series, so there is a significant outlier at that point. For this study, we set $REP = 40$ and $S = 20$. We allow $M = 2,000$ or $4,000$, and for each of these values we set $R = M$ or $2M$. For the rejection based particle filter algorithm, it only makes sense to take $M = R$ and so, when $R > M$, we repeat the calculations as if $M = R$. Finally, the rejection based method takes around twice as long as that of the SIR based particle filter when $M = R$.

Figure 13.4 shows that the fully adapted particle filter is considerably more accurate than the other particle filters. It also has the advantage of not depending on R. The auxiliary particle filter is more efficient than the plain particle filter, but the difference is small, reflecting the fact that for the SV model, the conditional likelihood is not very sensitive to the state.

Mixtures of normals

Suppose $f(\alpha_{t+1}|\alpha_t)$ is Gaussian, but the measurement density is a discrete mixture of normals $\sum_{j=1}^{P} \lambda_j f_j(y_{t+1}|\alpha_{t+1})$. We can perfectly sample from

$f(\alpha_{t+1}, k|\mathcal{F}_{t+1})$ by working with

$$f(\alpha_{t+1}, k, j|\mathcal{F}_{t+1}) \propto \lambda_j f_j(y_{t+1}|\alpha_{t+1}) f\left(\alpha_{t+1}|\alpha_t^k\right) = w_{j,k} f_j(\alpha_{t+1}|\alpha_t^k, y_{t+1}).$$

We sample from $f(\alpha_{t+1}, k, j|\mathcal{F}_{t+1})$ by selecting the index k, j with probability proportional to $w_{j,k}$, and then drawing from $f_j(\alpha_{t+1}|\alpha_t^k, y_{t+1})$. The disadvantage of this approach is that the complete enumeration and storage of $w_{j,k}$ involves PM calculations. This approach can be trivially extended to cover the case in which $f(\alpha_{t+1}|\alpha_t)$ is a mixture of normals. MCMC smoothing methods for state-space models with mixtures have been studied by, for example, (Carter and Kohn 1994) and (Shephard 1994). This relates to the mixture Kalman filter introduced by (Doucet 1998) and discussed at length by (Chen and Liu 2000).

13.4   Fixed-lag filtering

The auxiliary particle filter method can also be used when we update the estimates of the states not by a single observation but by a block of observations. This idea appeared in the working paper version of (Pitt and Shephard 1999b) circulated in 1997, but was taken out of the published paper. This was subsequently studied in (Clapp and Godsill 1999).

Again, suppose that we approximate the density of $\alpha_t|\mathcal{F}_t = (y_1, ..., y_t)'$ by a distribution with discrete support at the points $\alpha_t^1, ..., \alpha_t^M$, with mass $\left\{\frac{1}{M}\right\}$. Then the task will be to update this distribution to provide a sample from $\alpha_{t+1}, ..., \alpha_{t+p}|\mathcal{F}_{t+p}$. At first sight, this result seems specialised because it is not often that we have to update after the arrival of a block of observations. However, as well as solving this problem, it also suggests a way of reducing the bias caused by using the empirical prediction density as an approximation to $f(\alpha_{t+1}|\mathcal{F}_t)$. Suppose that, instead of updating p future observations simultaneously, we store $p-1$ observations and update them together with an empirical prediction density for $f(\alpha_{t-p+2}|\mathcal{F}_{t-p+1})$. This would provide us with draws from $f(\alpha_{t+1}|\mathcal{F}_{t+1})$ as required. We call this fixed-lag filtering. We know that the influence of errors in the empirical prediction density reduce at an exponential rate as it propagated p times through the transition density using results in (Künsch 2000).

We can carry out fixed-lag particle filtering using SIR, rejection sampling or MCMC, or by building in an auxiliary variable so that we sample from $\alpha_{t+1}, ..., \alpha_{t+p}, k|\mathcal{F}_{t+p}$. Typically, the gains from using the auxiliary approach are greater here, for as p grows, so naive implementations of the particle filter will become less and less efficient owing to an inability to adapt the sampler to the measurement density.

To illustrate this general setup consider the use of an auxiliary particle filter where we take

$$g(k|\mathcal{F}_{t+p}) \propto \int f(y_{t+p}|\mu_{t+p}^k)...f(y_{t+1}|\mu_{t+1}^k)dF(\alpha_{t+p}|\alpha_{t+p-1})...dF(\alpha_{t+1}|\alpha_t^k)$$

$$= \prod_{j=1}^{p} f(y_{t+j}|\mu_{t+j}^k),$$

and then sampling the index k with weights proportional to $g(k|\mathcal{F}_{t+p})$. Having selected the index k^j, we then propagate the transition equation p steps to produce a draw $\alpha_{t+1}^j, ..., \alpha_{t+p}^j$, $j = 1, ..., R$. These are then reweighted according to the ratio

$$\frac{f(y_{t+p}|\alpha_{t+p}^j)...f(y_{t+1}|\alpha_{t+1}^j)}{f(y_{t+p}|\mu_{t+p}^{k^j})...f(y_{t+1}|\mu_{t+1}^{k^j})}.$$

This approach has three main difficulties. First, it requires us to store p sets of observations and $p \times M$ mixture components. This is more expensive than the previous method and is also slightly more difficult to implement. Second, each auxiliary variable draw now involves $3p$ density evaluations and the generation of p simulated propagation steps. Third, the auxiliary variable method is based on approximating the true density of $f(k, \alpha_{t-p+1}, ..., \alpha_t|Y_t)$, and this approximation is likely to deteriorate as p increases. This suggests that the more sophisticated adaptive sampling schemes, discussed above, may be particularly useful at this point. Again however, this complicates the implementation of the algorithm.

We will illustrate the use of this sampler on an outlier problem at the end of the next section.

13.5   Reduced random sampling

13.5.1   Basic ideas

The generic auxiliary particle filter given in Section 13.3.2 has two sets of weighted bootstraps, and so introduces a large degree of randomness to the procedure. This is most stark if $f(y_{t+1}|\alpha_{t+1})$ does not depend on α_{t+1} making y_{t+1} uninformative about α_{t+1}. For such a problem, the first-stage weights are $\lambda_k \propto g(k|\mathcal{F}_{t+1}) \propto 1$, while the second-stage weights $w_j \propto 1$. The implication is that it would have been better simply to propagate every α_t^k through $f(\alpha_{t+1}|\alpha_t^k)$ once to produce a new α_{t+1}^k. This would produce a more efficient sample than our method, which samples with replacement from these populations twice, killing interesting particles for no good reason.

This observation has appeared in the particle filtering literature on a number of occasions. (Liu and Chen 1995) discuss carrying weights forward instead of resampling in order to keep alive particles which would otherwise die. (Carpenter et al. 1999a) think about the same issue using stratification ideas taken from sampling theory.

Here we use a method similar to an idea discussed by (Liu and Chen 1998) in this context. We resample from a population $\alpha_t^1, ..., \alpha_t^M$ with weights $\pi_t^1, ..., \pi_t^M$ to produce a sample of size R in the following way. We produce stratified uniforms $\tilde{u}_t^1, ..., \tilde{u}_t^R$ by writing

$$\tilde{u}_t^k = \frac{(k-1) + u_t^k}{R}, \quad \text{where} \quad u_t^k \overset{iid}{\sim} UID(0,1).$$

This is the scheme suggested by (Carpenter et al. 1999a). We then compute the cumulative probabilities

$$\tilde{\pi}_t^r = \sum_{s=1}^{r} \pi_t^s, \quad r = 1, ..., M.$$

We allocate n^k copies of the particle α_t^k to the new population, where n^k is the number of $\tilde{u}_t^1, ..., \tilde{u}_t^R$ in the interval

$$\left(\sum_{s=1}^{k-1} \pi_t^s, \sum_{s=1}^{k} \pi_t^s \right].$$

The computation of each of $\{\tilde{u}_t^k\}$, $\{\tilde{\pi}_t^r\}$ and $\{n^k\}$ is straightforward, so this type of stratified sampling is very fast. In fact, it is much faster than simple random sampling.

Although, as we noted above, this idea is not new, it is particularly useful in the context of our generic auxiliary particle filter suggestion, which has two weighted bootstraps – while a typical SIR based particle filter has only one. Hence, we might expect that the gains to be made over the original suggestion in (Pitt and Shephard 1999b) should be particularly large.

13.5.2 Simple outlier example

We tried random and stratified sampling using fixed-lag versions of SIR based particle and auxiliary particle filters on a difficult outlier problem in which the analytic solution is available via the Kalman filter. We assume the observations arise from an auto-regression observed with noise

$$\begin{aligned} y_t &= \alpha_t + \varepsilon_t, & \varepsilon_t &\sim NID(0, 0.707^2) \\ \alpha_{t+1} &= 0.9702\alpha_t + \eta_t, & \eta_t &\sim NID(0, 0.178^2), \end{aligned} \quad (13.5.8)$$

where ε_t and η_t are independent processes. The model is initialised by α_t's stationary prior while we used $n = 35$. We added to the simulated $y_{n/2}$ a shock 6.5×0.707, which represents a very significant outlier. Throughout, we set $M = R = 500$ and measure the precision of the filter by the log mean square error criteria (13.3.6), taking $REP = 30$ and $S = 20$. As the

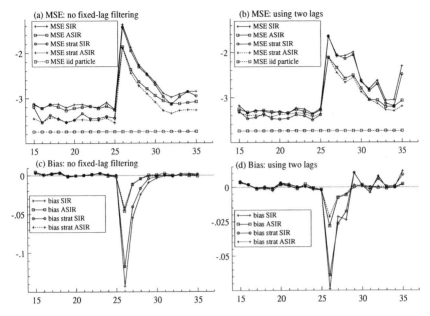

Figure 13.5. The mean square error (MSE), using a log10 scale, and bias of four different particle filters using no and two filtered lag filters. The x-axis is always time, but we only graph results for $t = T/4, T/4 + 1, ..., 3T/4$ in order to focus on the crucial aspects. The four particle filters are: SIR, ASIR, stratified SIR and stratified ASIR. The results are grouped according to the degree of fixed lag filtering. In particular: (a) shows the MSE when $p = 0$, (b) shows the MSE when $p = 2$. (c) shows the bias when $p = 0$, while (d) indicates the bias with $p = 2$. Throughout, we have taken $M = R = 500$.

problem is Gaussian, the Kalman filter's MSE divided by M provides a lower bound on the mean square error criteria.

Figure 13.5 shows the results from the experiment, recording the mean square errors and bias of the various particle filters. It is important to note that the mean square errors are drawn on the log10 scale. The main features of the graphs are: (i) when there is no outlier, all the particle filters are basically unbiased with stratification being important. The use of the auxiliary variable does not have much impact in this situation (although it is still better); (ii) during an outlier, the ASIR methods dominate both in terms of bias and MSE. Stratification makes very little difference in this situation; (iii) after the outlier, stratified ASIR continues to work well while ASIR returns to being less effective than stratified SIR. (iv) The introduction of fixed-lag filtering reduces the bias of all methods by an order of magnitude while the MSE reduces quite considerably.

In order to benchmark these results, we have repeated the experiment, but now with $M = R = 2500$. The results are given in Figure 13.6. This

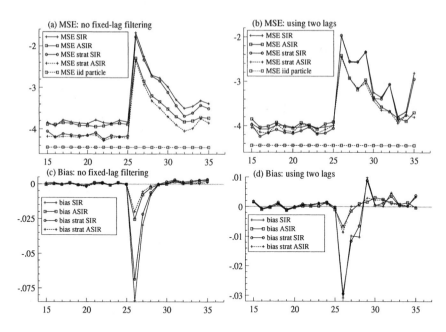

Figure 13.6. Repeat of figure 13.5, but with $M = R = 2500$.

picture is remarkably similar to Figure 13.5 but, with smaller bias and MSE. An important feature of this experiment is that the reduction in bias and MSE of a five fold increase in M and R produces around the same impact as the introduction of fixed-lag filtering.

13.6   Conclusion

This chapter has studied the weaknesses of the very attractive particle filtering method originally proposed by (Gordon et al. 1993). The SIR implementation of this method is not robust to outliers for two reasons: sampling efficiency, and the unreliability of the empirical prediction density in the tails of the distribution. We introduce an auxiliary variable into the particle filter to overcome the first of these problems, providing a direct generalisation of the SIR method, which allows more reliability and flexibility. We show that introducing the auxiliary variables into the problem allows many problems to be tackled efficiently by introducing adaption into the particle filter, which can bring about large efficiency gains. The fixed-lag filter partly tackles the second problem, which suggests a possible real improvement in the reliability of these methods.

13.7 Acknowledgements

Michael Pitt thanks the ESRC for its financial support through the grant "Modelling and analysis of econometric and financial time series." Neil Shephard's research is supported by the ESRC through the grant "Econometrics of trade-by-trade price dynamics," which is coded R000238391. We thank Neil Gordon for his particularly editorial comments.

14
Improved Particle Filters and Smoothing

Photis Stavropoulos
D.M. Titterington

14.1 Introduction

Exact recursive Bayesian inference is essentially impossible except in very special scenarios, such as the linear-Gaussian dynamic systems that are amenable to the Kalman filter and associated methods. Otherwise, some form of approximation is necessary. In some contexts, a parametric approximation might still be workable, as in (Titterington 1973)'s use of two-component Normal mixtures in a simple extremum-tracking problem (which we revisit later in this chapter), but nowadays it is more common to carry forward, as an estimate of the current distribution of the items of interest, what is claimed to be a simulated sample from that distribution, in other words, a particle filter.

A natural way of recursively generating such samples is to specify a prior distribution from which a sample can be simulated, and then to use weighted bootstrap techniques as described by (Smith and Gelfand 1992), with weights defined by the likelihood of the most recent data, to update that sample so that it better represents current information. However, repeated standard application of the weighted bootstrap suffers from a number of deficiencies if the number of updating stages becomes large. In particular, the sample being carried forward becomes 'impoverished', in that the number of distinct values it contains can decline dramatically. This phenomenon is referred to as 'sample attrition' by Liu and West (2001: this volume) and in its most extreme form as 'particle degeneracy' by Musso, Oudjane and Le Gland (2001: this volume); these two chapters contain much valuable material that complements the contents of this one.

The purpose of this chapter is to describe a number of approaches to this issue, and in general to refine the weighted bootstrap.

14.2   The methods

14.2.1   The smooth bootstrap

This method is a very straightforward extension of the weighted boot-strap; see, for example, (Gordon 1993, West 1993a, Givens and Raftery 1996, Gordon 1997). Suppose we have a random sample $\mathbf{x}(1), \ldots, \mathbf{x}(N)$ from an importance sampling function (ISF) g, and that our objective is inference about a distribution with probability density function (pdf) $\pi(\mathbf{x}) = \frac{f(\mathbf{x})}{\int f(\mathbf{x})d\mathbf{x}}$. We obtain a mixture approximation to π by assigning to each $\mathbf{x}(i)$ a normalised weight

$$q_i = \frac{f(\mathbf{x}(i))/g(\mathbf{x}(i))}{\sum_{j=1}^{N} f(\mathbf{x}(j))/g(\mathbf{x}(j))}$$

and taking

$$\hat{\pi}(\mathbf{x}) = \sum_{i=1}^{N} q_i K(\mathbf{x}; \mathbf{x}(i), B_N), \qquad (14.2.1)$$

where $K(\mathbf{x}; \mathbf{y}, B)$ is the value at $\mathbf{x}$ of a symmetric kernel K with mean $\mathbf{y}$ and variance matrix B. Sampling from $\hat{\pi}$ will be easy if sampling from K is easy. Since K is continuous, the samples generated by this smooth bootstrap method will not suffer from the impoverishment of the support of the sample that is inevitable with the ordinary bootstrap. We must also ensure that the samples from $\hat{\pi}$ converge to random samples from π as N goes to infinity. Intuitively, this will happen if we make the method resemble the weighted bootstrap as N becomes larger. This can be achieved if the variance parameters corresponding to the kernels decrease asymptotically to zero as functions of N, which explains the notation B_N in (14.2.1).

This is proved formally in (Stavropoulos and Titterington 1998). For reasons of completeness of exposition, we repeat our proof here for the case of univariate x and for the Normal kernel K. In this case the approximation to π becomes

$$\hat{\pi}(x) = \sum_{i=1}^{N} q_i N(x; x(i), b_N^2), \qquad (14.2.2)$$

where $N(x; a, b^2)$ is the value at x of the pdf of a Normal distribution with mean a and variance b^2.

All the limits that will be mentioned correspond to $N \longrightarrow \infty$. (Geweke 1989) gives conditions under which importance sampling estimates are asymptotically consistent and Normal. The same conditions are required for our proof to be valid, and they are expressed below:

- π is a proper pdf;

- $x(1), \ldots, x(N)$ are independently and identically distributed draws from g;

- the support of g contains that of π;

- the quantities $\mathbb{E}_g \left(\frac{f(x)}{g(x)} \right)$ and $\mathbb{E}_g \left[\left(\frac{f(x)}{g(x)} \right)^2 \right]$ exist and are finite.

- $b_N^2 \longrightarrow 0$.

Suppose that x^* is a draw from (14.2.2). Then,

$$
\begin{aligned}
\Pr(x^* \leq z | x(1), \ldots, x(N)) &= \sum_{i=1}^{N} q_i \Pr(x^* \leq z | x^* \sim N(x(i), b_N^2)) \\
&= \frac{\frac{1}{N} \sum_{i=1}^{N} \frac{f(x(i))}{g(x(i))} \Phi \left(\frac{z - x(i)}{b_N} \right)}{\frac{1}{N} \sum_{i=1}^{N} \frac{f(x(i))}{g(x(i))}},
\end{aligned}
$$

where Φ denotes the cumulative distribution function of the standard Normal distribution. According to the law of large numbers,

$$
\frac{1}{N} \sum_{i=1}^{N} \frac{f(x(i))}{g(x(i))} \longrightarrow \mathbb{E}_g \left(\frac{f(x)}{g(x)} \right) = \int_{-\infty}^{\infty} f(x) dx,
$$

in an appropriate sense. Furthermore, we define

$$
S_N = \frac{1}{N} \sum_{i=1}^{N} \frac{f(x(i))}{g(x(i))} \Phi \left(\frac{z - x(i)}{b_N} \right)
$$

and

$$
\mu_N = \mathbb{E}_g(S_N) = \mathbb{E}_g \left[\frac{f(x)}{g(x)} \Phi \left(\frac{z - x}{b_N} \right) \right] = \int_{-\infty}^{\infty} f(x) \Phi \left(\frac{z - x}{b_N} \right) dx.
$$

Then $\Phi \left(\frac{z - x}{b_N} \right) \longrightarrow I(x \leq z)$ for any x, where I is the indicator function, so that

$$
\mu_N \longrightarrow \mu_\infty = \int_{-\infty}^{\infty} f(x) I(x \leq z) dx = \int_{-\infty}^{z} f(x) dx.
$$

If we can show that $S_N \longrightarrow \mu_\infty$, then we will have shown that

$$
\Pr(x^* \leq z | x(1), \ldots, x(N)) \longrightarrow \frac{\int_{-\infty}^{z} f(x) dx}{\int_{-\infty}^{\infty} f(x) dx} = \int_{-\infty}^{z} \pi(x) dx = \Pr_\pi(x^* \leq z).
$$

$$(14.2.3)$$

However, for any positive real number ϵ, we have from Chebyshev's inequality that

$$\Pr(|S_N - \mu_\infty| > \epsilon) \leq \frac{\mathbb{E}_g[(S_N - \mu_\infty)^2]}{\epsilon^2} = \frac{\mathrm{var}_g(S_N) + (\mu_N - \mu_\infty)^2}{\epsilon^2}$$

$$= \frac{\sigma_N^2}{N\epsilon^2} + \frac{(\mu_N - \mu_\infty)^2}{\epsilon^2}, \qquad (14.2.4)$$

where

$$\sigma_N^2 = \mathrm{var}_g\left[\frac{f(x)}{g(x)}\Phi\left(\frac{z-x}{b_N}\right)\right] = \int_{-\infty}^{\infty} \frac{f^2(x)}{g(x)}\Phi^2\left(\frac{z-x}{b_N}\right)dx - \mu_N^2.$$

Therefore,

$$\sigma_N^2 \longrightarrow \sigma_\infty^2 = \int_{-\infty}^{z} \frac{f^2(x)}{g(x)}dx - \mu_\infty^2.$$

However, $\mathbb{E}_g\left[\left(\frac{f(x)}{g(x)}\right)^2\right]$ is assumed finite, so that $\int_{-\infty}^{z}\frac{f^2(x)}{g(x)}dx - \mu_\infty^2$ is finite too, and in consequence so is σ_∞^2. It follows that $\sigma_N^2/N \longrightarrow 0$. From (14.2.4) this leads to the conclusion that $\Pr(|S_N - \mu_\infty| > \epsilon) \longrightarrow 0$, and so (14.2.3) is true. The cumulative distribution function (cdf) of the values generated by smooth bootstrap converges to that of the target density π.

It is straightforward to construct a multivariate generalisation of the proof, or adapt it to the case of non-Normal K. Any symmetric uni-model kernel K is suitable for smooth bootstrap, provided its value decreases towards zero in the tails. The use of kernels may have the consequence that the convergence of smooth bootstrap is slower than that of weighted bootstrap.

The choice of B_N is also important because it is desirable that (14.2.1) should be a good approximation to π not only in the limit, but also for finite values of N. Kernel density estimation theory (Silverman 1986) offers valuable advice, and an obvious choice is given by

$$B_N = V \cdot b_N^2 \quad \text{with} \quad b_N = \left(\frac{4}{(n_x + 2)N}\right)^{\frac{1}{n_x + 4}}, \qquad (14.2.5)$$

where $V = \sum_{i=1}^{N} q_i(\mathbf{x}(i) - \sum_{j=1}^{N} q_j\mathbf{x}(j))(\mathbf{x}(i) - \sum_{j=1}^{N} q_j\mathbf{x}(j))^T$ is an importance sampling estimate of the variance matrix of π and n_x is the dimensionality of $\mathbf{x}$. This particular value for the bandwidth b_N would be asymptotically optimal if π were a Normal density and $\mathbf{x}(1), \ldots, \mathbf{x}(N)$ had been obtained from such a distribution. In view of this dependence, one can consider other bandwidth choices. (Gordon 1993), for example, derives his by minimising the mean integrated square error between π and $\hat{\pi}$, still relying on the assumption that π is Normal. Alternatively, if we believe that the bandwidth given by (14.2.5) is too large, we can divide it by some constant $c > 1$, to avoid oversmoothing.

Of concern also are the statistical properties of estimates derived from samples obtained by smooth bootstrap. For example, suppose that $l(\mathbf{x})$ is a univariate function of $\mathbf{x}$ with finite mean $\bar{l} = \int l(\mathbf{x})\pi(\mathbf{x})d\mathbf{x}$ and variance $\mathrm{var}_\pi(l(\mathbf{x})) = \int (l(\mathbf{x})-\bar{l})^2\pi(\mathbf{x})d\mathbf{x}$. If $\mathbf{x}^*(1),\dots,\mathbf{x}^*(n)$ is a weighted bootstrap sample from π, (Berzuini et al. 1997) prove that

$$\frac{\bar{l}^* - \bar{l}}{\sqrt{\frac{u_1}{N} + \frac{u_2}{n}}} \longrightarrow N(0,1) \quad \text{in distribution as} \quad n, N \longrightarrow \infty,$$

where $\bar{l}^* = \frac{1}{n}\sum_{i=1}^n l(\mathbf{x}^*(i))$, $u_1 = \mathbb{E}_g\left[(l(\mathbf{x}) - \bar{l})^2 \left(\frac{\pi(\mathbf{x})}{f(\mathbf{x})}\right)^2 \right]$ and, finally, $u_2 = \mathrm{var}_\pi(l(\mathbf{x}))$. This result is proven by considering the expectation of the conditional mean and variance of $\bar{l}^*$ given $\mathbf{x}(1),\dots,\mathbf{x}(N)$, the sample from g. These are

$$\mathbb{E}(\bar{l}^*|\mathbf{x}(1),\dots,\mathbf{x}(N)) = \sum_{i=1}^N q_i l(\mathbf{x}(i)) \quad \text{and}$$

$$\mathrm{var}(\bar{l}^*|\mathbf{x}(1),\dots,\mathbf{x}(N)) = \frac{1}{n}\sum_{i=1}^N q_i\{l(\mathbf{x}(i)) - \sum_{j=1}^N q_j l(\mathbf{x}(j))\}^2.$$

Suppose now that, starting with $\mathbf{x}(1),\dots,\mathbf{x}(N)$, we obtain $\mathbf{x}'(1),\dots,\mathbf{x}'(n)$ by sampling from (14.2.1), i.e. by smooth bootstrap. Let $\bar{l}'$ be the mean $\frac{1}{n}\sum_{i=1}^n l(\mathbf{x}'(i))$. If we can prove that

$$\mathbb{E}(\bar{l}'|\mathbf{x}(1),\dots,\mathbf{x}(N)) = \sum_{i=1}^N q_i l(\mathbf{x}(i)), \tag{14.2.6}$$

$$\mathrm{var}(\bar{l}'|\mathbf{x}(1),\dots,\mathbf{x}(N)) = \frac{1}{n}\sum_{i=1}^N q_i\{l(\mathbf{x}(i)) - \sum_{j=1}^N q_j l(\mathbf{x}(j))\}^2, \tag{14.2.7}$$

then the central limit theorem of (Berzuini et al. 1997) will be true here too. (West 1993a) mentions that if B_N is given by (14.2.5) and K is the Normal kernel the mean and variance of (14.2.1) are $\sum_{i=1}^N q_i\mathbf{x}(i)$ and $V(1 + b_N^2)$ respectively. If on the other hand we shrink $\mathbf{x}(1),\dots,\mathbf{x}(N)$ towards their mean and use $\hat{\mathbf{x}}(1),\dots,\hat{\mathbf{x}}(N)$, where $\hat{\mathbf{x}}(i) = \alpha\mathbf{x}(i)+(1-\alpha)\sum_{j=1}^N q_j\mathbf{x}(j)$, but retain the same B_N, then we obtain mean $\sum_{i=1}^N q_i\mathbf{x}(i)$ and variance matrix $(\alpha^2 + b_N^2)V$. Therefore $\alpha = \sqrt{1 - b_N^2}$ will give, for (14.2.1), the variance matrix V (West 1993a). Then (14.2.6) and (14.2.7) will be true for $l(\mathbf{x}) = \mathbf{x}$, and indeed for any linear function l. Consequently the central limit theorem will be true for such functions. However, we cannot extend the result to the case of a general function l because (14.2.1) is a continuous pdf. Of course, this does not mean that the central limit theorem does not hold in general for smooth bootstrap samples. On the contrary, it seems intuitively that it should, but we do not have a formal proof. Note also that the convergence of smooth bootstrap samples to samples from the target is not affected by shrinkage, since $\alpha \longrightarrow 1$ as $N \longrightarrow \infty$. (Givens and Raftery 1996) point out that shrinkage causes π not to be well represented by its mixture approximation if π has non-convex contours. We still believe, however, that with due caution smooth bootstrap can be employed with shrinkage because of its consequent close relationship to the weighted bootstrap.

14.2.2 Adaptive importance sampling

A crucial decision in resampling is which ISF should be used. Satisfactory ISF's resemble the shape of the target density, and the ratio N/n of sample sizes need not then be large. There are instances, however, when such ISF's cannot be found. Resampling will then be very inefficient, as N/n would have to be very large. Adaptive importance sampling (AIS) has been suggested as a way of avoiding this problem (West 1993a). If the ISF is poor, then (14.2.1), with or without shrinkage, will not be a good approximation to the target, although it can serve as a new ISF, which ought to be better than the previous one. The process may be iterated as often as necessary. The formal structure of AIS is as follows.

- Initially (stage 0) we choose ISF $g_0(\mathbf{x})$ and sample m_0 points $\mathbf{x}^{(0)}(1), \ldots, \mathbf{x}^{(0)}(m_0)$ from it. We calculate their normalised weights $q_1^{(0)}, \ldots, q_{m_0}^{(0)}$ and we form a mixture approximation to π as usual:

$$g_1(\mathbf{x}) = \sum_{i=1}^{m_0} q_i^{(0)} K(\mathbf{x}; \mathbf{x}^{(0)}(i), B_{m_0}).$$

We may use shrinkage if we wish. We then iterate the process for k stages by repeating the following.

- At stage j we obtain m_j points $\mathbf{x}^{(j)}(1), \ldots, \mathbf{x}^{(j)}(m_j)$ from $g_j(\mathbf{x})$ and we calculate their normalised weights,

$$q_i^{(j)} = \frac{f(\mathbf{x}^{(j)}(i))/g_j(\mathbf{x}^{(j)}(i))}{\sum_{r=1}^{m_j} f(\mathbf{x}^{(j)}(r))/g_j(\mathbf{x}^{(j)}(r))}.$$

We then form a new mixture approximation,

$$g_{j+1}(\mathbf{x}) = \sum_{i=1}^{m_j} q_i^{(j)} K(\mathbf{x}; \mathbf{x}^{(j)}(i), B_{m_j}).$$

At stage k we sample n points from $g_{k+1}(\mathbf{x})$, using these points as the sample from the target.

A small k ($k = 1$ or 2) will usually suffice. As a guide for stopping the process, (West 1993a) suggests monitoring the relative entropy of the normalised weights, which at stage j is

$$-\sum_{i=1}^{m_j} q_i^{(j)} \frac{\log q_i^{(j)}}{\log m_j}.$$

If the relative entropy approaches 1, there is not much scope for further improvement, so we can stop at stage j. If the relative entropy is much smaller than 1, we continue to iterate the smooth bootstrap. Alternatively, we can monitor the relative entropy and stop when this stops changing significantly between stages.

Adaptive importance sampling is a very appealing method. (Givens and Raftery 1996), however, disagree with the use of the same variance matrix B_{m_j} for all kernels at stage j. Simulation examples in (Givens and Raftery 1996) show that, because of the misrepresentation of the target caused by this practice and by shrinkage, the behaviour of AIS can be erratic, and sometimes even worse than weighted bootstrap. (Givens and Raftery 1996) favour AIS with 'local' variance matrices. More specifically, at stage j their g_{j+1} is

$$g_{j+1}(\mathbf{x}) = \sum_{i=1}^{m_j} q_i^{(j)} K(\mathbf{x}; \mathbf{x}^{(j)}(i), B_{m_j}(i)), \qquad (14.2.8)$$

where $B_{m_j}(i)$ is calculated as usual, but based only on the n_j points closest to $\mathbf{x}^{(j)}(i)$. They argue that (14.2.8) conforms better to the local behaviour of the target because the target's variance matrix can sometimes be misleading as to the target's shape. They suggest taking $n_j \approx 0.5m_j$, but mention that their experience has shown that any value from $0.2m_j$ to close to m_j will perform almost equally well. In their examples, they typically use up to about 5 stages, with each m_j of the order of 1000. They envisage that in practice there may plausibly be a constraint on the total $\sum_j m_j$, and they compare adaptive and non-adaptive approaches subject to that constraint.

There is also a computational problem with AIS, in that the evaluation of each $q_i^{(j)}$ requires us to calculate the values of m_j kernels K, and this can obviously be very time consuming. (West 1993a) suggests an algorithm that reduces the number of components of a mixture by amalgamating components with small weights with close neighbours that have larger weights; in his examples he typically reduces the number of components from several thousand down to a few hundred. While this does offer computational gains, it is computationally demanding. A similar component-merging algorithm appears in (Salmond 1990). (Oehlert 1998) has suggested using linear tensor splines to approximate the ISFs. They are easier to evaluate, but more difficult to update between stages of AIS.

The simulation experiments of (Givens and Raftery 1996) show that, if the sample we draw from an ISF is of sufficiently large size N, then we can use simple weighted or smooth bootstrap because it will outperform

AIS. If, however, N is not large, then it is better to split the sample into sub-samples to be obtained at successive stages of AIS.

14.2.3 The kernel sampler of Hürzeler and Künsch

(Hürzeler and Künsch 1998)'s method, motivated by rejection sampling rather than importance sampling, is specifically designed for application in Bayesian settings. It provides samples from the posterior distribution of the parameters of interest when we can sample from their prior.

Let $\mathbf{x}$ be the vector of parameters and let $\mathbf{x}(1), \ldots, \mathbf{x}(N)$ be a sample from its prior distribution, which has density $p(\mathbf{x})$. Based on this sample, we form a kernel density estimate (kde) of $p(\mathbf{x})$,

$$\hat{p}(\mathbf{x}) \propto \frac{1}{N} \sum_{i=1}^{N} K(V^{-1/2}(\mathbf{x} - \mathbf{x}(i))),$$

where V is an appropriate variance matrix and K is a radially symmetric kernel such that

$$K(\mathbf{x}) > 0 \quad \text{only if} \quad ||\mathbf{x}|| \leq 1. \tag{14.2.9}$$

If $\mathbf{y}$ denotes the observed data, and the likelihood function is $p(\mathbf{y}|\mathbf{x})$, then the posterior distribution of $\mathbf{x}$ is estimated by

$$\hat{p}(\mathbf{x}|\mathbf{y}) \propto \frac{1}{N} \sum_{i=1}^{N} p(\mathbf{y}|\mathbf{x}) K(V^{-1/2}(\mathbf{x} - \mathbf{x}(i))).$$

Suppose that we can find bounds c_i such that

$$p(\mathbf{y}|\mathbf{x}) \leq c_i \quad \text{for any} \quad \mathbf{x} \quad \text{that satisfies} \quad ||V^{-1/2}(\mathbf{x} - \mathbf{x}(i))|| \leq 1.$$

Notice that, if the kernel does not have a bounded support as indicated by (14.2.9), then all the bounds c_i would be equal to the maximised likelihood based on $\mathbf{y}$, which may not even exist. If, however, such bounds can be found, then we can write

$$\hat{p}(\mathbf{x}|\mathbf{y}) \propto \sum_{i=1}^{N} c_i \frac{p(\mathbf{y}|\mathbf{x})}{c_i} K(V^{-1/2}(\mathbf{x} - \mathbf{x}(i))).$$

This motivates the interpretation of $\hat{p}(\mathbf{x}|\mathbf{y})$ as the marginal pdf of $\mathbf{x}$ obtained from

$$\hat{p}(\mathbf{x}, i|\mathbf{y}) \propto c_i \frac{p(\mathbf{y}|\mathbf{x})}{c_i} K(V^{-1/2}(\mathbf{x} - \mathbf{x}(i))),$$

$i = 1, \ldots, N$, which in turn suggests the following rejection sampling algorithm. We draw $j, \mathbf{x}^*$ from a distribution with pdf

$$g(j, \mathbf{x}^*) \propto c_j K(V^{-1/2}(\mathbf{x}^* - \mathbf{x}(j))),$$

and u from a Uniform distribution $U(0,1)$. (To sample from $g(j, \mathbf{x}^*)$, we first choose j according to probabilities proportional to the c_j and then, for that j, take $\mathbf{x}^* = \mathbf{x}(j) + V^{-1/2}\mathbf{v}$, where $\mathbf{v}$ is sampled according to the pdf $K(\mathbf{v})$.) Then, if

$$u \le \frac{p(\mathbf{y}|\mathbf{x})}{c_j},$$

we can accept $\mathbf{x}^*$ as a sample from $p(\mathbf{x}|\mathbf{y})$. The existence of 'local' bounds c_i for the likelihood improves the performance of this method when compared with a rejection method, which would use a global bound for $p(\mathbf{y}|\mathbf{x})$. The existence of such bounds is not always guaranteed, but particular cases where they do exist are for linear models with certain noise distributions. If the mean of $\mathbf{y}$ given $\mathbf{x}$ is a linear function $\mathbf{h}^T\mathbf{x}$, and if the pdf $s(\cdot)$ of the noise has a single mode at 0, then, by defining

$$a_i^{\pm} = \mathbf{h}^T\mathbf{x}(i) \pm b_N(\mathbf{h}^T\Sigma\mathbf{h})^{1/2},$$

where Σ is the sample variance matrix of $\mathbf{x}(1), \ldots, \mathbf{x}(N)$, we can take

$$c_i = \begin{cases} s(a_i^+) & \text{if} & a_i^+ < 0 \\ s(a_i^-) & \text{if} & a_i^- > 0 \quad \text{and} \\ 0 & \text{otherwise.} \end{cases}$$

The advantage of the kernel sampler is that it produces samples from the target, or rather its estimate, for any sample size N, in contrast with the resampling methods, which only achieve this asymptotically. On the other hand, the number of draws required to obtain a sample from the target is random and unbounded, although see Musso, Oudjane and Le Gland (2001: this volume) for an effective way of estimating the mean number of draws. There is a connection with resampling, however. If $V = \mathbf{0}$, then the kernel sampler becomes weighted bootstrap. With $V \ne \mathbf{0}$ we ensure that the resulting sample consists of distinct values. We can use standard asymptotic formulae for bandwidths. For instance, for the Epanechnikov (quadratic) kernel scaled to satisfy (14.2.9), we can take

$$V = b_N^2 \cdot \Sigma \quad \text{with} \quad b_N = \left(8c_{n_x}^{-1}(n_x + 4)(2\sqrt{\pi})^{n_x}\right)^{1/(n_x+4)} N^{-1/(n_x+4)},$$

where c_{n_x} denotes the volume of the unit-radius n_x-dimensional hypersphere. (Hürzeler and Künsch 1998) suggest using a smaller value of b_N in order to make the method more effective for multi-modal targets.

14.2.4 Partially smooth bootstrap

We propose this method as a compromise between the weighted and the smooth bootstrap. The idea behind it is that we do not need to discard the weighted bootstrap entirely. We apply the weighted bootstrap to the sample $\mathbf{x}(1), \ldots, \mathbf{x}(N)$ from the ISF, but we also set things up for applying the smooth bootstrap; we choose a kernel K, and we calculate a bandwidth b_N, a shrinkage parameter α and the 'shrunk' values $\hat{\mathbf{x}}(1), \dots, \hat{\mathbf{x}}(N)$. We subsequently apply weighted bootstrap as usual. In the resulting sample, some $\mathbf{x}(i)$'s may appear more than once, especially if they correspond to large weights. We retain only one replicate of each such $\mathbf{x}(i)$. The rest are replaced by draws from the kernel K with mean $\mathbf{x}(i)$ or $\hat{\mathbf{x}}(i)$. By this device we retain all the advantages of smooth bootstrap while minimising the problem associated with shrinkage identified by (Givens and Raftery 1996). This is because one copy of each $\mathbf{x}(i)$ selected in the weighted bootstrap step is retained at its original 'pre-shrinking' location.

As a result of our method's connections with the weighted and smooth bootstraps, it is to be expected that the samples it produces also converge to samples from the target distribution π as N approaches infinity. For simplicity, we again prove this for the univariate case, with Normal kernel and no shrinkage, although the results generalise to other choices of kernel and to multivariate cases. Once again, all the limits are with respect to $N \longrightarrow \infty$.

Suppose k' ($k' < n$, k' possibly 0) values have already been generated. Also suppose, during the weighted bootstrap steps that led to their generation, k distinct values ($k \le k'$) were selected from among $x(1), \dots, x(N)$. We can re-label $x(1), \dots, x(N)$ and their respective weights so that $x(1), \dots, x(k)$ are the values that have already been selected. Let x^* be the next value to be generated by partially smooth bootstrap. Its cdf, given $x(1), \dots, x(N)$, is

$\Pr(x^* \le z | x(1), \dots, x(N))$

$$
= \sum_{j=k+1}^{N} q_j I(x(j) \le z) + \sum_{j=1}^{k} q_j \Pr(x^* \le z | x^* \sim N(x(j), b_N^2))
$$

$$
= \frac{\frac{1}{N} \sum_{j=1}^{N} \frac{f(x(j))}{g(x(j))} I(x(j) \le z)}{\frac{1}{N} \sum_{j=1}^{N} \frac{f(x(j))}{g(x(j))}} - \frac{\frac{1}{N} \sum_{j=1}^{k} \frac{f(x(j))}{g(x(j))} I(x(j) \le z)}{\frac{1}{N} \sum_{j=1}^{N} \frac{f(x(j))}{g(x(j))}}
$$

$$
+ \frac{\frac{1}{N} \sum_{j=1}^{k} \frac{f(x(j))}{g(x(j))} \Pr(x^* \le z | x^* \sim N(x(j), b_N^2))}{\frac{1}{N} \sum_{j=1}^{N} \frac{f(x(j))}{g(x(j))}}. \quad (14.2.10)
$$

From the law of large numbers we obtain that

$$
\frac{\frac{1}{N} \sum_{j=1}^{N} \frac{f(x(j))}{g(x(j))} I(x(j) \le z)}{\frac{1}{N} \sum_{j=1}^{N} \frac{f(x(j))}{g(x(j))}} \longrightarrow \Pr_\pi(x \le z).
$$

Moreover,

$$
\sum_{j=1}^{k} \frac{f(x(j))}{g(x(j))} \Pr(x^* \le z | x^* \sim N(x(j), b_N^2)) \longrightarrow \sum_{j=1}^{k} \frac{f(x(j))}{g(x(j))} I(x(j) \le z),
$$

and therefore the second and third terms of (14.2.10) go to 0. This proves that $\Pr(x^* \leq z|x(1),\ldots,x(N)) \longrightarrow \Pr_\pi(x^* \leq z)$.

A difference between partially smooth bootstrap and its two constituent methods is that the samples it generates do not consist of independent values, even when we condition on $x(1),\ldots,x(N)$. It is therefore difficult to investigate the statistical properties of its sample statistics. However, since the method converges to the weighted bootstrap as N goes to infinity, the asymptotic unbiasedness of the sample mean is clear. We believe that partially smooth bootstrap is a promising resampling technique.

14.2.5 Roughening and sample augmentation

The two methods described here differ from those presented earlier in that they are employed after weighted bootstrap has been applied. Suppose that $\mathbf{x}^*(1),\ldots,\mathbf{x}^*(N)$ is the sample produced by weighted bootstrap. (Gordon et al. 1993) suggest 'roughening' this sample by adding to each of its members a quantity drawn at random from a Normal distribution. Each n_x-dimensional $\mathbf{x}^*(i)$ is replaced by $\mathbf{x}^{**}(i)$, where

$$\mathbf{x}^{**}(i) = \mathbf{x}^*(i) + \mathbf{z}(i)$$

and $\mathbf{z}(i)$ is a draw from $N(0, J_{n_x})$. Here J_{n_x} is a $n_x \times n_x$ diagonal matrix with the jth element on its diagonal being

$$\sigma_j^2 = (kE_j N^{-1/n_x})^2,$$

where E_j is the difference between the maximum and minimum values of the jth component of $\mathbf{x}^*(1),\ldots,\mathbf{x}^*(N)$ and k is a positive tuning constant chosen subjectively by the user. (Gordon et al. 1993) do not give any guidelines concerning its choice. The term N^{-1/n_x} is the distance, in each Cartesian coordinate, between adjacent points when N points have been placed equidistantly on the unit rectangular n_x-dimensional grid.

(Sutherland and Titterington 1994) propose 'augmentation' of the weighted bootstrap sample. Since $\mathbf{x}^*(1),\ldots,\mathbf{x}^*(N)$ asymptotically come from the target distribution π we can form a kernel density estimate

$$\hat{\pi}(\mathbf{x}) = \frac{1}{N}\sum_{i=1}^{N} N(\mathbf{x}; \mathbf{x}^*(i), B_N)$$

and sample $\mathbf{x}^{**}(1),\ldots,\mathbf{x}^{**}(N)$ from it. This sample will replace the previous one. The choice of B_N is dictated by kernel density estimation theory as in Section 14.2.1.

Note that the two methods are very similar since roughening also samples from a kernel density estimate of π. Augmentation could be said to be more appealing from a theoretical point of view because roughening uses a bandwidth of order N^{-1/n_x} instead of the theoretical $N^{-1/(n_x+4)}$. However, the former could be beneficial when the target is multi-modal. Moreover, the constant k can be set to a high value if under-smoothing must be avoided.

Both methods share the disadvantage of their reliance on a weighted bootstrap sample. In severe cases, this sample may consist of very few distinct values, and hence it will provide a poor basis for estimation of the target and for obtaining $\mathbf{x}^{**}(1), \ldots, \mathbf{x}^{**}(N)$. In addition, the methods can be applied even in the limit when the weighted bootstrap sample comes from the target. The resulting sample will then come from a mixture of Normal densities with mixing weights derived from the target. Of course, this is principally a technical point because the user will usually know whether the weighted bootstrap sample is good, but it shows that these two methods in particular need to be applied with care.

Our opinion is that they offer no advantage over the methods described in the previous sections. This conclusion is reinforced by the results of numerical experiments reported in detail in (Stavropoulos and Titterington 1998). They considered two examples, a simple linear dynamic model and a simple two-dimensional imaging problem, concerning noisy observation of a rectangle rotating in a plane. The second problem is distinctly nonlinear. The methods compared in the first experiment were the weighted bootstrap, the smooth bootstrap, the partially smooth bootstrap, the weighted bootstrap roughened according to the method of (Gordon et al. 1993) and the weighted bootstrap roughened according to the method of (Sutherland and Titterington 1994).

For the first problem, highest posterior density intervals for unknown parameters can be compared with the exact intervals because of the linearity of the model, and it turned out that all methods were effective and performed equally well. They were not so effective at tracking certain features of the unobservable state variables, such as the correlations between components of the state vector, although the partially smooth bootstrap was moderately successful in this respect.

For the second, nonlinear problem, comparisons with exact distributional results can of course not be made, and the methods were judged on the basis of the extent to which the ranges of the simulated samples covered the corresponding true values. The (Gordon et al. 1993) roughening was arguably the most successful method, but neither it nor the (Sutherland and Titterington 1994) roughening was superior in practically meaningful terms. On the other hand, the weighted bootstrap was discarded because the corresponding samples very quickly became impoverished.

14.2.6   Application of the methods in particle filtering and smoothing

Particle filtering and smoothing do not pose greater difficulties for the methods presented here than they do for weighted bootstrap. The use of importance sampling and resampling, and associated issues such as the choice of ISF, have been presented elsewhere; see for example (Gordon 1993, Doucet 1998, Kitagawa 1996, Liu and Chen 1998). The method of (Hürzeler and Künsch 1998) can also be very straightforwardly adapted to filtering and we do not elaborate further on this here. We do, however, present their proposal for smoothing in detail.

Suppose that data $\mathbf{y}_1, \ldots, \mathbf{y}_T$ have been collected and that from each filtering distribution $p(\mathbf{x}_t|\mathbf{y}_{1:t})$, where $\mathbf{y}_{1:t} = (\mathbf{y}_1, \ldots, \mathbf{y}_t)$, we have a sample $\mathbf{x}_{t|t}(1), \ldots, \mathbf{x}_{t|t}(m)$. Each smoothing density $p(\mathbf{x}_t|\mathbf{y}_{1:T})$ can be expressed as

$$p(\mathbf{x}_t|\mathbf{y}_{1:T}) = \int \frac{p(\mathbf{x}_{t+1}|\mathbf{x}_t)p(\mathbf{x}_t|\mathbf{y}_{1:t})}{p(\mathbf{x}_{t+1}|\mathbf{y}_{1:t})} p(\mathbf{x}_{t+1}|\mathbf{y}_{1:T})d\mathbf{x}_{t+1}. \qquad (14.2.11)$$

The sampling from smoothing densities begins with $p(\mathbf{x}_T|\mathbf{y}_{1:T})$, from which a sample is of course already available. At time t we approximate (14.2.11) by

$$\hat{p}(\mathbf{x}_t|\mathbf{y}_{1:T}) \propto \frac{1}{N} \sum_{j=1}^{N} p(\mathbf{x}_{t+1|T}(j)|\mathbf{x}_t)\hat{p}(\mathbf{x}_t|\mathbf{y}_{1:t}),$$

where $\mathbf{x}_{t+1|T}(1), \ldots, \mathbf{x}_{t+1|T}(N)$ is the sample from $p(\mathbf{x}_{t+1}|\mathbf{y}_{1:T})$, while the quantity $\hat{p}(\mathbf{x}_t|\mathbf{y}_{1:t})$ is a kde based on $\mathbf{x}_{t|t}(1), \ldots, \mathbf{x}_{t|t}(N)$ and given by the formula

$$\hat{p}(\mathbf{x}_t|\mathbf{y}_{1:t}) \propto \frac{1}{N} \sum_{i=1}^{N} K(V^{-1/2}(\mathbf{x}_t - \mathbf{x}_{t|t}(i)));$$

here K is the same kernel that was used for filtering by (Hürzeler and Künsch 1998), and $V = b_N^2 \Sigma$ with Σ the sample variance matrix for $\mathbf{x}_{t|t}(1), \ldots, \mathbf{x}_{t|t}(N)$ and $b_N = \left(\frac{4}{(n_x+2)N}\right)^{1/(n_x+4)}$.
Therefore,

$$\hat{p}(\mathbf{x}_t|\mathbf{y}_{1:T}) \propto \sum_{j=1}^{N} \sum_{i=1}^{N} p(\mathbf{x}_{t+1|T}(j)|\mathbf{x}_t)K(V^{-1/2}(\mathbf{x}_t - \mathbf{x}_{t|t}(i))),$$

which can also be written as

$$\hat{p}(\mathbf{x}_t|\mathbf{y}_{1:T}) \propto \sum_{j=1}^{N} \sum_{i=1}^{N} c_{t,ij} \frac{p(\mathbf{x}_{t+1|T}(j)|\mathbf{x}_t)}{c_{t,ij}} K(V^{-1/2}(\mathbf{x}_t - \mathbf{x}_{t|t}(i))).$$

The quantities $c_{t,ij}$ are again 'local' upper bounds, such that

$$p(\mathbf{x}_{t+1|T}(j)|\mathbf{x}_t) \le c_{t,ij} \text{ for any } \mathbf{x}_t \text{ such that } ||V^{-1/2}(\mathbf{x}_t - \mathbf{x}_{t|t}(i))|| \le 1.$$

Rejection sampling is used to draw one point $\mathbf{x}_t$ from each mixture component density, which is proportional to

$$p(\mathbf{x}_{t+1|T}(j)|\mathbf{x}_t) \sum_{i=1}^{N} K(V^{-1/2}(\mathbf{x}_t - \mathbf{x}_{t|t}(i))).$$

Given $\mathbf{x}_{t+1|T}(j)$, we choose $\mathbf{x}_{t|t}(i)$ with probability proportional to $c_{t,ij}$, and next we draw $\mathbf{x}$ from a distribution with pdf proportional to $K(V^{-1/2}(\mathbf{x} - \mathbf{x}_{t|t}(i)))$. We then draw $u \sim U(0,1)$, and if

$$u \leq \frac{p(\mathbf{x}_{t+1|T}(j)|\mathbf{x})}{c_{t,ij}},$$

we accept $\mathbf{x}$ as a member of the sample from $p(\mathbf{x}_t|\mathbf{y}_{1:T})$.

Adaptive importance sampling techniques become more time consuming in filtering and smoothing problems. In filtering, for example, at any time point t and for any stage apart from the first, the weights are proportional to the quantity $p(\mathbf{y}_t|\mathbf{x}_t)p(\mathbf{x}_t|\mathbf{y}_{1:t-1})$, whose second factor is not available in closed form. We can estimate it by taking

$$\hat{p}(\mathbf{x}_t|\mathbf{y}_{1:t-1}) = \frac{1}{N}\sum_{i=1}^{N} p(\mathbf{x}_t|\mathbf{x}_{t-1|t-1}(i)),$$

but then we will need a further N density evaluations per draw. For this part of the computations, the method of (Oehlert 1998) cannot be of help because it only applies to the ISF's.

A final point of great importance for filtering is that the methods we have presented offer significant improvements in inference about unknown constants. It is well known that weighted bootstrap samples deteriorate with time. In filtering, the intervention of the system equation reintroduces variability into the samples of the system states, but not into the samples for the constants. By their design, the methods shown in this chapter overcome this deficiency.

14.3   Application of smooth bootstrap procedures to a simple control problem

14.3.1   Description of the problem

We illustrate our smooth bootstrap with reference to the simple extremum adaptation problem considered by (Titterington 1973). Noisy observations are available of a response curve or surface, the location of which changes randomly over time. The objective is to keep track of the extremum $\mathbf{x}_t$ of the true curve or function.

In the particular example we consider, x_t is the scalar minimum of a parabolic response curve, and x_t moves along the real line in a random manner. If at time t we estimate the minimum by u_t, then we observe

$$y_t = \frac{1}{2}(x_t - u_t)^2 + \epsilon_t.$$

We assume that the ϵ_t are independent, identically distributed $N(0, \sigma_\epsilon^2)$ random variables. Also, by choosing this setting for u_t, we incur a cost of $\frac{1}{2}(x_t - u_t)^2$, so that inaccuracy in estimation is detrimental to the 'system'.

Furthermore, we assume that x_1, the initial minimum of the curve, comes from a $N(0, \sigma_1^2)$ distribution and that the system equation governing the movement of x_t is a random walk:

$$x_t = x_{t-1} + \eta_t,$$

where the η_t are independent $N(0, \sigma_\eta^2)$ random variables.

Note that, even if the curve could be observed without noise, there would be ambiguity concerning the location of the minimum. At each instant t, there would be two candidates for its location, namely, the solutions $u_t - \sqrt{2y_t}$ and $u_t + \sqrt{2y_t}$ of the equation

$$y_t = \frac{1}{2}(x_t - u_t)^2.$$

This clearly complicates the noisy version of the problem as well. In addition, the nonlinearity of the observation equation and the probable ignorance about the variance parameters σ_ϵ^2 and σ_η^2 have the consequence that neat Kalman-filter formulae are not available from which x_t can be inferred and a sensible choice of u_t derived.

14.3.2 An approach to the continuous-time version of the problem

(Titterington 1973) uses the following ad-hoc approach to deal with the continuous-time version of this problem.

If x_1 has a-priori a uni-model distribution centred on the origin and we set $u_1 = 0$, then as long as u_t remains constant the posterior distribution of x_t will be symmetric about zero, and it will become bimodal, with its two humps becoming increasingly distinct, if x_t drifts away from the origin. Early on, the location of x_t will be uncertain because the observations are swamped by error, and it would not be sensible to try to change u. However, this will not matter much because x_t is still close to zero and little cost is incurred. The idea is to wait until x_t has reached the neighbourhood of a prescribed value κ in magnitude, where κ is chosen in such a way that by the time $|x_t|$ reaches it, the deterministic part of the observations will have become dominant. This will translate into two distinct and very concentrated modes for the posterior of x_t. When this happens u_t is set, at random, to κ or $-\kappa$ and a (desirably very short) test period ensues, during which one tests whether the choice of u_t was the correct one. On the basis of the test, the sign of u may be switched. During the test period it is expected that the true x_t should not have moved far; it is assumed that we are now effectively back in the initial configuration and the cycle begins again.

(Titterington 1973) establishes a recipe for choosing κ, balancing the cost incurred by letting x_t drift, untracked, against the uncertainty about x_t, if it is not allowed to drift. In view of the nonlinearities in the system, the exact posterior distribution of x_t is unavailable. (Titterington 1973) approximates it by a symmetric mixture of two Gaussian components and derives equations for the evolution of the common variance and magnitude of the means of the two components. This evolution is inevitably carried out in discrete time, but the calculation of a suitable value for κ refers specifically to the continuous-time scenario and does not have an obvious discrete-time analogue. However, the discrete-time version of the problem is a natural testbed for our sample-based methods, three of which we describe in the next sections.

To this end, we introduce some notation. Let $y_{1:t} = (y_1, \ldots, y_t)$ be the data collected to time t, and let $u_{1:t}$ and $x_{1:t}$ be similarly defined.

14.3.3   An adaptation of Titterington's method

As in the continuous-time case, we choose a value of κ such that by the time $|x_t|$ approaches its posterior distribution will have two well separated modes. When the modes of the posterior go beyond $[-\kappa, \kappa]$, we move u to one of the two modes, and this becomes u_{t+1}. If we choose the wrong mode, then this will become immediately apparent and will be corrected at the next stage, so there is no need for a formal test period. In the spirit of this book, of course, we represent the posterior distribution by samples, rather than by an approximate parametric model. We discuss the mechanism for producing the samples later. For more details of the procedure, see P. Stavropoulos's unpublished 1998 University of Glasgow doctoral dissertation.

14.3.4   Probabilistic criterion 1

Here we try to maximise the probability that the distance between x_{t+1} and u_{t+1} will be smaller than a specified quantity v. This probability is based on the distribution of x_{t+1} given the information up to time t. In other words we want u_{t+1} to maximise

$$\Pr\left(|X_{t+1} - u_{t+1}| \leq v | y_{1:t}, u_{1:t}\right).$$

If we define an indicator variable

$$I(x_{t+1}, u_{t+1}) = \begin{cases} 1 & \text{if } |x_{t+1} - u_{t+1}| \leq v \\ 0 & \text{otherwise} \end{cases},$$

then

$$\Pr\left(|X_{t+1} - u_{t+1}| \leq v | y_{1:t}, u_{1:t}\right) = \mathbb{E}\left[I(X_{t+1}, u_{t+1}) | y_{1:t}, u_{1:t}\right].$$

If we have a sample $x(1), \ldots, x(N)$ from $p(x_{t+1}|y_{1:t}, u_{1:t})$, then the probability above can be estimated by

$$\hat{\Pr}\left(|X_{t+1} - u_{t+1}| \le v|y_{1:t}, u_{1:t}\right) = \frac{1}{N}\sum_{i=1}^{N} I(x(i), u_{t+1}). \qquad (14.3.12)$$

Alternatively, we can write

$$\Pr\left(|X_{t+1} - u_{t+1}| \le v|y_{1:t}, u_{1:t}\right) = \int_{-\infty}^{\infty} p(x_t|y_{1:t}, u_{1:t})\left[\Phi\left(\frac{u_{t+1} + v - x_t}{\sigma_\eta}\right)\right.$$
$$\left. - \Phi\left(\frac{u_{t+1} - v - x_t}{\sigma_\eta}\right)\right] dx_t.$$

Also, if we have a sample $x(1), \ldots, x(N)$ from $p(x_t|y_{1:t}, u_{1:t})$, this probability can be estimated by

$$\hat{\Pr}\left(|X_{t+1} - u_{t+1}| \le v|y_{1:t}, u_{1:t}\right) = \frac{1}{N}\sum_{i=1}^{N}\left[\Phi\left(\frac{u_{t+1} + v - x(i)}{\sigma_\eta}\right)\right.$$
$$\left. - \Phi\left(\frac{u_{t+1} - v - x(i)}{\sigma_\eta}\right)\right]. \qquad (14.3.13)$$

14.3.5   Probabilistic criterion 2: working directly with the cost

An alternative goal is to have small cost y_{t+1} with high probability. After all, the cost is what we observe and we will know at each time point whether our target has been attained. We want a u_{t+1} that maximises

$$\Pr(Y_{t+1} \le b|y_{1:t}, u_{1:t}, u_{t+1}),$$

for specified b. It turns out that

$$\Pr(Y_{t+1} \le b|y_{1:t}, u_{1:t}, u_{t+1}) = \mathbb{E}\left[\Phi\left(\frac{b - \frac{1}{2}(X_{t+1} - u_{t+1})^2}{\sigma_\epsilon}\right)\bigg|y_{1:t}, u_{1:t}\right].$$

Again, if $x(1), \ldots, x(N)$ is a sample from $p(x_{t+1}|y_{1:t}, u_{1:t})$ we can estimate the probability by

$$\hat{\Pr}(Y_{t+1} \le b|y_{1:t}, u_{1:t}, u_{t+1}) = \frac{1}{N}\sum_{i=1}^{N}\Phi\left(\frac{b - \frac{1}{2}(x(i) - u_{t+1})^2}{\sigma_\epsilon}\right). \qquad (14.3.14)$$

14.3.6   Unknown variances

Parameter estimation So far, we have assumed that the variances $\sigma_1^2, \sigma_\eta^2$ and σ_ϵ^2 are known. This is not always the case in practice, but our methods are easily adapted to cope with this. We consider here two different scenarios.

- All variances, $\sigma_1^2, \sigma_\eta^2$ and σ_ϵ^2 are unknown.

- σ_1^2 is known and $\sigma_\eta^2 = \beta\sigma_\epsilon^2$ with β known but σ_ϵ^2 unknown.

In practice, we work with the precisions, ϕ, which are the reciprocals of the variances. The posteriors are now joint distributions for x and the unknown precisions. For Titterington's method, no modification is needed. When the posterior sample has been obtained, we only deal with its x component. For the other methods, slight changes are needed in the calculation of the probabilities involved. For instance, in the case of totally unknown variances,

$$\Pr\left(|X_{t+1} - u_{t+1}| \le v | y_{1:t}, u_{1:t}\right) = \mathbb{E}\left[\Phi\left((u_{t+1} + v - X_t)\phi_\eta^{1/2}\right)\right.$$
$$\left. - \Phi\left((u_{t+1} - v - X_t)\phi_\eta^{1/2}\right) | y_{1:t}, u_{1:t}\right].$$

If $(x_t(1), \phi_\eta(1)), \dots, (x_t(N), \phi_\eta(N))$ is a sample from $p(x_t, \phi_\eta | y_{1:t}, u_{1:t})$ we have the estimator

$$\hat{\Pr}\left(|X_{t+1} - u_{t+1}| \le v | y_{1:t}, u_{1:t}\right) = \frac{1}{N}\sum_{i=1}^{N}\left[\Phi\left((u_{t+1} + v - x_t(i))\phi_\eta^{1/2}(i)\right)\right.$$
$$\left. - \Phi\left((u_{t+1} - v - x_t(i))\phi_\eta^{1/2}(i)\right)\right] (14.3.15)$$

In the method involving Y_{t+1} we have

$$\Pr(Y_{t+1} \le b | y_{1:t}, u_{1:t}, u_{t+1}) = \mathbb{E}\left[\Phi\left((b - \frac{1}{2}(X_{t+1} - u_{t+1})^2)\phi_\epsilon^{1/2}\right) | y_{1:t}, u_{1:t}\right],$$

and, if $(x_{t+1}^*(1), \phi_\epsilon^*(1)), \dots, (x_{t+1}^*(N), \phi_\epsilon^*(N))$ is a sample from the distribution $p(x_{t+1}, \phi_\epsilon | y_{1:t}, u_{1:t})$ we have the estimator

$$\hat{\Pr}(Y_{t+1} \le b | y_{1:t}, u_{1:t}, u_{t+1}) = \frac{1}{N}\sum_{i=1}^{N}\left[\Phi\left((b - \frac{1}{2}(x_{t+1}^*(i) - u_{t+1})^2)\phi_\epsilon^*(i)^{1/2}\right)\right].$$
$$(14.3.16)$$

When only σ_ϵ^2 is unknown, we have $\phi_\epsilon = \beta\phi_\eta$ and therefore, while (14.3.16) remains unchanged, (14.3.15) becomes

$$\hat{\Pr}\left(|X_{t+1} - u_{t+1}| \le v | y_{1:t}, u_{1:t}\right) = \frac{1}{N}\sum_{i=1}^{N}\left[\Phi\left(\frac{(u_{t+1} + v - x_t(i))\phi_\epsilon^{1/2}(i)}{\sqrt{\beta}}\right)\right.$$
$$\left. - \Phi\left(\frac{(u_{t+1} - v - x_t(i))\phi_\epsilon^{1/2}(i)}{\sqrt{\beta}}\right)\right] (14.3.17)$$

based on samples of x and ϕ_ϵ.

14.3.7 Resampling implementation

In view of the dynamic nature of the problem, resampling techniques are appropriate, and we used smooth bootstrap techniques so as to not impoverish the samples. Depending on the degree of knowledge about the variances, different samples, and therefore different resampling weights, are required. For illustration, we give the details for the case of totally unknown variances.

The unknowns here are x, ϕ_1, ϕ_η and ϕ_ϵ, although ϕ_1 is not necessary in the bulk of the calculations. We leave the sampling from the first prior aside for the moment because of some peculiarities that have to be taken into account.

Suppose that, at time t, $(x_t^*(1), \phi_\eta^*(1), \phi_\epsilon^*(1)), \ldots , (x_t^*(N), \phi_\eta^*(N), \phi_\epsilon^*(N))$ are available from $p(x_t, \phi_\eta, \phi_\epsilon | y_{1:t-1}, u_{1:t-1})$. After y_t is observed at u_t, the weight of point $(x_t^*(i), \phi_\eta^*(i), \phi_\epsilon^*(i))$ is equal to

$$w_t(i) = \phi_\epsilon^*(i)^{1/2} \exp\left[-\frac{\phi_\epsilon^*(i)}{2} \left(y_t - \frac{1}{2}(x_t^*(i) - u_t)^2 \right)^2 \right].$$

We see that ϕ_η makes no contribution to the weight. The resampling creates $(x_t(1), \phi_\eta(1), \phi_\epsilon(1)), \ldots , (x_t(N), \phi_\eta(N), \phi_\epsilon(N))$ from the distribution $p(x_t, \phi_\eta, \phi_\epsilon | y_{1:t}, u_{1:t})$, and we obtain a sample from $p(x_{t+1}, \phi_\eta, \phi_\epsilon | y_{1:t}, u_{1:t})$ by setting

$$\begin{aligned} x_{t+1}^*(i) &= x_t(i) + z_i \\ \phi_\eta^*(i) &= \phi_\eta(i) \\ \phi_\epsilon^*(i) &= \phi_\epsilon(i), \end{aligned}$$

$i = 1, \ldots , N$, where $z_i \sim N(0, \phi_\eta(i)^{-1})$.

As we see, ϕ_η is essential in updating posterior samples, and we have to ensure that the initial pool of points of ϕ_η is good. This makes the sampling from the very first prior and the first resampling very crucial. If we use only the first observation y_1, any sampled point of ϕ_η can pass through the resampling. For this reason we have to consider $x_2, x_1, \phi_1, \phi_\eta, \phi_\epsilon$ as a block. Then, using observations y_1 and y_2, we will get a sample from $p(x_2, x_1, \phi_1, \phi_\eta, \phi_\epsilon | y_{1:2}, u_{1:2})$. The prior of this block of unknowns is

$$p(x_2, x_1, \phi_1, \phi_\eta, \phi_\epsilon) \propto p(\phi_1) p(\phi_\eta) p(\phi_\epsilon) p(x_1 | \phi_1) p(x_2 | x_1, \phi_\eta).$$

The priors for the precisions are Gamma, $p(x_1 | \phi_1)$ is the $N(0, \phi_1^{-1})$ density and $p(x_2 | x_1, \phi_\eta)$ is the $N(x_1, \phi_\eta^{-1})$ density. To get $(x_1(i), x_2(i), \phi_1(i), \phi_\eta(i), \phi_\epsilon(i))$ from the prior, we independently sample $\phi_1(i), \phi_\eta(i)$ and $\phi_\epsilon(i)$ from their corresponding priors, then $x_1(i)$ from $N(0, \phi_1(i)^{-1})$ and finally $x_2(i)$ from $N(x_1(i), \phi_\eta(i)^{-1})$. The corresponding weight is

$$w(i) = \phi_\epsilon(i) \exp\left\{ -\frac{\phi_\epsilon(i)}{2} \left[\left(y_1 - \frac{1}{2}(x_1(i) - u_1)^2 \right)^2 + \left(y_2 - \frac{1}{2}(x_2(i) - u_2)^2 \right)^2 \right] \right\}.$$

The resampling will favour 'good' pairs of x_1 and x_2 and therefore 'good' points for ϕ_η too. We do not care about 'good' points for ϕ_1 because immediately after the resampling ϕ_1 is no longer of interest. The size of the first prior sample is taken to be much larger than N. We usually set $u_1 = u_2 = 0$, which is the prior mean of x_1.

Occasionally, all the resampling weight attaches to a single point and subsequent samples 'lose' x. To counteract this, we amended the procedure, as we now explain for the unknown-variances case.

Irrespective of whether we know the ratio of noise and disturbance variances, the weight is a function of two variables, x and the noise precision, ϕ_ϵ:

$$w(x, \phi_\epsilon) = \phi_\epsilon^{1/2} \exp\left[-\frac{\phi_\epsilon}{2} \left(y - \frac{1}{2}(x - u)^2 \right)^2 \right].$$

The function has extrema in x at $x_A = u + \sqrt{2y}$ and $x_B = u - \sqrt{2y}$. When both these values fall outside the range of the x-values in the sample, the resampling tends to degenerate. In this event we instead generate x_t points from a mixture of two Normals with means x_A and x_B and standard deviation

$$\sigma = \frac{\sqrt{2y_t + 4\bar{\sigma}_\epsilon}}{2},$$

where

$$\bar{\sigma}_\epsilon = \frac{1}{n} \sum_{i=1}^{n} \frac{1}{\sqrt{\phi_\epsilon^*(i)}}.$$

With this amendment the samples never 'lost' x_t in any of our simulations.

14.3.8 Simulation results

We present and discuss the results of an experiment designed to give us more insight into the problem of choosing κ for Titterington's method, and to compare the performance of the methods in several situations. The criterion used to assess the performance is the mean rate of loss

$$E(\gamma(T)) = \frac{\mathbb{E}\left(\sum_{t=1}^{T} (x_t - u_t)^2 \right)}{T}.$$

We considered five different combinations of values for the variances:

(i) $\sigma_1^2 = \sigma_\eta^2 = 1, \sigma_\epsilon^2 = 0.01$

(ii) $\sigma_1^2 = \sigma_\eta^2 = 1, \sigma_\epsilon^2 = 1$

(iii) $\sigma_1^2 = \sigma_\eta^2 = 1, \sigma_\epsilon^2 = 4$

(iv) $\sigma_1^2 = \sigma_\eta^2 = 4, \sigma_\epsilon^2 = 1$

(v) $\sigma_1^2 = \sigma_\eta^2 = 4, \sigma_\epsilon^2 = 4.$

For each combination, we consider all three different situations concerning the degree of knowledge about the variances, including complete knowledge. Therefore, in all we compare the methods under fifteen settings. We took $T = 100$.

For each setting, we simulate the mean rate of loss using fifty simulated chains $x_1, \ldots, x_{100}$ of realisations of x. Each method is applied in turn, producing its own fifty chains of optimal control points $u_1, \ldots, u_{100}$. If x_{ti} and u_{ti} signify the true minimum and the control for time t and chain i, the mean rate of loss is estimated by

$$\hat{\gamma}(T) = \frac{\sum_{t=1}^{T} \sum_{i=1}^{50} (x_{ti} - u_{ti})^2}{50 \cdot T}, \quad T = 1, \ldots, 100.$$

In each setting we consider the two probabilistic criteria referred to as probabilistic criteria 1 and 2, and with criterion 1 we use both estimator (14.3.13) and also (14.3.12) based on the indicator variable. With criterion 1 we set $v = 1.5$, which is quite small for the values of σ_η considered. With criterion 2 we set $b = 1.5$ unless $\sigma_\epsilon = 0.1$, in which case we set $b = 0.15$. In all cases, if the resulting probability is above 0.9 we set $u_{t+1} = u_t$; otherwise we set u_{t+1} at the maximiser of the probability. (Maximisation was achieved using the nonlinear optimisation routine 'e04bbf' of the NAG Fortran subroutine library.) It very rarely happens that $u_{t+1} = u_t$. The prior distribution for x_1 when the variance σ_1^2 is known is just $N(0, \sigma_1^2)$. The precisions are given Gamma priors with means 1 and variances 10, unless the true precision is 100, in which case that is taken to be the prior mean as well. We always take $u_1 = 0$. The samples drawn are of size $N = 1000$, except that when the variances are unknown the sample from the very first prior is of size $N = 10000$. We also consider Titterington's method with three different values for κ, of which the largest is one that we believe is too large for these variance values.

Figure 14.1 displays the results for $\sigma_1^2 = \sigma_\eta^2 = \sigma_\epsilon^2 = 1$. It shows the evolution of $\hat{\gamma}_T$ for each method as T goes to 100. It shows typical behaviour in that the expected rates of loss stabilise quickly. The limits do not depend on the degree of knowledge about the variances, but they do depend on the system variances and, to a lesser extent, on the noise variance.

In particular, if $\sigma_\epsilon^2 = 1$, increasing the system variances from 1 to 4 quadruples the limit. When $\sigma_\epsilon^2 = 4$, the increase in the limit is somewhat less. Changes in the noise variance when the system variances remain constant have lesser effects on the limit. In his dissertation, P. Stavropoulos shows theoretically, for a noise-free version of the problem, that the large-sample limit of $E(\gamma_T)$ is indeed proportional to σ_η^2.

In all fifteen settings, except for scenario (iv) with known variances, Titterington's method with $\kappa = 0$ turns out to be the best in achieving the lowest values for the estimated $\gamma(T)$, and in all scenarios except (i) Titterington's method with $\kappa = 5$ was the poorest. Trials with even larger

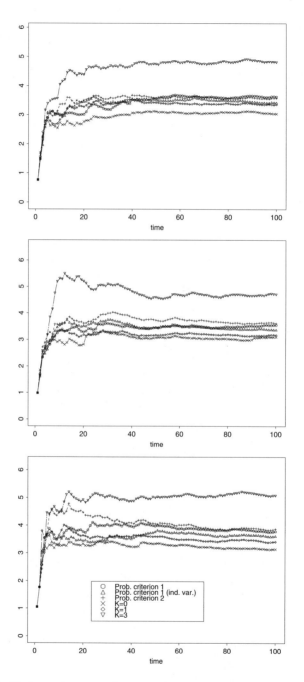

Figure 14.1. Estimated expected rates of loss. Top image: known variances $\sigma_1^2 = 1, \sigma_\eta^2 = 1, \sigma_\epsilon^2 = 1$. Middle image: only σ_1^2 and $\frac{\sigma_\eta^2}{\sigma_\epsilon^2}$ known. Bottom image: entirely unknown variances.

ratios of noise to disturbance variance have provided no concrete evidence that a value of κ other than 0 would be optimal.

14.3.9 Further work on this problem

We have described here only part of our work on this problem. In his thesis, P. Stavropoulos extends the degree of ignorance about the system, allowing for an unknown auto-regression parameter in the system equation and an unknown multiplicative parameter in the observation equation. He also considers the case in which the state variable is bivariate, meaning the response function is a surface rather than a curve. Finally, he applies versions of auxiliary variable particle filtering and stratified particle filtering to the univariate version, again using smooth bootstrap for all necessary resampling. We hope that more extensive details of this work will be reported publicly elsewhere, but in general the more sophisticated methods did not lead to noticeable improvement over the simpler versions of particle filtering applied above.

Part IV

Applications

15

Posterior Cramér-Rao Bounds for Sequential Estimation

Niclas Bergman

15.1 Introduction

The posterior filter density of most nonlinear recursive estimation problems cannot be described analytically by a finite number of parameters. Several examples of sub-optimal algorithms for practical sequential estimation have therefore appeared in the literature. Generally, these procedures approximate either the estimation model or the description of the posterior distribution. These inevitable approximations may seriously degrade the estimation performance when compared with the results that would have been obtained had Bayesian inference, based on the true posterior density, been carried out. It is of great practical interest to quantify this performance degradation, and measure the effect of the introduced approximations. A benchmark simulation evaluation against the optimal solution is not possible because it would require infinite computing power and unlimited memory to run the optimal algorithm. However, even if it is intractable to implement and run the optimal solution a lower bound on the performance of this solution can be obtained. The estimation error obtained with an optimal algorithm depends only on the fundamental properties of the model, e.g. signal-to-noise levels and prior assumptions on the sought parameters. Characteristics of the estimation error from the optimal solution define lower limits on the performance that can be achieved using any practical implementation. The characteristics of the sub-optimal estimation error achieved by an approximate implementation are revealed in simulations using the implemented procedure. The discrepancy from the lower bound gives an indication of the effect of the approximations introduced in the implemented algorithm. A relative comparison between the sub-optimal and the optimal algorithms is therefore possible even if it is intractable to implement the optimal solution. A lower bound on some property of the estimation error is convenient for evaluation purposes, but can also be used to shed light on the fundamental performance level that can be reached for the estimation problem.

In this chapter, we derive explicit expressions for the fundamental Cramér-Rao bound in sequential estimation. The Cramér-Rao bound sets a lower limit on the mean square estimation error in the sense that the difference between the expected mean square error correlation matrix and a matrix determined by the model of the estimation problem is positive semi-definite. The parametric and the Bayesian viewpoints on estimation give two different kinds of Cramér-Rao bounds. Usually, the Cramér-Rao bound is defined in the parametric framework connecting it to some asymptotic efficiency of maximum likelihood estimators. The parametric bound is based on the likelihood of the estimation problem and depends explicitly on the unknown but fixed parameters. In this work we consider the Bayesian, or posterior, Cramér-Rao bound for which the sought parameters are not fixed but random, and represented by their prior distribution. Unlike the parametric Cramér-Rao bound, the posterior bound also holds for estimators with unknown bias under the mild condition that the prior distribution should tend to zero at infinity. The definition of the posterior Cramér-Rao bound is very similar to the parametric bound. In the posterior bound, the joint density of parameters and observations plays the role of the likelihood in the parametric bound. Moreover, the posterior bound is computed by expectation with respect to the joint prior density of the states and the observations, while the parametric bound explicitly depends on the unknown parameter.

The posterior Cramér-Rao bound is reviewed in Section 15.2. Section 15.3 contains a review of Cramér-Rao bounds for sequential estimation and a derivation of the posterior Cramér-Rao bound for one-step-ahead sequential prediction. Section 15.4 illustrates an application of the bound to evaluation of sequential Monte Carlo methods for terrain navigation.

15.2   Review of the posterior Cramér-Rao bound

Consider a generic Bayesian inference problem. Let $\theta \in \mathbf{R}^n$ be a random parameter with prior density $p(\theta)$, such that $p(\theta) \to 0$ as θ tends to infinity. Furthermore, let $\hat{\psi}(z) : \mathcal{Z} \to \mathbf{R}^p$ be an estimate of an absolutely continuous function of this parameter, $\psi(\theta) : \mathbf{R}^n \to \mathbf{R}^p$, based on the observation $z \in \mathcal{Z}$. Then, we have the following Bayesian version of the Cramér-Rao bound. It is originally due to (Van Trees 1968), see also (Gill and Levit 1995) for different extensions and detailed regularity conditions.

Theorem 15.2.1. *The posterior Cramér Rao lower bound for estimating* θ *using* z *is given by*

$$\mathbf{E}_{p(z,\theta)}\left(\left(\hat{\psi}(z) - \psi(\theta)\right)\left(\hat{\psi}(z) - \psi(\theta)\right)^{\mathrm{T}}\right) \geq MPM^{\mathrm{T}} \qquad (15.2.1)$$

where

$$M^{\mathrm{T}} = \mathbf{E}_{p(z,\theta)}\left(\nabla_\theta \log p(z,\theta)\left(\hat{\psi}(z) - \psi(\theta)\right)^{\mathrm{T}}\right) = \int \nabla_\theta p(z,\theta)\left(\hat{\psi}(z) - \psi(\theta)\right)^{\mathrm{T}} dz d\theta$$

$$P^{-1} = \mathbf{E}_{p(z,\theta)}\left(\nabla_\theta \log p(z,\theta)\nabla_\theta^{\mathrm{T}} \log p(z,\theta)\right) = \int \nabla_\theta p(z,\theta)\nabla_\theta^{\mathrm{T}} \log p(z,\theta) dz d\theta.$$

Proof. Let $p(z,\theta)$ denote the joint density of the observation and the parameter. The correlation matrix of the random $(n+p)$-vector

$$\varphi = \left(\begin{array}{c} \hat{\psi}(z) - \psi(\theta) \\ \nabla_\theta \log p(z,\theta) \end{array}\right)$$

is by construction positive semi-definite and satisfies

$$\mathbf{E}_{p(z,\theta)}\left(\varphi\varphi^{\mathrm{T}}\right) = \left(\begin{array}{cc} C & M \\ M^{\mathrm{T}} & P^{-1} \end{array}\right) \geq 0. \qquad (15.2.2)$$

Under the assumption that $p(\theta)$ tends to zero as θ tends to infinity, integration by parts yields that

$$M^{\mathrm{T}} = \left[\left[\int p(z,\theta)\left(\hat{\psi}_j(z) - \psi_j(\theta)\right) dz d\theta_{-i}\right]_{\theta_i=-\infty}^{\theta_i=\infty}\right]_{ij} + \int p(z,\theta)\nabla_\theta \psi^{\mathrm{T}}(\theta) dz d\theta$$

$$= \mathbf{E}_{p(\theta)}\left(\nabla_\theta \psi^{\mathrm{T}}(\theta)\right)$$

where $\theta_{-i} = \theta_1, ..., \theta_{i-1}\theta_{i+1}, ..., \theta_n$. With the same argument we have that

$$P^{-1} = \left[\left[\int p(z,\theta)\frac{\partial}{\partial\theta_i} \log p(z,\theta) dz d\theta_{-i}\right]_{\theta_i=-\infty}^{\theta_i=\infty}\right]_{ij} - \int p(z,\theta)\nabla_\theta\nabla_\theta^{\mathrm{T}} \log p(z,\theta) dz d\theta$$

$$= \mathbf{E}_{p(z,\theta)}\left(-\Delta_\theta^\theta \log p(z,\theta)\right)$$

where $\Delta_\theta^\theta = \nabla_\theta\nabla_\theta^{\mathrm{T}}$ is the Laplacian operator. The Posterior Cramér-Rao Bound (PCRB) now follows from (15.2.2). $\qquad\square$

15.3   Bounds for sequential estimation

A general review of Cramér-Rao bounds for both continuous and discrete time nonlinear filtering is presented by (Kerr 1989). Although the continuous time case is of less practical interest, it has been studied in much greater detail than the discrete time case. The Cramér-Rao bound for discrete time filtering was initially studied by (Borobsky and Zakai 1975).

They consider the posterior bound with the restriction to scalar nonlinear models with additive Gaussian noise. Later, (Galdos 1980) extended the results to the multidimensional case. A more recent generalization using similar techniques is given by (Doerschuk 1995), (Shan and Doerschuk 1997) also points out some errors in the original paper by (Galdos 1980). The class of models considered in (Doerschuk 1995, Shan and Doerschuk 1997) are discrete time, state space representations of nonlinear autoregressive processes driven by Gaussian state dependent noise having full rank covariance matrix. The approach initiated in (Borobsky and Zakai 1975) and extended in (Galdos 1980, Doerschuk 1995) relies on a comparison between the information matrix of the nonlinear system and the information matrix of a suitable linear Gaussian system. According to (Kerr 1989), the results reported in (Borobsky and Zakai 1975) are somewhat imprecise and lack a proper definition of the equivalence between the linear and nonlinear models. Some additional limitations of this approach are also discussed in the general overview of lower bounds for nonlinear filtering provided in (Kerr 1989). The parametric bound has not been given the same amount of attention in the literature as the posterior bound. A parametric bound is derived by (Taylor 1979) for the case of continuous time deterministic state evolution and nonlinear discrete time measurements with additive Gaussian noise.

Lately, (Tichavský, Muravchik and Nehorai 1998) have presented general expressions for posterior Cramér-Rao bounds to discrete time nonlinear filtering using a different, and more general, approach than the one advocated in (Borobsky and Zakai 1975) and the references following it. Bounds for multidimensional state-space models without any limiting Gaussianity assumptions are given in (Tichavský et al. 1998). This work contains further extensions of these results that make them applicable to a larger class of nonlinear state-space models.

15.3.1   Estimation model

The states x_t are assumed to evolve according to a known prior distribution, defined by the discrete time dynamical state-space model

$$\begin{aligned} x_{t+1} &= f_t(x_t, w_t) \\ y_t &= h_t(x_t, e_t) \end{aligned} \qquad t = 0, 1, \ldots, \qquad (15.3.3)$$

where $x_t \in \mathbf{R}^n$, $w_t \in \mathbf{R}^m$ and both y_t and e_t are elements of $\mathbf{R}^p$. Let $\{w_t\}$ and $\{e_t\}$ be mutually independent i.i.d. sequences with known densities $p(w_t)$ and $p(e_t)$, respectively. The initial state is independent of both these noises at all times and has a known probability density function $p(x_0)$. The densities of the state and measurement noises, and the initial state together with the algebraic dependency induced by the model (15.3.3) determine the stochastic properties of the state and measurement sequences.

Apart from some regularity conditions such as continuity and differentiability, only one additional restriction is put on the functions in the state-space model (15.3.3). We assume that $h_t(x_t, \cdot) : \mathbf{R}^p \to \mathbf{R}^p$ is bijective for all t and all x_t. Denote the inverse of this function by $h_t^{-1}(x_t, \cdot)$. This assumption is perfectly reasonable from a modelling point of view because it assumes that there is no analytical redundancy between the elements in the measurement vector. The assumption serves to ensure that there are no Dirac delta measures in the likelihood function of y_t given x_t for all t and x_t,

$$p(y_t \mid x_t) = p_{e_t}\left(h_t^{-1}(x_t, y_t)\right). \qquad (15.3.4)$$

The lack of additional restrictions on the state transition equation allows us to include the case of models in which the conditional state transition density $p(x_{t+1} \mid x_t)$ is singular, i.e. involving some Dirac delta measures. This is a generalization compared with the model class considered in (Tichavský et al. 1998).

15.3.2 *Posterior Cramér-Rao bound*

One usually distinguishes among three different sequential estimation problems: filtering, prediction and smoothing. Further, smoothing can be decomposed into fixed lag, fixed interval and fixed point smoothing. In this text, we focus on the posterior Cramér-Rao bound for one step ahead prediction but the techniques we apply can straightforwardly be used to derive bounds both for filtering and for all kinds of smoothing in a similar fashion. The one step ahead prediction problem is defined as at time $t-1$ estimating x_t given the measurements $Y_{t-1} = \{y_i\}_{i=1}^{t-1}$. Straightforward application of the Cramér-Rao bound to a sequential problem would require forming the joint density $p(x_t, Y_{t-1})$. It is natural to consider sequences of bounds for increasing t in the sequential problem, which would thus result in an expression with increasing demand for both computational and memory resources. However, by utilizing the Markovian structure of the system model, the Cramér-Rao bound can be computed recursively, as shown in the following theorem.

Theorem 15.3.1. *Let $\hat{x}_t(Y_{t-1}) : \mathbf{R}^{(t-1)p} \to \mathbf{R}^n$ be any one-step-ahead predictor for the state vector x_t. Assume that given any $\varepsilon_t > 0$ there exists an $R_t < \infty$ such that*

$$\sup_{\|x_t\| \geq R_t} p(x_t) < \varepsilon_t \qquad \text{for every } t = 0, 1, \ldots.$$

Then, the correlation matrix of the estimation error, $\tilde{x}_t = \hat{x}_t(Y_{t-1}) - x_t$, is at any time $t = 0, 1, \ldots$ bounded from below by a matrix P_t,

$$\mathbb{E}\left(\tilde{x}_t \tilde{x}_t^T\right) \geq P_t \qquad t = 0, 1, \ldots$$

where P_t satisfies the recursion

$$P_{t+1} = F_t \left(P_t^{-1} + R_t^{-1}\right)^{-1} F_t^T + G_t Q_t G_t^{-1},$$

with

$$F_t^T = \mathbb{E}(\nabla_{x_t} f_t^T(x_t, w_t))$$
$$R_t^{-1} = \mathbb{E}(-\Delta_{x_t}^{x_t} \log p(y_t \mid x_t))$$
$$G_t^T = \mathbb{E}(\nabla_{w_t} f_t^T(x_t, w_t))$$
$$Q_t^{-1} = \mathbb{E}(-\Delta_{w_t}^{w_t} \log p(w_t)),$$

and the recursion is initiated by $P_0^{-1} = \mathbb{E}(-\Delta_{x_0}^{x_0} \log p(x_0))$. The statement holds under the additional assumption that the above mentioned differentiations, expectations and matrix inversions exist.

Proof. Consider the correlation matrix of the random vector

$$\xi^T = \begin{bmatrix} \tilde{x}_t^T & \tilde{x}_{t+1}^T & \nabla_{\begin{bmatrix} x_t \\ w_t \end{bmatrix}}^T \log p(x_t, w_t, Y_t) \end{bmatrix},$$

which by construction is positive semi definite. We thus have the matrix inequality

$$\mathbb{E}\left(\xi\xi^T\right) = \begin{bmatrix} C & M \\ M^T & K \end{bmatrix} \geq 0,$$

where

$$C = \mathbb{E}\left(\begin{bmatrix} \tilde{x}_t \\ \tilde{x}_{t+1} \end{bmatrix} \begin{bmatrix} \tilde{x}_t \\ \tilde{x}_{t+1} \end{bmatrix}^T\right) = \begin{bmatrix} C_{11} & C_{12} \\ C_{21} & C_{22} \end{bmatrix},$$

$$M = \begin{bmatrix} I & 0 \\ F_t & G_t \end{bmatrix}, \qquad K = \mathbb{E}\left(-\Delta_{\begin{bmatrix} x_t \\ w_t \end{bmatrix}}^{\begin{bmatrix} x_t \\ w_t \end{bmatrix}} \log p(x_t, w_t, Y_t)\right),$$

similarly to the proof of Theorem 15.2.1. The joint density

$$p(x_t, w_t, Y_t) = p(w_t) p(y_t \mid x_t) p(x_t, Y_{t-1})$$

induces a decomposition of the block K,

$$K = \begin{bmatrix} J_t + R_t^{-1} & 0 \\ 0 & Q_t^{-1} \end{bmatrix}$$

with

$$J_t = \mathbb{E}(-\Delta_{x_t}^{x_t} \log p(x_t, Y_{t-1})).$$

Assume that $\mathbb{E}(\tilde{x}_t \tilde{x}_t^T) \geq P_t$, which according to the posterior Cramér-Rao bound must satisfy $P_t^{-1} \geq J_t$. Therefore,

$$\begin{bmatrix} C_{11} & C_{12} & I & 0 \\ C_{21} & C_{22} & F_t & G_t \\ I & F_t^T & P_t^{-1} + R_t^{-1} & 0 \\ 0 & G_t^T & 0 & Q_t^{-1} \end{bmatrix} \geq 0,$$

and the matrix inequality

$$\begin{bmatrix} C_{11} & C_{12} \\ C_{21} & C_{22} \end{bmatrix} \geq \begin{bmatrix} I & 0 \\ F_t & G_t \end{bmatrix} \begin{bmatrix} \left(P_t^{-1} + R_t^{-1}\right)^{-1} & 0 \\ 0 & Q_t \end{bmatrix} \begin{bmatrix} I & F_t^T \\ 0 & G_t^T \end{bmatrix}$$

holds. The lower left block of this matrix yields that

$$\mathbb{E}(\tilde{x}_{t+1}\tilde{x}_{t+1}^T) = C_{22} \geq P_{t+1} = F_t \left(P_t^{-1} + R_t^{-1}\right)^{-1} F_t^T + G_t Q_t G_t^T$$

The result follows by induction over t noting that $J_0^{-1} = P_0$.   □

The result above provides an expression for the Cramér-Rao bound to the one step ahead estimator at all future time instants.

The recursive formula for P_t implicitly involves the computation of some quite complex expectations when forming the five matrices F_t, G_t, R_t, Q_t, and P_0. For nonlinear state space models with Gaussian additive noise, several of these matrices may be computed analytically, and can often be approximated within any given accuracy by direct Monte Carlo estimates. Convergence of these estimates are straightforwardly assessed since i.i.d. draws from the system model are easily generated. This is exemplified in Section 15.3.3.

There are several ways to use the bounds for recursive estimation presented here. Lower bounds on estimation performance can, for example be used to determine whether the imposed design specifications are realistic. Alternatively, plotting the bound against the range of different unknown but not estimated model parameters, can give valuable insight into the character of the estimation problem. In this work we use the bounds for evaluation of sub-optimal algorithms by means of Monte Carlo simulations. Such an evaluation can be used to determine the sub-optimal performance of the algorithm, and also to verify the correctness of the actual implementation of an optimal algorithm. Moreover, the Monte Carlo simulation can be used to determine the recursion for the bound P_t.

15.3.3 Relative Monte Carlo evaluation

Comparison of the performance of several algorithms on the basis of their Monte Carlo Root Mean Square (RMS) error is commonly used in signal processing applications. A relative comparison between different algorithms can be used to choose the most favorable approach for the problem at hand. On the other hand, the absolute RMS error only reveals the average performance of the algorithm under the simulation conditions used in the Monte Carlo evaluation. This error value will naturally depend on the subjective choice of noise levels and distributions used in the particular simulation study. By computing a lower bound against which to compare the results, a measure of the relative effectiveness of each algorithm is obtained. The discrepancy of the absolute RMS error from this lower bound reveals the effects of any sub-optimal approximations introduced in the current procedure.

In nonlinear estimation, the RMS error may depend on the region of the state-space in which the simulation is carried out. This is for example the case in the terrain navigation application, illustrated in Section 15.4, where simulations over flat terrain in general yield higher RMS errors than identical simulations performed over rough terrain. A statement about the average position error of such an algorithm in a simulation study therefore says nothing about the general performance of the algorithm. The statement needs to be accompanied by a measure of the terrain variation in the area where the simulations were carried out. The Cramér-Rao bound can be seen as a very natural way to quantise the terrain information in this particular application, and the optimal performance in a general sequential estimation problem. The relative difference between the bound and the algorithm performance is a sensible measure even if two different algorithms are evaluated over different terrain areas. Hence, the lower bound yields an opportunity to compare the Monte Carlo RMS performance between algorithms tested under different circumstances.

The posterior Cramér-Rao bound for sequential one step ahead estimation says that

$$\mathbb{E}\left((\hat{x}_t(Y_{t-1}) - x_t)(\hat{x}_t(Y_{t-1}) - x_t)^T\right) \geq P_t, \qquad (15.3.5)$$

where P_t is recursively defined in Theorem 15.3.1. It is more convenient to have a scalar inequality instead of the matrix inequality above. A suitable scalar measure is the mean square error given by the trace of (15.3.5),

$$\mathbb{E}\left((\hat{x}_t(Y_{t-1}) - x_t)^T(\hat{x}_t(Y_{t-1}) - x_t)\right) = \mathbb{E}\left(\|\hat{x}_t(Y_{t-1}) - x_t\|^2\right) \geq \mathrm{tr}P_t. \qquad (15.3.6)$$

Both the left hand side mean square error, and possibly the right hand side Cramér-Rao bound recursion, involve expectations that often lack simple analytical expressions, but may be estimated using Monte Carlo techniques.

Notice that the expectations are performed with respect to all random entities, i.e. both the state x_t and the set of measurements Y_{t-1}. A Monte Carlo simulation study should therefore generate M i.i.d. state trajectories of some fixed length N, $\{x_t^i\}_{t=0,i=1}^{N,M}$, by simulating the system model (15.3.3) M times starting at i.i.d. initial positions drawn from $p(x_0)$. The algorithm under inspection should then be applied to the corresponding measurements, $\{y_t^i\}_{t=0,i=1}^{N,M}$, yielding the set of estimates $\{\hat{x}_t^i\}_{t=0,i=1}^{N,M}$ where $\hat{x}_t^i$ is the estimate produced by the algorithm at time t in Monte

Carlo iteration i. Inserting the Monte Carlo estimate and taking the square root on both sides of (15.3.6) yields that

$$\sqrt{\frac{1}{M}\sum_{i=1}^{M}\|\hat{x}_t^i - x_t^i\|^2} \gtrsim \sqrt{\operatorname{tr}P_t} \qquad t = 0, 1, \ldots, N, \qquad (15.3.7)$$

where the notation $\gtrsim$ is used to denote that the inequality only holds approximately for finite M. Hence, the simulation root mean square (RMS) error in a Monte Carlo evaluation is bounded from below by the square root of the trace of the estimated Cramér-Rao bound. We conclude this chapter with an illustration of the proposed techniques using the terrain navigation application.

15.4 Example – terrain navigation

The terrain navigation application requires an on-line solution of a highly nonlinear recursive estimation problem and is therefore well suited for sequential Monte Carlo methods. The basic principle of terrain navigation is depicted in Figure 15.1, which shows how measurements of terrain elevation along the aircraft flight path are collected in real time. The aircraft altitude over mean sea-level is measured by a pressure meter, and the ground clearance distance is measured by a radar altimeter. The difference between these measurements is treated as a measurement of the terrain elevation beneath the aircraft. It is compared with a digital terrain elevation map, and the matching positions in the map are used to determine an aircraft position estimate. The matching procedure must naturally be carried out on-line, and high requirements for both performance and robustness must be met in this application. Detailed descriptions of the terrain navigation application, along with comparisons to alternative navigation notions, are provided in (Bergman 1999), and (Bergman, Ljung and Gustafsson 1999).

A Bayesian modelling of the estimation problem depicted in Figure 15.1 is needed in order to apply sequential Monte Carlo methods to the terrain navigation problem. Let $\mathbf{x}_t$ denote the aircraft two-dimensional position projected on the terrain map, and $\mathbf{y}_t$ denote the scalar terrain elevation measurement. An inertial navigation system outputs estimates of the distance travelled by the aircraft between two consecutive measurements. At time t, this estimate is denoted by $\mathbf{u}_t$, and $\mathbf{v}_t$ is the error drift of this estimate during one sample interval. A discrete time model of the recursive estimation problem in terrain navigation is given by

$$\begin{aligned} x_{t+1} &= x_t + u_t + v_t \\ y_t &= h(x_t) + e_t \end{aligned} \qquad t = 0, 1 \ldots . \qquad (15.4.8)$$

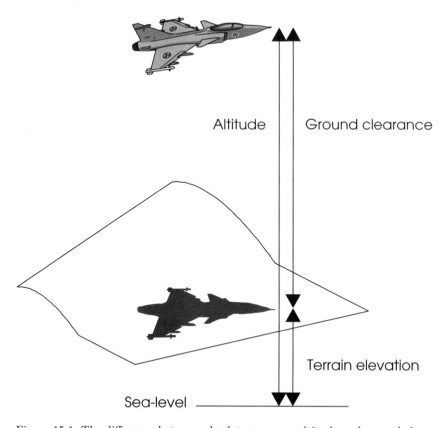

Figure 15.1. The difference between absolute pressure altitude and ground clearance distance is treated as a measurement of the terrain elevation beneath the aircraft.

Above, $h(\cdot)$ denotes the terrain elevation database equipped with an inherent interpolation scheme which can thus produce terrain elevation values intermediate to those actually stored in the database. The noise term e_t is the combined error in the measurements and the terrain database value. The noises v_t and e_t are assumed to be independent white sequences of random variables, and also independent of the initial state x_0. For simplicity, the initial state x_0, and the noises v_t and e_t are all set to be Gaussian distributed with known covariances.

The numerical parameters used in this simulation study are summarised in Table 15.1, where I_2 denotes the two-dimensional identity matrix. The resulting tracks generated from independent realizations of the system and measurement noises are depicted in Figure 15.2. The terrain map is shown as a contour plot behind the tracks. This is a commercial map depicting a central part of Sweden.

Initial covariance	$P_0 = 100^2 I_2$ m^2
State noise covariance	$Q = 5^2 I_2$ m^2
Measurement noise covariance	$\lambda = 16$ m^2
Track length	150 samples
Aircraft velocity	$u_t = [25, 25]^T$ m/sample
Monte Carlo runs	100

Table 15.1. Model parameter values used in the simulation evaluation.

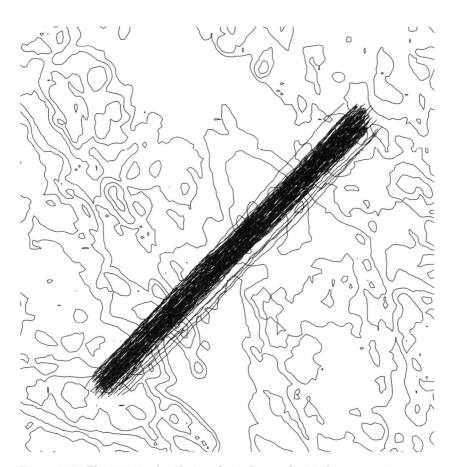

Figure 15.2. The 100 simulated aircraft tracks are depicted over a contour map of the commercial terrain map. The simulated aircraft tracks begin in the lower left corner and end in the upper right corner of the map.

Applying the Monte Carlo evaluation technique outlined in Section 15.3.3 and the result from Theorem 15.3.1 yields that (15.3.7) becomes

$$\sqrt{\frac{1}{100} \sum_{i=1}^{100} \|\hat{x}_t^i - x_t^i\|^2} \gtrsim \sqrt{\mathrm{tr} P_t} \qquad t = 0, 1, \ldots, N,$$

where P_t is given by the recursion

$$P_{t+1} = \left(P_t^{-1} + R_t^{-1}\right)^{-1} + Q \qquad R_t = \lambda^{-1} \frac{1}{100} \sum_{i=1}^{100} \nabla h(x_t^i)(\nabla h(x_t^i))^T,$$

where $\nabla h(x_t^i)$ denotes the gradient of $h(\cdot)$, evaluated at x_t^i.

Four algorithms were applied to this problem; three different Monte Carlo algorithms and one numerical integration filter based on a point-mass approach. The Monte Carlo algorithms under inspection were the Bayesian bootstrap filter (Gordon et al. 1993), the sequential importance sampling, and optimal importance sampling algorithms (Doucet 1998). The numerical integration algorithm was designed for this application, and is described in detail in (Bergman et al. 1999). Directly applying the Bayesian bootstrap filter to the model (15.4.8) yields the procedure summarised below.

Algorithm 1: Bayesian bootstrap

(i) *Set $t = 0$, and generate N samples $\{x_0^{(i)}\}_{i=1}^N$ from $\mathcal{N}(\hat{x}_0, P_0)$.*

(ii) *Compute the normalised weights*

$$w^{(i)} = \frac{N(y_t; h(x_t^{(i)}), \lambda)}{\sum_{j=1}^N N(y_t; h(x_t^{(j)}), \lambda)} \qquad i = 1, \ldots, N$$

and determine the estimate $\hat{x}_t = \sum_{i=1}^N w^{(i)} x_t^{(i)}$.

(iii) *Generate a new set $\{x_t^{(i\star)}\}_{i=1}^N$ by resampling with replacement N times from the set $\{x_t^{(j)}\}_{j=1}^N$ where $\Pr(x_t^{(i\star)} = x_t^{(j)}) = w^{(j)}$.*

(iv) *Generate $v_t^{(i)} \sim \mathcal{N}(0, Q)$ for $i = 1, \ldots, N$, and predict the particle cloud*

$$x_{t+1}^{(i)} = x_t^{(i\star)} + u_t + v_t^{(i)} \qquad i = 1, \ldots, N.$$

(v) *Set $t := t + 1$ and repeat from item 2.*

In this Monte Carlo algorithm, the likelihood directly affects the weight for each particle $x_t^{(i)}$. The weights will not be explicitly time dependent, and they are recalculated at each iteration of the algorithm. In sequential importance sampling with resampling, the degeneracy of the weights determines when to undertake the resampling step. The algorithm described below is the sequential importance sampling method with an importance function given by the prior distribution $p(\mathbf{x}_t \mid \mathbf{y}_{0:t-1})$, approximated by the particle cloud from the last iteration of the algorithm.

Algorithm 2: *Prior importance sampling*

(i) Set $t = 0$, $w_{-1}^{(i)} = \frac{1}{N}$ and generate N samples $\{\mathbf{x}_0^{(i)}\}_{i=1}^N$ from $\mathcal{N}(\hat{\mathbf{x}}_0, P_0)$.

(ii) Update the weights

$$w_t^{(i)} = w_{t-1}^{(i)} N(\mathbf{y}_t; h(\mathbf{x}_t^{(i)}), \lambda) \qquad i = 1, \ldots, N$$

and normalise them

$$w_t^{(i)} := \frac{w_t^{(i)}}{\sum_{j=1}^N w_t^{(j)}} \qquad i = 1, \ldots, N.$$

Compute the estimate $\hat{\mathbf{x}}_t = \sum_{j=1}^N w_t^{(i)} \mathbf{x}_t^{(i)}$.

(iii) If $\hat{N}_{\text{eff}} < N_{\text{thres}}$ resample with replacement from the set $\{\mathbf{x}_t^{(i)}\}_{i=1}^N$ where $w_t^{(i)}$ is the probability of resampling the state $\mathbf{x}_t^{(i)}$. Reset the weights $w_t^{(i)} = \frac{1}{N}$.

(iv) Generate $\mathbf{v}_t^{(i)} \sim \mathcal{N}(0, Q)$ for $i = 1, \ldots, N$, and predict the particle cloud

$$\mathbf{x}_{t+1}^{(i)} = \mathbf{x}_t^{(i)} + \mathbf{u}_t + \mathbf{v}_t^{(i)} \qquad i = 1, \ldots, N.$$

(v) Set $t := t + 1$ and repeat from item 2.

The effective sample size determines the approximate number of samples that actually contribute to the estimate $\hat{\mathbf{x}}_t$. It is defined as

$$\hat{N}_{\text{eff}} = \frac{1}{\sum_{i=1}^N \left(w_t^{(i)} \right)^2},$$

The threshold that determines when to carry out the resampling operation is a design parameter of the algorithm. In the simulations presented in this chapter, it was set to $N_{\text{thres}} = 2N/3$. The main difference between Algorithm 2 and Algorithm 1 is that the weights are computed sequentially.

Algorithm 2 uses the prior density $p(\mathbf{x}_t \mid \mathbf{y}_{0:t-1})$ as importance density in the sequential importance sampling algorithm. The use of the prior density as importance function may yield a sensitivity to outlier measurements. A more cleverly chosen importance function will naturally depend on the new observed measurement $\mathbf{y}_t$ as well. A structured way to choose the importance function is to pick the function that minimises the weight degeneracy in each iteration of the algorithm conditioned on the previously simulated trajectories. It can be shown that the optimal importance function in this sense is the density $p(\mathbf{x}_t \mid \mathbf{x}_{t-1}^{(i)}, \mathbf{y}_{0:t})$, see (Doucet 1998). However, for the terrain navigation problem, given the form (15.4.8), it is not possible to sample from this optimal importance distribution. In order to circumvent this problem, we use a local linearisation technique proposed by (Doucet 1998). Each sample candidate is then drawn from a distribution that is a local Gaussian approximation to the true optimal distribution conditioned on the current candidate trajectory.

Denote the terrain gradient evaluated at $\mathbf{x}_t^{(i)}$ by $h_t^{(i)} = \nabla h(\mathbf{x}_t^{(i)})$. With the local linearisation approach to optimal importance sampling applied to (15.4.8), the candidate $\mathbf{x}_t^{(i)}$ is generated from a Gaussian distribution with covariance and mean given by

$$\Sigma_t^{(i)} = Q - Q h_t^{(i)} (h_t^{(i)^T} Q h_t^{(i)} + \lambda)^{-1} h_t^{(i)^T} Q$$

$$\mu_t^{(i)} = \Sigma_t^{(i)} (Q^{-1}(\mathbf{x}_{t-1}^{(i)} + u_{t-1}) + h_t^{(i)}(\mathbf{y}_t - h(\mathbf{x}_t^{(i)}) + h_t^{(i)^T}(\mathbf{x}_{t-1}^{(i)} + u_{t-1}))\lambda^{-1}),$$

(15.4.9)

where and λ and Q are the noise covariances listed in Table 15.1. The derivation is straightforward and can be found in (Doucet 1998). Notice that the mean vector $\mu_t^{(i)}$ is affected by the measurement $\mathbf{y}_t$ such that the new candidate tends to be drawn to an area of higher likelihood than is the case with prior importance sampling.

The resulting importance filter using the linearisation approach to the optimal importance function follows.

Algorithm 3: *Optimal importance sampling*

(i) *Set $t = 0$, $w_{-1}^{(i)} = \frac{1}{N}$ and generate N samples $\{\mathbf{x}_0^{(i)}\}_{i=1}^N$ from $\mathcal{N}(\hat{\mathbf{x}}_0, P_0)$.*

(ii) *For each $i = 1, \ldots, N$ compute $\mu_t^{(i)}$ and $\Sigma_t^{(i)}$ using (15.4.9). Generate new samples*

$$\mathbf{x}_t^{(i)} \sim \mathcal{N}(\mu_t^{(i)}, \Sigma_t^{(i)}) \qquad i = 1, \ldots, N$$

(iii) *Update the weights*

$$w_t^{(i)} = w_{t-1}^{(i)} \frac{\mathrm{N}(\mathbf{y}_t; h(\mathbf{x}_t^{(i)}), \lambda)\mathrm{N}(\mathbf{x}_t^{(i)}; \mathbf{x}_{t-1}^{(i)} + u_{t-1}, Q)}{\mathrm{N}(\mathbf{x}_t^{(i)}; \mu_t^{(i)}, \Sigma_t^{(i)})} \qquad i = 1, \ldots, N$$

and normalise them

$$w_t^{(i)} := \frac{w_t^{(i)}}{\sum_{j=1}^N w_t^{(j)}} \qquad i = 1, \ldots, N.$$

Compute the estimate $\hat{\mathbf{x}}_t = \sum_{j=1}^N w_t^{(i)} \mathbf{x}_t^{(i)}$.

(iv) *If $\hat{N}_{eff} < N_{thres}$ resample with replacement from the set $\{\mathbf{x}_t^{(i)}\}_{i=1}^N$ where $w_t^{(i)}$ is the probability of resampling the state $\mathbf{x}_t^{(i)}$. Reset the weights $w_t^{(i)} = \frac{1}{N}$.*

(v) *Set $t := t + 1$ and repeat from item 2.*

Notice that this particular linearisation technique only works with Gaussian noises $\mathbf{e}_t$ and $\mathbf{v}_t$. In the terrain navigation application, these noises are usually not considered to be Gaussian distributed, so alternative approximation schemes must be applied in that case. See (Bergman 1999) for a more detailed discussion of the specific topics regarding terrain navigation.

As an alternative to the sequential Monte Carlo algorithms outlined above, we consider a direct numerical integration approach to solving the nonlinear estimation problem in terrain navigation. A numerical approximation to the optimal solution is obtained by limiting the continuous range of $\mathbf{x}_t$ to a discrete set of positions in a uniform mesh over the interesting part of the map. We adopt the name point-mass filter (PMF) for this numerical integration approach to terrain navigation. The point-mass set is represented by a matrix where each element corresponds to the probability of finding the aircraft in the current position given the observed measurements. In order to yield an efficient representation of the filter density,

the matrix is implemented as a sparse matrix that allows for large holes
of zero probability inside the rectangular support of the point-mass matrix. As the aircraft moves and measurement information is processed, the
grid mesh position and resolution must be updated accordingly. The implementation used determines the required number of grid points dynamically
by truncating small point masses. Details about this numerical integration
implementation are given in (Bergman 1999).

Algorithm 4: *Point-mass filter (PMF)*

(i) *Compute a discrete point-mass equivalent to the initial distribution*

$$p(\mathbf{x}_0 \mid \mathbf{y}_{0:-1}) = p(\mathbf{x}_0),$$

and set $t = 0$.

(ii) *Determine a point-mass equivalent to the likelihood $p_{e_t}(\mathbf{y}_t - h(\mathbf{x}_t))$, and compute the element-wise matrix multiplication between this point-mass matrix and the point-mass matrix of
$p(\mathbf{x}_t \mid \mathbf{y}_{0:t-1})$. Normalise the resulting point-mass approximation
of $p(\mathbf{x}_t \mid \mathbf{y}_{0:t})$ so that its weights sum to unity.*

(iii) *Compute the aircraft position estimate as the center of mass of
the current point-mass approximation.*

(iv) *Truncate all weights below some small probability number ε.*

(v) *Translate the grid mesh with the observed movement of the
aircraft.*

(vi) *Convolve the point-mass approximation with a discrete version
of the density of $p_{v_t}(\cdot)$. This will result in a point-mass approximation to $p(\mathbf{x}_{t+1} \mid \mathbf{y}_{0:t})$ which will be smooth and have slightly
larger support than prior to the time update convolution.*

(vii) *Set $t := t + 1$ and iterate from item 2.*

The numerical integration approach and the simulation-based approach
coincide in that they both represent the filter density by a discrete set
of weighted candidate trajectories, or aircraft positions. In the numerical
integration approach, these aircraft positions are determined by the design
of the algorithm grid mesh update, but are randomly generated by the
algorithm in the simulation-based approach.

Figure 15.3 summarises the resulting simulation comparison between all
four considered algorithms and the posterior Cramér-Rao bound. Notice

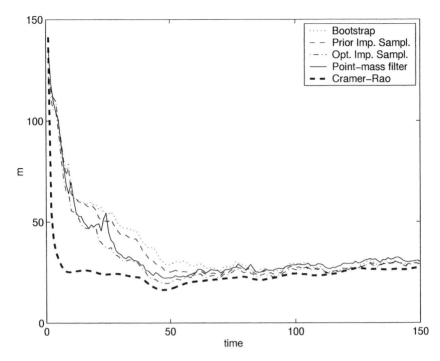

Figure 15.3. Monte Carlo RMS error and posterior Cramér-Rao bound, comparison between the three particle filters and a grid based numerical integration filter. The particle filters had sample size 400, while the point-mass filter had an adaptive grid which used on the average 339 grid points.

that all filters have similar performance, their Monte Carlo RMS errors all meet the posterior Cramér-Rao bound after less than half the simulation time. Neither the numerical integration nor any of the simulation-based filters experienced any divergence from the true aircraft track during these simulations. The prior importance sampling method yields the best performance for a given computational requirement, while Bayesian bootstrap and the point-mass filter demand a somewhat higher number of floating point operations. The optimal importance filter uses an approximation to the optimal importance function in order to limit the number of resampling steps. However, the inherent linearisations put substantially greater demands on the computational resources. A detailed analysis of the simulation results is provided in (Bergman 1999), see also (Bergman et al. 1999) for more background to the Bayesian approach in this application.

15.5   Conclusions

The posterior Cramér-Rao bound is a convenient tool for evaluating sequential algorithms. A general recursive expression with limited computational and memory requirements has been presented in this work. The bound gives a lower limit on the one step ahead mean square prediction error, and bounds for filtering and smoothing follow similarly. The posterior Cramér-Rao bound yields a natural approach to assess the relative performance in a Monte Carlo evaluation. This has been illustrated using three different simulation-based algorithms and one numerical integration algorithm applied to the terrain navigation problem. The terrain navigation application is well suited to a Bayesian approach in general, and for sequential Monte Carlo methods in particular. All four nonlinear filters yield close to the optimal performance predicted by the Cramér-Rao bound in the simulation evaluation conducted over a commercial terrain map.

16

Statistical Models of Visual Shape and Motion

Andrew Blake
Michael Isard
John MacCormick

16.1   Introduction

This paper[1] addresses some problems in the interpretation of visually observed shapes, both planar and three-dimensional, in motion. Mumford (1996), interpreting the Pattern Theory developed over a number of years by Grenander (1976), views images as pure patterns that have been distorted by a combination of four kinds of degradations. This view applies naturally to the analysis of static, two-dimensional images. The four degradations are given here, together with comments on how they need to be extended to take account of three-dimensional objects in motion.

(i) *Domain warping*, in which the domain of an image I is transformed by a mapping g:

$$I(\mathbf{r}) \to I(g(\mathbf{r})).$$

The three-dimensional nature of the world means that the warp g may be composed largely of projective or affine transformations. The dynamic nature of the problems addressed here will require time-varying warps $g(\mathbf{r}, t)$.

(ii) *Superposition:* objects may overlap, and in certain forms of imaging this may produce linear combinations. This is fortunate because such combinations can be analysed by linear spectral decomposition. In images of opaque,three dimensional objects, however, far surfaces are obscured by near ones.

---

[1] This is a modified version of (Blake, Bascle, Isard and MacCormick 1998), and we are grateful to the Royal Society for permission to reprint this material.

(iii) *Distortion and noise:* image measurements are corrupted by noise and blur:

$$I(\mathbf{r}) \to f(I(\mathbf{r}), \mathbf{n}).$$

Image degradations may be most effectively modelled as applying to certain image features obtained by suitable pre-processing of an image, rather than directly to an image itself.

(iv) *Observation failure:* disturbance of the observation process; often caused, in the work described here, by distracting background clutter.

A key idea in pattern theory is recognition by synthesis, in which predictions following from particular hypotheses play an important role. The predictions are generated and tested against the products of analysis of an image. Bayesian frameworks, which have gained significant influence in modelling perception (Knill, Kersten and Yuille 1996), seem to be a natural vehicle for this combination of analysis and synthesis. In the context of machine perception of shapes, we can state the problem as one of interpreting a *posterior* density function $p(\mathbf{X}|\mathbf{Z})$ for a shape $\mathbf{X}$ in some appropriate *shape-space* $\mathcal{S}$, given data $\mathbf{Z}$ from an image (or data $(\mathbf{Z}_1, \mathbf{Z}_2, \dots)$ from a sequence of images). The posterior density must be computed in terms of *prior* knowledge about $\mathbf{X}$ and inference about $\mathbf{X}$ based on the *observations* $\mathbf{Z}$. Bayes' formula expresses this as

$$p(\mathbf{X}|\mathbf{Z}) \propto p(\mathbf{Z}|\mathbf{X})p_0(\mathbf{X}), \qquad (16.1.1)$$

in which $p_0(\mathbf{X})$ is the prior density for $\mathbf{X}$ and the conditional density $p(\mathbf{Z}|\mathbf{X})$ conveys the range of likely observations to arise from a given shape $\mathbf{X}$. All this connects directly to the four degradations above. In particular, type 1 (warping) is represented in the prior p_0. Types 3 and 4 (noise and observation failure) are incorporated into the observation density $p(\mathbf{Z}|\mathbf{X})$.

The framework for Bayesian inference of visual shape and motion that forms the basis of this paper is set out in detail in (Blake and Isard 1998). Here, we aim to summarise that framework and introduce several new ideas. The organisation of this chapter is summarised by section, as follows.

16.2. *Statistical modelling of shape* — how to choose a suitable shape-space $\mathcal{S}$ and a prior p_0, or to learn them from a set of examples.

16.3. *Statistical modelling of image observations* — how to construct an effective observation density $p(\mathbf{Z}|\mathbf{X})$ that takes into account image intensities, both within the shape of interest and in the background.

16.4. *Sampling methods* — using random sample generation to construct an approximate representation of the posterior for $\mathbf{X}$, given that the complexity of $p(\mathbf{Z}|\mathbf{X})$ can render exact representation of the posterior infeasible.

16.5. *Modelling dynamics* — extending the Bayesian framework to deal with sequences of images demands priors for temporal sequences $\mathbf{X}_1, \mathbf{X}_2, \ldots$. These can either be constructed by hand or learned from examples.

16.6. *Learning dynamics* — the most effective way to set up dynamic models is to learn them from training sets.

16.7. *Particle filtering* — applied to the interpretation of shapes in motion.

16.8. *Dynamics with discrete states* — extending the dynamical repertoire to modelling of motion with several modes, for example walk–trot–canter–gallop.

16.2   Statistical modelling of shape

This section addresses the construction of a prior model $p_0(\mathbf{X})$ for a shape. This can be done in a somewhat general way if the dimensionality of the shape-space $\mathcal{S}$ is fixed in advance to be small, for example just translations in the plane. Then extended observation of the positions of moving objects in some area can be summarised as a histogram which serves as an approximate representation of the prior p_0 (Fernyhough, Cohn and Hogg 1996). In higher dimensional shape-spaces, involving three-dimensional rigid motion and deformation of shape, histograms are less practical. Here we focus on Gaussian distributions.

A Gaussian distribution is specified by its mean and covariance, and these can be estimated from a training sequence $\mathbf{X}_1, \mathbf{X}_2, \ldots$ of shapes by taking the sample mean $\overline{\mathbf{X}}$ and the sample covariance

$$\Sigma = \frac{1}{M} \sum_{k=1}^{M} (\mathbf{X}_k - \overline{\mathbf{X}})(\mathbf{X}_k - \overline{\mathbf{X}})^T.$$

Moreover, Principal Components Analysis (PCA) (Rao 1973) can be used to restrict the shape-space $\mathcal{S}$ to explain most of the variance in the training set while keeping the dimension of $\mathcal{S}$ small, in the interests of computational efficiency (Cootes, Taylor, Lanitis, Cooper and Graham 1993, Baumberg and Hogg 1994, Lanitis, Taylor and Cootes 1995, Beymer and Poggio 1995, Baumberg and Hogg 1995a, Vetter and Poggio 1996, Bascle and Blake 1998). An example is given in figure 16.1.

However, the resulting shape-space, though economical, is not especially easy to interpret because Principal Components need not be meaningful. More meaningful constructive shape-spaces can be generated by acknowledging three-dimensional projective effects and constructing affine spaces, for instance, whose components are directly related to rigid body transformations (Ullman and Basri 1991, Koenderink and van Doorn 1991). In

Figure 16.1. **PCA for faces.** *A shape-space of facial expressions is reduced here by PCA to the two-dimensional space that best covers the expressions in a certain training sequence.*

addition, named deformations can be included in a basis for $\mathcal{S}$ as key-frames (Blake and Isard 1994), as in figure 16.2.

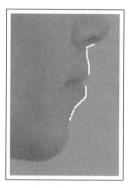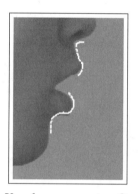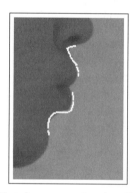

Template $\mathbf{Q}_0$ *Key-frame: opening* $\mathbf{Q}_1$ *Key-frame, protrusion* $\mathbf{Q}_2$

Figure 16.2. **Key-frames.** *Lips template followed by two key-frames, representing interactively tracked lips in characteristic positions. The key-frames are combined linearly with appropriate rigid degrees of freedom, to give a shape-space suitable for use in a tracker for non-rigid motion.*

A constructive shape-space S^c can be combined with PCA to give the best of both worlds. 'Residual PCA operates on a constructive shape-space that does not totally cover a certain dataset, supplying missing components by PCA. Then the constructive subspace retains its interpretation and only the residual components, covered by PCA, cannot be directly interpreted. This is done by constructing a projection operator E^c that maps $\mathcal{S}$ to $\mathcal{S}^c$

and applying PCA to the residual training-set vectors $\mathbf{X}_1^r, \mathbf{X}_2^r, \ldots$ where

$$\mathbf{X}^r = \mathbf{X} - E^c \mathbf{X}.$$

Full details of the algorithm are given in (Blake and Isard 1998) and an example of its application is shown in figure 16.3.

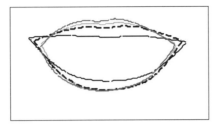

Figure 16.3. **Sampling from a prior for lip-shape, excluding translation** *Random sampling illustrates how a learned prior represents plausible lip configurations. Any rigid translations in the training set, due to head-motion, are separated out as a constructive shape-space in residual PCA.*

16.3 Statistical modelling of image observations

Gaussian distributions may often be acceptable as models of prior shape, but they are adequate as observation distributions only in the clutter-free case. Typically, in our framework, observations are made along a series of splines, normal to the hypothesised shape, as in figure 16.4. Consider the one-dimensional problem of observing, along a single spline, the set of feature positions $\{\mathbf{z} = (z_1, z_2, \ldots, z_M)\}$. Assuming a uniform distribution of background clutter, and a Gaussian model for error in measurement of the position of the true object edge, leads (Isard and Blake 1996) to the multi-modal observation density $p(\mathbf{z}|\mathbf{X})$ depicted in figure 16.5. The multiple peaks in the density are generated by clutter and cannot possibly be modelled by a single Gaussian. A mixture of Gaussians might be feasible, but a very efficient alternative is to use random sampling (next section).

The observation model above was based on the assumption that the observable contour is a "bent wire" resting on a cluttered background. This is not very realistic. It is highly desirable in practice to allow for object opacity and to distinguish between the textured interior of an object and its cluttered exterior. A probabilistic model that reflects this is based on the following assumptions.

Feature localisation error. It is assumed that the feature detector reports object outline position with an error whose density is $\mathcal{E}(\cdot)$, taken usually to be Gaussian.

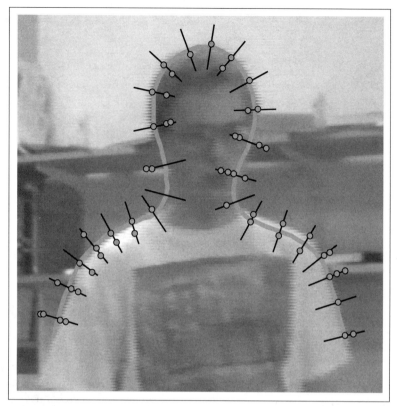

Figure 16.4. **Observation process.** *The thick line is a hypothesised shape, represented as a parametric spline curve. The spines are curve normals along which high-contrast features (white circles) are sought.*

Occlusion probability.   The possibility that the outline is missed by the feature detector is allowed, with probability q.

Clutter model.   Detection of clutter features is regarded as an i.i.d. random process on the portion of each measurement line that lies outside the object. The probability $\pi(n)$ that n clutter features are detected on a normal is generally taken to be uniform.

Interior model.   Interior features on a measurement line are modelled as uniformly distributed along the interior portion of the normal. The distribution $\rho(m)$ for the number m of interior features observed is taken to be Poisson with a known density parameter, which is actually learned by observing instances of the object.

A density $p(\mathbf{z}|\mathbf{X})$ based on these assumptions can be constructed and expressed as $p = \lambda D$ where λ is a constant and

$$D(\mathbf{X}) = p(\mathbf{z}|\mathbf{X})/p(\mathbf{z}|\text{no object present})$$

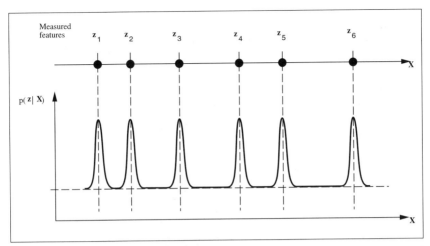

Figure 16.5. **Multi-modal observation density** *(one-dimensional illustration).* *A probabilistic observation model allowing for clutter and the possibility of missing the target altogether is specified here as a conditional density* $p(\mathbf{z}|\mathbf{X})$. *It has a peak corresponding to each observed feature.*

— the *contour discriminant*. This is a discriminant function (Duda and Hart 1973) in the form of a *likelihood ratio*. It has the attraction that, in addition to conveying the relative values of $p(\mathbf{z}|\mathbf{X})$, its absolute value is meaningful: $D(\mathbf{X}) > 1$ implies that the observed features $\mathbf{z}$ are more likely to have arisen from the object in location $\mathbf{X}$ than from clutter.

Lastly, densities $p(\mathbf{z}|\mathbf{X})$ for each normal need to be combined into a grand observation density $p(\mathbf{Z}|\mathbf{X})$, and this raises some issues about independence of measurements along an object contour. Details of the form and computation of the full observation density are given in (MacCormick and Blake 1998). Results of the evaluation of the contour discriminant on a real image are shown in figure 16.6. Analysis of the same image using the simpler bent wire observation model degrades the results, failing altogether to locate the leftmost of the three people. The explicit modelling of object opacity has clearly brought significant benefits.

16.4   Sampling methods

The next stage of the pattern recognition problem is to construct the posterior density $p(\mathbf{X}|\mathbf{Z})$ by applying Bayes' rule (16.1.1). In the previous section, it became plain that the observation density has a complex form in clutter. This means that direct evaluation of $p(\mathbf{X}|\mathbf{Z})$ is infeasible. However, iterative sampling techniques can be used (Geman and Geman 1984, Ripley and Sutherland 1990, Grenander, Chow and Keenan 1991, Storvik 1994). The *factored sampling* algorithm (Grenander et al. 1991) — also known

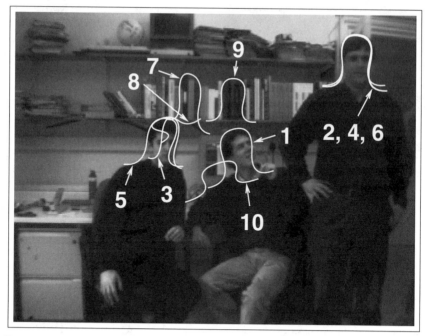

Figure 16.6. **Finding head-and-shoulders outlines in an office scene.** *The results of a sample of 1,000 configurations are shown ranked by value of their contour discriminant. The figure displays the cases in which $D > 1$, indicating a configuration that is more target-like than clutter-like.*

as sampling importance resampling, or SIR (Rubin 1988) — generates a random variate $\mathbf{X}$ from a distribution $\tilde{p}(\mathbf{X})$ that approximates the posterior $p(\mathbf{X}|\mathbf{Z})$. First a sample-set $\{\mathbf{s}^{(1)}, \ldots, \mathbf{s}^{(N)}\}$ is generated from the prior density $p(\mathbf{x})$ and then a sample $\mathbf{X} = \mathbf{X}_i$, $i \in \{1, \ldots, N\}$ is chosen with probability

$$\pi_i = \frac{p(\mathbf{Z}|\mathbf{X} = \mathbf{s}^{(i)})}{\sum_{j=1}^{N} p(\mathbf{Z}|\mathbf{X} = \mathbf{s}^{(j)})}.$$

Sampling methods have proved remarkably effective for recovering static objects from cluttered images. For such problems, $\mathbf{X}$ is a multi-dimensional set of parameters for curve position and shape. In that case the sample-set $\{\mathbf{s}^{(1)}, \ldots, \mathbf{s}^{(N)}\}$ is drawn from the posterior distribution of $\mathbf{X}$-values, as illustrated in figure 16.7.

16.5 Modelling dynamics

In order to be able to interpret moving shapes in sequences of images, it is necessary to supply a prior distribution, not just for shape, but also for

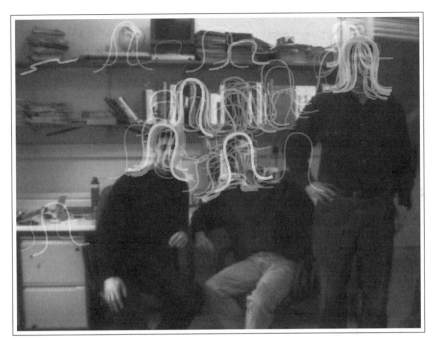

Figure 16.7. **Sample-set representation of posterior shape distribution** *for a curve with parameters* $\mathbf{X}$, *modelling a head outline. Each sample* $\mathbf{s}^{(n)}$ *is shown as a curve (of varying position and shape) with a mass proportional to the weight* $\pi^{(n)}$. *The prior is uniform over translation, with some constrained Gaussian variability in the remainder of its affine shape-space.*

the motion of that shape. Consider the problem of building an appropriate prior model for the position of a hand-mouse engaged in an interactive graphics task. A typical trace in the x-y plane of a finger drawing letters is given in figure 16.8. If the entire trajectory were treated as a training set, the methods discussed earlier could be applied to learn a Gaussian prior distribution for finger position. The learned prior is broad, spanning a sizeable portion of the image area, and places little constraint on the measured position at any given instant. Nonetheless, it is quite clear from the figure that successive positions are tightly constrained. Although the prior covariance ellipse spans about 300×50 pixels, successive sampled positions are seldom more than 5 pixels apart.

For sequences of images, then, a global prior $p_0(\mathbf{X})$ is not enough. What is needed is a conditional distribution $p(\mathbf{X}_k|\mathbf{X}_{k-1})$ giving the distributions of possibilities for the shape $\mathbf{X}_k$ at time $t = k\tau$ *given* the shape $\mathbf{X}_{k-1}$ at time $t = (k-1)\tau$ (where τ is the time-interval between successive images). This amounts to a first order Markov chain model in shape space in which, although in principle $\mathbf{X}_k$ may be correlated with all of $\mathbf{X}_1 \ldots \mathbf{X}_{k-1}$, only correlation with the immediate predecessor is explicitly acknowledged.

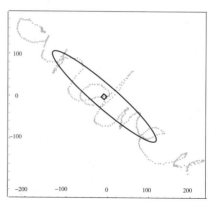

Figure 16.8. **The moving finger writes.** *The finger trajectory (left), which has a duration of about 10 seconds, executes a broad sweep over the plane. When the trajectory is treated as a training set, the learned Gaussian prior is broad, as the covariance ellipse (right) shows. Clearly, though, successive positions (individual dots represent samples captured every 20ms) are much more tightly constrained.*

For the sake of tractability, it is reasonable to restrict Markov modelling to linear processes. In principle and in practice it turns out that a first order Markov chain is not quite enough, generally, but second order suffices. The detailed arguments for this, addressing such issues as capacity to represent oscillatory signals and trajectories of inertial bodies, can be found in (Blake and Isard 1998, Chapter 9). Figure 16.9 illustrates the point for a practical example. A second order, Auto-Regressive Process (ARP) is most concisely expressed by defining a state vector

$$\mathcal{X}_k = \left(\begin{array}{c} \mathbf{X}_{k-1} \\ \mathbf{X}_k \end{array} \right), \tag{16.5.2}$$

and then specifying the conditional probability density $p(\mathcal{X}_k|\mathcal{X}_{k-1})$. In the case of a linear model, this can be done constructively as follows:

$$\mathcal{X}_k - \overline{\mathcal{X}} = A(\mathcal{X}_{k-1} - \overline{\mathcal{X}}) + B\mathbf{w}_k, \tag{16.5.3}$$

where

$$A = \left(\begin{array}{cc} 0 & I \\ A_2 & A_1 \end{array} \right), \quad \overline{\mathcal{X}} = \left(\begin{array}{c} \overline{\mathbf{X}} \\ \overline{\mathbf{X}} \end{array} \right) \quad \text{and} \quad B = \left(\begin{array}{c} 0 \\ B_0 \end{array} \right). \tag{16.5.4}$$

Each $\mathbf{w}_k$ is a vector of N_X independent random $\mathcal{N}(0,1)$ variables and $\mathbf{w}_k, \mathbf{w}_{k'}$ are independent for $k \neq k'$. This specifies the probable temporal evolution of the shape $\mathbf{X}$ in terms of parameters A, B, and covers multiple oscillatory modes and/or constant velocity motion. The constructive form is attractive because it is amenable to direct simulation, simply by supplying a realisation of the succession of random variates $\mathbf{w}_k$.

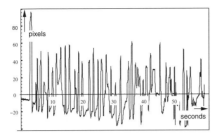

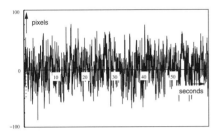 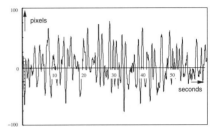

Figure 16.9. **Motion data from talking lips.** *Training sequence of 60 seconds duration (top). Random simulations of learned models — 1st order (left) and 2nd order (right). Only the 2nd order model captures the natural periodicity (around 1 Hz) of the training set, and spectrogram analysis confirms this.*

16.6   Learning dynamics

Motion parameters (A, B in this paper) can be set by hand to obtain desired effects, and a logical approach to this has been developed (Blake and Isard 1998, chapter 9). Experimentation allows these parameters to be refined by hand for improved tracking, but this is a difficult and unsystematic business. It is far more attractive to learn dynamic models on the basis of training sets. A number of alternative approaches have been proposed for learning dynamics, with a view to gesture-recognition — see for instance (Mardia, Ghali, Howes, Hainsworth and Sheehy 1993, Campbell and Bobick 1995, Bobick and Wilson 1995). The requirement there is to learn models that are sufficiently finely tuned to discriminate among similar motions. In our context of the problem of motion tracking, rather different methods are required so as to learn models that are sufficiently coarse to encompass all likely motions.

Initially, a hand-built model is used in a tracker to follow a training sequence that must be not be too hard to track. This can be achieved by allowing only motions that are not too fast and limiting background clutter, or eliminating it using background subtraction (Baumberg and Hogg 1994, Murray and Basu 1994, Koller, Weber and Malik 1994, Rowe and Blake 1996). Once a new dynamic model has been learned, it can be used to build a more competent tracker, one that is specifically tuned to the sort of motions it is expected to encounter. This can be used either to track the original training sequence more accurately, or to track a new and more demanding training sequence involving greater agility of motion. The cycle of learning and tracking is illustrated in figure 16.10. Typically two

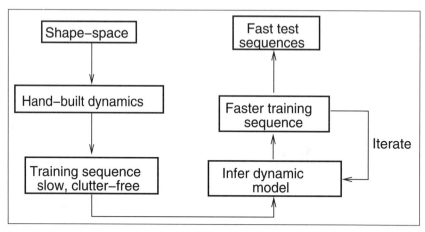

Figure 16.10. **Iterative learning of dynamics.** *The model acquired in one cycle of learning is installed in a tracker to interpret the training sequence for the next cycle. The process is initialised with a tracker whose prior is based on a hand-built dynamic model.*

or three cycles suffice to learn an effective dynamic model.

In mathematical terms, the general problem is to estimate the coefficients A_1, A_2, $\overline{\mathbf{X}}$ and B from a training sequence of shapes $\mathbf{X}_1, \ldots, \mathbf{X}_M$, gathered at the image sampling frequency. Known algorithms to do this are based on the Maximum Likelihood principle (Rao 1973, Kendall and Stuart 1979) and use variants of Yule-Walker equations for estimation of the parameters of auto-regressive models (Gelb 1974, Goodwin and Sin 1984, Ljung 1987). Suitable adaptations for multidimensional shape-spaces are given by (Blake and Isard 1994, Baumberg and Hogg 1995b, Blake, Isard and Reynard 1995), with a number of useful extensions in (Reynard, Wildenberg, Blake and Marchant 1996). One example is the scribble in figure 16.11, learned from the training-sequence in figure 16.8.

A more complex example consists of learning the motions of an actor's face, using the shape-space described earlier that covers both rigid and non-rigid motion. Figure 16.12 illustrates how much more accurately realistic facial motion can be represented by a dynamic model actually learned from examples.

The learning algorithms referred to above treat the training set as exact, whereas in fact it is inferred from noisy observations. Dynamics can be learned directly from the observations using Expectation-Maximisation (EM) (Dempster, Laird and Rubin 1977). Learning dynamics by EM is

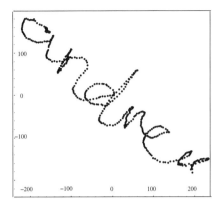 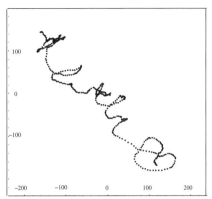

Figure 16.11. **Scribbling: simulating a learned model for finger-writing.** *A training set (left) consisting of six handwritten letters is used to learn a dynamical model for finger motion. A random simulation from the model (right) exhibits reasonable gross characteristics.*

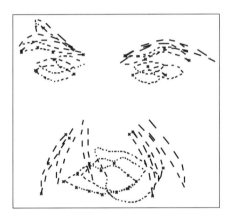 

Figure 16.12. **Trained dynamics for facial motion.** *Hand-built dynamics, exhibited here by random simulation (left) are just good enough, when used in tracking, to gather a training sequence. Trained dynamics (right) however, capture more precisely the constraints of realistic facial motion.*

suggested by Ljung (1987) and the detailed algorithm is given in (North and Blake 1997). It is related to the Baum-Welch algorithm used to learn speech models (Huang, Arika and Jack 1990, Rabiner and Bing-Hwang 1993) but with additional complexity because the state-space is continuous rather than discrete. In practice, accuracy of the learned dynamics is significantly improved when EM is used, especially in the case of more coherent oscillations.

An extension of the basic algorithm for *classes* of objects, dealing independently with motion and with variability of mean shape/position over the class, is described in (Reynard et al. 1996). The same algorithm is also used for modular learning — the aggregation of training sets for which a joint dynamic model is to be constructed.

16.7   Particle filtering

The principles of particle filtering have been developed earlier in the book. Particle filtering has been introduced into the practice of Computer Vision (Isard and Blake 1996, Isard and Blake 1998a) over the past few years, where it is known as the CONDENSATION algorithm, and has been applied to great effect for tracking moving objects in image sequences.

Each time-step of the particle filter generates a weighted, time-stamped sample-set, denoted $\mathbf{s}_k^{(n)}$, $n = 1, \ldots, N$ with weights $\pi_k^{(n)}$, representing approximately the conditional state-density $p(\mathcal{X}_k|\mathbf{Z}_k)$ at time $t = k\tau$, where $\mathbf{Z}_k = (\mathbf{Z}_1, \ldots, \mathbf{Z}_k)$, the history of observations. How is this sample-set obtained? Clearly, the process must begin with a prior density, and the effective prior for time-step k should be $p(\mathcal{X}_k|\mathbf{Z}_{k-1})$. This prior is of course multi-modal in general and no functional representation of it is available. It is derived from the representation as a sample set $\{(\mathbf{s}_{k-1}^{(n)}, \pi_{k-1}^{(n)}),\ n = 1, \ldots, N\}$ of $p(\mathcal{X}_{k-1}|\mathbf{Z}_{k-1})$, the output from the previous time-step, to which prediction must then be applied.

The iterative process applied to the sample-sets is depicted in figure 16.13. At the top of the diagram, the output from time-step $k - 1$ is the weighted sample-set $\{(\mathbf{s}_{k-1}^{(n)}, \pi_{k-1}^{(n)}),\ n = 1, \ldots, N\}$. The aim is to maintain, at successive time-steps, sample sets of fixed size N, so that the algorithm can be guaranteed to run within a given computational resource. The first operation therefore is to sample (with replacement) N times from the set $\{\mathbf{s}_{k-1}^{(n)}\}$, choosing a given element with probability $\pi_{k-1}^{(n)}$. Some elements, especially those with high weights, may be chosen several times, leading to identical copies of elements in the new set. Others with relatively low weights may not be chosen at all.

Each element chosen from the set is now subjected to a predictive step, using an ARP dynamic model, as in equation (16.5.3). This involves sampling a value of $\mathcal{X}_k$ randomly from the conditional density $p(\mathcal{X}_k|\mathcal{X}_{k-1})$ to form a new set member $\mathbf{s}_k^{(n)}$. Since the predictive step includes a random component, identical elements may now split as each undergoes its own independent random motion step. At this stage, the sample set $\{\mathbf{s}_k^{(n)}\}$ for the new time-step has been generated but, as yet, without its weights; it is approximately a fair random sample from the effective prior density $p(\mathcal{X}_k|\mathbf{Z}_{k-1})$ for time-step $t = k\tau$. Finally, the observation step from factored sampling is applied, generating weights from the observation density $p(\mathbf{Z}_k|\mathcal{X}_k)$ to obtain the sample-set representation $\{(\mathbf{s}_k^{(n)}, \pi_k^{(n)})\}$ of state-density for time t.

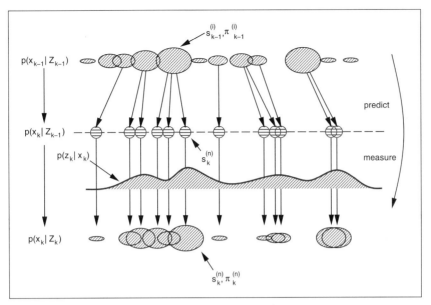

Figure 16.13. **One time-step in the** CONDENSATION **algorithm.** *Blob centres represent sample values and sizes depict sample weights.*

In general sequential importance sampling, almost any "importance distribution" can be used to propagate particles from one time-step to the next, provided the weights are calculated so as to counteract the effects of the importance distribution. The dynamical model $p(\mathcal{X}_k|\mathcal{X}_{k-1})$ used to propagate members of the sample set here is a particular choice of importance distribution that is effective for the visual tracking problems to be addressed. Of course this distribution cannot guarantee that the sample weights have low variance — indeed, the optimal distribution for this purpose is $p(\mathcal{X}_k|\mathbf{Z}_k, \mathcal{X}_{k-1})$ (Doucet, Godsill and Andrieu 2000), but this distribution is unknown and intractable in the problems considered here. Using $p(\mathcal{X}_k|\mathcal{X}_{k-1})$ as the importance distribution has two advantages: it can be learnt by the methods described in the last section, and samples can be drawn from it easily.

A good deal of experimentation has been conducted in applying the CONDENSATION algorithm to the tracking of visual motion, including moving hands and dancing figures. Perhaps one of the most stringent tests involved the tracking of a leaf on a bush, in which the foreground leaf is effectively camouflaged against the background. Results are shown in figure 16.14 and experimental details can be found in (Isard and Blake 1996).

Figure 16.14. **Tracking with camouflage.** *Stills depict mean contour configurations, with preceding tracked leaf positions plotted at 40 ms intervals to indicate motion.*

16.8 Dynamics with discrete states

A recent development of the dynamic models already described is to append to the state variable $\mathcal{X}$ a discrete state y_k to make a mixed state

$$\mathcal{X}_k^+ = \begin{pmatrix} \mathcal{X}_k \\ y_k \end{pmatrix}, \qquad (16.8.5)$$

where $y_k \in \{1, \ldots, N_S\}$ is drawn from a finite set of discrete states with integer labels. Each discrete state represents a mode of motion such as stroke, rest and shade for a hand engaged in drawing. Corresponding to each state $y_{k-1} = i$ there is a dynamical model $p_i(\mathcal{X}_k | \mathcal{X}_{k-1})$, which, in the case of the drawing hand, is likely to be an ARP as in (16.5.3). The stroke model, for instance, might represent constant velocity motion, whereas shading would be oscillatory. In addition, and independently, state transitions are governed by

$$P(y_k = j | y_{k-1} = i) = T_{i,j},$$

a transition matrix following usual practice for Markov chains. More generally, transition probabilities could be made sensitive to the context $\mathcal{X}_{k-1}$ in state-space, so that

$$P(y_k = j | y_{k-1} = i, \mathcal{X}_{k-1}) = T_{i,j}(\mathcal{X}_{k-1}).$$

For example this could be used to express an enhanced probability of transition into the resting state when the hand is moving slowly.

The incorporation of mixed states into the CONDENSATION algorithm is straightforward. It involves using the extended state $\mathcal{X}_k^+$ in place of the original $\mathcal{X}_k$, so that a sample $s_k^{(n)}$ is now a value of the extended state. The prediction step, which generates a new sample $s_k^{(n)}$ from an old one $s_{k-1}^{(n)}$, requires a discrete step and a continuous one. First, the discrete state y_k for the new sample is $y_k = j$, chosen randomly, with probability $T_{i,j}$, where i is the discrete state of the old sample. Then the continuous state is chosen by sampling randomly from a continuous density, as in the original algorithm, but now one of several possible densities $p_i(\mathcal{X}_k | \mathcal{X}_{k-1})$, where again i is the discrete state of the old sample.

Experiments with a three-state model for drawing have been described in detail elsewhere (Isard and Blake 1998b). In addition to enhancing tracking performance, there is the bonus that the current discrete state y_k can be estimated at each time $t = k\tau$, effectively carrying out gesture recognition as a side-effect. One interesting variation on the mixed-state theme uses continuous conditional densities $p_i(\mathcal{X}_k | \mathcal{X}_{k-1})$ which are not ARP models. Consider the example of a moving ball, which may occasionally bounce. This could be represented using two states $\{1, 2\}$ in which $i = 1$ stands for the free ballistic motion of the ball, and $i = 2$ is the bounce event. A suitable transition matrix would be

$$T = \begin{pmatrix} 1 - \epsilon & \epsilon \\ 1 & 0 \end{pmatrix},$$

in which $0 < \epsilon \ll 1$ so that ballistic motion has a mean duration τ/ϵ between bounces. The fact that $T_{2,2} = 0$ ensures that the model always returns to ballistic motion after a bounce — bouncing at each of two consecutive time-steps is disallowed. Now $p_1(\ldots | \ldots)$ is an ARP for ballistic motion but $p_2(\ldots | \ldots)$ models the instantaneous reversal of velocity normal to the reflecting surface. Details of experiments with such a model are in (Isard and Blake 1998b), but the results are illustrated in figure 16.15. The use of mixed discrete/continuous states in a particle filter has been proposed independently by (Semerdjiev, Jilkov and Angelove 1998).

16.9 Conclusions

A high-speed tour has been given of a framework for probabilistic modelling of shapes in motion, and of their visual observation. The key points are that

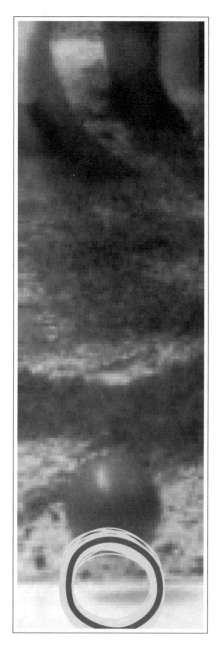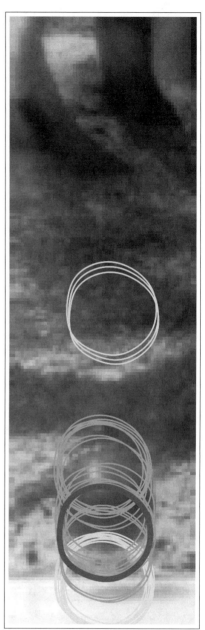

Figure 16.15. **Mixed states tighten constraints in dynamic models.** *A conventional, continuous-state ARP model (left) used to track ballistic motion fails unrecoverably as the ball bounces. Introducing an explicit discrete state for the bounce allows the sample set to split, so that a significant proportion are able to track the bounce.*

visual clutter makes motion analysis difficult, and demands sophisticated probabilistic mechanisms to handle the resulting uncertainty. Further, prior models of motion and observation provide powerful constraints, especially so when the models are learned. A more detailed treatment is given in (Blake and Isard 1998). Since that account, several new modelling tools have been developed. First, the contour discriminant is a new observational model expressed as a likelihood ratio taking opacity of objects into account. Second, complex models for combined rigid and nonrigid motion have been constructed, with a new algorithm for decomposing the two components. Third, extending dynamic states to include discrete labels can significantly enhance their power to constrain perceptual interpretation of shape.

Many interesting questions remain to be addressed. One of these is whether sampling methods for object localisation can be fused elegantly with the CONDENSATION algorithm, to allow robust handling of birth and death (Grenander and Miller 1994) processes in which objects enter and leave the scene. A second is to extend mixed-state models to give reliable gesture recognition on the fly, in a manner that is integrated with the tracking process. A third is to develop algorithms, based on EM, to learn dynamical models from sequences tracked by CONDENSATION, using the full richness of its probabilistic representation, both for continuous and mixed state systems. Finally, another important goal is the extension of particle filtering to allow its use in many dimensions and complex configuration spaces. Such techniques are emerging all the time, and recent contributions include the partitioned sampling of (MacCormick 1999) and the layered sampling of (Sullivan, Blake, Isard and MacCormick 1999).

Acknowledgements

The authors would like to acknowledge the support of the EPSRC.

17

Sequential Monte Carlo
Methods for Neural Networks

N de Freitas
C Andrieu
P Højen-Sørensen
M Niranjan
A Gee

17.1 Introduction

Many problems, arising in science and engineering, require the estimation
of nonlinear, time-varying functions that map a set of input signals to
a corresponding set of output signals. Some examples include: finding the
relation between an input pressure signal and the movement of a pneumatic
control valve; using past observations in a time series to predict future
events; and using a group of biomedical signals to carry out diagnoses and
prognoses. These problems can be reformulated in terms of a generic one
of estimating the parameters of a suitable neural network on-line as the
input-output data becomes available.

For our purposes, neural networks will simply consist of large and in-
tricate combinations of parameterised basis functions. In particular, we
shall focus on two types of architecture: multi-layer perceptrons (MLPs)
and radial basis function networks (RBFs). Neural networks of this type
have been shown to have the capacity to approximate any continuous function arbitrarily well as the number of neurons (basis functions) increases
without bound (Cybenko 1989, Poggio and Girosi 1990). They have also
been successfully applied to many and varied complex problems, including
speech recognition, hand-written digit recognition, financial modelling and
medical diagnosis, among others. Some of the key surveys of the topic are
(Bishop 1995, Ripley 1996).

The measurements are, typically, corrupted by noise. It is, therefore,
important to model this noise so that the network does not overfit the data.
An estimation (training) method that allows the network to fit the data

appropriately without fitting the noise is said to generalise well. To achieve this property, one needs to adopt both adequate probabilistic models and efficient stochastic estimation algorithms. We will show in this chapter that sequential Monte Carlo (SMC) methods provide a powerful solution to this problem. Readers unfamiliar with SMC are encouraged to read the introductory chapter of this volume. More extensive treatments of some of the ideas presented in this chapter can be found in (Andrieu, de Freitas and Doucet 1999a, de Freitas 1999, de Freitas, Niranjan, Gee and Doucet 2000). Finally, we ought to mention that the on-line methods presented in this chapter are applicable to any parametric regression or classification strategies.

17.2   Model specification

To model the relation between the measured input and output signals, we adopt the following Markov, nonlinear, dynamic state-space representation

$$\text{Prior model: } p(\boldsymbol{\theta}_0) \tag{17.2.1}$$

$$\text{Transition model: } p(\boldsymbol{\theta}_t|\boldsymbol{\theta}_{t-1}) \tag{17.2.2}$$

$$\text{Observation model: } p(\mathbf{y}_t|\mathbf{x}_t, \boldsymbol{\theta}_t) \tag{17.2.3}$$

where $\mathbf{x}_t \in \mathbb{R}^{n_x}$ denotes the input signals at time t, $\mathbf{y}_t \in \mathcal{Y}$ represents the output signals distributed according to $p(\mathbf{y}_t|\mathbf{x}_t, \boldsymbol{\theta}_t)$ and $\boldsymbol{\theta}_t \in \mathbb{R}^{n_\theta}$ corresponds to the parameters of the neural network $\mathbf{f}(\mathbf{x}_t, \boldsymbol{\theta}_t)$. The parameters are assumed to follow a random walk $\boldsymbol{\theta}_t = \boldsymbol{\theta}_{t-1} + \mathbf{u}_t$. The process noise could, for instance, be Gaussian $\mathbf{u}_t \sim \mathcal{N}(\mathbf{0}, \delta_t^2 \mathbf{I}_{n_\theta})$, where $\mathbf{I}_{n_\theta}$ denotes the identity matrix of size $n_\theta \times n_\theta$. Other noise models are, of course, possible. To complete the specification of the model, the initial state is $\boldsymbol{\theta}_0 \sim \mathcal{N}(\mathbf{0}, \delta_0^2 \mathbf{I}_{n_\theta})$. Note that the approach discussed in this paper can be used to model noise in the input $\mathbf{x}_t$. Moreover, our approach applies to time varying transition and observation densities.

17.2.1   MLP models for regression and classification

MLPs are one of the most popular choices of neural network architectures because of their simplicity, approximating power, relation to biological systems, as well as for various historical reasons. Figure 17.1 shows a typical two hidden layer MLP with logistic basis functions (also known as sigmoidal activation functions) in the hidden layers, and a single output linear neuron. Networks of this type can be represented mathematically as

$$\mathbf{f}_t = \phi\left(\boldsymbol{\varphi}_{t,n_y}\left(\cdots\mathcal{S}\left(\boldsymbol{\varphi}_{t,n_{h2}}\mathcal{S}\left(\boldsymbol{\varphi}_{t,n_{h1}}\mathbf{x}_t + \mathbf{b}_{t,n_{h1}}\right) + \mathbf{b}_{t,n_{h2}}\right)\cdots\right) + \mathbf{b}_{t,n_y}\right),$$

where $\boldsymbol{\varphi}_{t,n_{h1}} \in \mathbb{R}^{n_{h1} \times n_x}$ denotes the first hidden layer's weights for a network with n_{hi} hidden neurons in the ith layer, $\boldsymbol{\varphi}_{t,n_{hi}} \in \mathbb{R}^{n_{hi} \times n_{h(i-1)}}$ denotes the weights between the $(i-1)^{th}$ and i^{th} layers and $\mathbf{b}_{t,n_{hi}} \in \mathbb{R}^{n_{hi}}$ are known as the biases in the ith layer. The function $\mathcal{S} : \mathbb{R}^{n_{hi}} \mapsto \mathbb{R}^{n_{hi}}$ is a component-wise mapping, whose entries are logistic functions. For instance, the mapping for the jth component is

$$\mathcal{S}_j(\mathbf{z}_j) = \frac{1}{1 + \exp(-\mathbf{z}_j)}.$$

Notice that we can eliminate the need for the variables $\mathbf{b}_t$ by treating them as extra weights with input 1. The overall parameter vector will then include the MLP biases and weights $\boldsymbol{\theta}_t \triangleq \{\mathbf{b}_t, \boldsymbol{\varphi}_t\}$. The output basis functions $\phi(\cdot)$ will be linear for regression tasks and logistic for classification tasks. The latter choice allows us to interpret the outputs of the network as probabilities of class membership (Bishop 1995).

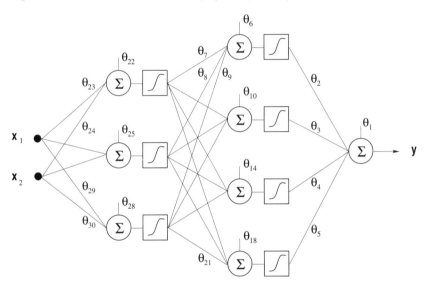

Figure 17.1. Typical multi-layer perceptron architecture with two sigmoidal hidden layers and a single output linear neuron. Here $n_y = 1$, $n_x = 2$ and $n_\theta = 30$.

We adopt the following model to carry out multivariate regression from $\mathbb{R}^{n_x}$ to $\mathcal{Y} = \mathbb{R}^{n_y}$

$$\mathbf{y}_t = f(\mathbf{x}_t, \boldsymbol{\theta}_t) + \mathbf{v}_t, \qquad (17.2.4)$$

where the output basis function is linear and the measurement noise $\mathbf{v}_t$ can be any specified distribution. For example, it can be Gaussian $\mathbf{v}_t \sim \mathcal{N}(\mathbf{0}, \sigma_t^2 \mathbf{I}_{n_y})$, in which case

$$p(\mathbf{y}_t|\mathbf{x}_t, \boldsymbol{\theta}_t) = \left(2\pi\sigma_t^2\right)^{-\frac{n_y}{2}} \exp\left(-\frac{1}{2\sigma_t^2}\left(\mathbf{y}_t - \mathbf{f}(\mathbf{x}_t, \boldsymbol{\theta}_t)\right)'\left(\mathbf{y}_t - \mathbf{f}(\mathbf{x}_t, \boldsymbol{\theta}_t)\right)\right).$$

Alternatively, if we believe that the data contains outliers, we could assume that the noise is t-distributed $\mathbf{v}_t \sim \mathcal{T}_{n_y}\left(\nu, \mathbf{0}, \sigma_t^2 \mathbf{I}_{n_y}\right)$, that is

$$p(\mathbf{y}_t|\mathbf{x}_t, \boldsymbol{\theta}_t) \propto \left(\nu\pi\sigma_t^2\right)^{-\frac{n_y}{2}} \left[1 + \frac{1}{\nu\sigma_t^2}\left(\mathbf{y}_t - \mathbf{f}(\mathbf{x}_t, \boldsymbol{\theta}_t)\right)'\left(\mathbf{y}_t - \mathbf{f}(\mathbf{x}_t, \boldsymbol{\theta}_t)\right)\right]^{-\frac{(\nu+n_y)}{2}}.$$

For binary classification, the output data are assumed to be noiseless class labels, that is $\mathcal{Y} = \{0, 1\}$. We want the network output $\mathbf{f}(\mathbf{x}_t, \boldsymbol{\theta}_t)$ to represent the probability of class membership $\Pr(1|\mathbf{x}_t)$ of class $\{1\}$. The posterior probability of class $\{0\}$ is then given by $1 - \mathbf{f}(\mathbf{x}_t, \boldsymbol{\theta}_t)$. Considering this, the likelihood of the observations should be given by the binomial (Bernoulli) distribution

$$p(\mathbf{y}_t|\mathbf{x}_t, \boldsymbol{\theta}_t) = f(\mathbf{x}_t, \boldsymbol{\theta}_t)^{y_t}\left(1 - f(\mathbf{x}_t, \boldsymbol{\theta}_t)\right)^{1-y_t}. \tag{17.2.5}$$

If one assumes that the outputs of the hidden layer neurons are exponentially distributed, then it follows that the output probabilities of class membership are logistic functions of the hidden layer outputs (Bishop 1995). We therefore use a logistic output basis function to carry out classification. This classification scheme can be extended to more output classes by adopting softmax basis functions and multinomial likelihood distributions (Bishop 1995).

It is possible to extend our framework to recurrent networks by the addition of multiple feedback connections or tapped delay lines (de Freitas, Macleod and Maltz 1999, Narendra and Parthasarathy 1990, Puskorius and Feldkamp 1994).

17.2.2 Variable dimension RBF models

RBF networks tend to be more tractable than MLPs. In these models, the training of the parameters corresponding to different layers is largely decoupled. We shall consider an approximation scheme consisting of a mixture of k RBFs and a linear regression term (Holmes and Mallick 1998). The unknown number of basis functions will be estimated from the data. Thus, unless the data is nonlinear, the model collapses to a standard linear model. More precisely, the linear-RBF model $\mathcal{M}$ at time t is given by

$$\mathcal{M}_{t,0}: \qquad \mathbf{y}_t = \boldsymbol{b}_t + \boldsymbol{\beta}_t' \mathbf{x}_t + \mathbf{v}_t \qquad\qquad k_t = 0$$

$$\mathcal{M}_{t,k_t}: \quad \mathbf{y}_t = \sum_{j=1}^{k_t} \mathbf{a}_{j,t}\phi(\|\mathbf{x}_t - \boldsymbol{\mu}_{j,t}\|) + \boldsymbol{b}_t + \boldsymbol{\beta}_t' \mathbf{x}_t + \mathbf{v}_t \quad k_t \geq 1,$$

where $\|\cdot\|$ denotes a distance metric (usually Euclidean or Mahalanobis), $\boldsymbol{\mu}_{j,t} \in \mathbb{R}^{n_x}$ denotes the jth RBF centre for a model with k_t RBFs, $\mathbf{a}_{j,t} \in \mathbb{R}^{n_y}$ denotes the jth RBF amplitude and $\boldsymbol{b}_t \in \mathbb{R}^{n_y}$ and $\boldsymbol{\beta}_t \in \mathbb{R}^{n_x \times n_y}$ denote the linear regression parameters. The noise sequence $\mathbf{v}_t \in \mathbb{R}^{n_y}$ is assumed to be zero-mean Gaussian; its variance changes over time. Notice that $\boldsymbol{b}_t$, $\boldsymbol{\beta}_t$ and $\mathbf{v}_t$ are affected by the value of k_t.

Figure 17.2 depicts the approximation model for $k = 3$, $n_y = 2$ and $n_x = 2$. Depending on our *a priori* knowledge about the smoothness of

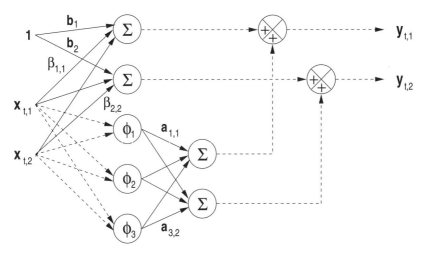

Figure 17.2. Linear-RBF approximation model with three radial basis functions, two inputs and two outputs. The solid lines indicate weighted connections.

the mapping, we can choose different types of basis functions (Poggio and Girosi 1990). The most common choices are

- Linear: $\phi(z) = z$.

- Cubic: $\phi(z) = z^3$.

- Thin plate spline: $\phi(z) = z^2 \ln(z)$.

- Multi-quadric: $\phi(z) = (z^2 + \lambda^2)^{1/2}$

- Gaussian: $\phi(z) = \exp(-\lambda z^2)$.

For the last two choices of basis functions, λ will be treated as a user-set parameter. Nevertheless, the Monte Carlo estimation strategies described in this paper can treat the choice of basis functions as a model selection problem. It is possible to place a prior distribution on the basis functions and allow the Monte Carlo algorithms to decide which of them provide a better solution (Andrieu et al. 1999b, Djurić 1999).

For convenience, the approximation model is expressed in vector-matrix form[1]

$$\mathbf{y}_t = \mathbf{D}(\boldsymbol{\mu}_{1:k_t,t}, \mathbf{x}_t)\boldsymbol{\alpha}_{1:1+n_x+k_t,t} + \mathbf{n}_t,$$

where

$$
\begin{aligned}
\mathbf{y}_t &= [\mathbf{y}_{1,t} \cdots \mathbf{y}_{n_y,t}] \\
\mathbf{D}(\boldsymbol{\mu}_{1:k_t,t}, \mathbf{x}_t) &= [1 \; \mathbf{x}_{1,t} \cdots \mathbf{x}_{n_x,t} \; \phi(\mathbf{x}_t, \boldsymbol{\mu}_{1:n_x,1,t}) \cdots \phi(\mathbf{x}_t, \boldsymbol{\mu}_{1:n_x,k_t,t})] \\
\boldsymbol{\alpha}_{1:n_y,1:1+n_x+k_t,t} &=
\begin{bmatrix}
b_{1,t} & \beta_{1,1,t} \cdots \beta_{n_x,1,t} & a_{1,1,t} \cdots a_{1,k_t,t} \\
b_{2,t} & \beta_{1,2,t} \cdots \beta_{n_x,2,t} & a_{2,1,t} \cdots a_{2,k_t,t} \\
& \vdots & \\
b_{n_y,t} & \beta_{1,n_y,t} \cdots \beta_{n_x,n_y,t} & a_{n_y,1,t} \cdots a_{n_y,k_t,t}
\end{bmatrix}',
\end{aligned}
$$

and, finally, the noise process is assumed to be normally distributed as follows

$$\mathbf{n}_t \sim \mathcal{N}\left(\mathbf{0}_{c \times 1}, diag\left(\sigma_{1,t}^2, \ldots, \sigma_{n_y,t}^2\right)\right).$$

The transition equations for the RBF are given by

$$
\begin{aligned}
k_{t+1} &\sim \Pr(k_{t+1}|k_t) \\
\boldsymbol{\mu}_{1:k_{t+1},t+1} &= \boldsymbol{\mu}_{1:k_{t+1},t} + \epsilon_\mu \\
\boldsymbol{\alpha}_{1:m_{t+1},t+1} &= \boldsymbol{\alpha}_{1:m_{t+1},t} + \epsilon_\alpha \\
\log(\sigma_{t+1}^2) &= \log(\sigma_t^2) + \epsilon_\sigma,
\end{aligned}
$$

where $\Pr(k_{t+1}|k_t)$ is a specified discrete distribution, while the diffusion processes are sampled from normal distributions

$$
\begin{aligned}
\epsilon_\mu &\sim \mathcal{N}(0, \delta_\mu^2 \mathbf{I}_{n_x}) \\
\epsilon_\alpha &\sim \mathcal{N}(0, \delta_\alpha^2 \mathbf{I}_{m_t}) \\
\epsilon_\sigma &\sim \mathcal{N}(0, \delta_\sigma^2 \mathbf{I}_{n_y}),
\end{aligned}
$$

where $m_t = k_t + n_x + 1$. Notice that, if k_t changes, the dimensions of $\boldsymbol{\alpha}_{1:m_t,t}$ and $\boldsymbol{\mu}_{1:k_t,t}$ also change accordingly. When k_t increases, new basis centres $\boldsymbol{\mu}_t$ are sampled from a Gaussian distribution centred at the data $\mathbf{x}_t$. That is, one places basis functions where there is more information in a

---

[1] The notation $\mathbf{y}_{1:n_y,t} \triangleq (\mathbf{y}_{1,t}, \mathbf{y}_{2,t}, \ldots, \mathbf{y}_{n_y,t})'$ is used to denote all the output observations at time t. To simplify this notation, $\mathbf{y}_t$ is equivalent to $\mathbf{y}_{1:n_y,t}$. That is, if one index does not appear, it is implied that we are referring to all its possible values.

stochastic way. On the other hand, if k_t decreases, one samples uniformly from the existing set of basis functions to decide which of the bases should be deleted.

We use multivariate normal distributions to represent the priors $p(\boldsymbol{\mu}_{k_0,0})$ and $p(\boldsymbol{\alpha}_{k_0,0})$, a uniform distribution for $p(\sigma^2_{k_0,0})$ and a discrete uniform distribution for $p(k_0)$. The parameters of these priors should reflect our degree of belief in the initial values of the various quantities of interest (Bernardo and Smith 1994). Figure 17.3 shows the graphical model representation of the joint distribution for the first four time steps.

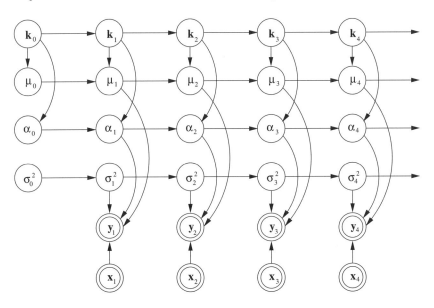

Figure 17.3. Dynamic graphical model. The circles represent unknowns, while the double circles correspond to the data. The hyper-parameters δ^2_μ, δ^2_α and δ^2_σ have been omitted for clarity.

17.3   Estimation objectives

When considering MLPs, our goal will be to approximate the posterior distribution $p(\boldsymbol{\theta}_{0:t}|\mathbf{d}_{1:t})$ and one of its marginals, the filtering density $p(\boldsymbol{\theta}_t|\mathbf{d}_{1:t})$, where $\mathbf{d}_{1:t} = \{\mathbf{x}_{1:t}, \mathbf{y}_{1:t}\}$. By computing the filtering density recursively, we do not need to keep track of the complete history of the parameters. We can also augment the state-space to approximate the joint posterior of the parameters and hyper-parameters. For example, for Gaussian noise, we might be interested in estimating the distribution $p(\boldsymbol{\theta}_{0:t}, \delta^2_{0:t}, \sigma^2_{0:t}|\mathbf{d}_{1:t})$.

With RBFs, our aim will be to estimate the full posterior distribution $p(k_{0:t}, \boldsymbol{\xi}_{k_{0:t},0:t} | \mathbf{d}_{1:t})$, where $\boldsymbol{\xi}_{k_t,t} \triangleq \{\boldsymbol{\alpha}_{1:m_t,t}, \boldsymbol{\mu}_{1:n_x,1:k_t,t}, \boldsymbol{\sigma}_{1:n_y,t}^2\}$, defined on the space $\boldsymbol{\Omega}_{0:t} = \bigcup_{l=1}^{(k_{max}+1)^{t+1}} \boldsymbol{\Omega}_{k_{0:t}^{[l]},0:t} \times \{k_{0:t}^{[l]}\}$, recursively in time. Note that $k_{0:t}^{[l]}$ represents the lth possible model trajectory for an arbitrary ordering.

It is, on most occasions, impossible to derive closed-form analytical expressions to obtain the posterior distribution and its features of interest, such as the posterior model probabilities $p(k_{0:t} | \mathbf{d}_{1:t})$. To compute these distributions in a reasonable time, we need to resort to sequential Monte Carlo methods. In some situations, however, one can combine analytical methods such as the Kalman filter, HMM filter or junction tree algorithm (Doucet, de Freitas, Murphy and Russell 2000) to compute some of the variables exactly, and sample the rest. For example, in the RBF case, it is clear that, conditional on the basis centres $\boldsymbol{\mu}$, the noise variance $\boldsymbol{\sigma}^2$ and the model order k, we end up with a linear Gaussian system with states $\boldsymbol{\alpha}$. It is therefore possible to integrate out $\boldsymbol{\alpha}$ (using a bank of Kalman filters) and only carry out particle filtering on the remaining variables. The variance of such an estimator (Rao-Blackwellised particle filter) can be shown to be smaller than the variance of the standard particle filter (Doucet et al. 1999, Propositions 1 and 2). Notice that it is also possible to integrate the output weights of the MLP in the same way.

In the next section, we shall present a generic SMC algorithm for MLPs and variable dimension RBFs. We shall group the variables that are sampled into the vector $\mathbf{r}$. Thus, for MLPs $\mathbf{r}_{0:t} \triangleq \{\boldsymbol{\theta}_{0:t}, \boldsymbol{\delta}_{0:t}^2, \boldsymbol{\sigma}_{0:t}^2\}$, while for RBFs $\mathbf{r}_{0:t} \triangleq \{k_{0:t}, \boldsymbol{\mu}_{0:t}, \boldsymbol{\sigma}_{0:t}^2\}$. We also group the variables that are integrated out into the vector $\mathbf{s}$. Hence, for RBFs we have $\mathbf{s}_{0:t} \triangleq \{\boldsymbol{\alpha}_{0:t}\}$.

17.4   General SMC algorithm

Given N particles (samples) $\{(\mathbf{r}_{0:t-1}^{(i)}, \mathbf{s}_{0:t-1}^{(i)}); i \in \{1, \ldots, N\}\}$ at time $t-1$, approximately distributed according to $p(\mathbf{r}_{0:t-1}, \mathbf{s}_{0:t-1} | \mathbf{d}_{1:t-1})$, SMC allow us to compute N particles $(\mathbf{r}_{0:t}^{(i)}, \mathbf{s}_{0:t}^{(i)})$, approximately distributed according to the posterior $p(\mathbf{r}_{0:t}, \mathbf{s}_{0:t} | \mathbf{d}_{1:t})$, at time t. This is accomplished by sampling from the importance function $q(\mathbf{r}_t | \mathbf{r}_{0:t-1}, \mathbf{d}_{1:t})$. We shall now present the general algorithm and, subsequently, discuss its main steps.

Generic Sequential Monte Carlo

(i) Importance sampling step

- For $i = 1, \ldots, N$, sample:

$$\widehat{\mathbf{r}}_t^{(i)} \sim q(\mathbf{r}_t | \mathbf{r}_{0:t-1}^{(i)}, \mathbf{d}_{1:t})$$

and set:

$$\widehat{\mathbf{r}}_{0:t}^{(i)} \triangleq \left(\widehat{\mathbf{r}}_t^{(i)}, \mathbf{r}_{0:t-1}^{(i)}\right)$$

- For $i = 1, \ldots, N$, evaluate the importance weights up to a normalising constant:

$$w_t^{(i)} = \frac{p(\widehat{\mathbf{r}}_{0:t}^{(i)} | \mathbf{d}_{1:t})}{q(\widehat{\mathbf{r}}_t^{(i)} | \mathbf{r}_{0:t-1}^{(i)}, \mathbf{d}_{1:t}) p(\mathbf{r}_{0:t-1}^{(i)} | \mathbf{d}_{1:t-1})}$$

- For $i = 1, \ldots, N$, normalise the importance weights:

$$\widetilde{w}_t^{(i)} = w_t^{(i)} \left[\sum_{j=1}^{N} w_t^{(j)}\right]^{-1}$$

(ii) Selection step

- Multiply/suppress samples $\widehat{\mathbf{r}}_{0:t}^{(i)}$ with high/low importance weights $\widetilde{w}_t^{(i)}$, respectively, to obtain N random samples $\widetilde{\mathbf{r}}_{0:t}^{(i)}$ approximately distributed according to $p(\widetilde{\mathbf{r}}_{0:t}^{(i)} | \mathbf{d}_{1:t})$.

(iii) MCMC step

- Apply a Markov transition kernel with invariant distribution given by $p(\mathbf{r}_{0:t}^{(i)} | \mathbf{d}_{1:t})$ to obtain $\mathbf{r}_{0:t}^{(i)}$.

(iv) Exact step

- Compute the sufficient statistics of $\mathbf{s}_t^{(i)}$.

17.4.1   Importance sampling step

If we restrict ourselves to importance functions of the form

$$q\left(\mathbf{r}_{0:t} | \mathbf{d}_{1:t}\right) = q\left(\mathbf{r}_0\right) \prod_{k=1}^{t} q\left(\mathbf{r}_k | \mathbf{d}_{1:k}, \mathbf{r}_{1:k-1}\right), \tag{17.4.6}$$

we can obtain recursive formulae to evaluate $w\left(\mathbf{r}_{0:t}\right) = w\left(\mathbf{r}_{0:t-1}\right) w_t$ and thus $\widetilde{w}_{1:t}$, w_t, being given by

$$w_t \propto \frac{p\left(\mathbf{y}_t | \mathbf{x}_t, \mathbf{d}_{1:t-1}, \mathbf{r}_{0:t}\right) p\left(\mathbf{r}_t | \mathbf{r}_{t-1}\right)}{q\left(\mathbf{r}_t | \mathbf{d}_{1:t}, \mathbf{r}_{0:t-1}\right)} \tag{17.4.7}$$

There are infinitely many possible choices for $q\left(\mathbf{r}_{0:t} | \mathbf{d}_{1:t}\right)$, however we must make sure that its support includes the one of $p\left(\mathbf{r}_{0:t} | \mathbf{d}_{1:t}\right)$. A possible strategy for choosing the importance function is to minimise the variance of the importance weights given $\mathbf{r}_{0:t-1}$ and $\mathbf{d}_{1:t}$. According to this strategy, the following proposition establishes the optimal importance distribution.

Proposition 17.4.1. [Proposition 3 of (Doucet et al. 1999)] *The distribution* $p\left(\mathbf{r}_t | \mathbf{d}_{1:t}, \mathbf{r}_{0:t-1}\right)$ *minimises the variance of the importance weights conditional upon* $\mathbf{r}_{0:t-1}$ *and* $\mathbf{d}_{1:t}$.

For simplicity, we use the transition prior $p\left(\mathbf{r}_t | \mathbf{r}_{t-1}\right)$ as importance distribution for the MLPs. In this case, the importance weights are equal to the likelihood $p\left(\mathbf{y}_t | \mathbf{x}_t, \mathbf{r}_t\right)$. However, we also demonstrate that one can use the results of a sub-optimal deterministic algorithm to construct an importance sampling distribution that incorporates the latest observations. In particular, we use the extended Kalman filter's (EKF) approximation to the posterior distribution of the network weights as proposal distribution. In doing so, the importance weights become

$$w_t \propto \frac{p\left(\mathbf{y}_t | \mathbf{x}_t, \mathbf{d}_{1:t-1}, \mathbf{r}_{0:t}\right) p\left(\boldsymbol{\theta}_t | \boldsymbol{\theta}_{t-1}\right)}{p_{EKF}\left(\boldsymbol{\theta}_t | \mathbf{d}_{1:t}, \mathbf{r}_{0:t-1}\right)}. \tag{17.4.8}$$

For RBFs, we sample from the transition priors and, consequently, the importance weights w_t are given by the one-step-ahead (innovations) densities $p(\mathbf{y}_t | \boldsymbol{\mu}_{1:k_{0:t}, 0:t}, \boldsymbol{\sigma}_{0:t}^2, k_{0:t}, \mathbf{x}_t, \mathbf{d}_{1:t-1})$.

The following proposition shows that, for importance functions of the form (17.4.6), the variance of $w\left(\mathbf{r}_{0:t}\right)$ can only increase (stochastically) over time. The proof of this proposition is an extension of a Kong-Liu-Wong (Kong et al. 1994, p. 285) theorem to the case of an importance function of the form (17.4.6).

Proposition 17.4.2. [Proposition 4 of (Doucet et al. 1999)] *The unconditional variance (that is, with the observations* $\mathbf{d}_{1:t}$ *being interpreted as random variables) of the importance weights* $w\left(\mathbf{r}_{0:t}\right)$ *increases over time.*

In practice, the degeneracy caused by the variance increase can be observed by monitoring the importance ratios. Typically, what we observe is that, after a few iterations, one of the normalised importance ratios tends to one, while the remaining ratios tend to zero.

17.4.2   Selection step

To avoid the degeneracy of the sequential importance sampling simulation method, a selection (resampling) stage may be used to eliminate samples with low importance ratios and multiply samples with high importance ratios. A selection scheme associates to each particle $\mathbf{r}_{0:t}^{(i)}$ a number of "children", say $N_i \in \mathbb{N}$, such that $\sum_{i=1}^{N} N_i = N$. Several selection schemes have been proposed in the literature. These schemes satisfy $\mathbb{E}\left(N_i\right) = N\widetilde{w}_t^{(i)}$ but their performance varies in terms of the variance of the number of particles $var(N_i)$. Examples of these selection schemes include multinomial sampling (Doucet 1998, Pitt and Shephard 1999b), residual resampling (Higuchi 1997, Liu and Chen 1998) and systematic sampling (Carpenter et al. 1999a, Kitagawa 1996). Their computational complexity is $\mathcal{O}(N)$.

17.4.3 MCMC Step

After the selection scheme at time t, we obtain N particles distributed marginally approximately according to $p(\mathbf{r}_{0:t}|\mathbf{d}_{1:t})$. The discrete nature of the approximation can lead to a skewed importance distribution. That is, many particles have no children ($N_i = 0$), whereas others have a large number of children, the extreme case being $N_i = N$ for a particular value i. A strategy to avoid this depletion of samples involves introducing MCMC steps with invariant distribution $p(\mathbf{r}_{0:t}|\mathbf{d}_{1:t})$ on each particle (Andrieu et al. 1999b, MacEachern et al. 1999, Gilks and Berzuini 1999). Essentially, if the particles are distributed according to the posterior $p(\widetilde{\mathbf{r}}_{0:t}|\mathbf{y}_{1:t})$, then applying a Markov chain transition kernel $\mathcal{K}(\mathbf{r}_{0:t}|\widetilde{\mathbf{r}}_{0:t})$, with invariant distribution $p(\mathbf{r}_{0:t}|\mathbf{y}_{1:t})$ such that $\int \mathcal{K}(\mathbf{r}_{0:t}|\widetilde{\mathbf{r}}_{0:t})p(\widetilde{\mathbf{r}}_{0:t}|\mathbf{y}_{1:t}) = p(\mathbf{r}_{0:t}|\mathbf{y}_{1:t})$, still results in a set of particles distributed according to the posterior of interest. However, the new particles might have been moved to more interesting areas of the state-space. In fact, by applying a Markov transition kernel, the total variation of the current distribution with respect to the invariant distribution can only decrease (Gilks and Berzuini 1999). Note that we can incorporate any of the standard MCMC methods, such as the Gibbs sampler and Metropolis Hastings algorithms, into the on-line estimation framework, but we no longer require the kernel to be ergodic. The MCMC move step can also be interpreted as sampling from the finite mixture distribution $N^{-1}\sum_{i=1}^{N}\mathcal{K}(\mathbf{r}_{0:t}|\widetilde{\mathbf{r}}_{0:t}^{(i)})$. In the case of MLPs, we can use an MH step as follows.

MH step

- Sample the proposal candidate $\boldsymbol{\theta}_t^{\star(i)} \sim p(\boldsymbol{\theta}_t|\boldsymbol{\theta}_{t-1}^{(i)})$

- If $v \sim \mathcal{U}_{[0,1]} \leq \min\left\{1, \dfrac{p(\mathbf{y}_t|\mathbf{x}_t, \boldsymbol{\theta}_t^{\star(i)})}{p(\mathbf{y}_t|\mathbf{x}_t, \widetilde{\boldsymbol{\theta}}_t^{(i)})}\right\}$

    - then accept move: $\boldsymbol{\theta}_{0:t}^{(i)} = \left(\widetilde{\boldsymbol{\theta}}_{0:t-1}^{(i)}, \boldsymbol{\theta}_t^{\star(i)}\right)$.
    - else reject move: $\boldsymbol{\theta}_{0:t}^{(i)} = \widetilde{\boldsymbol{\theta}}_{0:t}^{(i)}$.

  End If.

It is also possible to adopt more complex proposals. For instance, we could benefit from the EKF approximation to the posterior, as below.

EKF MH step

- Compute the Jacobian $F_t^{\star(i)}$ of $\mathbf{f}(\mathbf{x}_t, \boldsymbol{\theta}_t)$.

- Update the states in the direction of the gradient with the EKF:

$$\overline{\boldsymbol{\theta}}_{t|t-1}^{\star(i)} = \widetilde{\boldsymbol{\theta}}_{t-1}^{(i)}$$

$$P_{t|t-1}^{\star(i)} = \widetilde{P}_{t-1}^{(i)} + Q_t$$

$$K_t = P_{t|t-1}^{\star(i)} F_t^{\star'(i)} [R_t + F_t^{\star(i)} P_{t|t-1}^{\star(i)} F_t^{\star'(i)}]^{-1}$$

$$\boldsymbol{\theta}_t^{\star(i)} = \overline{\boldsymbol{\theta}}_{t|t-1}^{\star(i)} + K_t(\mathbf{y}_t - \mathbf{f}(\mathbf{x}_t, \overline{\boldsymbol{\theta}}_{t|t-1}^{\star(i)}))$$

$$P_t^{\star(i)} = P_{t|t-1}^{\star(i)} - K_t F_t^{\star(i)} P_{t|t-1}^{\star(i)}$$

- Sample the candidate $\boldsymbol{\theta}_t^{\star(i)} \sim q(\boldsymbol{\theta}_t | \widetilde{\boldsymbol{\theta}}_{0:t-1}^{(i)}, \mathbf{d}_{1:t}) = \mathcal{N}(\overline{\boldsymbol{\theta}}_t^{\star(i)}, P_t^{\star(i)})$

- If $v \sim \mathcal{U}_{[0,1]} \leq \min\left\{1, \dfrac{p(\mathbf{y}_t|\mathbf{x}_t, \boldsymbol{\theta}_t^{\star(i)}) p(\boldsymbol{\theta}_t^{\star(i)} | \widetilde{\boldsymbol{\theta}}_{t-1}^{(i)}) q(\widetilde{\boldsymbol{\theta}}_t | \widetilde{\boldsymbol{\theta}}_{0:t-1}^{(i)}, \mathbf{d}_{1:t})}{p(\mathbf{y}_t|\mathbf{x}_t, \widetilde{\boldsymbol{\theta}}_t^{(i)}) p(\widetilde{\boldsymbol{\theta}}_t^{(i)} | \widetilde{\boldsymbol{\theta}}_{t-1}^{(i)}) q(\boldsymbol{\theta}_t^{\star(i)} | \widetilde{\boldsymbol{\theta}}_{0:t-1}^{(i)}, \mathbf{d}_{1:t})}\right\}$

  - then accept: $\boldsymbol{\theta}_{0:t}^{(i)} = (\widetilde{\boldsymbol{\theta}}_{0:t-1}^{(i)}, \boldsymbol{\theta}_t^{\star(i)})$ and $P_{0:t}^{(i)} = (\widetilde{P}_{0:t-1}^{(i)}, P_t^{\star(i)})$
  - else reject: $\boldsymbol{\theta}_{0:t}^{(i)} = \widetilde{\boldsymbol{\theta}}_{0:t}^{(i)}$ and $P_{0:t}^{(i)} = \widetilde{P}_{0:t}^{(i)}$

  End If.

In this algorithm, the EKF noise covariance parameters (R and Q) should be set so that the tails of the proposal distribution are heavier than the tails of the target distribution. It is possible to improve on this approximation by replacing the EKF with the unscented filter (van der Merwe, Doucet, de Freitas and Wan 2000).

Standard MCMC methods are not able to jump between subspaces $\Omega_{k_{0:t}, 0:t}$ of different dimension. However, Green has introduced a flexible class of MCMC samplers, the so-called reversible jump MCMC, capable of jumping between subspaces of different dimensions (Green 1995). Here, the chain must move across subspaces of different dimensions, and therefore the proposal distributions are more complex: see (Andrieu et al. 1999a) for details. For our problem, the following moves have been selected.

(i) Birth of a new basis, that is, proposing a new basis function.

(ii) Death of an existing basis, that is, removing a basis function chosen randomly.

(iii) Update the RBF centres.

These moves are defined by heuristic considerations, the only condition to be fulfilled being to maintain the correct invariant distribution. A particular choice will only have influence on the convergence rate of the algorithm. The birth and death moves allow the network to grow from k_t to $k_t + 1$ and decrease from k_t to $k_t - 1$ respectively. The main algorithm for the MCMC step is as follows (details are presented in (Andrieu et al. 1999a)).

MCMC Step for Bayesian Model Selection in RBFs

- If $\left(u \sim \mathcal{U}_{[0,1]} \leq \Pr(k_t - 1|k_t)\right)$
    - then perform a "birth" move.
    - else if $(u \leq \Pr(k_t - 1|k_t) + \Pr(k_t + 1|k_t))$ then perform a "death" move.
    - else update $\widetilde{\mu}_{1:\widetilde{k}_t,t}$ and $\widetilde{k}_t$.

    End If.

17.4.4   Exact step

The RBF state-space model for $\boldsymbol{\alpha}_{1:m_t,t}$ is conditionally linear Gaussian given k_t, σ_t^2 and $\boldsymbol{\mu}_t$. It is, therefore, possible to integrate out $\boldsymbol{\alpha}_{1:m_t,t}$, thereby reducing the variance of the estimates (Doucet et al. 1999, Propositions 1 and 2). The result corresponds to a bank of Kalman filters for each network output (Anderson and Moore 1979). A more detailed explanation of this step is presented in (de Freitas 1999). We shall now concentrate on a few examples that demonstrate the theory discussed so far.

17.5   On-line classification

Sequential classification problems arise in a few areas of technology, including condition monitoring and real-time decision systems (Melvin 1996, Penny, Roberts, Curran and Stokes 1999). For example, when monitoring patients, we might wish to decide whether they require an increase in drug intake at several intervals in time. Here, it is shown that particle filters provide an efficient and elegant probabilistic solution to this problem. We shall consider both a synthetic problem and a fault detection problem using data from a marine diesel engine. In both cases, the target classes change with time.

17.5.1 Simple classification example

In this synthetic example, the data were generated from two bivariate, overlapping and time-varying Gaussian clusters as depicted in Figure 17.4. If we look at all the generated data (bottom right of Figure 17.4), we see that non-sequential classification strategies would fail here.

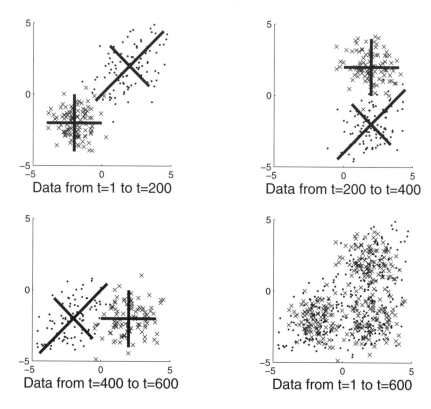

Figure 17.4. Data generated for the classification problem.

An MLP with 4 hidden logistic functions and an output logistic function was applied to classify the data. The network was trained, sequentially, with an SMC algorithm by proposing from the transition prior. A smoothing MCMC step was used to reduce the resampling variance. The number of particles was set to 200, the prior network weights were sampled from $\mathcal{N}(\mathbf{0}, 10\mathbf{I}_2)$, while their diffusion parameter was set to 0.2.

Figure 17.5 shows the one-step-ahead predicted class probabilities, the output labels obtained by thresholding the output at 0.5 and the cumulative mis-classification errors. Figure 17.6 depicts the evolution of the probabilities of class membership. Despite the change in the target distributions at $t = 200$ and $t = 400$, the algorithm recovers very quickly.

17.5.2   An application to fault detection in marine diesel engines

In this section, we apply the proposed on-line classifier to monitor the exhaust value condition in a marine diesel engine. The ability to detect valve burn-through, or leakage, in marine diesel engines is of great practical interest because it makes it possible to plan costly maintenance more efficiently. For instance, in the case of a minor leakage, it is possible to get the valve reconditioned and re-installed when the leakage is detected early. However, an undetected leakage will continue to aggravate the exhaust value condition and hence the performance of the motor owing to a reduction of pressure and power. If the leakage continues undetected, the damage will become too serious to allow for reconditioning. Therefore, the goal is to detect the leakage before the motor's performance becomes unacceptable, or irreversible damage occurs.

The non-intrusive monitored measurements consisted of vibrations and structure-borne stress waves, also known as acoustic emission (AE), acquired during various load conditions (25%, 50%, 75% and 100%) and

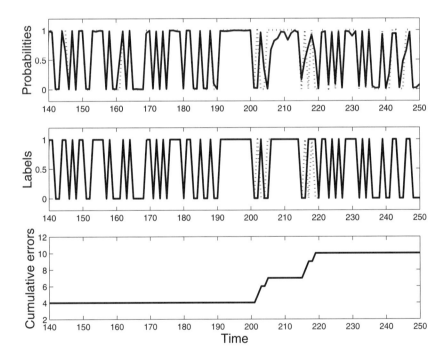

Figure 17.5. The top plot shows the true labels [· · ·] and one-step-ahead predicted probabilities of class membership [—]. The middle plot shows the predicted labels using a threshold of 0.5. The bottom plot shows the cumulative number of misclassification errors.

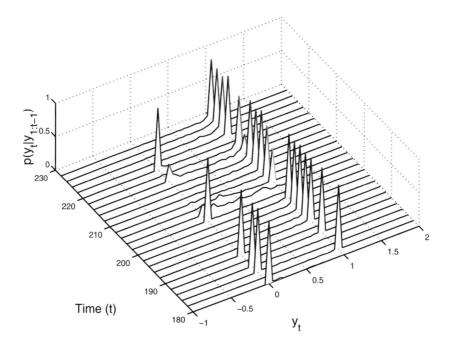

Figure 17.6. Probabilities of class membership for the classification problem.

valve conditions (normal, small leakage and large leakage). The measurements were carried out using a stationary plant where the propeller was simulated by a water brake. The RMS AE time-series was trigger sampled using a shaft timing signal obtained from an angle encoder pulse signal yielding 2048 angle positions per piston cycle. Additional dimensionality reduction of the input-space was obtained by using principal component analysis. In this paper, we focused on classifying normal motor condition against large value leakage. Figure 17.7 shows the data projected on the two first principal components. This figure suggest that the two classes are approximately linearly separable. Because of the non-dynamic data acquisition, a temporal dynamic motor operation was simulated by collecting all data acquired with the same motor load, and with the load level increasing in time (that is, 25%, 50%, 75% and 100%). Finally, for each fixed load level, the data was shuffled randomly.

An MLP with two hidden unit and five input nodes was used as to increase the modelling flexibility, even though Figure 17.7 suggests that a linear discriminant on the data projected on the two first principal components is sufficient to separate the two classes. Figure 17.8 shows the true label and one-step-ahead predicted probabilities of class membership obtained using 500 particles. We anticipated that an initial convergence time

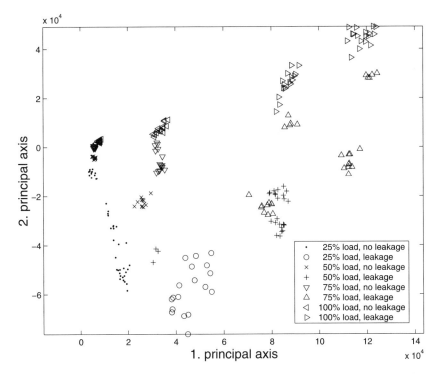

Figure 17.7. Trigger sampled AE measurements projected on the two first principal components. The two classes are approximately linearly separable

would be required for the classifier to start tracking the changes in the operating condition.

17.6   An application to financial time series

Derivatives are financial instruments whose value depends on some basic underlying cash product, such as interest rates, equity indices, commodities, foreign exchange and bonds. An option is a specific type of derivative that gives the holder the right to do something. For example, a call option allows the holder to buy a cash product, at a specified date in the future, for a price determined in advance. The price at which the option is exercised is known as the strike price, whereas the date at which the option lapses is referred to as the maturity time. Put options, on the other hand, allow the holder to sell the underlying cash product. Research results in this field seem to provide clear evidence that there is a nonlinear and non-stationary relation between the option price and the cash product price, maturity time, strike price, interest rates and variance of the returns on the cash product (volatility).

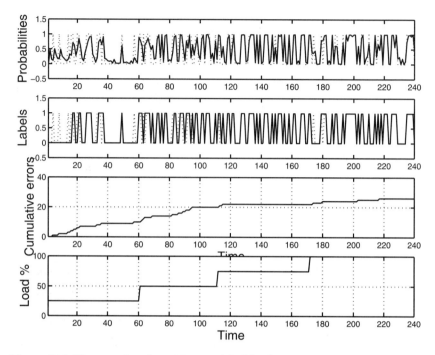

Figure 17.8. The top plot shows the true label [· · ·] and one-step-ahead predicted probabilities of class membership [—]. The next plot shows the predicted classes using a threshold of 0.5. The next plot shows the cumulative number of misclassification errors. The bottom plot shows the operating condition of the motor as a function of time.

Here, we employ neural networks to map the stock's price to the options' price. This approach, in conjunction with the use of volatility smiles and other strategic financial information, might reveal whether an option in the market is either overpriced or underpriced. However, our main interest is to simply demonstrate the use of our methods on a typical non-stationary, nonlinear time series problem. The experiment used a call option contract on the FTSE100 index (daily close prices on the London Stock Exchange from February 1994 to December 1994) to evaluate the pricing algorithms. The sampling algorithms all used 500 samples. The MLP architecture consisted of five sigmoidal hidden units and a linear output neuron. We considered the following algorithms:

Trivial : This method simply involves using the current value of the option as the next prediction.

MLP-EKF : MLP trained with the EKF algorithm.

MLP-SMC-G : MLP trained with plain SMC (no MCMC step) under the assumptions of Gaussian process and measurement noises.

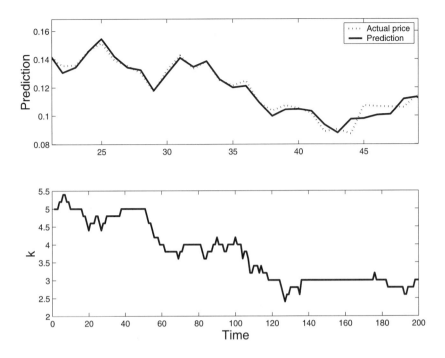

Figure 17.9. One step ahead predictions of the options prices (above) and posterior mean model order (below) for the RBF network on the options pricing problem.

MLP-SMC-G-EKF : MLP with Gaussian assumptions and EKF proposal. It also incorporated an EKF-MCMC step.

MLP-SMC-T : MLP with Gaussian process noise and t-distributed measurement noise.

MLP-SMC-TT : MLP with t-distributed process and measurement noises.

RBF-SMC : RBF with thin plate spline basis functions trained with the model selection algorithm.

The results obtained, over 50 runs, are shown in Table 17.1. One can draw several conclusions from these results

- Most of the methods will on the average perform better than the trivial approach.

- Using heavy tailed distributions resulted in more accurate sampling methods. This is not surprising considering the nature of the data and standard results from importance sampling.

Method	RMS mean	RMS std. dev.
Trivial	0.0064	0
MLP-EKF	0.0037	0.0001
MLP-SMC-G	0.0055	0.0019
MLP-SMC-T	0.0052	0.0009
MLP-SMC-TT	0.0048	0.0007
MLP-SMC-G-EKF	0.0049	0.0004
RBF-SMC	0.0039	0.0002

Table 17.1. Means and standard deviations for the root mean square errors obtained over 50 simulations in the options pricing problem.

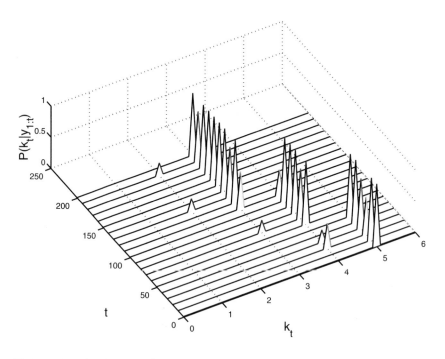

Figure 17.10. Model order (number of neurons) probabilities for the RBF network on the options pricing problem.

- SMC with EKF proposal distributions can lead to better results than standard SMC.

- The RBF model selection strategy performed very well (see also Figures 17.9 and 17.10). This demonstrates that model selection and Rao-Blackwellisation are two good strategies for dealing with high-dimensional problems.

- Although the EKF results look better, we should not forget that the sampling methods are far more general. They allow us to compute noise variances, model order and other important quantities. In many applications, such as finance or tracking, these quantities are the main goal of the estimation process. If we also consider the fact that the EKF often results in poor estimates of the state's covariance (van der Merwe et al. 2000), we see that there is a real need for SMC methods.

17.7   Conclusions

In this chapter, we discussed various SMC strategies for neural networks. We considered both problems of parameter estimation and model selection. We also pointed out some techniques, such as Rao-Blackwellisation, for dealing with high-dimensionality. The experiments show that SMC methods perform well, especially when the posterior distribution is non-Gaussian and time-varying. Compared with standard algorithms, SMC methods are computationally expensive. Yet, they are far more general and flexible. In addition, they allow for the computation of many quantities of interest, among them noise parameters and model order.

Acknowledgments

We are very grateful to Arnaud Doucet for his help during many stages of this work. We would also like to thank Bill Fitzgerald and David Lowe for their valuable comments.

18

Sequential Estimation of Signals under Model Uncertainty

Petar M. Djurić

18.1 Introduction

Two important areas of signal processing are parameter estimation of signals and signal detection. In standard textbooks they are usually addressed separately, although in many practical problems they are applied jointly. In estimation theory, it is almost always assumed that the model of the signal is known, and the objective is to estimate its parameters from noisy signal data. For example, if the observed data vector is $\mathbf{y}$, where $\mathbf{y} \in \mathbb{R}^{n_y}$, and the model of the signal is represented by a function whose analytical form is known, the goal is to estimate the signal parameters $\mathbf{x}$ from $\mathbf{y}$, where $\mathbf{x} \in \mathbb{R}^{n_x}$.[1] In cases when it is unclear whether there is a signal in the data, one resorts to estimation of the signal parameters under the assumption that the data contain a signal, thereafter applying a detection scheme to decide if the signal is indeed in the data.

One approach to estimation of the signal parameters is to apply the Bayesian methodology (Bernardo and Smith 1994, O'Hagan 1994). Suppose that the data are described by

$$\mathbf{y} = g(\mathbf{x}, \mathbf{v}),$$

where $\mathbf{v} \in \mathbb{R}^{n_v}$ is a noise vector, and $g(\cdot)$ is a known function of the signal and noise vectors. In addition, let $\mathbf{v} \sim p(\mathbf{v}|\boldsymbol{\psi})$, where the mathematical form of the probability density function $p(\cdot)$ is known but the parameters

---

[1] The multidimensional real spaces $\mathbb{R}^{n_y}$ and $\mathbb{R}^{n_x}$ in many problems are often replaced by complex spaces, $\mathbb{C}^{n_y}$ and $\mathbb{C}^{n_x}$, or $\mathbb{R}^{n_x}$ is replaced by real and complex spaces that correspond to real and complex parameters, respectively, which altogether reflects that the data and/or the signal parameters may be complex. In this chapter, all the parameter and data spaces are real.

$\psi \in \Psi \subset \mathbb{R}^{n_\psi}$ may be unknown. Then all the information about the signal parameters after observing $\mathbf{y}$ is in the a posteriori density of $\mathbf{x}$ given by

$$p(\mathbf{x}|\mathbf{y}) \propto \int_\Psi p(\mathbf{y}|\mathbf{x}, \psi)\pi(\mathbf{x})\pi(\psi)d\psi, \qquad (18.1.1)$$

where $\pi(\mathbf{x})$ and $\pi(\psi)$ are the densities of $\mathbf{x}$ and ψ, respectively. Notice that in (18.1.1), it is assumed that the variables $\mathbf{x}$ and ψ are independent.

In some problems, data arrive sequentially, and one observes independent data vectors $\mathbf{y}_1, \mathbf{y}_2, \cdots, \mathbf{y}_t$. The signal and noise parameters of each data record, $\mathbf{x}_t$ and ψ_t respectively, are random and drawn independently from *unknown* distributions $\pi(\mathbf{x})$ and $\pi(\psi)$, the main objective being to estimate $\pi(\mathbf{x})$ by the posterior density $p(\mathbf{x}|\mathbf{y}_{1:t})$, where $\mathbf{y}_{1:t} \triangleq \{\mathbf{y}_1, \mathbf{y}_2, \cdots, \mathbf{y}_t\}$. With the arrival of the data record $\mathbf{y}_t$, the scheme for finding $p(\mathbf{x}|\mathbf{y}_{1:t})$ can be implemented by modifying the density $p(\mathbf{x}|\mathbf{y}_{1:t-1})$. With the independence assumption

$$p(\mathbf{y}_t|\mathbf{y}_{1:t-1}, \mathbf{x}, \psi) = p(\mathbf{y}_t|\mathbf{x}, \psi),$$

and one can write

$$p(\mathbf{x}|\mathbf{y}_{1:t}) \propto \int_\Psi p(\mathbf{y}_t|\mathbf{x}, \psi)p(\mathbf{x}, \psi|\mathbf{y}_{1:t-1})d\psi, \qquad (18.1.2)$$

where $p(\mathbf{x}, \psi|\mathbf{y}_{1:t-1})$ is the joint posterior density of $\mathbf{x}$ and ψ given the data $\mathbf{y}_{1:t-1}$. The expression in (18.1.2) can be rewritten as

$$p(\mathbf{x}|\mathbf{y}_{1:t}) = \frac{p(\mathbf{y}_t|\mathbf{x}, \mathbf{y}_{1:t-1})}{p(\mathbf{y}_t|\mathbf{y}_{1:t-1})} p(\mathbf{x}|\mathbf{y}_{1:t-1}),$$

which formally shows the update of the density from $p(\mathbf{x}|\mathbf{y}_{1:t-1})$ to $p(\mathbf{x}|\mathbf{y}_{1:t})$.

In many applications, the assumption that the signal is always present in the data is not valid. In other words, the data may have been generated by one of two hypotheses, for example, one with a signal and another without a signal. Which hypothesis is true, of course, is not known, and the problem is to estimate the signal parameters by optimally processing the available data. For instance, in processing of synaptic currents measured in studies in neurology, synaptic events can be parameterised and the interest is to estimate the distribution of the signal parameters from patch clamp recordings of the synaptic events. There can be many recordings, and each recording does not necessarily contain an event. And if it does, the parameters that describe these events are different from the parameters of the events from other recordings. The challenge that arises in this setup is to optimally combine the information from the various recordings into a unique posterior of the signal parameters. A somewhat different scenario appears in communications, when one of several known signals is transmitted through an unknown and time-varying channel. The performance of the communication system can degrade considerably if the channel varies rapidly and the changes are not taken into account, and therefore the channel characteristics must be estimated as new data continue to arrive. The estimation of the channel is carried out under the uncertainty of the transmitted signal.

We show that, in this problem of estimation under model uncertainty, the posterior is a mixture density whose components are weighted by the posterior probabilities of the hypotheses. The number of component densities is equal to 2^t, where t is the number of processed data records. This clearly creates a major computational problem, because, as new data records continue to be collected, the computations of the posterior become slower and slower. Also, the optimal solution requires various integrations of multidimensional functions, which are often impossible to carry out analytically. To overcome these two difficulties, we use sequential sampling schemes based on the idea of approximating the posteriors with mixtures whose number of component densities does not change with time, and by propagating and updating these component densities as new data are observed. In this approach, the number of computations per observation is constant and the proposed schemes are easily parallelisable.

In Section 18.2, a problem formulation is provided, and in Section 18.3, two important propositions are proved and a solution to the problem given. In Section 18.4, sequential sampling schemes are presented that provide a practical implementation for computing the posterior. A simple example that illustrates the performance of one of the schemes is shown in Section 18.5. In Section 18.6, some brief remarks and conclusions are made.

18.2   The problem of parameter estimation under uncertainty

In this section, we formally define the problem of parameter estimation under uncertainty. Let the observed data vector at time t be $\mathbf{y}_t \in \mathbb{R}^{n_y}$, and assume that it is generated according to one of the following two hypotheses:

$$\begin{aligned}
\mathcal{H}_1 : \quad \mathbf{y}_t &= g(\mathbf{x}_t, \mathbf{v}_t) \\
\mathcal{H}_0 : \quad \mathbf{y}_t &= \mathbf{v}_t,
\end{aligned} \qquad (18.2.3)$$

where the noise vector $\mathbf{v}_t \in \mathbb{R}^{n_v}$, $n_v = n_y$, the signal parameters $\mathbf{x}_t \in \mathbb{R}^{n_x}$, and $g(\cdot)$ is some known function of $\mathbf{x}_t$ and $\mathbf{v}_t$. The signal parameters are random variables whose probability density function $\pi(\mathbf{x})$ is *unknown*. In other words, if $\mathbf{y}_t$ has a signal, the parameters of the signals are generated according to

$$\mathbf{x}_t \sim \pi(\mathbf{x}). \qquad (18.2.4)$$

Also, the parameters of the signals in the various data vectors are independent and identically distributed. For example, if $\mathbf{y}_1$ and $\mathbf{y}_2$ contain signals, $\mathbf{x}_1$ and $\mathbf{x}_2$ are independently drawn from $\pi(\mathbf{x})$. The noise vector is distributed according to a known probability density function $p(\mathbf{v}_t|\psi)$, where $\psi \in \mathbb{R}^{n_\psi}$ is a vector of known noise parameters. We present the case of known noise parameters to ease the presentation. When the noise parameters are unknown, the solution to the problem follows the same line of argument. To simplify the notation, we suppress the conditioning on ψ when writing densities. For example, instead of $p(\mathbf{y}_t|\mathbf{x}, \psi)$, we write $p(\mathbf{y}_t|\mathbf{x})$.

It is important to observe that the true hypothesis at time t can be different from the true hypotheses at times $t-1$, $t-2$, and so on. This is consequential and to emphasise it, we write $\mathcal{M}_t$ for the true hypothesis at time t and say that $\mathcal{M}_t = 1$ if $\mathcal{H}_1$ is true, and $\mathcal{M}_t = 0$ if $\mathcal{H}_0$ is true. Therefore, (18.2.3) formally becomes

$$\mathbf{y}_t = \begin{cases} g(\mathbf{x}_t, \mathbf{v}_t), & \mathcal{M}_t = 1 \\ \mathbf{v}_t, & \mathcal{M}_t = 0 \end{cases} \qquad (18.2.5)$$

and

$$p(\mathcal{M}_t|\mathcal{M}_{t-1}, \mathcal{M}_{t-2}, \cdots) = p(\mathcal{M}_t).$$

The probabilities $p(\mathcal{M}_t = 1)$ and $p(\mathcal{M}_t = 0)$ are assumed known. It is supposed that these probabilities do not change with time, although modifications can easily be made to accommodate such a possibility.

At time $t = 0$, all we know about $\mathbf{x}$ is given by the prior of $\mathbf{x}$, $\pi_a(\mathbf{x})$, which in general is *different* from $\pi(\mathbf{x})$. With the above assumptions and with the arrival of new data records $\mathbf{y}_t$, the objective is to update sequentially the estimate of $\pi(\mathbf{x})$, that is, the goal is to find a recursive algorithm that computes $p(\mathbf{x}|\mathbf{y}_{1:t})$ from $p(\mathbf{x}|\mathbf{y}_{1:t-1})$ and $\mathbf{y}_t$.

18.3   Sequential updating of the solution

At the beginning, before any data record has been received, all the information about $\mathbf{x}$ is summarised by $\pi_a(\mathbf{x})$. When the first data record becomes available, our knowledge about $\mathbf{x}$ is modified to $p(\mathbf{x}|\mathbf{y}_1)$ according to

$$\begin{aligned} p(\mathbf{x}|\mathbf{y}_1) &= p(\mathbf{x}|\mathbf{y}_1, \mathcal{M}_1 = 1)p(\mathcal{M}_1 = 1|\mathbf{y}_1) \\ &+ p(\mathbf{x}|\mathbf{y}_1, \mathcal{M}_1 = 0)p(\mathcal{M}_1 = 0|\mathbf{y}_1). \end{aligned} \qquad (18.3.6)$$

In writing this expression, the total probability theorem was used. For the factor of the second term on the right-hand side of (18.3.6), $p(\mathbf{x}|\mathbf{y}_1, \mathcal{M}_1 = 0)$, one can write

$$p(\mathbf{x}|\mathbf{y}_1, \mathcal{M}_1 = 0) = \pi_a(\mathbf{x}),$$

which simply reflects the fact that, if there is no signal in $\mathbf{y}_1$, there is nothing to learn about $\mathbf{x}$ from $\mathbf{y}_1$, implying that the knowledge about $\mathbf{x}$ is unchanged. If for the posterior probability mass function of $\mathcal{M}_1$ we use the notation

$$c_{i,1} = p(\mathcal{M}_1 = i|\mathbf{y}_1), \quad i = 0, 1$$

the posterior $p(\mathbf{x}|\mathbf{y}_1)$ in (18.3.6) becomes

$$p(\mathbf{x}|\mathbf{y}_1) = c_{1,1}p(\mathbf{x}_1|\mathbf{y}_1, \mathcal{M}_1 = 1) + c_{0,1}\pi_a(\mathbf{x}). \tag{18.3.7}$$

Since $c_{1,1}$ and $c_{0,1}$ are constants, where $c_{1,1} + c_{0,1} = 1$, it follows that the posterior of $\mathbf{x}$, after observing only one data record, is a mixture of two component densities. One of the components is the prior $\pi_a(\mathbf{x})$, and the other is

$$p(\mathbf{x}|\mathbf{y}_1, \mathcal{M}_1 = 1) = \frac{p(\mathbf{y}_1|\mathbf{x}, \mathcal{M}_1 = 1)\pi_a(\mathbf{x})}{p(\mathbf{y}_1|\mathcal{M}_1 = 1)}.$$

For the mixing weights of the components, we can write formally

$$\begin{aligned} c_{i,1} &= p(\mathcal{M}_1 = i|\mathbf{y}_1) \\ &= \frac{p(\mathbf{y}_1|\mathcal{M}_1 = i)p(\mathcal{M}_1 = i)}{p(\mathbf{y}_1)}, \end{aligned}$$

where $i = 0, 1$, and

$$p(\mathbf{y}_1|\mathcal{M}_1 = 1) = \int p(\mathbf{y}_1|\mathbf{x}, \mathcal{M}_1)\pi_a(\mathbf{x})d\mathbf{x}.$$

Next, the second data vector $\mathbf{y}_2$ is received, and its prior of $\mathbf{x}$ is given by (18.3.7). Again, note that, if there is a signal in $\mathbf{y}_2$, its parameters $\mathbf{x}_2$ are independently drawn according to (18.2.4). Then

$$p(\mathbf{x}|\mathbf{y}_{1:2}) = p(\mathbf{x}|\mathbf{y}_{1:2}, \mathcal{M}_2 = 1)p(\mathcal{M}_2 = 1) + p(\mathbf{x}|\mathbf{y}_{1:2}, \mathcal{M}_2 = 0)p(\mathcal{M}_2 = 0),$$

which further simplifies to

$$p(\mathbf{x}|\mathbf{y}_{1:2}) = p(\mathbf{x}|\mathbf{y}_{1:2}, \mathcal{M}_2 = 1)p(\mathcal{M}_2 = 1) + p(\mathbf{x}|\mathbf{y}_1)p(\mathcal{M}_2 = 0), \tag{18.3.8}$$

and where

$$p(\mathbf{x}|\mathbf{y}_{1:2}, \mathcal{M}_2 = 1) \propto p(\mathbf{y}_2|\mathbf{x}, \mathcal{M}_2 = 1)p(\mathbf{x}|\mathbf{y}_1). \tag{18.3.9}$$

The last two expressions entail the posterior $p(\mathbf{x}|\mathbf{y}_{1:2})$ being a mixture of four densities. To see this, first notice that the second term on the right-hand side of (18.3.8) is a mixture with two components. But the first term, too, is a mixture with two components, which can easily be deduced from (18.3.9). There is another way of writing the posterior $p(\mathbf{x}|\mathbf{y}_{1:2})$, which directly leads to the conclusion that $p(\mathbf{x}|\mathbf{y}_{1:2})$ is a mixture composed of four components, and that is

$$\begin{aligned} p(\mathbf{x}|\mathbf{y}_{1:2}) = \quad & c_{0,2}p(\mathbf{x}|\mathbf{y}_{1:2}, \mathcal{M}_1 = 0, \mathcal{M}_2 = 0) \\ + \quad & c_{1,2}p(\mathbf{x}|\mathbf{y}_{1:2}, \mathcal{M}_1 = 1, \mathcal{M}_2 = 0) \\ + \quad & c_{2,2}p(\mathbf{x}|\mathbf{y}_{1:2}, \mathcal{M}_1 = 0, \mathcal{M}_2 = 1) \\ + \quad & c_{3,2}p(\mathbf{x}|\mathbf{y}_{1:2}, \mathcal{M}_1 = 1, \mathcal{M}_2 = 1), \end{aligned} \tag{18.3.10}$$

where the coefficients $c_{i,2}$ are defined by

$$
\begin{aligned}
c_{i,2} &= p(\mathcal{M}_1 = i_1, \mathcal{M}_2 = i_2 | \mathbf{y}_{1:2}) \\
&= \frac{p(\mathbf{y}_{1:2} | \mathcal{M}_1 = i_1, \mathcal{M}_2 = i_2) p(\mathcal{M}_1 = i_1, \mathcal{M}_2 = i_2)}{p(\mathbf{y}_{1:2})},
\end{aligned}
$$

with the indices $i_1, i_2 \in \{0, 1\}$, and $i = i_1 + 2i_2$.

Now we generalise this result, but first we introduce new notation. Suppose that, until time t, the sequence of models $\mathcal{M}_1 = i_1$, $\mathcal{M}_2 = i_2$, $\cdots$, $\mathcal{M}_t = i_t$ has generated the data $\mathbf{y}_1, \mathbf{y}_2, \cdots, \mathbf{y}_t$, where $i_l \in 0, 1$ for $l = 1, 2, \cdots, t$. Then, a brief notation for this sequence of models will be $\mathcal{M}_{k,t}$, with $k = \sum_{l=1}^{t} i_l 2^{l-1}$. Next we state the proposition that generalises the result in (18.3.10).

Proposition 18.3.1. Let $\mathbf{y}_1, \mathbf{y}_2, \cdots, \mathbf{y}_t$, be observed data generated according to (18.2.5) under conditions formulated in Section 18.2. The posterior density of $\mathbf{x}$ is given by a mixture density with 2^t components and is of the form

$$
p(\mathbf{x} | \mathbf{y}_{1:t}) = \sum_{k=0}^{K_t} c_{k,t} p_{k,t}(\mathbf{x}), \tag{18.3.11}
$$

where $K_t = 2^t - 1$, the mixing weights $c_{k,t}$ satisfy $\sum_k c_{k,t} = 1$ and are defined by

$$
c_{k,t} = p(\mathcal{M}_{k,t} | \mathbf{y}_{1:t}), \quad k = \sum_{l=1}^{t} i_l 2^{l-1}, \quad i_l \in \{0, 1\} \tag{18.3.12}
$$

and the components $p_{k,t}(\mathbf{x})$ are given by

$$
p_{k,t}(\mathbf{x}) = p(\mathbf{x} | \mathbf{y}_{1:t}, \mathcal{M}_{k,t}). \tag{18.3.13}
$$

The proof of the proposition follows straightforwardly by applying the total probability theorem.

For the mixing weights of the components in (18.3.11) given by (18.3.12), we can say that they are joint posterior probabilities of specific sequences of models. Notice that each component, given by (18.3.13), represents the posterior of $\mathbf{x}$ if a particular sequence of models has generated the data $\mathbf{y}_{1:t}$. The mixture is then formed by multiplying each of these components with coefficients equal to the posterior probabilities of occurrence of the model sequences corresponding to the components. For example, if the sequence is $\mathcal{M}_l = 0$, for $l = 1, 2, \cdots, t$, then the component is

$$
p(\mathbf{x} | \mathbf{y}_{1:t}, \mathcal{M}_{0,t}) = \pi_a(\mathbf{x}),
$$

with a weighting coefficient $c_{0,t}$ given by

$$
c_{0,t} = p(\mathcal{M}_{0,t} | \mathbf{y}_{1:t}) = \frac{\prod_{l=1}^{t} p(\mathbf{y}_l | \mathcal{M}_l = 0) p(\mathcal{M}_l = 0)}{p(\mathbf{y}_{1:t})}.
$$

Next we investigate the sequential update of the posterior $p(\mathbf{x}|\mathbf{y}_{1:t})$, given that we know $p(\mathbf{x}|\mathbf{y}_{1:t-1})$. It is not difficult to show that

$$p(\mathbf{x}|\mathbf{y}_{1:t}) = \frac{p(\mathbf{y}_t|\mathbf{y}_{1:t-1}, \mathbf{x})}{p(\mathbf{y}_t|\mathbf{y}_{1:t-1})} p(\mathbf{x}|\mathbf{y}_{1:t-1}). \qquad (18.3.14)$$

This expression is interesting because it helps in establishing a direct relationship between the components of $p(\mathbf{x}|\mathbf{y}_{1:t})$ and $p(\mathbf{x}|\mathbf{y}_{1:t-1})$ and their weights. If we write

$$
\begin{aligned}
p(\mathbf{y}_t|\mathbf{y}_{1:t-1}, \mathbf{x}) &= p(\mathbf{y}_t|\mathbf{y}_{1:t-1}, \mathbf{x}, \mathcal{M}_t = 1)p(\mathcal{M}_t = 1) \\
&+ p(\mathbf{y}_t|\mathbf{y}_{1:t-1}, \mathbf{x}, \mathcal{M}_t = 0)p(\mathcal{M}_t = 0) \quad (18.3.15)
\end{aligned}
$$

and use it in (18.3.14), we readily see how the components of $p(\mathbf{x}|\mathbf{y}_{1:t})$ are formed from the components of $p(\mathbf{x}|\mathbf{y}_{1:t-1})$. The next proposition provides the exact relationship among them.

Proposition 18.3.2. Let $\mathbf{y}_1$, $\mathbf{y}_2$, $\cdots$, $\mathbf{y}_t$ be the observed data that are generated according to (18.2.5) under conditions formulated in Section 18.2. Let

$$p(\mathbf{x}|\mathbf{y}_{1:t-1}) = \sum_{k=0}^{2^{t-1}-1} c_{k,t-1} p_{k,t-1}(\mathbf{x})$$

be the posterior density of $\mathbf{x}$, which is a mixture density composed of 2^{t-1} components, after processing $t-1$ data records. Upon receiving the data record $\mathbf{y}_t$, the mixture $p(\mathbf{x}|\mathbf{y}_{1:t-1})$ is modified to $p(\mathbf{x}|\mathbf{y}_{1:t})$, where each $p_{k,t-1}(\mathbf{x})$ and $c_{k,t-1}$ produce two new components and mixing coefficients of $p(\mathbf{x}|\mathbf{y}_{1:t})$ according to

$$p_{k,t}(\mathbf{x}) = p_{k,t-1}(\mathbf{x}) \qquad (18.3.16)$$

$$p_{2^{t-1}+k,t}(\mathbf{x}) = \frac{p(\mathbf{y}_t|\mathbf{x}, \mathcal{M}_t = 1)}{p(\mathbf{y}_t|\mathbf{y}_{1:t-1}, \mathcal{M}_{2^{t-1}+k,t})} p_{k,t-1}(\mathbf{x}) \qquad (18.3.17)$$

$$c_{k,t} = \frac{p(\mathbf{y}_t|\mathcal{M}_t = 0)p(\mathcal{M}_t = 0)}{p(\mathbf{y}_t|\mathbf{y}_{1:t-1})} c_{k,t-1} \qquad (18.3.18)$$

$$c_{2^{t-1}+k,t} = \frac{p(\mathbf{y}_t|\mathbf{y}_{1:t-1}, \mathcal{M}_{2^{t-1}+k,t})p(\mathcal{M}_t = 1)}{p(\mathbf{y}_t|\mathbf{y}_{1:t-1})} c_{k,t-1} \qquad (18.3.19)$$

and

$$
\begin{aligned}
p_{0,1}(\mathbf{x}) = \pi_a(\mathbf{x}) \quad & p_{1,1}(\mathbf{x}) = p(\mathbf{x}|\mathbf{y}_1, \mathcal{M}_{1,1}) \\
c_{0,1} = p(\mathcal{M}_{0,1}|\mathbf{y}_1) \quad & c_{1,1} = p(\mathcal{M}_{1,1}|\mathbf{y}_1).
\end{aligned}
$$

Proof: First it is important to notice that, provided the first $t-1$ models are represented by the sequence $\mathcal{M}_{k,t-1}$, at time t the model sequence will become either $\mathcal{M}_{k,t}$ or $\mathcal{M}_{2^{t-1}+k,t}$, depending on whether the model at time t is $\mathcal{M}_t = 0$ or $\mathcal{M}_t = 1$, respectively. This implies that the component $p_{k,t-1}(\mathbf{x})$ of $p(\mathbf{x}|\mathbf{y}_{1:t-1})$ will produce the components $p_{k,t}(\mathbf{x})$ and $p_{2^{t-1}+k,t}(\mathbf{x})$ that are part of $p(\mathbf{x}|\mathbf{y}_{1:t})$. Similarly, the associated mixing coefficients $c_{k,t}$ and $c_{2^{t-1}+k,t}$ will be obtained from $c_{k,t-1}$. To find the exact expressions that relate them, we use (18.3.14) and expand $p(\mathbf{y}_t|\mathbf{y}_{1:t-1}, \mathbf{x})$ according to (18.3.15). Then we multiply it by the remaining factor on the right-hand side of (18.3.14), where we also use the representation (18.3.11). The result is

$$
\begin{aligned}
p(\mathbf{x}|\mathbf{y}_{1:t}) \;=\;& \frac{1}{p(\mathbf{y}_t|\mathbf{y}_{1:t-1})} \left(p(\mathbf{y}_t|\mathbf{y}_{1:t-1}, \mathbf{x}, \mathcal{M}_t = 1)p(\mathcal{M}_t = 1) \right. \\
&+ \left. p(\mathbf{y}_t|\mathbf{y}_{1:t-1}, \mathbf{x}, \mathcal{M}_t = 0)p(\mathcal{M}_t = 0) \right) \sum_{k=0}^{K_{t-1}} c_{k,t-1} p_{k,t-1}(\mathbf{x}),
\end{aligned}
$$

where $K_{t-1} = 2^{t-1} - 1$. The relationships (18.3.16)–(18.3.19) now follow straightforwardly using the identities

$$
p(\mathbf{y}_t|\mathbf{y}_{1:t-1}, \mathbf{x}, \mathcal{M}_t = 1) = p(\mathbf{y}_t|\mathbf{x}, \mathcal{M}_t = 1)
$$

and

$$
\int p(\mathbf{y}_t|\mathbf{x}, \mathcal{M}_t = 1)p(\mathbf{x}|\mathbf{y}_{1:t-1}, \mathcal{M}_{k,t-1})d\mathbf{x} = p(\mathbf{y}_t|\mathbf{y}_{1:t-1}, \mathcal{M}_{2^{t-1}+k,t}).
$$

$\diamondsuit$

It is interesting to observe how some component densities of the mixture are formed. For example, we can write $p_{k,t}(\mathbf{x}) = p_{k,t-1}(\mathbf{x})$, and as a special case deduce that

$$
p_{0,t}(\mathbf{x}) = p_{0,t-1}(\mathbf{x}) = \pi_a(\mathbf{x}). \tag{18.3.20}
$$

The last expression does make sense because it is saying that the posterior of $\mathbf{x}$ after observing t data records, which are all generated by the model $\mathcal{M} = 0$, is the same as the prior. This indeed should be the case as the data then do not contain any new information about $\mathbf{x}$. Similarly,

$$
p_{1,t}(\mathbf{x}) = p_{1,t-1}(\mathbf{x}) = p(\mathbf{x}|\mathbf{y}_1, \mathcal{M}_{1,1}),
$$

which means that the prior $\pi_a(\mathbf{x})$ has been updated after the arrival of the first data record, and ever since has been unchanged. Also,

$$
p_{2^{t-1},t}(\mathbf{x}) = \frac{p(\mathbf{y}_t|\mathbf{x}, \mathcal{M}_t = 1)}{p(\mathbf{y}_t|\mathbf{y}_{1:t-1}, \mathcal{M}_{2^{t-1},t})} p_{0,t-1}(\mathbf{x})
$$

and since from (18.3.20), $p_{0,t-1}(\mathbf{x}) = \pi_a(\mathbf{x})$, it follows that

$$
p_{2^{t-1},t}(\mathbf{x}) = p(\mathbf{x}|\mathbf{y}_t, \mathcal{M}_t = 1),
$$

that is, it is just the posterior of $\mathbf{x}$ obtained by modifying the prior using the information from the data record $\mathbf{y}_t$.

The above results show that as new data records continue to be received, the posterior of $\mathbf{x}$ becomes more and more complex. Even when all the densities required for sequential updating of the posterior are evaluated analytically, the number of components and mixing coefficients to be determined grows very quickly, and the computations become prohibitively expensive. With 10 data records, the mixture of the posterior has 1024 components, and after 20 data records, it is larger than one million. Therefore, the above results are almost only of theoretical interest.

18.4 Sequential algorithm for computing the solution

The analysis from the previous section showed that the computation of the posterior density $p(\mathbf{x}|\mathbf{y}_{1:t})$ is infeasible because, with the arrival of the new data records, there is an ever-increasing number of mixture components. Even if the large number of density components was not a problem, there are difficulties due to inherent integrations needed for computation of the coefficients $c_{k,t}$. Notice that in

$$c_{k,t} = \frac{\int p(\mathbf{y}_{1:t}|\mathbf{x}, \mathcal{M}_{k,t})\pi_a(\mathbf{x})d\mathbf{x} \ p(\mathcal{M}_{k,t})}{p(\mathbf{y}_{1:t})}$$

the numerator has a multidimensional integral, which in general cannot be solved analytically. So is the case with (18.3.19), if it is used for computation of $c_{k,t}$. This would imply that, with the arrival of every data record, a method for numerical integration would have to be applied.

A sequential method that can provide an approximate solution to our problem is based on a two-step procedure: first the posterior probabilities of the hypotheses are computed and second, depending on the obtained values of the probabilities, the density of $\mathbf{x}$ is updated or kept unaltered (Titterington, Smith and Makov 1985). In particular, if after time $t-1$ the posterior density of $\mathbf{x}$ is $p(\mathbf{x}|\mathbf{y}_{1:t-1})$, first the probabilities $p(\mathcal{M}_t = 1|\mathbf{y}_{1:t})$ and $p(\mathcal{M}_t = 0|\mathbf{y}_{1:t})$ are found using

$$p(\mathcal{M}_t = 1|\mathbf{y}_{1:t}) = \frac{\int p(\mathbf{y}_t|\mathbf{x}, \mathcal{M}_t = 1)p(\mathbf{x}|\mathbf{y}_{1:t-1})d\mathbf{x} \ p(\mathcal{M}_t = 1)}{p(\mathbf{y}_t|\mathbf{y}_{1:t-1})} \quad (18.4.21)$$

and

$$p(\mathcal{M}_t = 0|\mathbf{y}_{1:t}) = \frac{p(\mathbf{y}_t|\mathcal{M}_t = 0)p(\mathcal{M}_t = 0)}{p(\mathbf{y}_t|\mathbf{y}_{1:t-1})}. \quad (18.4.22)$$

Once these probabilities are obtained, there are several strategies that can be applied in the second step. One is to use a cut-off rule, such that if $p(\mathcal{M}_t = 1|\mathbf{y}_{1:t}) > p(\mathcal{M}_t = 0|\mathbf{y}_{1:t})$, the posterior $p(\mathbf{x}|\mathbf{y}_{1:t})$ is updated by assuming that $\mathcal{M}_t = 1$ is true. If $p(\mathcal{M}_t = 1|\mathbf{y}_{1:t}) < p(\mathcal{M}_t = 0|\mathbf{y}_{1:t})$, the posterior remains unchanged, i.e., $p(\mathbf{x}|\mathbf{y}_{1:t}) = p(\mathbf{x}|\mathbf{y}_{1:t-1})$. A "probabilistic" rule would choose randomly $\mathcal{M}_t = 1$ or $\mathcal{M}_t = 0$ according to the computed probabilities from (18.4.21) and (18.4.22) and proceed as in the previous rule with step 2. There are also other schemes, which in effect approximate the mixture with a single distribution.

18.4.1   A Sequential-Importance-Resampling scheme

A more accurate procedure to cope with the computational obstacles of this problem is based on a sequential sampling scheme known as sequential-importance-resampling (SIR). The main idea behind it is to approximate the posterior densities by samples (particles) drawn from these densities and propagate them as new data records become available. Thus, instead of keeping a direct track of the densities $p(\mathbf{x}|\mathbf{y}_{1:t})$, we maintain a set of particles that represent these densities.

Recall that the density of interest is $p(\mathbf{x}|\mathbf{y}_{1:t})$. Suppose that $\mathbf{x}_t^{(i)}$, $i = 1, 2, \cdots, N$, are particles from this density, each with probability mass $w_t^{(i)}$, where $\sum_{i=1}^{N} w_t^{(i)} = 1$. Notice that the subscript t in $\mathbf{x}_t^{(i)}$ simply denotes that the particle is associated with $p(\mathbf{x}|\mathbf{y}_{1:t})$ and $\mathbf{x}_0^{(i)}$ with $\pi_a(\mathbf{x})$. The set of particles $\{\mathbf{x}_t^{(i)}; i = 1, 2, \cdots, N\}$ represents the support of $p(\mathbf{x}|\mathbf{y}_{1:t})$, and with their weights $w_t^{(i)}$, they form a random measure that approximates the density $p(\mathbf{x}|\mathbf{y}_{1:t})$. The random measure is denoted as $\{\mathbf{x}_t^{(i)}, w_t^{(i)}; i = 1, 2, \cdots, N\}$. As an estimate of $p(\mathbf{x}|\mathbf{y}_{1:t})$, we then use

$$\hat{p}(\mathbf{x}|\mathbf{y}_{1:t}) = \sum_{i=1}^{N} w_t^{(i)} \delta(\mathbf{x} - \mathbf{x}_t^{(i)}),$$

where $\delta(\cdot)$ is the Dirac's delta function. In particular, when we compute expectations with respect to $p(\mathbf{x}|\mathbf{y}_{1:t})$, for instance,

$$\mathbb{E}(f(\mathbf{x})|\mathbf{y}_{1:t}) = \int f(\mathbf{x}) p(\mathbf{x}|\mathbf{y}_{1:t}) d\mathbf{x},$$

where $f(\mathbf{x})$ is some known function of $\mathbf{x}$, we use

$$\mathbb{E}(f(\mathbf{x})|\mathbf{y}_{1:t}) \simeq \sum_{i=1}^{N} w_t^{(i)} f(\mathbf{x}_t^{(i)}). \tag{18.4.23}$$

Under weak assumptions and if the particles $\mathbf{x}_t^{(i)}$ are independently drawn from the distribution $p(\mathbf{x}|\mathbf{y}_{1:t})$, the summation in (18.4.23) converges to the expectation $\mathbb{E}(f(\mathbf{x})|\mathbf{y}_{1:t})$ (Geweke 1989). Although the principles of SIR methodology are described in other chapters of this book, we briefly outline some of its basics for reasons of coherent presentation. SIR is based on a notion that involves the generation of particles from a proposal density, the particles being subsequently modified so that they become particles from a desired density (Rubin 1988). For example, if our desired density is $p(\mathbf{x}_t|\mathbf{y}_{1:t})$, the particles approximating it are obtained by carrying out three steps (Doucet, Godsill and Andrieu 2000),(Liu and Chen 1998).

(i) Generate candidate particles $\mathbf{x}_t^{*(i)}$ from a proposal density $q(\mathbf{x}_t)$.

(ii) Compute the weights of the candidate particles according to

$$w_t^{(i)} = \frac{p(\mathbf{x}_t^{*(i)}|\mathbf{y}_{1:t})/q(\mathbf{x}_t^{*(i)})}{\sum_{n=1}^{N} p(\mathbf{x}_t^{*(n)}|\mathbf{y}_{1:t})/q(\mathbf{x}_t^{*(n)})}. \qquad (18.4.24)$$

(iii) Resample N times with replacement from the discrete distribution

$$p(\mathbf{x} = \mathbf{x}_t^{*(i)}) = w_t^{(i)}, \quad i = 1, 2, \cdots, N.$$

In implementing SIR, an important issue is the choice of the proposal density $q(\mathbf{x}_t)$. Since, often, sequential sampling methods are applied in a scenario where the unknown signal parameters *evolve* in time according to

$$\mathbf{x}_t = h_t(\mathbf{x}_{t-1}, \mathbf{u}_t), \qquad (18.4.25)$$

where $h_t(\cdot)$ is some known function, and $\mathbf{u}_t \in \mathbb{R}^{n_u}$ is a noise vector with a known distribution, and the observations, in general, are generated by

$$\mathbf{y}_t = g_t(\mathbf{x}_t, \mathbf{v}_t),$$

the choice of $q(\mathbf{x}_t)$ depends on the probability densities of $\mathbf{u}_t$ and $\mathbf{v}_t$ as well as the functions $h_t(\cdot)$ and $g_t(\cdot)$. Ideally, $q(\mathbf{x}_t)$ should be $p(\mathbf{x}_t|\mathbf{x}_{0:t-1}, \mathbf{y}_{1:t})$. Quite often, however, it is impossible to sample from the ideal density, so various alternatives are used (Doucet, Godsill and Andrieu 2000). They include methods that try to approximate the optimal proposal density, for example, by using Markov chain Monte Carlo methods (Berzuini et al. 1997), (Liu and Chen 1998), or by employing a procedure where the new particles are generated by

$$\mathbf{x}_t^{(i)} = h_t(\mathbf{x}_{t-1}^{(i)}, \mathbf{u}_t^{(i)}),$$

where $\mathbf{u}_t^{(i)}$, $i = 1, 2, \cdots, N$ are noise samples drawn from the noise density (Gordon et al. 1993). The resampled particles $\mathbf{x}_t^{(i)}$ have weights $1/N$, and they approximate the density $p(\mathbf{x}_t|\mathbf{y}_{1:t})$. Note that some of the particles from the original set $\{\mathbf{x}_t^{*(i)}; i = 1, 2 \cdots, N\}$, in general, appear more than once in the modified particle set $\{\mathbf{x}_t^{(i)}; i = 1, 2 \cdots, N\}$. It is not difficult to see that, as the recursion proceeds, the number of distinct particles that represent a density decreases, which altogether leads to a poor approximation of the density (Berzuini et al. 1997), (Carpenter et al. 1999b), (Gordon et al. 1993), (Pitt and Shephard 1999b). This sample impoverishment is a serious defect of the SIR method for which several remedies have been proposed. Some of these include jittering (Gordon et al. 1993) and particle boosting (Rubin 1987b). The former adds a small amount of Gaussian noise to each particle drawn in the update step. The latter proposes kN particle candidates rather than N, where $k > 1$, and the resampling is carried out by drawing N out of the kN samples.

Getting back to our problem, we realise that the parameters $\mathbf{x}$ do evolve as in a special case of (18.4.25), i.e. $\mathbf{x}_t = \mathbf{u}_t$, where the probability density function of $\mathbf{u}_t$ is identical to $\pi(\mathbf{x})$. An SIR procedure can therefore be implemented. Assuming no uncertainty of signal being present in the data, the procedure would proceed as described. First a set of particles would be generated from the prior $\pi_a(\mathbf{x})$, and they would be the set of candidates for particles that would approximate $p(\mathbf{x}|\mathbf{y}_1)$. The weights would be computed as soon as the observation vector $\mathbf{y}_1$ is known and, based on these weights, a resampling carried out and a new random measure, $\{\mathbf{x}_1^{(i)}, 1/N; i = 1, 2, \cdots, N\}$, obtained. Since there is no evolution of the parameters as in the Markovian model (18.4.25), the particles from this random measure are candidates for particles of the next random measure that would approximate $p(\mathbf{x}|\mathbf{y}_{1:2})$. The procedure for selecting them is identical to that already used for choosing the $\mathbf{x}_1^{(i)}$'s. When the particles $\mathbf{x}_2^{(i)}$, $i = 1, 2, \cdots, N$, are determined, as before, they are the candidates for the next random measure, and they are sampled with replacement by using weights obtained according to (18.4.24).

When the hypothesis that generated the data is unknown, the above scheme is implemented with some modifications. Suppose that the current random measure is $\{\mathbf{x}_{t-1}^{(i)}, 1/N; i = 1, 2, \cdots, N\}$, and $\mathbf{y}_t$ becomes available. If we want to replicate the method from Section 18.3, an important difference from the procedure just described would be that the posterior probabilities of each hypothesis must be computed. If $p(\mathcal{M}_t = 0|\mathbf{y}_{1:t}) = 1$, there is no need to change the random measure and so we have $\{\mathbf{x}_t^{(i)} = \mathbf{x}_{t-1}^{(i)}, 1/N; i = 1, 2, \cdots, N\}$. If $p(\mathcal{M}_t = 1|\mathbf{y}_{1:t}) = 1$, then the random measure $\{\mathbf{x}_{t-1}^{(i)}, 1/N; i = 1, 2, \cdots, N\}$ is modified to $\{\mathbf{x}_t^{(i)}, 1/N; i = 1, 2, \cdots, N\}$ as already explained. In case the posterior probabilities have values between 0 and 1, the resampling is carried out by using particles from $\{\mathbf{x}_{t-1}^{(i)}; i = 1, 2, \cdots, N\}$ and $\{\mathbf{x}_t^{(i)}; i = 1, 2, \cdots, N|\mathcal{M}_t = 1\}$, where the meaning of the notation for the latter set signifies that it was obtained under the assumption $\mathcal{M}_t = 1$. The number of particles selected from each of the sets is proportional to the posterior probabilities $p(\mathcal{M}_t = 0|\mathbf{y}_{1:t})$ and $p(\mathcal{M}_t = 1|\mathbf{y}_{1:t})$.

An important problem that must be solved before applying this procedure is the computation of the posterior probabilities. For $p(\mathcal{M}_t = 0|\mathbf{y}_{1:t})$, we can write

$$p(\mathcal{M}_t = 0|\mathbf{y}_{1:t}) = \frac{p(\mathbf{y}_t|\mathcal{M}_t = 0)p(\mathcal{M}_t = 0)}{p(\mathbf{y}_t|\mathbf{y}_{1:t-1})} \qquad (18.4.26)$$

and for $p(\mathcal{M}_t = 1|\mathbf{y}_{1:t})$,

$$p(\mathcal{M}_t = 1|\mathbf{y}_{1:t}) = \frac{\int p(\mathbf{y}_t|\mathbf{x}, \mathcal{M}_t = 1)p(\mathbf{x}|\mathbf{y}_{1:t-1})d\mathbf{x} \ p(\mathcal{M}_t = 1)}{p(\mathbf{y}_t|\mathbf{y}_{1:t-1})}. \quad (18.4.27)$$

In general, the solution of the integral cannot be obtained analytically, so we resort to numerical integration. Since the random measure $\{\mathbf{x}_{t-1}^{(i)}, 1/N; i = 1, 2, \cdots, N\}$ is composed of particles that come from $p(\mathbf{x}|\mathbf{y}_{1:t-1})$, the integral in the numerator of (18.4.27) can be approximated by

$$\int p(\mathbf{y}_t|\mathbf{x}, \mathcal{M}_t = 1)p(\mathbf{x}|\mathbf{y}_{1:t-1})d\mathbf{x} \simeq \frac{1}{N}\sum_{i=1}^{N} p(\mathbf{y}_t|\mathbf{x}_{t-1}^{(i)}, \mathcal{M}_t = 1).$$

In summary, with the arrival of $\mathbf{y}_t$, the SIR approach to updating the posterior $p(\mathbf{x}|\mathbf{y}_{1:t-1})$, which is approximated by $\{\mathbf{x}_{t-1}^{(i)}, 1/N; i = 1, 2, \cdots, N\}$, can be implemented according to the following procedure.

(i) Set $\mathbf{x}_t^{*(i)} = \mathbf{x}_{t-1}^{(i)}$.

(ii) Compute the weights $w_t^{*(i)} = p(\mathbf{y}_t|\mathbf{x}_t^{*(i)}, \mathcal{M}_t = 1)$, normalise them by

$$w_t^{(i)} = \frac{w_t^{*(i)}}{\sum_{n=1}^{N} w_t^{*(i)}}$$

and associate them with the particles $\{\mathbf{x}_t^{*(i)}; i = 1, 2, \cdots N\}$. These weights are the probability masses of $\mathbf{x}_t^{*(i)}, i = 1, 2, \cdots, N$.

(iii) Estimate the probabilities $p(\mathcal{M}_t = 1|\mathbf{y}_{1:t})$ and $p(\mathcal{M}_t = 0|\mathbf{y}_{1:t})$ by first computing

$$I_t = \frac{1}{N}\sum_{i=1}^{N} p(\mathbf{y}_t|\mathbf{x}_{t-1}^{(i)}, \mathcal{M}_t = 1) \quad (18.4.28)$$

and then $p(\mathbf{y}_t|\mathcal{M}_t = 0)$. The required probabilities $p(\mathcal{M}_t = 1|\mathbf{y}_{1:t})$ and $p(\mathcal{M}_t = 0|\mathbf{y}_{1:t})$ can then be easily evaluated.

(iv) Resample from the two random measures $\{\mathbf{x}_{t-1}^{(i)}, 1/N; i = 1, 2, \cdots, N\}$ and $\{\mathbf{x}_t^{*(i)}, w_t^{(i)}; i = 1, 2, \cdots, N\}$ according to the estimated probabilities $p(\mathcal{M}_t = 0|\mathbf{y}_{1:t})$ and $p(\mathcal{M}_t = 1|\mathbf{y}_{1:t})$. For example, first choose the random measure according to these probabilities, and then follow up with sampling using the appropriate weights. The obtained particles form the random measure $\{\mathbf{x}_t^{(i)}, 1/N; i = 1, 2, \cdots, N\}$.

It bears mentioning that the number of operations needed to update a posterior when a new data record is available remains constant. It should also be obvious that this scheme can easily be implemented in parallel. Furthermore, since the number of distinct particles in the random measure may decrease rapidly, special care must be taken to avoid their depletion.

The sequential update can be carried out by avoiding the computation of the integral in (18.4.28). The point is to propagate the joint posterior $p(\mathbf{x}, \mathcal{M}_{1:t}|\mathbf{y}_{1:t})$, instead of $p(\mathbf{x}|\mathbf{y}_{1:t})$, and use the joint posterior to obtain $p(\mathbf{x}|\mathbf{y}_{1:t})$ (Kotecha and Djurić 2000). In our case, we can write

$$p(\mathbf{x}, \mathcal{M}_{1:t}|\mathbf{y}_{1:t}) = \frac{p(\mathbf{y}_t|\mathbf{x}, \mathcal{M}_t)p(\mathcal{M}_t|\mathcal{M}_{1:t-1})}{p(\mathbf{y}_t|\mathbf{y}_{1:t-1})} p(\mathbf{x}, \mathcal{M}_{1:t-1}|\mathbf{y}_{1:t-1})$$

and observe that a formal way to proceed is to use the N particles that approximate $p(\mathbf{x}, \mathcal{M}_{1:t-1}|\mathbf{y}_{1:t-1})$, and form two sets $\{\mathbf{x}_t^{*(i)}, \mathcal{M}_t = 1; i = 1, 2, \cdots, N\}$ and $\{\mathbf{x}_t^{*(i)}, \mathcal{M}_t = 0; i = 1, 2, \cdots, N\}$. When $\mathbf{y}_t$ becomes available, we compute the weights of each of the $2N$ particles, and resample accordingly. Since $p(\mathcal{M}_t|\mathcal{M}_{1:t-1}) = p(\mathcal{M}_t)$, we do not really need to keep track of the models that go with the particles. In fact, the procedure for updating $p(\mathbf{x}|\mathbf{y}_{1:t-1})$ with $\mathbf{y}_t$ amounts to the following steps.

(i) Set $\mathbf{x}_t^{*(i)} = \mathbf{x}_{t-1}^{(i)}$.

(ii) Compute the weights $w_t^{*(i)1} = p(\mathbf{y}_t|\mathbf{x}_t^{*(i)}, \mathcal{M}_t = 1)$ and $w_t^{*(i)0} = p(\mathbf{y}_t|\mathcal{M}_t = 0)$, where the latter weights are identical for $i = 1, 2, \cdots, N$. Form a random measure with the N particles, $\mathbf{x}_t^{*(i)}$, $i = 1, 2, \cdots, N$, with weights obtained by

$$w_t^{(i)} = \frac{w_t^{*(i)}}{\sum_{n=1}^{N} w_t^{*(i)}}$$

where

$$w_t^{*(i)} = w_t^{*(i)0} p(\mathcal{M}_t = 0) + w_t^{*(i)1} p(\mathcal{M}_t = 1).$$

(iii) Resample from the random measure $\{\mathbf{x}_t^{*(i)}, w_t^{(i)}; i = 1, 2, \cdots, N\}$.

The proposed schemes use the Rao-Blackwellisation technique. Recall that Rao-Blackwellisation is a method that reduces the variance of estimators (Casella and Robert 1996). The first method estimates the probabilities of the models by numerically integrating out the signal parameters, and the second, integrates out the models that generate the data to obtain the sample weights. Notice that the joint posterior $p(\mathbf{x}, \mathcal{M}_{1:t}|\mathbf{y}_{1:t})$ is of no interest, and instead it is $p(\mathbf{x}|\mathbf{y}_{1:t})$, and so at every time instant the second method marginalises $\mathcal{M}_{1:t}$. It updates the desired posterior according to

$$p(\mathbf{x}|\mathbf{y}_{1:t}) \propto p(\mathbf{y}_t|\mathbf{y}_{t-1}, \mathbf{x})p(\mathbf{x}|\mathbf{y}_{1:t-1}),$$

where the factor $p(\mathbf{y}_t|\mathbf{y}_{t-1}, \mathbf{x})$ is

$$p(\mathbf{y}_t|\mathbf{y}_{t-1}, \mathbf{x}) = p(\mathbf{y}_t|\mathbf{x}, \mathcal{M}_t = 1)p(\mathcal{M}_t = 1) + p(\mathbf{y}_t|\mathcal{M}_t = 0)p(\mathcal{M}_t = 0).$$
$$(18.4.29)$$

It is clear that the computation of the particle weights in the second step of this method amounts to using (18.4.29).

18.4.2 Sequential sampling scheme based on mixtures

Here we use a scheme that carries on the approximation of the posterior density with a mixture density whose number of components remains constant throughout the processing of the data. If the component densities are Gaussians, the method is similar to one of Gaussian sum filters introduced in the early seventies (Sorenson and Alspach 1971), (Alspach and Sorenson 1972), (Anderson and Moore 1979). However, the idea of Gaussian sum filters here is combined with the SIR scheme described in the previous subsection.

The basic premise is to approximate all the densities of interest by mixture densities (many of these densities are mixtures anyway). For example, the prior $\pi_a(\mathbf{x})$ is approximated by

$$\pi_a(\mathbf{x}) = \sum_{i=1}^{N} w_0^{(i)} p_0^{(i)}(\mathbf{x}),$$

where the component densities $p_0^{(i)}(\mathbf{x})$ are some adequately chosen densities, for example Gaussians. If the components are indeed Gaussian, they are individually defined by their means $\mathbf{x}^{(i)}$ and covariance matrices $\mathbf{C}^{(i)}$. The component densities form a random measure, which in essence is of the same type as the random measures from the previous subsections. There we could think of the component densities as Dirac delta impulses located at the values of the particles. If the component densities are Gaussians, which is assumed in the sequel, we not only have the values of the particles that define the component density, but also covariance matrices that are associated with them. We formally denote this random measure by $\{\mathbf{x}_0^{(i)}, \mathbf{C}_0^{(i)}, w_0^{(i)}; i = 1, 2, \cdots, N\}$. The idea now is to propagate the component densities as new data arrive so that the newly obtained component densities continue to approximate the posterior $p(\mathbf{x}|\mathbf{y}_{1:t})$.

When there is no model uncertainty, that is, it is known that a signal is always present in the data, once $\mathbf{y}_1$ is received, all the weights associated with the component densities are updated according to

$$w_1^{*(i)} = w_0^{(i)} \int p(\mathbf{y}_1|\mathbf{x})p_0^{(i)}(\mathbf{x})d\mathbf{x}$$

and

$$w_1^{(i)} = \frac{w_1^{*(i)}}{\sum_{n=1}^{N} w_1^{*(i)}}.$$

Then the mean vectors $\mathbf{x}_0^{(i)}$ and covariance matrices $\mathbf{C}_0^{(i)}$ of the components are modified to include information about $\mathbf{x}$ from $\mathbf{y}_1$. After this step is completed, a new random measure is formed, $\{\mathbf{x}_1^{(i)}, \mathbf{C}_1^{(i)}, w_1^{(i)}; i = 1, 2, \cdots, N\}$. With the next data record, $\mathbf{y}_2$, these steps are repeated, and the random measure $\{\mathbf{x}_2^{(i)}, \mathbf{C}_2^{(i)}, w_2^{(i)}; i = 1, 2, \cdots, N\}$ that approximates $p(\mathbf{x}|\mathbf{y}_{1:2})$ is obtained. The process is replicated along the same lines with the arrival of every data record. A problem that has to be resolved is the update of the mean vectors and covariance matrices. In the case of linear models, the updates can readily be accomplished by the equations that describe the Kalman filter. When the model is nonlinear, a possibility is to use the extended Kalman filter (Anderson and Moore 1979). If the model is nonlinear or the noise is non-Gaussian, the mean vectors and the covariance matrices can be computed by Monte Carlo integrations, for example, by using importance sampling. As new data are processed, the variance of the weights may grow, which would indicate that some of the component densities are negligible and others become dominating. Then, sampling from the mixture density will produce a new random measure in which all the weights are the same. Some strategies for doing this are described in (West 1993b).

Now, we return to the problem of estimating the unknowns under model uncertainty. Let the random measure be given by $\{\mathbf{x}_{t-1}^{(i)}, \mathbf{C}_{t-1}^{(i)}, 1/N; i = 1, 2, \cdots, N\}$ at time $t-1$. Notice that the weights of the components are $1/N$; the reason for this will be clear shortly. As with the SIR scheme, it is important to compute the random measure as if the true model were $\mathcal{M}_t = 1$. This can be done as described in the previous paragraph, and let the result be $\{\mathbf{x}_t^{(i)}, \mathbf{C}_t^{(i)}, w_t^{(i)}; i = 1, 2, \cdots, N|\mathcal{M}_t = 1\}$. If the measure is conditioned on $\mathcal{M}_t = 0$, we have $\{\mathbf{x}_t^{(i)} = \mathbf{x}_{t-1}^{(i)}, \mathbf{C}_t^{(i)} = \mathbf{C}_{t-1}^{(i)}, w_t^{(i)} = 1/N; i = 1, 2, \cdots, N|\mathcal{M}_t = 0\}$. The next task is to estimate the probabilities $p(\mathcal{M}_t = 1|\mathbf{y}_{1:t})$ and $p(\mathcal{M}_t = 0|\mathbf{y}_{1:t})$. This can be done by using (18.4.26)-(18.4.27), where the integral in (18.4.27) is computed by

$$\int p(\mathbf{y}_t|\mathbf{x}, \mathcal{M}_t = 1)p(\mathbf{x}|\mathbf{y}_{1:t-1})d\mathbf{x} \simeq \frac{1}{M}\sum_{m=1}^{M} p(\mathbf{y}_t|\tilde{\mathbf{x}}_{t-1}^{(m)}, \mathcal{M}_t = 1)$$

and the samples $\tilde{\mathbf{x}}_{t-1}$ used in the computation of the integral are drawn from the mixture density at $t-1$, i.e.,

$$\tilde{\mathbf{x}}_{t-1} \sim \frac{1}{N} \sum_{i=1}^{N} p_{t-1}^{(i)}(\mathbf{x}).$$

After computing the posterior probabilities of the models, we resort to re-sampling. We choose component densities from the two measures according to the computed probabilities. The component densities from the corre-sponding measures are drawn according to the weights of the measures. The resulting measure is $\{\mathbf{x}_t^{(i)}, \mathbf{C}_t^{(i)}, 1/N; i = 1, 2, \cdots, N\}$.

Again, it is obvious that we do not have the problem of component weights with large variance. In fact, at every t the weights are all equal. However, since we resample with replacement, we may have repeated com-ponents in the mixture. If the number of repeated components becomes larger than a predefined threshold, the mixture density can be used to produce a new random measure.

18.5   Example

We provide a simple example that shows the performances of two sequential estimators, one based on the proposed scheme that uses mixture densi-ties and the other, based on the cut-off rule described at the beginning of Section 18.4. Recall that with the cut-off rule we either update the poste-rior of $\mathbf{x}$ (if $p(\mathcal{M}_t = 1|\mathbf{y}_{1:t}) > p(\mathcal{M}_t = 0|\mathbf{y}_{1:t})$, or keep it unchanged (if $p(\mathcal{M}_t = 1|\mathbf{y}_{1:t}) < p(\mathcal{M}_t = 0|\mathbf{y}_{1:t})$). The data were generated according to

$$y_t = \begin{cases} x_t + v_t, & \mathcal{M}_t = 1 \\ v_t, & \mathcal{M}_t = 0 \end{cases},$$

where

$$x_t \sim \mathcal{N}(2, 1),$$

or $\pi(x) = \mathcal{N}(2, 1)$. The probabilities of $\mathcal{M}_t = 1$ and $\mathcal{M}_t = 0$ were the same. Thus, a succinct way of writing the model that generated the data is

$$y_t = b_t x_t + v_t,$$

where b_t is a Bernoulli random variable with $b_t \in \{0, 1\}$ and $P(b_t = 1) = P(b_t = 0) = 1/2$. The noise samples v_t were independent and identically distributed, and $v_t \sim \mathcal{N}(0, 3)$.

The number of components in the mixture was $N = 100$, and the prior density of x was assumed $\pi_a(x) = \mathcal{N}(0, 10)$. The component densities had means that were drawn from a Gaussian density with mean zero and variance 8. Thus, the component variances were all equal and set to 2.

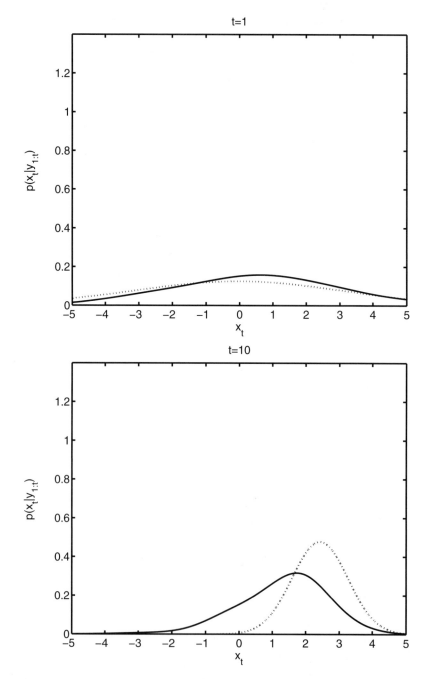

Figure 18.1. Comparison of the posterior densities $p(x|y_{1:t})$ obtained by the mixture (solid line) and cut-off (dotted line) methods. Top figure: $t = 1$; bottom figure: t=10.

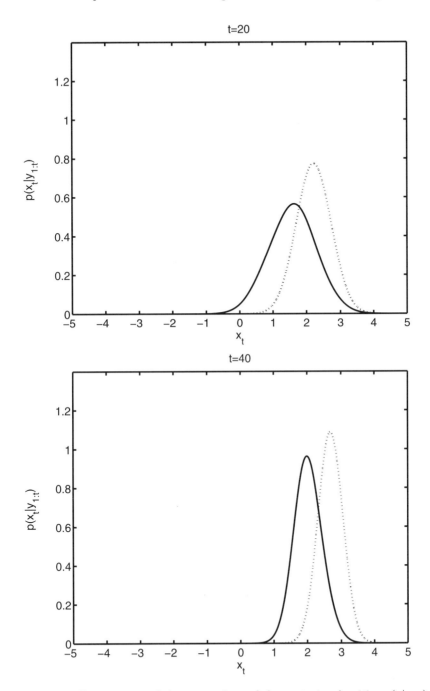

Figure 18.2. Continuation of the comparison of the posterior densities $p(x|y_{1:t})$ obtained by the mixture (solid line) and cut-off (dotted line) methods. Top figure: $t = 20$; bottom figure: t=40.

The results of the experiment are presented in the four figures, which progressively show the performance of the two methods as the number of observations increases. The first figure displays the posteriors $p(x|\mathbf{y}_{1:1})$, where the solid line corresponds to the mixture estimator, and the dotted line, to the cut-off method. After only one sample, there is obviously not much difference in performance. The second figure shows the results after 10 samples were received, and the third and fourth figures, after 20 and 40 samples, respectively. It is apparent that the posterior obtained by the mixture approach centers on the true mean, whereas the cut-off method provides a biased estimate. Repeated experiments showed similar performance.

18.6   Conclusions

In this chapter we addressed the problem of sequential estimation of signal parameters carried out under uncertainty. The uncertainty relates to the fact that the signal may not be present in the observed data. It was shown that the posterior density of the signal parameters is a mixture with 2^t components, where t is the number of observations. This result can be generalised to include a scenario with more than two hypotheses at each time instant. Then, the number of components in the mixture is L^t, where L is the number of hypotheses at every t. We also showed how to compute the weights of the components of the mixture at time t using the weights of the mixture at time $t-1$. The sequential updating of the mixture quickly becomes a computationally impossible task, so sequential importance sampling methods were invoked. Several schemes were discussed, all based on the notion of sampling-importance-resampling. Simulation results that compare the performance of a method based on Gaussian mixtures and a cut-off based rule approach that approximates posteriors with a single density were also presented. Their comparison showed that estimates based on sequential importance updates using mixtures are much better than those based on single densities.

18.7   Acknowledgment

Many fruitful discussions with Jayesh Kotecha are very much appreciated. This work was supported by the National Science Foundation under Award No. CCR-9903120.

19

Particle Filters for Mobile Robot Localization

Dieter Fox
Sebastian Thrun
Wolfram Burgard
Frank Dellaert

19.1   Introduction

This chapter investigates the utility of particle filters in the context of mobile robotics. In particular, we report results of applying particle filters to the problem of mobile robot localization, which is the problem of estimating a robot's pose relative to a map of its environment. The localization problem is a key one in mobile robotics, because it plays a fundamental role in various successful mobile robot systems; see e.g., (Cox and Wilfong 1990, Fukuda, Ito, Oota, Arai, Abe, Tanake and Tanaka 1993, Hinkel and Knieriemen 1988, Leonard, Durrant-Whyte and Cox 1992, Rencken 1993, Simmons, Goodwin, Haigh, Koenig and O'Sullivan 1997, Weiß, Wetzler and von Puttkamer 1994) and various chapters in (Borenstein, Everett and Feng 1996) and (Kortenkamp, Bonasso and Murphy 1998). Occasionally, it has been referred to as "the most fundamental problem to providing a mobile robot with autonomous capabilities" (Cox 1991).

The mobile robot localization problem takes different forms. The simplest localization problem—which has received by far the most attention in the literature—is *position tracking*. Here the initial robot pose is known, and localization seeks to correct small, incremental errors in a robot's odometry. More challenging is the *global localization problem*, where a robot is not told its initial pose, but instead has to determine it from scratch. The global localization problem is more difficult, since the robot's localization error can be arbitrarily large. Even more difficult is the *kidnapped robot problem* (Engelson and McDermott 1992), in which a well-localised robot is teleported to some other position without being told. This problem differs from the global localization problem in that the robot might firmly believe to be somewhere else at the time of the kidnapping. The kidnapped robot problem is often used to test a robot's ability to recover autonomously from catastrophic localization failures. Finally, there also exists the *multi-robot localization problem*, in which a team of robots seeks to localise themselves. The multi-robot localization problem is particularly interesting if robots are able to perceive each other, which introduces non-trivial statistical dependencies in the individual robots' estimates.

The beauty of particle filters is that they provide solutions to all the problems above. Even the most straightforward implementation of particle filters exhibits excellent results for the position tracking and the global localization problem. Extensions of the basic algorithm have led to excellent results on the kidnapped robot and the multi-robot localization problem.

The power of particle filters relative to these problems has many sources. In contrast to the widely used Kalman filters, particle filters can approximate a large range of probability distributions, not just normal distributions. Once a robot's belief is focused on a subspace of the space of all poses, particle filters are computationally efficient, because they focus their resources on regions in state-space with high likelihood. Particle filters are also easily implemented as any-time filters (Dean and Boddy 1988, Zilberstein and Russell 1995), by dynamically adapting the number of samples based on the available computational resources. Finally, particle filters for localization are remarkably easy to implement, which also contributes to their popularity.

This article describes a family of methods, known as *Monte Carlo localization (MCL)* (Dellaert, Burgard, Fox and Thrun 1999, Fox, Burgard, Dellaert and Thrun 1999). The MCL algorithm is a particle filter combined with probabilistic models of robot perception and motion. Building on this, we will describe a variation of MCL that uses a different proposal distribution (a mixture distribution) that facilitates fast recovery from global localization failures. As we shall see, this proposal distribution has a range of advantages over that used in standard MCL, but it comes with the price that it is more difficult to implement and requires an algorithm for sampling poses from sensor measurements, which might be difficult to obtain. Lastly, we will present an extension of MCL to cooperative multi-robot localization of robots that can perceive each other during localization. All these approaches have been tested thoroughly in practice. Experimental results are provided to demonstrate their relative strengths and weaknesses in practical robot applications.

19.2 Monte Carlo localization

19.2.1 Bayes filtering

Particle filters have already been discussed in the introductory chapters of this book. For the sake of consistency, let us briefly repeat the basics, beginning with Bayes filters. Bayes filters address the problem of estimating the state x of a dynamic system from sensor measurements. For example, in mobile robot localization the dynamical system is a mobile robot and its environment, the state is the robot's pose therein (often specified by a position in a two-dimensional Cartesian space and the robot's heading direction θ), and measurements may include range measurements, camera images, and odometry readings. Bayes filters assume that the environment is *Markov*, that is, past and future data are (conditionally) independent if one knows the current state.

The key idea of Bayes filtering is to estimate the posterior probability density over the state-space conditioned on the data. In the robotics and AI literature, this posterior is typically called the *belief*. Throughout this chapter, we will use the following notation:

$$Bel(x_t) \quad = \quad p(x_t \mid d_{0...t}).$$

Here x denotes the state, x_t is the state at time t, and $d_{0...t}$ denotes the data starting at time 0 up to time t. For mobile robots, we distinguish two types of data: *perceptual data* such as laser range measurements, and *odometry data* or *controls*, which carries information about robot motion. Denoting the former by y and the latter by u, we have

$$Bel(x_t) \quad = \quad p(x_t \mid y_t, u_{t-1}, y_{t-1}, u_{t-2} \ldots, u_0, y_0). \qquad (19.2.1)$$

Without loss of generality, we assume that observations and actions occur in an alternating sequence. Notice that the most recent perception in $Bel(x_t)$ is y_t, whereas the most recent controls/odometry reading is u_{t-1}.

Bayes filters estimate the belief *recursively*. The *initial* belief characterises the *initial* knowledge about the system state. In the absence of such knowledge (e.g., global localization), it is typically initialised by a *uniform distribution* over the state-space.

To derive a recursive update equation, we observe that Expression (19.2.1) can be transformed by Bayes rule to

$$Bel(x_t) \quad = \quad \frac{p(y_t \mid x_t, u_{t-1}, \ldots, y_0) \, p(x_t \mid u_{t-1}, \ldots, y_0)}{p(y_t \mid u_{t-1}, \ldots, y_0)}$$

$$= \quad \frac{p(y_t \mid x_t, u_{t-1}, \ldots, y_0) \, p(x_t \mid u_{t-1}, \ldots, y_0)}{p(y_t \mid u_{t-1}, d_{0...t-1})}. \qquad (19.2.2)$$

The *Markov assumption* states that measurements y_t are conditionally independent of past measurements and odometry readings given knowledge of the state x_t

$$p(y_t \mid x_t, u_{t-1}, \ldots, y_0) \;=\; p(y_t \mid x_t).$$

This allows us to conveniently simplify Equation (19.2.2)

$$Bel(x_t) \;=\; \frac{p(y_t \mid x_t)\, p(x_t \mid u_{t-1}, \ldots, y_0)}{p(y_t \mid u_{t-1}, d_{0\ldots t-1})}.$$

To obtain our final recursive form, we now have to integrate out the pose x_{t-1} at time $t-1$, which yields

$$\frac{p(y_t \mid x_t)}{p(y_t \mid u_{t-1}, d_{0\ldots t-1})} \int p(x_t \mid x_{t-1}, u_{t-1}, \ldots, y_0) p(x_{t-1} \mid u_{t-1}, \ldots, y_0) dx_{t-1}.$$

The *Markov assumption* also implies that, given knowledge of x_{t-1} and u_{t-1}, the state x_t is conditionally independent of past measurements $y_1 \ldots, y_{t-1}$ and odometry readings $u_1 \ldots, u_{t-2}$ up to time $t-2$, that is

$$p(x_t \mid x_{t-1}, u_{t-1}, \ldots, y_0) \;=\; p(x_t \mid x_{t-1}, u_{t-1}).$$

Using the definition of the belief Bel, we obtain a recursive estimator known as *Bayes filter*

$$
\begin{aligned}
Bel(x_t) \;&=\; \frac{p(y_t \mid x_t)}{p(y_t \mid u_{t-1}, d_{0\ldots t-1})} \int p(x_t \mid x_{t-1}, u_{t-1})\, Bel(x_{t-1})\, dx_{t-1} \\
&=\; \eta\, p(y_t \mid x_t) \int p(x_t \mid x_{t-1}, u_{t-1})\, Bel(x_{t-1})\, dx_{t-1}, \quad (19.2.3)
\end{aligned}
$$

where η is a normalizing constant. This equation is of central importance, as it is the basis for various MCL algorithms studied here.

19.2.2   Models of robot motion and perception

In the context of mobile robot localization, Bayes filters are also known as *Markov localization* (Burgard, Fox, Hennig and Schmidt 1996, Fox, Burgard and Thrun 1999, Kaelbling, Cassandra and Kurien 1996, Koenig and Simmons 1996, Nourbakhsh, Powers and Birchfield 1995, Simmons and Koenig 1995, Thrun 1998). To implement Markov localization, one needs to know three distributions: the initial belief $Bel(x_0)$ (e.g., uniform), the next state probability $p(x_t \mid x_{t-1}, u_{t-1})$ (called the *motion model*), and the perceptual likelihood $p(y_t \mid x_t)$ (called the *perceptual model*). The specific shape of these probabilities depends on the robot's odometry, and the type of sensors used for localization. Both of these models are time-invariant; we will henceforth omit the time index t.

A specific motion model (for an RWI B21 robot) is shown in Figure 19.1. This figure shows the probabilistic outcome of two example motion commands indicated by the lines. The grey-scale corresponds to $p(x' \mid x, a)$, projected into 2D. This specific model is the result of convolving conventional robot kinematics with two independent zero-mean random variables,

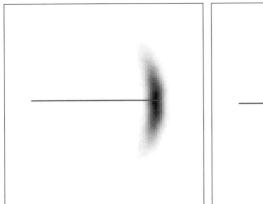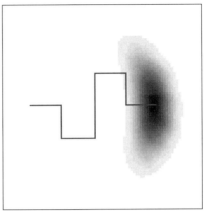

Figure 19.1: The density $p(y \mid x)$ after moving 40 meters (left diagram) and 80 meters (right diagram). The darker a pose, the more likely it is.

one of which models noise in rotation, and one models translational noise. The model is easily coded in 20 lines of C code.

The perceptual model $p(y \mid x)$ depends on the specific sensor. If y are raw camera images, computing $p(y \mid x)$ is related to the computer graphics problem in that the appearance of an image y at pose x has to be predicted. However, $p(y \mid x)$ is considerably simpler when one uses range finders for perception. Such sensors measure the distance of the robot to nearby obstacles, using sound or structured laser light. Figure 19.2 illustrates the model of robot perception for a planar 2D laser range finder, which is commonly used in mobile robotics. Figure 19.2a shows a laser scan and a map. The specific density $p(y \mid x)$ is computed in two stages. First, the measurement in an ideal, noise-free environment is computed. For laser range finders, this is easily done using ray-tracing in a geometric map of the environment, such as that shown in Figure 19.2a. Second, the desired density $p(y \mid x)$ is obtained as a mixture of random variables, composed of one that models the event of getting the correct reading (convolved with small Gaussian-distributed measurement noise), one for receiving a max-range reading (which occurs frequently), and one that models random noise and is exponentially distributed. Figure 19.2b shows a picture of $p(y \mid x)$, and Figure 19.2c plots $p(y \mid x)$ for the specific sensor scan y shown in Figure 19.2a.

19.2.3   Implementation as particle filters

If the state-space is continuous, as is the case in mobile robot localization, implementing the belief update equation (19.2.3) is *not* a trivial matter—particularly if one is concerned about efficiency. The idea of MCL (and other particle filter algorithms) is to represent the belief $Bel(x)$ by a set of

(a) laser scan and map

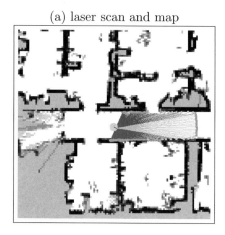

(b) sensor model $p(y \mid x)$

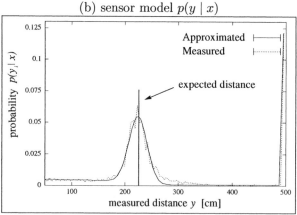

(c) probability distribution for different poses

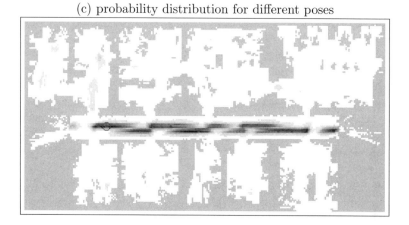

Figure 19.2: (a) Laser range scan, projected into a map. (b) The density $p(y \mid x)$. (c) $p(y \mid x)$ for the scan shown in (a). Based on a single sensor scan, the robot assigns high likelihood for being somewhere in the main corridor.

m weighted samples distributed according to $Bel(x)$:

$$Bel(x) \;=\; \{x^{(i)}, p^{(i)}\}_{i=1,\ldots,m}.$$

Here each $x^{(i)}$ is a sample (a state), and $p^{(i)}$ are non-negative numerical factors called *importance factors*, which sum to one. As the name suggests, the importance factors determine the weight (=importance) of each sample.

In global mobile robot localization, the *initial* belief is a set of poses drawn according to a uniform distribution over the robot's universe, annotated by the uniform importance factor $\frac{1}{m}$.

The recursive update is effectuated in three steps, computing the expression in (19.2.3) *from the right to the left*.

(i) Sample a state x_{t-1} from $Bel(x_{t-1})$, by drawing a random $x_{t-1}^{(i)}$ from the sample set representing $Bel(x_{t-1})$ according to the (discrete) distribution defined through the importance factors $p_{t-1}^{(i)}$.

(ii) Use the sample $x_{t-1}^{(i)}$ and the action u_{t-1} to sample $x_t^{(j)}$ from the distribution $p(x_t \mid x_{t-1}, u_{t-1})$. The predictive density of $x_t^{(j)}$ is now given by the product $p(x_t \mid x_{t-1}, u_{t-1})Bel(x_{t-1})$.

(iii) Finally, weight the sample $x_t^{(j)}$ by the (non-normalised) importance factor $p(y_t \mid x_t^{(j)})$, the likelihood of the sample $x_t^{(j)}$ given the measurement y_t.

After the generation of m samples, the new importance factors are normalised so that they sum to 1 and hence define a probability distribution. The reader should quickly see that this procedure in fact implements (19.2.3), using an (approximate) sample-based representation. Obviously, our algorithm is just one possible implementation of the particle filtering idea; other sampling schemes exist that further reduce variance (Kitagawa 1996). Detailed convergence results can be found in Chapters 2 and 3 of this book.

Further below, notice that, in this version of MCL, the proposal distribution for approximating $Bel(x_t)$ via importance sampling is given by

$$q \;:=\; p(x_t \mid x_{t-1}, u_{t-1})Bel(x_{t-1}), \qquad (19.2.4)$$

which is used to approximate the desired posterior

$$\frac{p(y_t \mid x_t)\, p(x_t \mid u_{t-1}, x_{t-1})\, Bel(x_{t-1})}{p(y_t \mid d_{0\ldots t-1}, u_{t-1})}. \qquad (19.2.5)$$

Figure 19.3: Two of the robots used for testing: RHINO (left) and MINERVA (right), which successfully guided thousands of people through crowded museums.

Consequently, the importance factors are given by the quotient

$$[p(x_t \mid x_{t-1}, u_{t-1}) Bel(x_{t-1})]^{-1} \frac{p(y_t \mid x_t) \, p(x_t \mid u_{t-1}, x_{t-1}) \, Bel(x_{t-1})}{p(y_t \mid d_{0...t-1}, u_{t-1})}$$

$$\propto \quad p(y_t \mid x_t). \tag{19.2.6}$$

19.2.4   Robot results

MCL has been at the core of our robot navigation software. It is more efficient and accurate than any of our previous algorithms. We thoroughly tested MCL in a range of real-world environments, applying it to at least three different types of sensors (cameras, sonar, and laser proximity data). Our experiments were carried out using several B21, B18 Pioneer, Scout, and XR4000 robots, two of which are shown in Figure 19.3. These robots were equipped with arrays of sonar sensors (from 7 to 24), one or two laser range finders, and in the case of Minerva, the robot shown in center and right of Figure 19.3, a B/W camera pointed at the ceiling.

A typical example of MCL is shown in Figure 19.4. This example illustrates MCL in the context of localizing a mobile robot globally in an office environment. This robot is equipped with sonar range finders, and it is

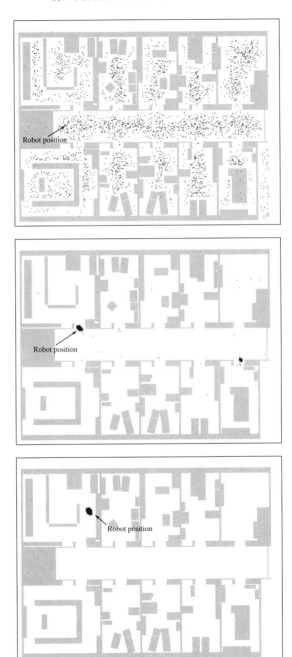

Figure 19.4: Global localization of a mobile robot using MCL (10,000 samples).

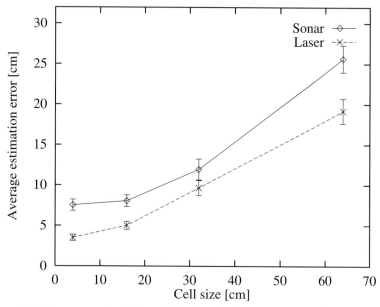

Figure 19.5: Accuracy of grid-based Markov localization using different spatial resolutions.

also given a map of the environment. In Figure 19.4a, the robot is globally uncertain; hence the samples are spread uniformly trough the free-space (projected into 2D). Figure 19.4b shows the sample set after approximately 1 meter of robot motion, at which point MCL has disambiguated the robot's position up to a single symmetry. Finally, after another 2 meters of robot motion the ambiguity is resolved, and the robot knows where it is. The majority of samples is now dispersed tightly around the correct position, as shown in Figure 19.4c.

19.2.5   Comparison to grid-based localization

To elucidate the advantage of particle filters over alternative representations, we are particularly interested in grid-based representations, which are at the core of an alternative family of Markov localization algorithms (Fox, Burgard and Thrun 1998). The algorithm described in (Fox et al. 1998) relies on a fine-grained grid approximation of the belief $Bel(\cdot)$, using otherwise identical sensor and motion models. Figure 19.5 plots the localization accuracy for grid-based localization as a function of the grid resolution. Notice that the results in Figure 19.5 were not generated in real-time. As shown there, the accuracy increases with the resolution of the grid, both for sonar (solid line) and for laser data (broken line). However, grid sizes be-

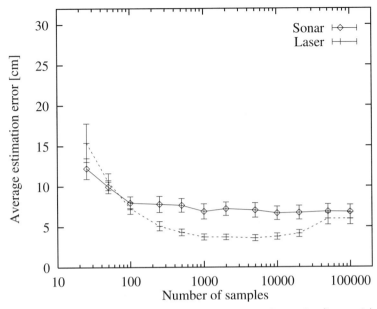

Figure 19.6: Accuracy of MCL for different numbers of samples (log scale).

yond 8 cm do not permit updating in real-time, even when highly efficient, selective update schemes are used (Fox et al. 1998).

Results for MCL with fixed sample set sizes are shown in Figure 19.6. These results have been generated under real-time conditions, where large sample sizes (> 1,000 samples) result in loss of sensor data because of time constraints. Here very small sample sets are disadvantageous, since they infer too large an error in the approximation. Large sample sets are also disadvantageous, since processing them requires too much time and fewer sensor items can be processed in real-time. The "optimal" sample set size, according to Figure 19.6, is somewhere between 1,000 and 5,000 samples. Grid-based localization, to reach the same level of accuracy, has to use grids with 4cm resolution—which is infeasible even given our fastest computers.

In contrast, the grid-based approach, with a resolution of 20 cm, requires almost exactly ten times as much memory when compared to MCL with 5,000 samples. During global localization, integrating a single sensor scan requires up to 120 seconds using the grid-based approach, whereas MCL consumes consistently less than 3 seconds under otherwise equal conditions. This illustrates that particle filters are clearly superior over grid-based representations, which previously were among the most effective algorithms for the global localization problem.

Similar results were obtained using a camera as the primary sensor for localization (Dellaert, Fox, Burgard and Thrun 1999). To test MCL

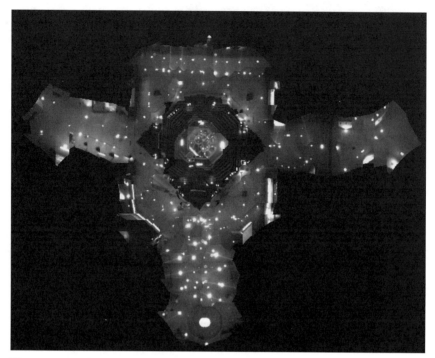

Figure 19.7: Ceiling map of the National Museum of American History, which was used as the perceptual model in navigating with a vision sensor.

under extreme circumstances, we evaluated it using data collected in a populated museum. During a two week exhibition, our robot Minerva (Figure 19.3) was employed as a tour-guide in the Smithsonian's Museum of Natural History, during which it traversed more than 44km (Thrun, Bennewitz, Burgard, Cremers, Dellaert, Fox, Hähnel, Rosenberg, Roy, Schulte and Schulz 1999). To aid localization, Minerva is equipped with a camera pointed towards the ceiling. Using this camera, the brightness of a small patch of the ceiling directly above the robot is measured, and compared with a large-scale mosaic of the museum's ceiling obtained beforehand (Dellaert, Thorpe and Thrun 1999), shown in Figure 19.7. This constitutes the likelihood model. The data used here is among the most difficult data sets in our possession, as the robot travelled with speeds of up to 163 cm/sec. Whenever it entered or left the carpeted area in the center of the museum, it crossed a 2cm bump that introduced significant errors in the robot's odometry.

When *only* using vision information, grid-based localization fatally failed to track the robot. This is because the enormous computational overhead makes it impossible to sufficiently incorporate many images. MCL, however, succeeded in globally localizing the robot and tracking its position.

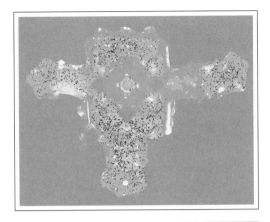

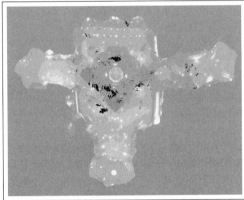

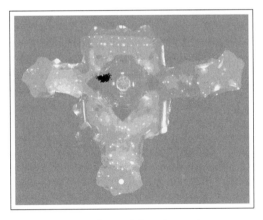

Figure 19.8: Global localization of a mobile robot using a camera pointed at the ceiling.

Figure 19.8 shows an example of global localization with MCL. In the begin-

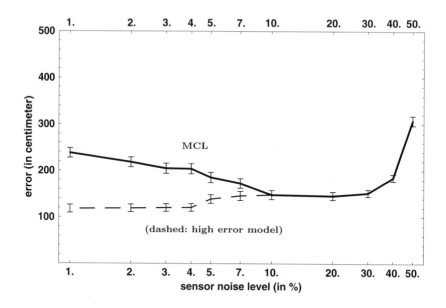

Figure 19.9: Solid curve: error of MCL after 100 steps, as a function of the sensor noise. 95% confidence intervals are indicated by the bars. Notice that this function is *not* monotonic, as one might expect. Dashed curve: Same experiment with high-error model.

ning, the robot starts with 2,000 uniformly distributed samples representing the absolute uncertainty about the robot's position. After incorporating 15 images (first diagram), the samples were still scattered over the whole area, but were starting to concentrate on several locations. After incorporating 38 images, most of the ambiguities were resolved, and the samples concentrated on a small number of peaks (second diagram). Finally, after 126 iterations, the robot was highly certain about its position (third diagram), which is represented by a concentration of the samples of the true location of the robot.

19.3 MCL with mixture proposal distributions

19.3.1 The need for better sampling

As recognised by several authors (Doucet 1998, Lenser and Veloso 2000, Liu and Chen 1998, Pitt and Shephard 1999b), the basic particle filter performs poorly if the proposal distribution, which is used to generate samples, places too few samples in regions where the desired posterior $Bel(x_t)$ is large.

This problem has indeed great practical importance in the context of MCL, as the following example illustrates. The solid curve in Figure 19.9 shows the accuracy MCL achieves after 100 steps, using $m = 1,000$ samples. These results were obtained in simulation, enabling us to vary the amount of perceptual noise from 50% (on the right) to 1% (on the left); in particular, we simulated a mobile robot localizing an object in 3D space from mono-camera imagery. It appears that MCL works best for 10% to 20% perceptual noise. The degradation of performance towards the right, when there is *high* noise, barely surprises. The less accurate a sensor, the larger an error one should expect. However, MCL also performs poorly when the noise level is too small. In other words, MCL with accurate sensors may perform *worse* than MCL with inaccurate sensors. This finding is counter-intuitive because it suggests that MCL only works well in specific situations, namely those where the sensors possess the right amount of noise.

At first glance, one might attempt to fix the problem by using a perceptual likelihood $p(y_t \mid x_t)$ that overestimates the sensor noise. In fact, such a strategy partly alleviates the problem: the dashed curve in Figure 19.9b shows the accuracy if the error model assumes a fixed 10% noise (shown there only for smaller "true" error rates). While the performance is better, this is hardly a principled way of fixing the problem. The overly pessimistic sensor model is inaccurate, throwing away precious information in the sensor readings. In fact, the resulting belief is not any longer a posterior, even if infinitely many samples were used. As we shall see below, a mathematically sound method exists that produces much better results.

To analyze the problem more thoroughly, we first notice that the true goal of Bayes filtering is to calculate the product distribution specified in Equation (19.2.5). Thus, the optimal proposal distribution would be this product distribution. However, sampling from this distribution directly is too difficult. As noticed above, MCL samples instead from the proposal distribution q defined in Equation (19.2.4), and uses the importance factors (19.2.6) to account for the difference. It is well-known from the statistical literature (Doucet 1998, Pitt and Shephard 1999b, Liu and Chen 1998, Tanner 1993) that the divergence between (19.2.5) and (19.2.4) determines the convergence speed. This difference is accounted for by the perceptual density $p(y_t \mid x_t)$: if the sensors are entirely uninformative, this distribution is flat and (19.2.5) equals (19.2.4). For low-noise sensors, however, $p(y_t \mid x_t)$ is typically quite narrow, hence MCL converges slowly. Thus, the error in Figure 19.9 is in fact caused by two different types of errors: one arising from the limitation of the sensor data (=noise), and the other arising from the mismatch of (19.2.5) and (19.2.4) in MCL. This suggests the use of different proposal distributions for sampling that can accommodate highly accurate sensors.

19.3.2 *An alternative proposal distribution*

To remedy this problem, one can use a different proposal distribution, one that samples according to the most recent sensor measurement y_t (see also (Lenser and Veloso 2000, Thrun, Fox and Burgard 2000)). The key idea is to sample x_t directly from a distribution that is proportional to the perceptual likelihood $p(y_t \mid x_t)$

$$\bar{q} \; := \; \frac{p(y_t \mid x_t)}{\pi(y_t)} \quad \text{with} \quad \pi(y_t) \; = \; \int p(y_t \mid x_t) \, dx_t. \quad (19.3.7)$$

This new proposal distribution possesses orthogonal limitations from the one described above, in that it generates samples that are highly consistent with the most recent sensor measurement, but ignorant of the belief $Bel(x_{t-1})$ and the control u_{t-1}.

The importance factors for these samples can be calculated in three ways. Recall that our goal is to sample from the product distribution

$$\frac{p(y_t|x_t)p(x_t|u_{t-1}, x_{t-1})Bel(x_{t-1})}{p(y_t|d_{0...t-1}, u_{t-1})} = \frac{p(y_t|x_t)p(x_t|d_{0...t-1}, u_{t-1})}{p(y_t|d_{0...t-1}, u_{t-1})} \quad (19.3.8)$$

Approach 1

(proposed by Arnaud Doucet, personal communication): The idea is to draw random pairs $\langle x_t^{(i)}, x_{t-1}^{(i)} \rangle$ by sampling $x_t^{(i)}$ as described above, and $x_{t-1}^{(i)}$ by drawing from $Bel(x_{t-1})$. Obviously, the combined proposal distribution is then given by

$$\frac{p(y_t \mid x_t^{(i)})}{\pi(y_t)} \times Bel(x_{t-1}^{(i)}), \quad (19.3.9)$$

and hence the importance factors are given by the quotient

$$\left[\frac{p(y_t \mid x_t^{(i)})}{\pi(y_t)} \times Bel(x_{t-1}^{(i)}) \right]^{-1} \frac{p(y_t \mid x_t^{(i)}) \, p(x_t^{(i)} \mid u_{t-1}, x_{t-1}^{(i)}) \, Bel(x_{t-1}^{(i)})}{p(y_t \mid d_{0...t-1}, u_{t-1})}$$

$$= \frac{p(x_t^{(i)} \mid u_{t-1}, x_{t-1}^{(i)}) \, \pi(y_t)}{p(y_t \mid d_{0...t-1}, u_{t-1})}$$

$$\propto \; p(x_t^{(i)} \mid u_{t-1}, x_{t-1}^{(i)}). \quad (19.3.10)$$

This approach is mathematically more elegant than the two options described below, in that it avoids the need to transform sample sets into densities (which will be the case below). We have not yet implemented this approach, and hence cannot comment on how well it works. However, in the context of global mobile robot localization, we suspect the importance factor $p(x_t^{(i)} \mid u_{t-1}, x_{t-1}^{(i)})$ will be zero for many pose pairs $\langle x_t^{(i)}, x_{t-1}^{(i)} \rangle$.

Approach 2

Another approach uses forward sampling and kd-trees to generate an approximate density of $p(x_t \mid d_{0...t-1}, u_{t-1})$. This density is then used in a second phase to calculate the desired importance factor. More specifically, Equations (19.3.7) and (19.3.8) suggest that the importance factors of a sample $x_t^{(i)}$ can be written as

$$\left[\frac{p(y_t \mid x_t^{(i)})}{\pi(y_t)}\right]^{-1} \frac{p(y_t \mid x_t^{(i)}) \, p(x_t^{(i)} \mid d_{0...t-1}, u_{t-1})}{p(y_t \mid d_{0...t-1}, u_{t-1})}$$

$$\propto \quad p(x_t^{(i)} \mid d_{0...t-1}, u_{t-1}). \tag{19.3.11}$$

Computing these importance factors is *not* trivial, since $Bel(x_{t-1})$ is represented by a set of samples. The trick here is to employ a two-staged approach that first approximates $p(x_t \mid d_{0...t-1}, u_{t-1})$ and then uses this approximate density to calculate the desired importance factors.

The following algorithm implements this alternative importance sampler.

(i) Generate a set of samples $x_t^{(j)}$, first by sampling from $Bel(x_{t-1}^{(j)})$ and then sampling from $p(x_t^{(j)} \mid u_{t-1}, x_{t-1}^{(j)})$ as described above. Obviously, these samples approximate $p(x_t^{(j)} \mid d_{0...t-1}, u_{t-1})$.

(ii) Transform the resulting sample set into a kd-tree (Bentley 1980, Moore 1990). The tree generalises samples to arbitrary poses $x_t^{(j)}$ in pose space, which is necessary to calculate the desired importance factors.

(iii) Lastly, sample $x_t^{(i)}$ from our proposal distribution $\frac{p(y_t \mid x_t^{(i)})}{\pi(y_t)}$. Weight each such sample by an importance factor that is proportional to its probability under the previously generated density tree.

This approach avoids the danger of generating pairs of poses $\langle x_t^{(i)}, x_{t-1}^{(i)} \rangle$ with zero probability under $p(x_t^{(i)} \mid u_{t-1}, x_{t-1}^{(i)})$. However, it involves an explicit forward sampling phase.

Approach 3

The third approach combines the best of both worlds, in that it avoids the explicit forward-sampling phase of the second approach, but also generates importance factors that are large. In particular, this approach transforms the initial belief $Bel(x_{t-1})$ into a kd-tree. It then generates samples $x_t^{(i)}$

according to $\frac{p(y_t|x_t)}{\pi(y_t)}$. For each such sample $x_t^{(i)}$, it generates a sample $x_{t-1}^{(i)}$ according to

$$\frac{p(x_t^{(i)} \mid u_{t-1}, x_{t-1})}{\pi(x_t^{(i)} \mid u_{t-1})}, \tag{19.3.12}$$

where

$$\pi(x_t^{(i)} \mid u_{t-1}) = \int p(x_t^{(i)} \mid u_{t-1}, x_{t-1}) \, dx_{t-1}. \tag{19.3.13}$$

Each of these combined samples $\langle x_t^{(i)}, x_{t-1}^{(i)} \rangle$ is, thus, sampled from the joint distribution

$$\frac{p(y_t \mid x_t^{(i)})}{\pi(y_t)} \times \frac{p(x_t^{(i)} \mid u_{t-1}, x_{t-1}^{(i)})}{\pi(x_t^{(i)} \mid u_{t-1})}. \tag{19.3.14}$$

The importance factor is calculated as follows:

$$\left[\frac{p(y_t \mid x_t^{(i)})}{\pi(y_t)} \times \frac{p(x_t^{(i)} \mid u_{t-1}, x_{t-1}^{(i)})}{\pi(x_t^{(i)} \mid u_{t-1})} \right]^{-1} \frac{p(y_t \mid x_t^{(i)}) \, p(x_t^{(i)} \mid x_{t-1}^{(i)}, u_{t-1}) \, Bel(x_{t-1}^{(i)})}{p(y_t \mid d_{0...t-1})}$$

$$= \frac{\pi(y_t) \, \pi(x_t^{(i)} \mid u_{t-1}) \, Bel(x_{t-1}^{(i)})}{p(y_t \mid d_{0...t-1})}$$

$$\propto \pi(x_t^{(i)} \mid u_{t-1}) \, Bel(x_{t-1}^{(i)}), \tag{19.3.15}$$

where $Bel(x_{t-1}^{(i)})$ is calculated using the kd-tree representing this belief density. The only complication arises from the need to calculate $\pi(x_t^{(i)} \mid u_{t-1})$, which depends on both $x_t^{(i)}$ and u_{t-1}. Luckily, in mobile robot localization, $\pi(x_t^{(i)} \mid u_{t-1})$ can safely be assumed to be a constant—even though this assumption is not valid in general. This leads to the following Monte Carlo algorithm.

(i) Sample a pose $x_t^{(i)}$ from a proposal distribution that is proportional to $P(y_t \mid x_t)$.

(ii) For this $x_t^{(i)}$, sample a pose $x_{t-1}^{(i)}$ from a distribution that is proportional to $P(x_t^{(i)} \mid u_{t-1}, x_{t-1})$.

(iii) Set the importance factor to a value proportional to the posterior probability of $x_{t-1}^{(i)}$ under the density tree that represents $Bel(x_{t-1})$.

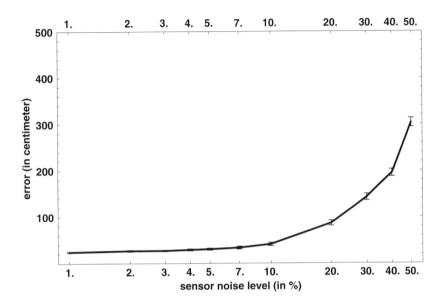

Figure 19.10: Error of MCL with mixture proposal distribution as a function of the sensor noise. Compare this curve with that in Figure 19.9.

19.3.3   The mixture proposal distribution

Neither proposal distribution alone—the original distribution q described in (19.2.4) and the alternative distribution $\bar{q}$ given in (19.3.7)—is satisfactory. The original MCL proposal distribution fails if the perceptual likelihood is too peaked. The alternative proposal distribution, however, only considers the most recent sensor measurement, hence being prone to failure when the sensors err.

A mixture of both proposal distributions gives excellent results:

$$(1 - \phi)q \; + \; \phi\bar{q}. \tag{19.3.16}$$

Here ϕ (with $0 \leq \phi \leq 1$) denotes the *mixing ratio* between regular and dual MCL. Figure 19.10 shows performance results of MCL using this mixture proposal distribution, using a fixed mixing ratio $\phi = 0.1$. All data points are averaged over 1,000 independent experiments. Comparison with Figure 19.9 suggests that this proposal distribution is uniformly superior to regular MCL, and in certain cases reduces the error by more than one order of magnitude.

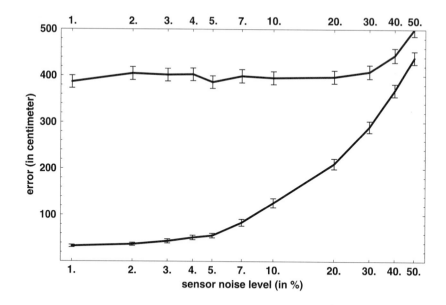

Figure 19.11: Error of MCL (top curve) and hybrid MCL (bottom curve) with 50 samples (instead of 1,000) for each belief state.

These results have been obtained with the third method for calculating importance factors described in the previous section. In our simulation experiments, we found that the second approach yielded less favourable results, but the difference was not significant at the 95% confidence level. As we said above, we have not yet implemented the first approach. In our robot results below, we use the second method for calculating importance factors.

19.3.4   Robot results

A series of experiments was conducted, carried out both in simulation and using physical robot hardware, to elucidate the difference between MCL with the standard and mixture proposal distribution. We found that the modified proposal distribution scales much better to small sample set sizes than conventional MCL. Figure 19.11 plots the error of both MCL algorithms for different error levels, using $m = 50$ samples only. With 50 samples, the computational load is 0.126% on a 500MHz Pentium Computer—meaning that the algorithm is approximately 800 times

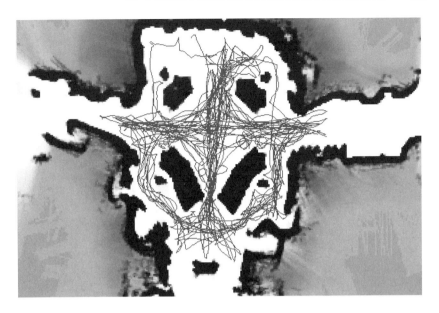

Figure 19.12: Part of the map of the Smithsonian's Museum of National History, and path of the robot.

faster than real-time. While MCL with the standard proposal distribution fails under this circumstances to track the robot's position, our extended approach gives excellent results, which are only slightly worse than those obtained with 1,000 sample.

The following experiment evaluates MCL with mixture proposal distribution in the context of the kidnapped robot problem. This MCL algorithm addresses the issue of recovery from a kidnapping, in that it generates samples that are consistent with momentary sensor readings. Our approach was tested using laser range data recorded during the two-week deployment of the robot Minerva. Figure 19.12 shows part of the map of the museum and the path of the robot used for this experiment. To enforce the kidnapped robot problem, we repeatedly introduced errors into the odometry information. These errors made the robot lose track of its position with probability of 0.01 when advancing one meter. These errors where synthetic; however, they accurately modelled the effect of kidnapping a robot to a random location.

Figure 19.13 shows results for three different approaches. The error is measured by the percentage of time, during which the estimated position deviates more than 2 meters from the reference position. Obviously, using the mixture proposal distribution yields significantly better results, even if the basic proposal distribution is mixed with 5% random samples (as suggested in (Fox, Burgard, Dellaert and Thrun 1999) to alleviate the

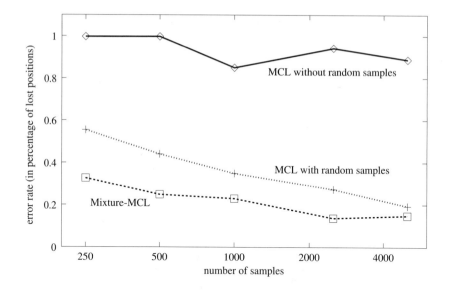

Figure 19.13: Performance of MCL with the conventional (top curve) and mixture proposal distribution (bottom curve), evaluated for the kidnapped robot problem in the Smithsonian museum. The middle curve reflects the performance of MCL with a small number of random samples added in the resampling step, as suggested in (Fox, Burgard, Kruppa and Thrun 2000) as a means to recover from localization failures. The error rate is measured in percentage of time during which the robot lost track of its position.

kidnapped robot problem). The mixture proposal distribution reduces the error rate of localization by as much as 70% more than MCL if the standard proposal distribution is employed; and 32% when compared with the case in which the standard proposal distribution is mixed with a uniform distribution. These results are significant at the 95% confidence level, evaluated over actual robot data.

We also compared MCL with different proposal distributions in the context of visual localization, using only camera imagery obtained with the robot Minerva during public museum hours. The specific image sequence is of extremely poor quality, as people often intentionally covered the camera with their hand and placed dirt on the lens. Figure 19.14 shows the localization error obtained when using vision only (calculated using the localization results from the laser as ground truth). The data cover a period of approximately 4,000 seconds, during which MCL processes a total of 20,740 images. After approximately 630 seconds, a drastic error in the robot's odometry leads to a loss of the position (which is an instance of the kidnapped robot problem). As the two curves in Figure 19.14 illustrate, the regular MCL sampler (dashed curve) is unable to recover from

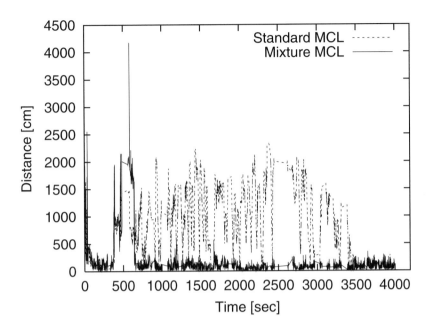

Figure 19.14: MCL with the standard proposal distribution (dashed curve) compared to MCL with the new mixture distribution (solid line). Shown here is the error for a 4,000-second episode of camera-based localization in the Smithsonian museum.

this event, whereas MCL with mixture proposal distribution (solid curve) recovers quickly. These results are not statistically significant because only a single run is considered, but they confirm our findings with laser range finders. Taken together, our results suggest that the mixture distribution drastically increases the robustness of the statistical estimator for mobile robot localization.

19.4   Multi-robot MCL

19.4.1   Basic considerations

The final section of this chapter briefly addresses the multi-robot localization problem. As mentioned in the introduction, multi-robot localization involves a team of robots that simultaneously seek to determine their poses in a known environment. This problem is particularly interesting if robots can sense each other during localization. The ability to detect each other can significantly speed up learning; however, it also creates dependencies in the pose estimates of individual robots that pose major challenges for the design of the estimator.

Formally speaking, the multi-robot localization problem is the problem of estimating a posterior density over a product space $X = \bigotimes_{i=1}^{N} X^i$, where X^i describes the position of the ith robot. Every time a robot senses, it obtains information about the relative poses of all other robots, either by detecting nearby robots, or by not detecting them, which also provides information about other robots' poses. Let $r_t^{i,j}$ denote the random variable that models the detection of robot j by robot i at time t ($i \neq j$). Thus, the variable $r_t^{i,j}$ either takes the value *not detected* or it contains a relative distance and bearing of robot j relative to robot i. The multi-robot localization problem, thus, extends the single robot localization problem by additional observations $r_t^{i,j}$; which are modelled using a time-invariant sensor model $p(x^i \mid r^{i,j}, x^j)$ (time index omitted as above).

The first and most important thing to notice is that the multi-robot localization problem is very difficult and, in fact, we only know of a rudimentary solution which, while exhibiting reasonable performance in practice, possesses clear limitations. What makes this problem so difficult is the fact that the random variables $r_t^{i,j}$ introduce dependencies in the robots' beliefs. Thus, ideally one would like to estimate the posterior over the joint distribution $X = \bigotimes_{i=1}^{N} X^i$. However, such calculations cannot be carried out locally (a desirable property of autonomous robots) and, more importantly, the size of X increases exponentially with the number of robots N. The latter is not much of a problem if all robots are well-localised; however, during global localization large subspaces of X would have to be populated with samples, rendering particle filters inapplicable to this difficult problem.

Our approach ignores these non-trivial interdependencies and instead represents the belief at time t by the product of its marginals

$$Bel(x_t) = \prod_{i=1}^{N} Bel(x_t^i). \tag{19.4.17}$$

Thus, our representation effectively makes a (false) independence assumption; see (Boyen and Koller 1998) for a way to overcome this independence assumption while still avoiding the exponential death of the full product space.

When a robot detects another robot, the observation is folded into a robot's current belief, and the result is used to update the belief of the other robots. More specifically, suppose robot i detects robot j at time t. Then j's belief is updated according to

$$Bel(x_t^j) = \int p(x_t^j \mid x_{t-1}^i, r_{t-1}^{i,j}) \, Bel(x_{t-1}^i) \, dx_{t-1}^i \, Bel(x_{t-1}^j). \tag{19.4.18}$$

The derivation of this formula is analogous to the derivation of Bayes filters, above, and can be found in (Fox et al. 2000). By symmetry, the same detection is be used to constrain the ith robot's position based on the belief of the jth robot.

Clearly, our update rule assumes independence. Hence, when applied more than once it can lead to repetitive use of the same evidence, which will make our robots more confident than warranted by the data. Unfortunately, we are not aware of a good solution to this problem that would maintain the same computational efficiency as our approach. To reduce this effect, our current algorithm only processes positive sightings, that is, the event of *not* seeing another robot has no effect. Additionally, repetitive sightings in short time intervals are ignored. Nevertheless, the occasional transfer from one robot to another can have a substantial effect on each robot's ability to localise.

The implementation of the multi-robot localization algorithm as a distributed particle filter requires some thought. This is because, under our factorial representation, each robot maintains its own, local sample set. When one robot detects another, both sample sets have to be synchronised according to Equation (19.4.18). Note that this equation requires the multiplication of two *densities*, which means that we have to establish a correspondence between the individual samples of robot j and the density representing robot i's belief about the position of robot j. However, both of these densities are themselves represented by sample sets, and with probability one no two samples in these sets are the same. To solve this problem, our approach transforms sample sets into density functions using *density trees* (Koller and Fratkina 1998, Moore, Schneider and Deng 1997, Omohundro 1991). Density trees are continuations of sample sets that approximate the underlying density using a variable-resolution piecewise constant density.

Figure 19.16 shows such a tree, which corresponds to a robot's estimate of another robot's location. Together with Figure 19.15, it shows a map of our testing environment along with a sample set obtained during global localization. The resolution of the tree is a function of the densities of the samples: the more samples there are in a region of space, the more fine-grained the tree representation. The tree enables us to integrate the detection into the sample set of the detected robot using importance sampling for each individual sample $\langle x, w \rangle$:

$$w \quad \longleftarrow \quad \alpha \int p(x_t^j \mid x_{t-1}^i, r_{t-1}^i) \, Bel(x_{t-1}^i) \, dx_{t-1}^i. \quad (19.4.19)$$

19.4.2 Robot results

Multi-robot MCL has been tested using two RWI Pioneer robots, equipped with a camera and a laser range finder for detection (see (Fox et al. 2000)

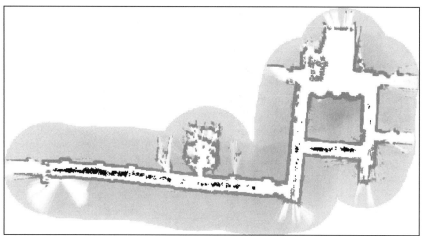

Figure 19.15: Sample set representing a robot's belief.

for details). In particular, our implementation detects robots visually, and uses a laser range finder to determine the relative distance and bearing. The perceptual models $p(x^i \mid r^{i,j}, x^j)$ were estimated from data collected in a separate training phase, where the exact location of each robot was known. After training, the mean error of the distance estimation was 48.26 cm, and the mean angular error was 2.2 degree. Additionally, there was a 6.9% chance of erroneously detecting a robot (false positive).

Figure 19.17 plots the localization error as a function of time, averaged over ten experiments involving physical robots in the environment shown in Figure 19.15. The ability of the robots to detect each other clearly reduces the time required for global localization. Obviously, the overuse of evidence, while theoretically present, appears not to harm the robots' ability to localise themselves. We attribute this finding to the fact that our multi-robot MCL is highly selective when incorporating relative information. These findings were confirmed in systematic simulation experiments (Fox et al. 2000) involving larger groups of robots in a range of different environments.

19.5 Conclusion

This chapter has surveyed a family of particle filters for mobile robot localization, commonly known as *Monte Carlo localization* (MCL). MCL algorithms provide efficient and robust solutions for a range of mobile robot localization problems, such as position tracking, global localization, robot kidnapping, and multi-robot localization.

This chapter investigated three variants of the basic algorithm: the basic MCL algorithm, which has been applied with great success to global lo-

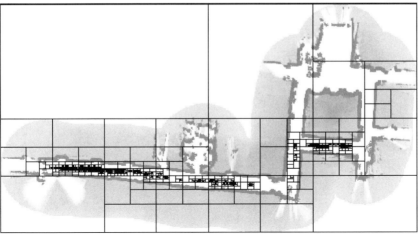

Figure 19.16: Tree representation extracted from the sample set.

calization and tracking, followed by an extension that uses a more sensible proposal distribution, which overcomes certain limitations of MCL such as poor performance when sensors are too accurate, and sub-optimal recovery from robot kidnapping. Finally, the paper proposed an extension to multi-robot localization, where a distributed factorial representation was employed to estimate the joint posterior.

For all these algorithms, we obtained favorable results in practice. In fact, an elaborate experimental comparison with our previous best method, a version of Markov localization that uses fine-grained grid representations (Fox, Burgard and Thrun 1999), showed that MCL is one order of magnitude more efficient and accurate than the grid-based approach.

The derivation of all these algorithms is based on a collection of independence assumptions, ranging from a static world assumption to the assumption that the joint belief space of multiple robots can be factorised into independent components that are updated locally on each robot. Clearly, in most application domains all these independence assumptions are violated. Robot environments, for example, are rarely static. Relaxing those assumptions is a goal of current research, with enormous potential benefits for robot practitioners.

Acknowledgments

We would like to thank Hannes Kruppa for his contributions to the multi-robot MCL approach, as well as the members of CMU's Robot Learning Lab for stimulating discussions that have influenced this research. We are also indebted to Nando de Freitas and Arnaud Doucet for their insightful comments on an earlier draft of this paper, comments which substantially improved the presentation of the material.

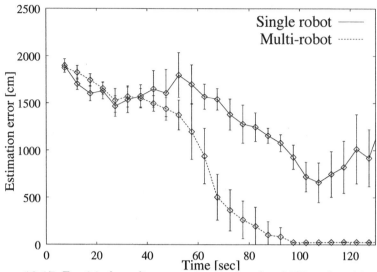

Figure 19.17: Empirical results comparing single robot MCL and multi robot MCL.

This research is sponsored by the National Science Foundation (CAREER grant number IIS-9876136 and regular grant number IIS-9877033), and by DARPA-ATO via TACOM (contract number DAAE07-98-C-L032) and DARPA-ISO via Rome Labs (contract number F30602-98-2-0137), which we gratefully acknowledge. The views and conclusions contained in this document are our own and should not be interpreted as necessarily representing official policies or endorsements, either expressed or implied, of the United States Government or any of the sponsoring institutions.

20

Self-Organizing Time Series Model

Tomoyuki Higuchi

20.1 Introduction

20.1.1 Generalised state-space model

The generalised state-space model (GSSM) that we deal with in this study
is defined by a set of two equations,

$$
\begin{array}{llll}
\textit{system model} & \mathbf{x}_t & = & f(\mathbf{x}_{t-1}, \mathbf{v}_t) & (20.1.1) \\
\textit{observation model} & \mathbf{y}_t & \sim & r(\ \cdot\ |\mathbf{x}_t, \boldsymbol{\theta}_{obs}), & (20.1.2)
\end{array}
$$

where $\mathbf{x}_t$ is an $n_x \times 1$ vector of unobserved sate variables, and $\mathbf{y}_t$ is an n_y
dimensional vector observation. $f : \mathbb{R}^{n_x} \times \mathbb{R}^{n_v} \to \mathbb{R}^{n_x}$ is a given function.
$\{\mathbf{v}_t\}$ is an independent and identically distributed (i.i.d.) random process
with $\mathbf{v}_t \sim q(\mathbf{v}|\boldsymbol{\theta}_{sys})$. r is the conditional distribution of $\mathbf{y}_t$ given $\mathbf{x}_t$. $q(\cdot|\cdot)$
and $r(\cdot|\cdot)$ are, in general, non-Gaussian densities specified by the unknown
parameter vectors, $\boldsymbol{\theta}_{sys}$ and $\boldsymbol{\theta}_{obs}$, respectively. In this study, we set $\boldsymbol{\theta} = [\boldsymbol{\theta}'_{sys}, \boldsymbol{\theta}'_{obs}]'$. The initial state $\mathbf{x}_0$ is distributed according to the density
$p_0(\mathbf{x})$.

The GSSM includes the nonlinear non-Gaussian state-space model (Kitagawa 1987, Kitagawa 1991, Tanizaki 1993, Gordon et al. 1993) as a special
case

$$
\begin{array}{llll}
\textit{system model} & \mathbf{x}_t & = & f(\mathbf{x}_{t-1}, \mathbf{v}_t) & (20.1.3) \\
\textit{observation model} & \mathbf{y}_t & = & h(\mathbf{x}_t, \mathbf{e}_t), & (20.1.4)
\end{array}
$$

where $\{\mathbf{e}_t\}$ is an i.i.d. random process that follows $\mathbf{e}_t \sim r(\mathbf{e}|\boldsymbol{\theta}_{obs})$. $h : \mathbb{R}^{n_x} \times \mathbb{R}^{n_e} \to \mathbb{R}^{n_y}$ is a given function. The simplest version of the nonlinear non-
Gaussian state-space model is the well-known linear Gaussian state-space
model (Anderson and Moore 1979), which is given by

$$
\begin{array}{lll}
\textit{system model} & \mathbf{x}_t = F\mathbf{x}_{t-1} + G\mathbf{v}_t, & \mathbf{v}_t \sim N(0, Q) \quad (20.1.5) \\
\textit{observation model} & \mathbf{y}_t = H\mathbf{x}_t + \mathbf{e}_t, & \mathbf{e}_t \sim N(0, R), \quad (20.1.6)
\end{array}
$$

where F, G, and H are $n_x \times n_x$, $n_x \times n_v$, and $n_y \times n_x$ matrices, respectively. Q and R are the covariance matrices of $\mathbf{v}$ and $\mathbf{e}$.

The GSSM also includes a model frequently used for the analysis of discrete valued time series (West et al. 1985, Kitagawa 1987, Kitagawa 1991, Kitagawa and Gersch 1996, West and Harrison 1997, Higuchi 1999)

$$\textit{system model} \quad \mathbf{x}_t = F\mathbf{x}_{t-1} + G\mathbf{v}_t, \quad \mathbf{v}_t \sim N(0, Q) \qquad (20.1.7)$$

$$\textit{observation model} \quad \mathbf{y}_t \sim \exp(\alpha_t' \mathbf{y}_t - b(\alpha_t) + c(\mathbf{y}_t)), \quad \alpha_t = H\mathbf{x}_t, \quad (20.1.8)$$

where F, G, and H are properly defined matrices, Q is the covariance matrix, and $b(\cdot)$ and $c(\cdot)$ are properly defined functions. This type of distribution, called the exponential family of distributions, can cover a broad class of distributions frequently used in statistical analysis, such as the Poisson distribution and the binomial distribution. The model specified by (20.1.7) and (20.1.8) is called the Dynamic Generalised Linear Model (DGLM) (West et al. 1985, West and Harrison 1997). This type of model has the favourable property that the filtering distribution is uni-model and close to symmetric. By using this property and properly treating the difference from the normal distribution, computationally efficient and precise estimators have been proposed (West et al. 1985, Fahrmeir 1992, Schnatter 1992, Frühwirth-Schnatter 1994, Durbin and Koopman 1997).

20.1.2 Monte Carlo filter

The GSSM is very flexible and suitable for a wide variety of time series. It allows one to derive recursive formulas for estimating the state vector given observations (Kitagawa 1987). However, we have to solve difficult integrations over a state-space that increases enormously with respect to the state dimension n_x (Carlin et al. 1992, Fahrmeir 1992, Frühwirth-Schnatter 1994, Schnatter 1992). A Monte Carlo method for filtering and smoothing, called Monte Carlo Filter (MCF) has been proposed (Kitagawa 1996) to overcome this numerical problem. The bootstrap filter (Gordon et al. 1993, Doucet, Barat and Duvaut 1995) is a similar algorithm.

To review MCF, suppose that $p(\mathbf{x}_t|\mathbf{y}_{1:t-1})$ and $p(\mathbf{x}_t|\mathbf{y}_{1:t})$ are approximated by the N realizations

$$X_{t|t-1} \equiv \{\mathbf{x}_{t|t-1}^{(i)}|i = 1, \dots, N\} \quad \text{and} \qquad (20.1.9)$$

$$X_{t|t} \equiv \{\mathbf{x}_{t|t}^{(i)}|i = 1, \dots, N\}, \qquad (20.1.10)$$

respectively. It can be shown that these realisations (particles) can be generated recursively by the following algorithm.

(i) For $i = 1, \ldots, N$ generate the n_x-dimensional random number $\mathbf{x}_{0|0}^{(i)} \sim p_0(\mathbf{x})$.

(ii) Repeat the following steps for $t = 1, \ldots, T$. For (a), (b) and (c), repeat N times independently for $i = 1, \ldots, N$.

   (a) Generate the following n_v-dimensional system random vector $\mathbf{v}_t^{(i)} \sim q(\mathbf{v}|\boldsymbol{\theta}_{sys})$.

   (b) Compute $\mathbf{x}_{t|t-1}^{(i)} = f(\mathbf{x}_{t-1|t-1}^{(i)}, \mathbf{v}_t^{(i)})$.

   (c) Compute $\widetilde{w}_t^{(i)} = r(\mathbf{y}_t|\mathbf{x}_t = \mathbf{x}_{t|t-1}^{(i)}, \boldsymbol{\theta}_{obs})$.

   (d) Obtain $X_{t|t}$ by sampling with replacement from $X_{t|t-1}$ with sampling probabilities proportional to $\widetilde{w}_t^{(1)}, \cdots, \widetilde{w}_t^{(N)}$.

A significant advantage of this Monte Carlo filter is that it can be applied to almost any type of high dimensional nonlinear and non-Gaussian state-space models. This filtering algorithm can be extended to handle smoothing by storing the past particles and resampling the vector of particles $(\mathbf{x}_{t|t-1}^{(i)}, \mathbf{x}_{t-1|t-1}^{(i)}, \cdots, \mathbf{x}_{t-L|t-L}^{(i)})$ rather than the single particle $\mathbf{x}_{t|t-1}^{(i)}$. Other smoothing algorithms appear in (Clapp and Godsill 1999).

In this Monte Carlo filter the likelihood is approximated by

$$p(\mathbf{y}_t|\mathbf{y}_{1:t-1}, \boldsymbol{\theta}) = \int r(\mathbf{y}_t|\mathbf{x}_t, \boldsymbol{\theta}_{obs})p(\mathbf{x}_t|\mathbf{y}_{1:t-1}, \boldsymbol{\theta})d\mathbf{x}_t \cong \frac{1}{N}\sum_{i=1}^{N} \widetilde{w}_t^{(i)}. \quad (20.1.11)$$

Hence, the log-likelihood in the MCF, $l^{\mathrm{MCF}}(\boldsymbol{\theta})$, is given by

$$\log p(\mathbf{y}_{1:T}|\boldsymbol{\theta}) = \sum_{t=1}^{T} \log p(\mathbf{y}_t|\mathbf{y}_{1:t-1}, \boldsymbol{\theta})$$

$$\cong \sum_{t=1}^{T} \log \left(\sum_{i=1}^{N} \widetilde{w}_t^{(i)} \right) - T \log N = l^{\mathrm{MCF}}(\boldsymbol{\theta}). \quad (20.1.12)$$

The log-likelihood calculated by MCF, l^{MCF}, suffers from an error inherent in the Monte Carlo approximation, which gives rise to difficulties in parameter estimation based on maximising the log-likelihood. This error somewhat reduces the applicability of MCF in practical data analysis. A different procedure for estimating a parameter vector is therefore required, unless a very large number of particles is used or an average of the approximated log-likelihoods is computed by the massive parallel use of many Monte Carlo filters.

The organization of this paper is as follows. In Section 20.2, we explain two types of self-organizing time series model able to diminish the difficulties associated with parameter estimation. One is a Genetic algorithm filter based on a strong parallelism of the Monte Carlo filter to a Genetic algorithm (GA). The other is a self-organizing state-space model that is an extension of the GSSM. In Section 20.3, we discuss a resampling procedure that plays an important role in the filtering of the MCF. It is shown in Section 20.2 that the resampling is identical to the selection procedure in GA, which can be implemented in a variety of ways. The typical three selection methods are realised in the filtering of the MCF, then performance is assessed by applying them to a time series smoothing problem. In Section 20.4, we show an application of the self-organizing state-space model. In this study, we focus on estimation of a time-dependent frequency that changes rapidly over a relatively short interval. In particular, we deal with an analysis of the small count time series for which a time-dependent mean exhibits a wavy behaviour whose frequency evolves in time. For this case, conventional methods of identifying a time-varying frequency are incapable of estimating it precisely. A benefit of considering the self-organizing state-space model for this problem is illustrated by applying it to actual data. Finally, we offer some concluding remarks in Section 20.5.

20.2   Self–organizing time series model

20.2.1   Genetic algorithm filter

Parallelism to genetic algorithm

The genetic algorithm (GA) is the adaptive search procedure inspired by evolution, and one of the most popular among the population-based optimization techniques (Holland 1975, Goldberg 1989, Davis 1991, Whitley 1994). GA is characterised by keeping the N candidates for optimal solution at each iteration composed of three steps: crossover, mutation, and selection (or reproduction). Each candidate at the tth iteration is called a string at the tth generation. It has been pointed out that there exists a close relationship between MCF and GA, and that an essential structure involved in MCF is quite similar to that in GA (Higuchi 1997). The correspondence between MCF's and GA's terminology is summarised in Table 20.1. Here we give a brief explanation of the strong parallelism of MCF to GA.

$\mathbf{x}_{t|t-1}^{(i)}$ and $\mathbf{x}_{t|t}^{(i)}$, which are identified as particles in the MCF, are considered the strings in GA. The filtering procedure of MCF exactly corresponds to the selection of GA by regarding $r(\cdot|\boldsymbol{\theta}_{obs})/N$ as the evaluation function in GA. Accordingly, in the GA the importance weight $w_t^{(i)} = r(\mathbf{y}_t|\mathbf{x}_{t|t-1}^{(i)}, \boldsymbol{\theta}_{obs})/ \sum_{j=1}^{N} r(\mathbf{y}_t|\mathbf{x}_{t|t-1}^{(j)}, \boldsymbol{\theta}_{obs})$ can be called the fitness of the string $\mathbf{x}_{t|t-1}^{(i)}$. It should be noticed that, whereas $r(\mathbf{y}_t|\mathbf{x}_{t|t-1}^{(i)}, \boldsymbol{\theta}_{obs})/N$ is obviously dependent on time due to the presence of data $\mathbf{y}_t$, the evaluation function in GA without any modification such as a scaling (stretching) (e.g., (Goldberg 1989)) is independent of t. $\mathbf{x}_{t-1|t-1}^{(i)}$ and $\mathbf{x}_{t|t-1}^{(i)}$ are regarded as the parent and offspring, respectively, because $\mathbf{x}_{t-1|t-1}^{(i)}$ creates $\mathbf{x}_{t|t-1}^{(i)}$ which then undergoes selection. The maximum likelihood principle is interpreted as a rule to choose the model that maximises the family fitness under a circumstance of given data $\mathbf{y}_{1:T}$.

Monte Carlo filter (MCF)	$\Longleftrightarrow$	Genetic algorithm (GA)
t: time	—	generation
$\mathbf{x}_{t|t-1}^{(i)}$ and $\mathbf{x}_{t|t}^{(i)}$: particle	—	string
$f(\cdot, \mathbf{v}_t = 0)$	—	genetic drift
system noise	—	crossover, mutation
filtering	—	selection
$r(\mathbf{y}_t|\mathbf{x}_t, \boldsymbol{\theta}_{obs})/N$	—	evaluation function
$w_t^{(i)}$	—	fitness
$p(\mathbf{y}_t|\mathbf{y}_{1:t-1}, \boldsymbol{\theta})$	—	population fitness
$p(\mathbf{y}_{1:T}|\boldsymbol{\theta})$	—	family fitness
$\mathbf{y}_t$: observation	—	environment
$\mathbf{y}_{1:T}$: data	—	history of environment

Table 20.1. Analogy between the MCF and the GA

The crossover and mutation in GA may be interpreted from the Bayesian point of view. In particular, a fluctuation caused by a mutation is exactly regarded as the system noise stemming from a non–Gaussian probability density function. When we represent a model using a binary code and use the canonical GA, a closed form of this density function can be obtained easily. In the MCF there is an apparent movement of the population at each step that results from the systematic behaviour of the particle driven by a nonlinear function f with $\mathbf{v}_t \equiv 0$ in the system model. The GA penalises any drift in model space that is independent of any performance of optimization, this being called genetic drift.

GA filter

Based on the close relationship between MCF and GA, a new algorithm, in which the prediction step in MCF is replaced by the mutation and crossover operators in GA, has been proposed to avoid an estimation of $\boldsymbol{\theta}_{sys}$ involved in the system model by maximising $l^{\mathrm{MCF}}(\boldsymbol{\theta})$. While within MCF the stochastic behaviour of the particles is determined by an outer force independent of the particles, the crossover yields strong interactions between particles.

This means that a description of the resulting stochastic behaviour cannot take a simple form as in MCF. There may occur a gradual change in inertial force driving the population through the crossover. The time series model involving the genetic operators is, therefore, no longer the same model as one expressed in terms of GSSM. We call the GSSM involving these genetic operators GA filter. Examples of applications of the GA filter including a seasonal adjustment can be found (Higuchi 1997).

20.2.2   Self-organizing state-space model

As explained in Section 20.1.2, the parameter estimation of $\boldsymbol{\theta}$ by maximising the likelihood happens to be rendered impractical by the sampling error of $l^{\mathrm{MCF}}(\boldsymbol{\theta})$. To remedy such difficulty, a self-organizing state-space model has been proposed (Kitagawa 1998). The general form of the self-organizing state-space model is obtained by augmenting the state vector $\mathbf{x}_t$ with the parameter vector $\boldsymbol{\theta}$ as $\mathbf{z}_t = [\mathbf{x}_t', \boldsymbol{\theta}_t']'$. The state-space model for this augmented state vector $\mathbf{z}_t$ is given by

$$\text{system model} \quad \mathbf{z}_t \;=\; F^*(\mathbf{z}_{t-1}, \mathbf{v}_t) \qquad (20.2.13)$$

$$\text{observation model} \quad \mathbf{y}_t \;\sim\; r(\cdot|\mathbf{z}_t), \qquad (20.2.14)$$

where the nonlinear function F^* is defined by

$$F^*(\mathbf{z}, \mathbf{v}) = \begin{bmatrix} f(\mathbf{x}, \mathbf{v}) \\ \boldsymbol{\theta} \end{bmatrix}. \qquad (20.2.15)$$

f and r are given in (20.1.1) and (20.1.2), respectively. Another approach to the problem of learning about time-varying state vectors and fixed model parameters is discussed by Liu and West (2001: this volume).

Here we consider Bayesian estimation of $\boldsymbol{\theta}$, instead of using the maximum likelihood method. Once Monte Carlo smoothing is carried out for this self-organizing state-space model, the posterior distribution of $\boldsymbol{\theta}$, $p(\boldsymbol{\theta}|\mathbf{y}_{1:T})$, is simply approximated by $p(\boldsymbol{\theta}|\mathbf{y}_{1:T}) \cong \int p(\mathbf{z}_T|\mathbf{y}_{1:T})d\mathbf{x}_T$. Similarly, the posterior distribution of the state vector $\mathbf{x}_t$ is defined by a marginal posterior distribution given by $p(\mathbf{x}_t|\mathbf{y}_{1:T}) = \int p(\mathbf{z}_t|\mathbf{y}_{1:T})d\boldsymbol{\theta}_t$.

When we need to define an optimal $\boldsymbol{\theta}$, several methods are available based on $p(\boldsymbol{\theta}|\mathbf{y}_{1:T})$. In this study, the optimal $\boldsymbol{\theta}$ is determined by calculating a median of the marginal distribution of $p(\boldsymbol{\theta}|\mathbf{y}_{1:T})$ for each variable. The successful use of this parameter estimation procedure relies on the setting of its initial distribution, $p_0(\boldsymbol{\theta})$. Usually, it is recommended that we adopt a uniform distribution over a range covering the possible values of an optimal $\boldsymbol{\theta}$. For situations in which there is prior information, the distribution can be tailored.

With time-varying parameters, $\boldsymbol{\theta} = \boldsymbol{\theta}_t$, a model that accounts for time-changes of the parameter $\boldsymbol{\theta}_t$ is necessary. For example, we may use the random walk model $\boldsymbol{\theta}_t = \boldsymbol{\theta}_{t-1} + \boldsymbol{\varepsilon}_t$, where $\boldsymbol{\varepsilon}_t$ is a white noise with density function $\phi(\varepsilon|\boldsymbol{\xi})$. In this case, the nonlinear function F^* is defined by

$$F^*(\mathbf{z}_{t-1}, \mathbf{u}_t) = \begin{bmatrix} f(\mathbf{x}_{t-1}, \mathbf{v}_t) \\ \boldsymbol{\theta}_{t-1} + \varepsilon_t \end{bmatrix}, \tag{20.2.16}$$

where the system noise is defined by $\mathbf{u}_t = [\mathbf{v}'_t, \varepsilon'_t]'$. Therefore, there is no formal difference between the generalised state space model and the self-organizing state-space model. The essential difference is that the self-organizing state-space model contains the parameter in the state vector. In the self-organizing state space models, the parameter vector, $\boldsymbol{\theta}_t$, is automatically determined as the estimate of the state vector. The parameter to be estimated by the maximum likelihood method, $\boldsymbol{\xi}$, is expected to be a one or two dimensional vector.

20.3 Resampling scheme for filtering

20.3.1 Selection scheme

As mentioned above, the resampling form appearing in the filtering of the MCF is identical to the one in the selection procedure of the GA. Consequently, we can choose the resampling scheme best suited for estimating the log-likelihood from several different ways of carrying out selection (Goldberg 1989, Whitley 1994). Here we compare three typical selection methods from the ones classified by Brindle (Brindle 1981):

(i) Stochastic sampling with replacement (Roulette wheel selection).

(ii) Remainder stochastic sampling with replacement.

(iii) Deterministic sampling.

To explain them, we imagine a roulette wheel on which each particle $\mathbf{x}_{t|t-1}^{(i)}$ is represented by a space that proportionally corresponds to $w_t^{(i)}$. The stochastic sampling is very simple because we choose N particles by repeatedly spinning this roulette wheel. The remainder stochastic sampling begins by calculating the integer and fractional portions of $N w_t^{(i)}$, $k_t^{(i)}$ and $\xi_t^{(i)}$, respectively. We denote the sum of $k_t^{(i)}$ by $N_{int} = \sum_{i=1}^{N} k_t^{(i)}$. Secondly, we get $k_t^{(i)}$ copies of $\mathbf{x}_{t|t-1}^{(i)}$ for each particle. Finally, we choose the $N_{frac} = N - N_{int}$ particles by repeatedly spinning the "remainder roulette wheel" on which each particle has a space in proportion to $\xi_t^{(i)}$. This sampling method can be realised in an efficient way by adopting the resampling method known as stochastic universal sampling (Baker 1987). Deterministic sampling does not employ this roulette wheel, but instead chooses the N_{frac} particles in a deterministic way in that we order $\xi_t^{(i)}$ and take the N_{frac} largest particles.

20.3.2   Comparison of performance: simulation study

We discuss three resampling methods by considering a problem of time series smoothing. Figure 20.1(a) shows the artificially generated data ob-

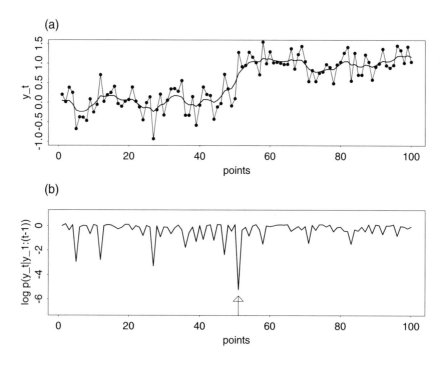

Figure 20.1. (a) Data and smoothed data by KFS. (b) log-likelihood $p(y_t|y_{1:t-1})$.

tained by adding i.i.d. Gaussian white noise with a variance of 0.1 to the step function with a jump of 1 at $t = 51$, where $T = 100$. To smooth the given data, we consider the simplest model defined by

$$\begin{aligned} \textit{system model} \quad \mu_t &= \mu_{t-1} + v_t, \quad v_t \sim N(0, \tau^2) \\ \textit{observation model} \quad y_t &= \mu_t + e_t, \quad e_t \sim N(0, \sigma^2), \end{aligned} \qquad (20.3.17)$$

where μ_t is a trend component at time t, thereby setting the one-dimensional state vector $\mathbf{x}_t = [\mu_t]$, and $\mathbf{v}_t$ is the one-dimensional white noise sequence. For this case, the Kalman filter yields a "true" value of the likelihood. The true value of the log-likelihood based on the Kalman filter is specified by l^{true} henceforth, from which we omit $\boldsymbol{\theta}$ for simplicity. In this case, the maximum likelihood estimate $\boldsymbol{\theta}^* = [\sigma^{2*}, \tau^{2*}]'$ can be easily obtained because the likelihood function has a very simple form as a function of $\lambda^2 = \tau^2/\sigma^2$, the ratio of the variance of the system noise to that of the observation noise. The solid curve is obtained by applying the fixed interval smoother with $\lambda^{2*} = 0.25$.

Figure 20.1(b) shows the log–likelihood of $\mathbf{y}_t$ for the model with λ^{2*}, $\log p(\mathbf{y}_t|\ \mathbf{y}_{1:t-1}, \boldsymbol{\theta}^*)$, which is denoted by l_t^{true}. It is seen that l_t^{true} takes a minimum value at $t = 51$, indicated by the arrow, where there is a discontinuous change in the given trend. An approximation by MCF to l_t^{true} is denoted by $l_t^{\text{MCF},[j]}$, where $[j]$ indicates the jth trial.

(a) Deterministic   (b) Roulette Wheel   (c) Stochastic Universal

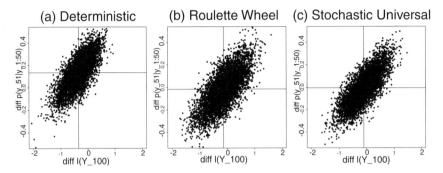

Figure 20.2. Plots of $N_p = 5,000$ trials of $\Delta l^{[j]}$ versus $\Delta l_{51}^{[j]}$ for (a) deterministic, (b) roulette wheel, and (c) stochastic universal sampling methods.

We concentrate on the discrepancy of $l^{\text{MCF},[j]}$ from l^{true}, $\Delta l^{[j]}$, as well as $\Delta l_{51}^{[j]} = l_{51}^{\text{MCF},[j]} - l_{51}^{\text{true}}$. We pay special attention to $\Delta l_{51}^{[j]}$ because a sparse distribution of $\mathbf{x}_{51|50}^{(i)}$ around y_{51} could result in poor approximation of $l_{51}^{\text{MCF},[j]}$ to l_{51}^{true}. Figure 20.2(a) shows a distribution of $(\Delta l^{[j]}, \Delta l_{51}^{[j]})$ for a case in which the deterministic sampling is adopted to implement the resampling in the filtering procedure of MCF. The number of trials is $N_p = 5,000$ and the particle number is fixed $N = 1,000$ for each trial. The vertical and horizontal lines indicate the mean values of $\Delta l^{[j]}$ and $\Delta l_{51}^{[j]}$, $\overline{\Delta l^{[j]}}$ and $\overline{\Delta l_{51}^{[j]}}$, respectively. We call $\overline{\Delta l^{[j]}}$ the bias. An apparent positive correlation seen in this figure (correlation coefficient $\rho = 0.737$) supports our conjecture that the poor approximation of the MCF to l_{51}^{true} is significant in controlling $\Delta l^{[j]}$. The straight line obtained by the least squares fit to these plots has a slope of 0.265. From the positive value of slope smaller than 1, it suggests that there remains a systematic error in approximation of $l_{t'}^{\text{MCF}}$ for $51 < t'$. Namely, $X_{51|50}$ with negative $\Delta l_{51}^{[j]}$ may lead $X_{52|51}$ to give negative $\Delta l_{52}^{[j]}$ through $X_{51|51}$.

Figures 20.2(b) and (c) demonstrate a distribution of $(\Delta l^{[j]}, \Delta l_{51}^{[j]})$ for the roulette wheel selection and for the stochastic universal sampling. As in Figure 20.2(a), we can see the same tendency that $\Delta l^{[j]}$ depends mainly on $\Delta l_{51}^{[j]}$. Whereas the sign of the bias is common to three cases $(\overline{\Delta l^{[j]}} < 0)$, its magnitude for the deterministic sampling is apparently larger than that of the others. It is clearly seen that plots of $(\Delta l^{[j]}, \Delta l_{51}^{[j]})$ for the roulette wheel selection show a broader distribution compared with plots for the stochastic universal sampling. Table 20.2 summarises the bias, standard deviation of $\Delta l^{[j]}$, correlation coefficient, and slope of the line using a least squares fit, for each resampling scheme. Based on these statistics, we suggest that the stochastic universal sampling is the most suitable for the resampling in the filtering procedure of the MCF.

Sampling scheme	$\overline{\Delta l^{[j]}}$	s.d. of $\Delta l^{[j]}$	ρ	slope
Deterministic	-0.344	0.392	0.737	0.265
Roulette wheel	-0.127	0.496	0.723	0.234
Stochastic universal	-0.095	0.455	0.776	0.249

Table 20.2. Comparison of selection schemes: $N_p = 5,000$ trials.

A reduction of standard deviation can be achieved using the resampling suitable for the filtering, but Figure 20.2 demonstrates that there still remains a larger variance un-favourable to choosing an optimal parameter value. Of course, we can make a drastic reduction of variance by increasing the particle number N by as much as the computer memory space permits. Even an increase in N for the predictive distribution $p(\mathbf{x}_t|\mathbf{y}_{1:t-1})$ would lead to a significant reduction of variance (Gordon et al. 1993). There are interesting resampling schemes designed to reduce this variance (Crisan et al. 1999, Crisan and Grunwald 1999).

20.4   Application

20.4.1   Time-varying frequency wave in small count data

In this chapter, we analyse the nearly cyclic (wave-like) signal that can be characterised by repeating a similar pattern successively with a gradual change in its wave form in the time domain. Most of the attention in actual data analysis is usually paid to an estimation of a time–dependence of its frequency (or, equivalently, wavelength) and/or amplitude among quantities specifying its time–dependent structure. In particular, when the wave–like signal seems to consist of a sinusoid corrupted by observation noise, a good estimation of time–dependence of frequency may lead to further insights into a system that generates the cyclic behaviour. For example, investigating the time-varying structure of a power spectrum with a few peaks plays an important role in destructive tests of airplane wings. Here, eigen-oscillations of the wing show a different behaviour as the wind velocity increases towards the point at which the wing could be destroyed

(Ikoma 1996). In this study, we deal with these monotonic sinusoidal waves, whose frequency changes as time elapses.

In general, an attempt to estimate the time-varying frequency of the wave is achieved by searching a significant peak over an instantaneous power spectra (IPS). This can be done in several ways. One way to calculate IPS is to fit a time-varying coefficient AR model (TVCAR) to data, giving an IPS that is based on the relation between a stationary AR process and the theoretical spectrum of the process (Kitagawa and Gersch 1985, Gersch and Kitagawa 1988). One reason to develop a method for an analysis of the time-varying frequency wave is that, when an identification of the time-varying frequency wave is difficult owing to significant observation noise, TVCAR is likely to provide no peak in IPS. This occurs even when a signature of the time-varying frequency wave can be obviously detected by visual inspection of plots of data. Another reason stems from a demand that we treat the time–varying frequency as the state variable (Ikoma 1996) because a linear increase or decrease in the time–varying frequency is very often encountered in data analysis.

Our attention here is focused on time series involving small counts, which are frequently found in many fields, such as the biomedical statistics (for example, monthly numbers of polio incidences (Zeger 1988)) and astrophysics. Recent progress in instruments for measuring faint light from star brightness, such as CCD cameras, has led to a drastic change in the obtained light intensity data, which is usually allowed to take a continuous value, to a photon count number. When we want to investigate the behaviour of targets with a small time scale, the sampling time, within which photons arriving at the instrument are counted, is shortened and gives us the small count data time series. Thus it is important to model small count data in astrophysics.

We deal with small count data in terms of a statistical view point because the time series involving relatively larger counts can be well analyzed by means of a time series model without regard to the fact that observed data follow a discrete distribution, and does not require more sophisticated models in practice. Actually recent Bayesian approaches to time series modelling have paid considerable attention to the analysis of small count data. (West et al. 1985, Fahrmeir 1992, Frühwirth-Schnatter 1994, Chan and Ledolter 1995, Kashiwagi and Yanagimoto 1992, Higuchi 1999). A practical example of the applications of the Monte Carlo filter to various nonlinear problems can be found in (Carpenter et al. 1999b).

20.4.2 Self-organizing state-space model for time-varying frequency wave

We assume that the observation is generated from a Poisson distribution with time–varying mean λ_t: $\mathbf{y}_t \sim Poisson(\lambda_t)$. $\log \lambda_t$ is, in this study, decomposed into two factors: a trend component μ_t, and time–varying frequency wave component s_t, $\log \lambda_t = \mu_t + s_t$. In other words, we deal with the non–stationary Poisson model in which the time–varying mean is expressed in the multiplicative form $\lambda_t = \exp(\mu_t)\exp(s_t)$.

We assume that μ_t follows a first order trend model given by $\mu_t = \mu_{t-1} + v_{t,\mu}$, $v_{t,\mu} \sim C(0, \tau_\mu^2)$, where $C(0, \tau_\mu^2)$ represents the Cauchy distribution concentrated around 0 with dispersion parameter τ_μ^2. As mentioned above, we want to estimate the time–varying frequency. Accordingly, we denote the time-varying frequency at time t by f_t and treat it as the state variable. The time–varying frequency wave component, s_t, with a time-varying frequency f_t, is described by the second order difference equation

$$s_t = 2 \cos (2\pi f_t) s_{t-1} - s_{t-2}. \qquad (20.4.18)$$

Since f_t is allowed to take a value between 0 to $1/2$, a nonlinear transformation based on a sigmoid function, $f_t = 0.5/(1 + \exp(-\beta_t))$ $(-\infty < \beta_t < +\infty)$, is performed and β_t is used as a state variable instead of f_t. We assume that β_t follows a first order trend model given by $\beta_t = \beta_{t-1} + v_{t,\beta}$, $v_{t,\beta} \sim N(0, \tau_\beta^2)$. The parameter vector, $[\log_{10} \tau_\mu^2, \log_{10} \tau_\beta^2]'$, is also included in the state vector to optimise it by means of the self-organizing state-space model. These models can be represented by the state-space model form

system model

$$
\mathbf{x}_t =
\begin{bmatrix}
\mu_t \\
s_t \\
s_{t-1} \\
\beta_t \\
\log_{10} \tau_{t,\mu}^2 \\
\log_{10} \tau_{t,\beta}^2
\end{bmatrix}
=
\begin{bmatrix}
\mu_{t-1} + v_{t,\mu} \\
2 \cos \left(\frac{\pi}{1+\exp(-\beta_{t-1})} \right) s_{t-1} - s_{t-2} \\
s_{t-1} \\
\beta_{t-1} + v_{t,\beta} \\
\log_{10} \tau_{t-1,\mu}^2 \\
\log_{10} \tau_{t-1,\beta}^2
\end{bmatrix},
\qquad (20.4.19)
$$

where $v_{t,\mu} \sim C(0, \tau_{t,\mu}^2)$ and $v_{t,\beta} \sim N(0, \tau_{t,\beta}^2)$.

20.4.3   Results

To demonstrate the applicability of our method to the small count time-varying frequency wave data, a simple test is conducted on a simulated data set. f_t is given beforehand to show a linear increase and decrease during a relatively short interval. The dotted line in Figure 20.3(b) indicates this dependence of f_t on time. The next step to generate simulated data is to define an amplitude and phase in s_t. A phase is determined by integrating given f_t. The time-varying amplitude is taken to present a gradual increase and decrease, as indicated by the dotted line in Figure 20.3(a). The given fixed value of μ_t ($\exp(\mu_t) = 20$) together with given s_t produces λ_t by $\lambda_t = \exp(\mu_t + s_t)$. Lastly, $\mathbf{y}_t$ is obtained by drawing a random number from $Poisson(\lambda_t)$. The total data number is $T = 100$. The solid straight line in Figure 20.3(a) shows simulated data obtained by the aforementioned procedure.

To illustrate the good performance of our model in terms of accuracy of estimation of the time-varying frequency, the TVCAR procedure is first applied to compute a spectral peak frequency within a frequency range of $f = 0 \sim 0.2$ from IPS. The TVCAR with varying M_{AR} in a range of $M_{AR} = 2, 4, \ldots, 10$ for $N_{obs} = 1, 2, 5, 10, 20$ is applied in this study, where M_{AR} and N_{obs} are the numbers of the AR coefficients and of successive data points for which the time-varying AR coefficients are set to be constant (for detail, see (Kitagawa and Gersch 1985, Gersch and Kitagawa 1988)). In fact, any combination of (N_{obs}, M_{AR}) fails to define a peak frequency because of the absence of a peak in IPS.

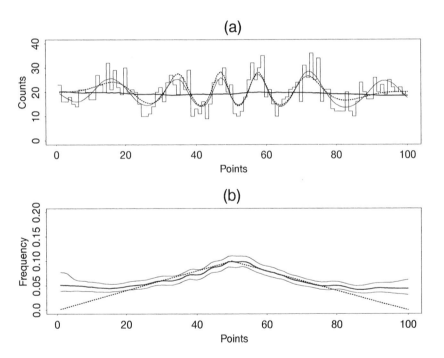

Figure 20.3. Simulation results. (a) Given data and estimated $\lambda_t = \exp(\mu_t + s_t)$. The thick line shows the estimated $\exp(\mu_t)$. (b) Estimated f_t and its confidence interval $\pm\sigma$.

We apply the self-organizing state-space model explained in Section 20.4.2 to this data set. For the initial state $\mathbf{x}_0$, it is assumed, for example, that $\mu_0 \sim N(2.56, 1)$, $s_{-1} \sim N(0, 0.5)$, $s_0 \sim (0, 0.5)$, $\beta_0 \sim U([-2.60, -0.81])$, $\log_{10} \tau_{0,\mu}^2 = U([-7, -4])$, $\log_{10} \tau_{0,\beta}^2 = U([-3, 0])$, where

$U([\ ,\ ])$ denotes the uniform distribution. The initial range of β_0 corresponds to the range of f_0 between 0.0346 and 0.154, which is inferred from a visual inspection of the given data. The self-organizing state-space model estimates the values of $\log_{10} \tau^2_{T,\mu}$ and $\log_{10} \tau^2_{T,\beta}$: $\log_{10} \hat{\tau}^2_{T,\mu} = -5.36(-5.99, -4.48)$, $\log_{10} \hat{\tau}^2_{T,\beta} = -1.83(-2.14, -1.48)$. The two values in the parentheses indicate the confidence interval $\pm\sigma$.

The obtained optimal values for $\hat{\tau}^2_{\mu}$ and $\hat{\tau}^2_{\beta}$ allow us to conduct the MCF for the generalised state-space model, which is simply defined by setting $\tau^2_{\mu} = 10^{-5.36}$ and $\tau^2_{\beta} = 10^{-1.83}$ and taking $\mathbf{x}_t = [\mu_t, s_t, s_{t-1}, \beta_t]'$ as the state vector. The marginal posterior distribution of f_t, $p(f_t|\mathbf{y}_{1:T})$, is determined through $p(\mathbf{x}_t|\mathbf{y}_{1:T})$, and its median is shown by a thick line in Figure 20.3(b), as well as, its confidence interval $\pm\sigma$ (two thin lines). Of course, the self-organizing state space model can produce the posterior distribution of f_t, $p(f_t|\mathbf{y}_{1:T})$, which can be obtained from the marginal posterior distribution of β_t, $p(\beta_t|\mathbf{y}_{1:T})$, but we are interested in examining how much of a difference arises between the two estimates. In fact, the estimate obtained by the generalised state-space model is only shown in this study, because the difference between the two estimates is invisible.

The estimated $\lambda_t = \exp(\mu_t + s_t)$ and $\exp(\mu_t)$ are in Figure 20.3(a) denoted by the thin and thick lines. Although a good agreement of the estimate with the given true curve of λ_t can be seen for the interval between $t = 20$ and 80, a significant discrepancy is found for the outside intervals, where the given amplitude is smaller than that for the middle interval. Poor agreement of the estimated λ_t to the given one for those intervals turns out to be made apparent by a comparison of the estimated f_t to the given one (dotted straight line) in Figure 20.3(b). Such poor performance is explained by two reasons. One is due to the fact that the given amplitude for those intervals is smaller than that for the middle interval. The system model (20.4.18) itself becomes close to the second order trend model as f_t decreases. Therefore, an estimation of the time-varying frequency is intrinsically difficult by means of the system model (20.4.18) for a case with $\cos(2\pi f_t) \sim 1$.

Here we show one example of an application of the above mentioned procedure to an actual data set. The data set to be used for an illustration of our method is the signature of the spiral density waves obtained by the Voyager photopolarimeter (PPS) during a stellar occultation of Saturn's rings. The original Voyager PPS data is also a small count time series. The discussion of instruments and operation of Voyager PPS data can be seen in (Horn, Showalter and Russell 1996). In fact, we use the simulated data, shown in Figure 20.4(a), which is generated to resemble the typical density wave, named Prometheus 7:6. The reason we have not used the original data is that the data set we have at hand has already undergone the detrending procedure, and hence is no longer count data. The simulated data is obtained by taking an average over every three points of the original

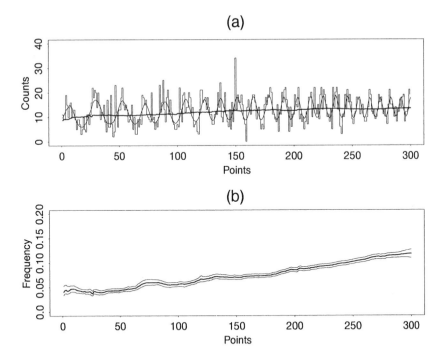

Figure 20.4. Analysis of the Voyager PPS data. (a) Given data and estimated $\lambda_t = \exp(\mu_t + s_t)$. The thick line shows the estimated $\exp(\mu_t)$. (b) Estimated f_t and its confidence interval $\pm\sigma$. The dotted line shows the given f_t.

data and quantizing it in an attempt to generate count data. The behaviour displayed in this figure is one of the strongest density waves and the signature of the time-varying frequency wave is clearly visible in the data. The Voyager data revealed a wide variety of ring structure's of Saturn; waves and wakes. The understanding of this structure plays an important role in studying the generation of planetary ring structure. A detailed explanation of the spiral density waves is found in (Horn et al. 1996).

It has been reported that a theory of the spiral density wave suggests a linear increase in f_t (Horn et al. 1996). We apply the self-organizing state-space model as used in the simulated data set. The solid and thick lines in Figure 20.4(a) show the estimated λ_t and $\exp(\mu_t)$, respectively, which are based on the marginal posterior distribution $p(\mathbf{x}_t|\mathbf{y}_{1:T})$. Figure 20.4(b) shows an estimated f_t and its confidence interval $\pm\sigma$. The tendency of a linear increase in f_t, which is expected from a theoretical model based approach, is clearly identified.

20.5   Conclusions

In this study, we introduced two types of self-organizing time series models: the genetic algorithm filter and the self-organizing state-space model. The former has been developed on the basis of the strong parallelism between the genetic algorithm and the MCF. The latter is an extension of the GSSM which is formed by augmenting the state vector with a parameter vector. Both methods remedy the difficulty of parameter estimation by maximising the likelihood associated with the sampling error. The self-organizing time series model makes the model selection procedure practical and facilitates the automatic processing of time series data. Obviously, the required amount of computation to realise the self-organizing time series model increases considerably. However, considering the rapid progress in computing capacity, thereby obviating the need for human calculation, the future of time series modelling seems assured.

21

Sampling in Factored Dynamic Systems

Daphne Koller
Uri Lerner

21.1 Introduction

In many real-world situations, we are interested in monitoring the evolution of a complex situation over time. For example, we may be monitoring a patient's vital signs in an intensive care unit (Dagum and Galper 1995), analyzing a complex freeway traffic scene with the goal of controlling a moving vehicle (Huang, Koller, Malik, Ogasawara, Rao, Russell and Weber 1994), localizing a robot in a complex environment (Fox, Burgard and Thrun 1999) (see also Murphy and Russell (2001: this volume)), or tracking motion of non-rigid objects in a cluttered visual scene (Isard and Blake 1998a). We treat such systems as being in one of a possible set of states, where the state changes over time. We model the states as changing at discrete time intervals, so that $\mathbf{x}_t$ is the state of the system at time t. In most systems, we model the system states as having some internal structure: the system state is typically represented by some vector of *variables* $\mathbf{X} = (X_1, \ldots, X_n)$, where each X_i takes on values in some space $Dom[X_i]$. The possible states $\mathbf{x}$ are assignments of values to the variables $\mathbf{X}$. In a traffic surveillance application, the state might contain variables such as the vehicle position, its velocity, the weather, and more.

In most cases, the system dynamics are unpredictable. We model such systems as a stochastic dynamic system, where we encode a probability distribution over all possible trajectories. By making the assumptions that the system is Markovian and time-invariant, we can represent this distribution using two components: a *prior probability distribution* π_0, which represents the distribution over the initial state of the system; and a *transition model*

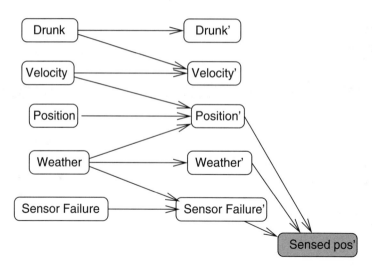

Figure 21.1. A highly simplified DBN for traffic surveillance.

$p(\mathbf{X}' \mid \mathbf{X})$, which represents the probability that the system will transition from state $\mathbf{X}$ to state $\mathbf{X}'$ in a single step.[1]

Dynamic Bayesian networks (DBNs) (Dean and Kanazawa 1989) are a representation language for stochastic dynamic systems that allows us to represent complex systems in a compact and natural way. DBNs exploit certain assumptions about locality of interaction that are very common in real-world systems. For example, in a freeway traffic surveillance domain, we might reasonably assume that a vehicle's position at time t depends only on its position at time $t-1$ and its velocity at time $t-1$. The vehicle's forward velocity at time t might depend on the road's condition, and on the clearance to the vehicle in front at time $t-1$. Like a Bayesian network (BN) (Pearl 1988), a DBN utilises a graphical notation to represent the direct dependencies between the variables in the model. The transition model of the system — $p(\mathbf{X}' \mid \mathbf{X})$ — is represented in a factorised way using a network fragment such as that shown in Figure 21.1, appropriately annotated with probabilities.

DBNs provide a convenient and compact representation that allows us to model even very large and complex systems, including very large discrete systems and systems that include both discrete and continuous variables. Hidden Markov models are a very simple special case of DBNs, as are linear Gaussian systems (Kalman 1960, Bar-Shalom and Fortman 1987). DBNs have been used for a variety of applications, including freeway surveil-

---

[1]We use boldface to distinguish between a set of random variables $\mathbf{X}$ and particular variable X. We use upper-case letters X (or $\mathbf{X}$) to denote random variables, such as the (unknown) state of the process at a given time, and lower-case letters x (or $\mathbf{x}$) to denote particular instantiations to these variables, such as a particular state of the system.

lance (Huang et al. 1994), complex factories (Jensen, Kjærulff, Olesen and Pedersen 1989), robotics (Nicholson and Brady 1994), medical monitoring (Dagum and Galper 1995), speech recognition (Zweig and Russell 1998), and more.

Of course, modelling complex systems is only the first step; we also need to use these models for inference. A common goal in dynamic systems is *tracking* (also called *filtering* or *monitoring*). At each time point, we get an observation $\mathbf{y}_t$, which is a partial observation of the state of the system. We wish to keep track of the state of the system, based on a partial observation $\mathbf{y}_t$. Because our observation does not completely determine the true state, the best we can do is to keep track of $p(\mathbf{X}_t \mid \mathbf{y}_1, \dots, \mathbf{y}_t)$. We call this distribution the *belief state at time t*.

In static Bayesian networks (that is, those that represent a probability distribution over a fixed finite set of variables), the model structure can be exploited to allow effective inference. Unfortunately, we can show that the same phenomenon does not apply to dynamic systems. Even in highly structured DBNs, the belief state is entirely unstructured (except in degenerate cases) (Boyen and Koller 1998). Hence, we can only carry out exact tracking by maintaining our belief state as a full joint distribution over the possible states. In discrete DBNs, the cost is exponential in the number of state variables. In continuous and hybrid DBNs, the situation is even worse. In most systems (Kalman filters being the only notable exception), the complexity of the belief state grows unboundedly over time, preventing any closed-form representation.

There have been several approximate inference techniques proposed for DBNs (Ghahramani and Jordan 1997, Boyen and Koller 1998), but they are designed primarily for discrete domains (or, at best, a very limited class of hybrid domains). Sequential Monte Carlo methods (Handschin and Mayne 1969, Gordon et al. 1993), originally proposed for DBNs under the name *survival of the fittest* (Kanazawa, Koller and Russell 1995) are currently the only approach that allows us to perform tracking in general purpose hybrid DBN models.

The remainder of the paper is structured as follows. In Section 21.2, we review the representation and semantics of DBN models. In Section 21.3, we show how particle filtering can be applied to general DBN models. In Section 21.4, we present experimental results for particle filtering in two large-scale DBN models, one discrete and the other hybrid. We conclude in Section 21.5 with some discussion and directions for future work.

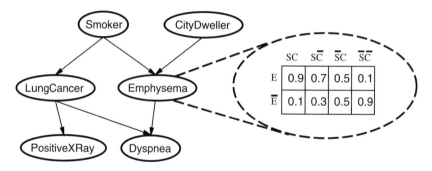

Figure 21.2. A simple Bayesian network, with a CPD for one of the nodes, specifying the conditional probability of each possible value of the variable *Emphysema*, given each possible combination of values of the parent nodes *Smoker* and *CoalMiner*.

21.2   Structured probabilistic models

21.2.1   Bayesian networks

Bayesian networks are a representation for probability distributions over complex domains. We consider probability spaces defined as the set of possible assignments to some set of random variables $X_1, \ldots, X_n$, each of which has a domain $Dom[X_i]$ of possible values. For example, we might have that the domain of the discrete variable *Weather* is {*clear, rain, snow*}. The domain of the variable *Velocity* can be $\mathbb{R}$, or it can be discretised into some appropriate partition. The goal is to represent a joint probability distribution over these variables.

Unfortunately, in discrete networks, an explicit description of the joint distribution requires a number of parameters that is exponential in n, the number of variables. Bayesian networks (Pearl 1988) derive their power from the ability to represent conditional independencies among variables, which allows them to take advantage of the "locality" of causal influences. Intuitively, a variable is independent of its indirect causal influences given its direct causal influences. In Figure 21.2, for example, the outcome of the X-ray does not depend on whether the patient is a smoker because we know that the patient has lung cancer. If each variable has at most k other variables that directly influence it, then the total number of required parameters is linear in n and exponential in k. This enables the compact representation of fairly large problems.

Formally, a Bayesian network is defined by a directed acyclic graph together with a local probabilistic model for each node. There is a node in the graph for each random variable $X_1, \ldots, X_n$. The edges in the graph denote direct dependency of a variable X_i on its parents $Parents(X_i)$. More formally, the graphical structure encodes a set of conditional independence assumptions: each node X_i is conditionally independent of its non-descendants given its parents.

The local probabilistic model for a node in the graph encodes the quantitative relationship between the node and its parents. It consists of a *conditional probability distribution* (CPD) $p(X_i \mid Parents(X_i))$, which specifies the distribution over the values of X_i given any possible assignment of values to its parents.

The qualitative independence assumptions implied by the network structure, combined with the CPDs associated with the nodes, are enough to specify a full joint distribution through the following equation, known as the *chain rule for Bayesian networks*:

$$p(X_1 \ldots X_n) = \prod_{i=1}^{n} p(X_i \mid Parents(X_i)). \qquad (21.2.1)$$

21.2.2 *Hybrid networks*

While the CPD is often represented as a simple table, as in Figure 21.2, it can also be represented implicitly as a parameterised function from values of $Parents(X_i)$ to probability distributions over X_i. With implicitly represented CPDs, a network can include continuous as well as discrete variables.

As an example, we might represent a conditional distribution of a continuous node with a discrete parent as a conditional Gaussian. Formally, for a variable X with parent set $\mathbf{U}$, we can specify a CPD as follows: for every value $\mathbf{u} \in Dom[\mathbf{U}]$, the CPD has a parameter $\mu_{\mathbf{u}}$ and $\sigma_{\mathbf{u}}^2$; the conditional distribution is then

$$p(X \mid \mathbf{u}) = N(\mu_{\mathbf{u}}; \sigma_{\mathbf{u}}^2).$$

For a continuous node X with continuous parents $Y_1, \ldots, Y_k$, we might choose to model X as a linear Gaussian of its parents. In this case, we would specify the CPD by means of parameters $a_0, \ldots, a_k$ and σ^2, and have

$$p(X \mid y_1, \ldots, y_k) = N(a_0 + a_1 y_1 + \cdots + a_k y_k; \sigma^2).$$

The linear Gaussian model is a very natural model; it is also very useful because there is an equivalence between BNs with linear Gaussian models and multivariate Gaussian distributions: every continuous BN all of whose CPDs are linear Gaussian defines a multivariate Gaussian distribution; conversely, every multivariate Gaussian distribution can be represented as a continuous BN with linear Gaussian CPDs (Shachter and Kenley 1989).

We can extend a linear Gaussian CPD to allow the continuous node to have discrete parents. The idea is the same as the one used above. If a node X has continuous parents $\mathbf{Y}$ and discrete parents $\mathbf{U}$, we could have a set of different linear Gaussians for X as a function of $\mathbf{Y}$ — one linear Gaussian (with its own parameters) for each value $\mathbf{u}$ of $\mathbf{U}$. This dependence model is called a *conditional linear Gaussian*. A BN whose discrete nodes have only discrete parents and whose continuous nodes have conditional linear Gaussian CPDs induces a joint distribution that defines a multivariate Gaussian distribution over the continuous variables for every instantiation to the discrete network variables.

We can also accommodate the case of a discrete child with a continuous parent. One reasonable model is based on the *softmax* density (Bishop 1995). Intuitively, the *softmax CPD* (Koller, Lerner and Angelov 1999) defines a set of R regions (for some parameter R of our choice). The regions are defined by a set of R linear functions over the continuous variables. A region is characterised as that part of the space where one particular linear function is bigger than all the others. Each region is also associated with some distribution over the values of the discrete child; this distribution is the one used for the variable within this region. The actual CPD is a continuous version of this region-based idea, allowing for smooth transitions between the distributions in neighboring regions of the space.

More precisely, let U be a discrete variable, with continuous parents $\mathbf{Y} = \{Y_1, \ldots, Y_k\}$. Assume that U has m possible values, $\{u_1, u_2, \ldots, u_m\}$. Each of the R regions is defined by means of two vectors of parameters $\alpha^r, \mathbf{p}^r$. The vector α^r is a vector of weights $\alpha_0^r, \alpha_1^r, \ldots, \alpha_k^r$ specifying the linear function associated with the region. The vector $\mathbf{p}^i = \{p_1^r, \ldots, p_m^r\}$ is the probability distribution over $u_1, \ldots, u_m$ associated with the region (i.e., $\sum_{j=1}^m p_j^r = 1$). The CPD is now defined as

$$p(U = u_j \mid \mathbf{Y}) \;=\; \sum_{r=1}^{R} w^r p_j^r$$

$$w^r \;=\; \frac{\exp(\alpha_0^r + \sum_{i=1}^k \alpha_i^r Y_i)}{\sum_{q=1}^R \exp(\alpha_0^q + \sum_{i=1}^k \alpha_i^q Y_i)}.$$

In other words, the distribution is a weighted average of the region distributions, where the weight of each region depends exponentially on how big the value of its defining linear function is, relative to the rest. The choice of α^i determines both the regions and the slope of the transitions between them; the choice of $\mathbf{p}^i$ determines the distribution defining each region. Like the conditional linear Gaussian CPD, if U also has discrete parents, our softmax CPD will have a separate component for each instantiation of the discrete parents.

The power to choose the number of regions R to be as large as we wish is the key to the rich expressive power of the generalised softmax CPD. Figure 21.3 demonstrates this expressivity. In Figure 21.3(a), we present an example CPD for a binary variable with $R = 4$ regions. In Figure 21.3(b), we show how this CPD can be used to represent a simple classifier. Here, U is a sensor with three values: low, medium and high. The probability of each

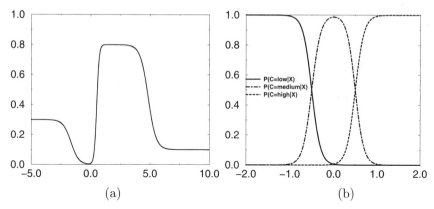

Figure 21.3. Expressive power of a softmax CPD.

of these values depends on the value of the continuous parent X. Notice that we can easily accommodate a variety of noise models for the sensor: we can make it less reliable in borderline situations by making the transitions between regions more moderate; we can make it inherently more noisy by having the probabilities of the different values in each of the regions be farther away from 0 and 1.

We have presented a small set of CPD models that are useful across a wide range of applications. Of course, there is an unlimited range of representations. For example, the model of (Fox, Burgard and Thrun 1999) for robot localization uses a linear Gaussian CPD to represent the distribution of the sonar sensor value given the position of a known obstacle, and a very different parameterised CPD (an exponential distribution) to represent the distribution if there is no known obstacle. In principle, we can use parametric representation for a function of the appropriate type; we are limited only by how rich we choose to make our language for specifying CPDs. (The BUGS system (Gilks et al. 1996) provides a very rich language for this purpose.)

21.2.3   Dynamic Bayesian networks

A *dynamic Bayesian network* uses the same ideas as a BN for modelling dynamic systems. As stated in the introduction, we typically assume that the process is *Markovian* so that the probability of being in state $\mathbf{x}'$ at time $t + 1$ depends only on the state of the world at time t. We also typically assume that the process is time invariant, so that $p(\mathbf{X}_{t+1} \mid \mathbf{X}_t)$ does not depend on t. Thus, it can be specified using a single *transition model* which holds for all time points.

In a DBN, the state of the process at any given time is specified in terms of a set of state variables $X_1, \dots, X_n$. We specify both the prior distribution π_0 and the transition model $p_\to(\mathbf{X'} \mid \mathbf{X})$ using a compact graphical representation.

Definition 21.2.1.: A *2-time-slice Bayesian network (2-TBN)* for a process whose state variables are $\mathbf{X}$ is a Bayesian network *fragment* defined as follows. The graph structure is a directed acyclic graph whose nodes are $\mathbf{X} \cup \mathbf{X'}$, where nodes in $\mathbf{X}$ have no parents. The nodes in $\mathbf{X'}$ are annotated with CPDs, $p(X' \mid Parents(X'))$. The 2-TBN represents a *conditional distribution* $p_\to(\mathbf{X'} \mid \mathbf{X})$ via the *chain rule for 2-TBNs*:

$$p_\to(\mathbf{X'} \mid \mathbf{X}) = \prod_{i=1}^{n} p(X_i' \mid Parents(X_i')).$$

∎

A 2-TBN for a simple vehicle surveillance problem was shown in Figure 21.1. Notice that a 2-TBN represents a conditional distribution; hence, it only specifies a probabilistic model for the variables $\mathbf{X'}$: for each variable X_i', it specifies a set of parents $Parents(X_i')$, which can be variables either in the previous time slice or the current one, and a CPD. The 2-TBN model allows us to encode stronger independence assumptions than the standard Markov property: a variable X_i' depends only on its immediate parents, i.e., it is conditionally independent of all other variables in the past given values for its parents.

Definition 21.2.2.: A *dynamic Bayesian network* is a pair $\langle \mathcal{B}_0, \mathcal{B}_\to \rangle$, where $\mathcal{B}_0$ is a Bayesian network over $\mathbf{X}_0$, representing the initial distribution π_0 over the initial state, and $\mathcal{B}_\to$ is a 2-TBN for the process representing the transition model $p_\to$. For any T, the distribution over $\mathbf{X}_0, \dots, \mathbf{X}_T$ is defined as:

$$p(\mathbf{X}_0 = \mathbf{x}_0, \dots, \mathbf{X}_T = \mathbf{x}_T) = \pi_0(\mathbf{x}_0) \cdot \prod_{t=1}^{T} p_\to(\mathbf{X'} = \mathbf{x}_t \mid \mathbf{X} = \mathbf{x}_{t-1}).$$

∎

In most dynamic systems, we have an *observation model* in addition to a transition model. The observation model tells us the probability that we observe $\mathbf{y}$ given that the current state is $\mathbf{x}$. In a DBN, it is convenient to represent the observation model as part of the 2TBN. We simply define a subset $\mathbf{Y} \subset \mathbf{X}$ to be the variables that are observable. We define $\mathbf{y}_t$ to be the instantiation of values for these variables observed at time t.

DBNs provide a representation language for dynamic systems that allows us to represent complex distributions in a factored way and that accommodates both discrete and continuous variables. It accommodates, as special cases, the important classes of *hidden Markov models* (HMMs) (Rabiner

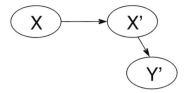

Figure 21.4. The representation of an HMM or Kalman filter as a DBN.

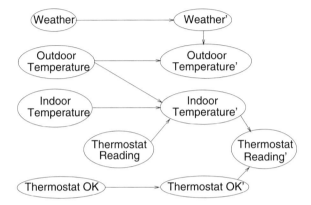

Figure 21.5. The thermostat network.

and Juang 1986) and *Kalman filters*. Both can be viewed as a very simple DBN with two variables X and Y, where the variable X represents the current (hidden) state of the system, and the variable Y the observation. The structure of both models is shown in Figure 21.4. In the case of an HMM, X is discrete, and typically so is Y. In the case of a Kalman filter, X and Y are usually vector-valued continuous variables, and the CPDs are linear Gaussians. Other, more expressive, variants of an HMM (such as the *factorial HMM* of (Ghahramani and Jordan 1997)) can also be viewed as a DBN (Smyth, Heckerman and Jordan 1997). [2]

As an example for a 2-TBN using various types of CPDs we consider the thermostat network, shown in Figure 21.5. This 2-TBN describes a simple process with two phases. A thermostat measures the (continuous) temperature inside the house. It can be operational, or it can malfunction. If it is working properly it gives a reading which is a softmax classification of the room temperature into one of three readings: *Cold*, *Normal* and *Warm*. If the thermostat malfunctions, it gives one of the possible readings with

---

[2]We note that a discrete DBN can also be viewed as an HMM by agglomerating all of the variables into a single state variable, and defining the transition model for that single variable using $p_{\rightarrow}$. However, the size of the resulting representation is exponential in the number of state variables, and it loses all of the regularity encoded in the DBN structure.

a uniform distribution. The room temperature is a conditional Gaussian that depends on the previous room temperature, the outside temperature and the thermostat. If the thermostat shows *Cold*, it turns on a heating device and the room temperature is likely to go up; if the thermostat shows *Warm*, it turns on the air conditioning. Finally the outside temperature is a conditional Gaussian that depends on the weather.

21.3   Particle filtering for DBNs

Particle filtering is a general purpose Monte Carlo scheme for tracking in dynamic systems. It maintains the belief state at time t as a set of *particles* $\mathbf{x}_t^{(1)}, \dots, \mathbf{x}_t^{(N_t)}$, where each $\mathbf{x}_t^{(i)}$ is a full instantiation of $\mathbf{X}_t$. To generate a new (time $t + 1$) particle $\mathbf{x}_{t+1}^{(j)}$ from a (time t) particle $\mathbf{x}_t^{(i)}$, we simply sample it. Ideally, we want to sample it from the distribution $p(\mathbf{X}' \mid \mathbf{X} = \mathbf{x}_t^{(i)}, \mathbf{Y}' = \mathbf{y}_{t+1})$. However, this distribution is rarely one that we can compute efficiently. Hence, we use importance sampling: we sample from one distribution, and use an importance weight to make up for the difference.

The sampling algorithm for generating a new particle and an importance weight in a DBN is roughly as follows. The algorithm traverses the network in an order consistent with the partial order implied by the edges. It picks a value for every variable X' in turn. A value for an unobserved variable is sampled according to the appropriate conditional distribution, given the values already selected for its parents. A variable X' whose value is observed as evidence is not sampled; rather it is simply instantiated to its observed value x', and we update the weight of the new particle to reflect the probability that X *would* have been sampled as x', given the current instantiation of its parents. It is straightforward to verity that, while our algorithm only generates points $\mathbf{x}'$ that are consistent with our observations, the expected weight for any such $\mathbf{x}'$ (i.e., the probability with which it is generated times its weight when it is) is exactly its probability $p(\mathbf{x}')$. Thus, our weighted samples are an unbiased estimator of the (unnormalised) distribution over the states and the observations. The formal version of the algorithm is shown in Figure 21.6; we assume that $X_1', \dots, X_n'$ is an ordering which is consistent with the edges in the 2TBN, in that if $i < j$ then X_j' cannot be an ancestor of X_i'.

Notice that there are several distinctions between this algorithm and a naive application of particle filtering. In most applications of particle filtering, it is customary to sample $\mathbf{z}'$ from the prior distribution $p(\mathbf{Z}' \mid \mathbf{x}^{(i)})$, where $\mathbf{Z}' = \mathbf{X}' - \mathbf{Y}'$, and then use $p(\mathbf{y}' \mid \mathbf{z})$ as an importance weight. In DBNs, we can do better than that provided the evidence is not always a leaf in the network. For example, in the Thermostat network of Figure 21.5, the weather is an observable that is not a leaf. The DBN sampling algorithm

```
SampleParticle(x, y)
      /*x is the state at time t
        y is the observation at time t + 1 */
   w := 1
   Set x' to be empty
   For i := 1 to n
     Let u be the assignment to Parents(X'_i) in x, x'
     If X'_i is not in Y'
        Sample x'_i from p(X'_i | Parents(X'_i) = u)
     Else
        Set x'_i to be X_i's observed value in y
        Set w := w · p(x'_i | Parents(X'_i) = u)
   Return(x', w)
```

Figure 21.6. Algorithm for sampling a particle in a DBN.

also samples a particle without explicitly generating the entire transition model $p(\mathbf{X}' \mid \mathbf{X})$.

This algorithm is called with a particle $\mathbf{x}_t^{(i)}$ at time t, and it generates a new particle $\mathbf{x}_{t+1}^{(j)}$ for time $t + 1$, and an associated weight w. It can be used with any of the many variants of particle filtering. For example, we can use it several times to generate several samples from the same particle. If we maintain particles that have weights, we can define the weight of the new particle $\mathbf{x}'$ to be the weight of the old particle $\mathbf{x}$ times w.

At a high level, the variant of particle filtering used in our experiments is standard *bootstrap sampling*: At time t, we start out with an approximate belief state represented as a set of weighted particles $\mathbf{x}_t^{(1)}, \ldots, \mathbf{x}_t^{(N_t)}$. We bootstrap sample from the belief state: we randomly select the desired number of particles from among $\mathbf{x}_t^{(1)}, \ldots, \mathbf{x}_t^{(N_t)}$, where each $\mathbf{x}_t^{(i)}$ is selected according to its relative weight. The result is a set of *unweighted* time-t particles: we have already compensated for the relative weights of the original time-t particles by sampling each one according to its weight. Each of the new time-t particles is propagated forward using the SampleParticle procedure. The resulting time-$t+1$ particles, with their weights, induce the next approximate belief state, which is the starting point for the next phase.

We modify the most basic implementation of this scheme by using a simple form of density estimation. Our time t samples are necessarily very sparse. In a discrete setting, this implies that many entries in the joint distribution representing the approximate belief state will be estimated to have probability zero, even when their probability in the true belief state is positive. This type of behavior can cause significant problems, as samples at time $t+1$ are only generated based on our existing samples at time t. If the process is not very stochastic, i.e., if there are parts of the state-space that only transition to other parts with very low probability, parts of the space that are not represented in our particles will not be explored. Unfortunately, the parts of the space that are not represented may have non-negligible probability; our sampling process may simply have missed them earlier, or they may be the results of trajectories that appeared unlikely in earlier time slices because of misleading evidence.

We can use simple density estimation to address this concern. In the continuous case, we simply place a small Gaussian kernel around each of our particles. In the discrete case, we are trying to estimate a density over the joint probability space of $\mathbf{X}_{t+1}$. We smooth the distribution as follows. Let $w_{t+1}^{(1)}, \ldots, w_{t+1}^{(N_{t+1})}$ be the weights of the particles generated at time $t+1$. Let $W_{t+1} = \sum_i w_{t+1}^{(i)}$. We add a certain fraction of W to each entry in the joint consistent with the evidence. In other words, our new approximate belief state — the one from which bootstrap particles will be generated — is defined as follows: for each $\mathbf{x}'$ consistent with $\mathbf{y}_{t+1}$, we define

$$\pi_t(\mathbf{x}') = \frac{1}{Z} \sum_{i \,:\, \mathbf{x}_{t+1}^{(i)} = \mathbf{x}'} w_{t+1}^{(i)} + \alpha,$$

where α is some small fraction of W_t and Z is a normalizing constant. This effect is equivalent to doing Bayesian density estimation over the belief state, using a simple Dirichlet prior (DeGroot 1970). We mention that, even though $\pi_t(\mathbf{x}') > 0$ for every $\mathbf{x}'$, we need only represent explicitly those states that have materialised in our sampling algorithm. All the others have the same weight α. We can easily sample from this distribution using a simple two-phase process. Let $W = \sum_i w_{t+1}^{(i)}$; then $Z = W + \alpha M$, where M is the total number of states consistent with $\mathbf{y}_{t+1}$. [3] With probability $(M\alpha)/Z$, we select some state consistent with the evidence; each such state is then selected uniformly at random. With probability W/Z, we use our standard bootstrap sampling process from the set of time-t samples. Thus, the cost of using the smoothed belief state is no higher than that of maintaining our original sparse set of samples.

The use of density estimation, and the associated smoothing effect, serves to spread out some of the probability mass over unobserved states,

---

[3] M can easily be computed when all CPDs do not contain zero entries. If CPDs do contain zero entries, the computation becomes more difficult. In general, deciding which instantiations are consistent with the evidence is an NP-hard problem, so we either have to resign ourselves to having a small positive probability even for some states that are inconsistent with our current evidence, or we must resort to some other smoothing scheme.

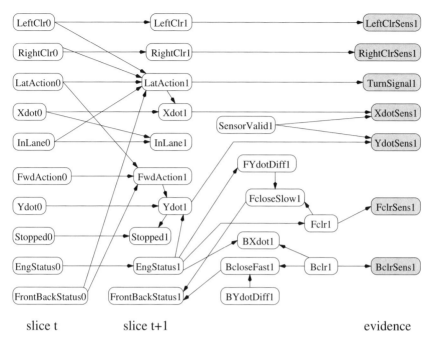

Figure 21.7. The BAT network.

increasing the amount of exploration done for unfamiliar regions of the space.

21.4   Experimental results

We have tested particle filtering on two large DBNs. The first network is the discrete BAT network (Huang et al. 1994), used for monitoring freeway traffic (see Figure 21.7). The network describes a car driving on a freeway, by modelling the car velocity and acceleration both on the X axis (lane changes) and on the Y axis (forward movement). The network also models relations between the car and nearby cars on the freeway. The belief state of this network includes 10 variables; altogether, there are 66,528 possible states in the belief state.

Our first experiment with this network was to see how well we can track it using particle filtering. To do so, we randomly generated an observation sequence from the network. At each time step, we compared the results of running particle filtering with the correct distributions computed using exact inference (which is feasible for a model of this size).

We used two different error metrics: KL-distance (Cover and Thomas 1991) and L_1-norm. For each of these metrics, we computed the error on the entire joint distribution of the belief state and also the average error on the marginals of all the variables in the belief state. Obviously, estimating the marginals is much easier than estimating the full joint distribution, therefore the average error over the marginals is much smaller.

Figure 21.8 shows the error as a function of the number of samples. In both cases we report the average error over the entire run. The results were averaged over 10 runs. Figure 21.8(a) shows the KL error on the entire belief state. Figure 21.8(b) shows the average L1-norm error over the marginals of the random variables. As expected, the error drops sharply initially and then the improvements become smaller and smaller. Notice that the drop-off occurs at around 8000 samples, substantially less than the number of states in the belief state.

Figure 21.9 shows how the error behaves over the entire run. We present the KL-error, averaged over 10 runs with identical evidence (L1-norm has a similar behavior) with 5000 samples per time step. It is interesting to see that the error changes dramatically over the sequence. Generally speaking, the spikes in the error correspond to unlikely evidence, in which case the samples become less reliable. This phenomenon suggests that it may be advantageous to use a more adaptive sampling scheme, e.g., by using a different number of samples at different time steps.

To further investigate the quality of the approximation, we focused on one of the variables in the belief state: the forward velocity (*Ydot1* in Figure 21.7). In this case we used 5000 samples per time step. The results are given in Figure 21.10. As can be seen, particle filtering gives a very good estimate of the true distribution over this variable. We point out that the error for any particular value of the variable is tiny; the error of 0.06 suggested by Figure 21.9 is the result of summing over all possible values of the variable.

The second network we experimented with is a hybrid network that models a process which is commonly used as a benchmark in the fault diagnostics community (Mosterman and Biswas 1999). The physical system is shown in figure Figure 21.11 and the DBN is shown in Figure 21.12.

The system contains two water tanks which are connected by a pipe. Each of the tanks has a pipe from which water flows out of the tank. The first tank also has a constant flow of water going into it. We get noisy measurements of the flows in each one of the three pipes.

There are three possible types of failures in the system:

- **Measurement failures:** Usually measurements are quite reliable, but in case of a measurement failure, the measurement becomes extremely noisy. In this case, the value of the measurement has almost nothing to do with the actual flow; thus, to track the system correctly, it is necessary to identify these faulty measurements and ignore them.

- **Pipe bursts:** A pipe can suddenly change its resistance to some unknown value.

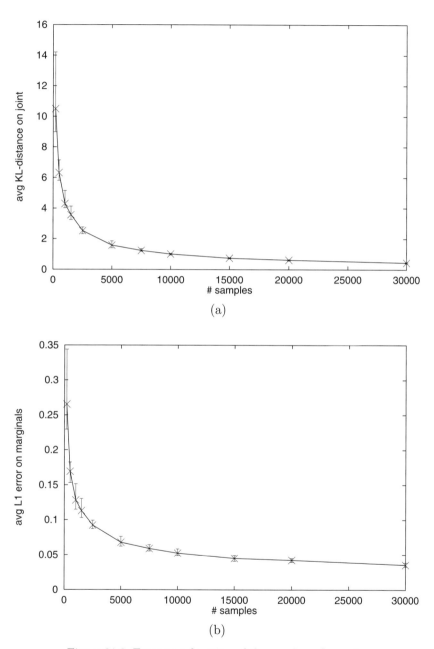

Figure 21.8. Error as a function of the number of samples

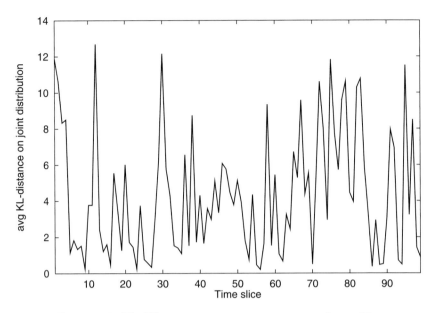

Figure 21.9. The KL error over a sequence, averaged over 10 runs

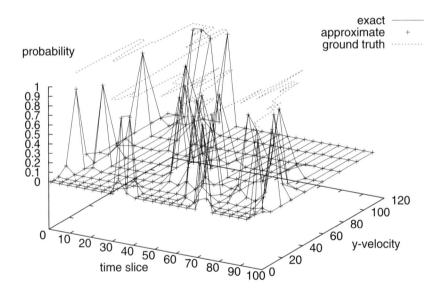

Figure 21.10. The exact and approximate belief states for the *Ydot1* variable, and the ground truth used to generate the trajectory.

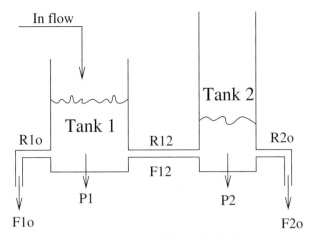

Figure 21.11. The Two Water Tanks System.

- **Drifts:** The resistance of the pipe can drift, which gradually increases or decreases the pipe's resistance.

The probability of a measurement failure is 0.01. The probability of a burst is 0.003 and the probability of a drift (positive or negative) is also 0.003. Once a drift begins, it persists with probability 0.991.

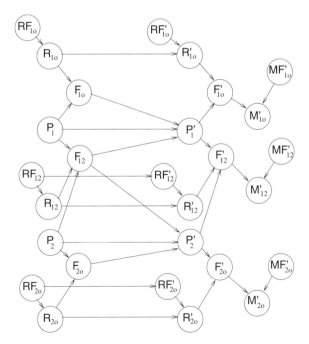

Figure 21.12. The *two-tank* DBN.

The DBN which models the two-tank system is shown in Figure 21.12. The nodes labelled by RF indicate faults in the resistances of the pipes (drifts or bursts) and the nodes labelled by MF indicate measurement failures. The nodes labelled P, F, and R are continuous and indicate pressures, flows, and pipe resistances.

By explicitly modelling the pipe resistances, we accurately model the physical system. However, since the flow is the ratio between the pressure and the resistance, the system is not a linear system. Dealing with ratios is difficult, especially when the values are close to zero. Therefore, instead of modelling the resistances, we choose to model the conductances (the conductance is defined as the reciprocal of the resistance). This transformation results in products rather than ratios. The system is still nonlinear, but this does not pose a problem for the particle filtering algorithm, which is general enough to deal with nonlinear CPDs.

Notice the difference between the measurement failure nodes and the pipe faults. The pipe faults are persistent, and therefore appear both at time t and at time $t + 1$. In fact, the pipe faults are a part of the belief state. Measurement failures, on the other hand, are transient and therefore appear only at time $t + 1$, so need not be included in the belief state. As a result, the network has six pipe fault variables and three measurement failure variables, leading to 32,768 different discrete states.

Unfortunately, this network is far too complicated for us to be able to use exact inference. In fact, the belief state at time t is a mixture of Gaussians, with a number of mixture components that grows exponentially with t, and with the number of discrete variables in each time slice. Hence, we examine how well the algorithm tracks the value of one particular random variable in a complex sequence of failures.

Our sequence includes the following failures.

- At $t = 12$ a negative drift begins in R_{1o} (the output pipe of the first tank).

- At $t = 23$, R_{12} bursts.

- At $t = 31$ we have a measurement failure for F_{12}.

- At $t = 42$ the negative drift for R_{1o} ends.

- At $t = 57$ we have another measurement failure, this time for F_{1o}.

- At $t = 70$ and at $t = 81$ we have bursts at R_{1o} and R_{2o}, respectively.

The most interesting variable to track during this sequence is R_{1o} which experienced all types of failures: a drift, a measurement failure, and then a burst. The results, based on 10 runs with 50000 samples in each run, are shown in Figure 21.13. The results show the true value of R_{1o}, the mean of the samples for R_{1o} and the confidence interval (taken to be plus or minus two standard deviations from the mean).

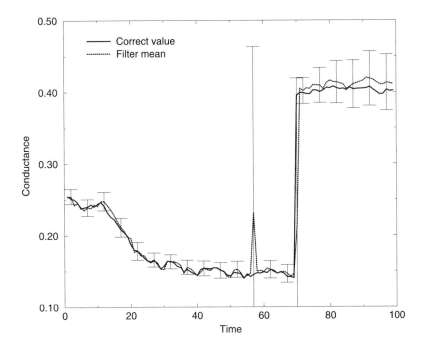

Figure 21.13. Tracking R_{1o}

We can see that we are able to track the resistance quite well. The algorithm correctly finds the negative drift and tracks the resistance accordingly. The probability that the algorithm assigns to a negative drift increases very gradually, since it takes quite a bit of evidence to support this unlikely event. When the drift begins at $t = 12$ the probability of a negative drift is only 0.008. This increases to 0.145 at $t = 16$, 0.86 at $t = 21$ and 0.993 at $t = 25$. When the drift ends at $t = 42$, the probability gradually decreases.

The big increase in uncertainty at $t = 57$ is due to the measurement error. At this point, the algorithm considers two hypotheses — either a burst or a measurement error. As more evidence is collected, the hypothesis of a measurement failure dominates, and the algorithm again tracks the resistance correctly. At $t = 70$ there is a similar increase in uncertainty, but this time the burst hypothesis is consistent with the subsequent evidence, and the algorithm continues to track the resistance correctly (although with a somewhat greater uncertainty over the true value).

21.5   Conclusions

Dynamic Bayesian networks are a rich modelling language that allows us to represent very complex stochastic systems, including those involving both discrete and continuous variables. Particle filtering provides a general-purpose inference algorithm that can be applied to virtually any DBN (we only require that the representation of the CPDs admits sampling). Thus, it allows us to deal, at least in principle, with a wide variety of dynamic systems.

The ability to represent very large-scale systems raises the question of whether particle filtering approaches scale up to very high-dimensional spaces. In this paper, we have examined this issue by looking at two realistic systems — one discrete and one hybrid. Our results show that particle filters are surprisingly robust, even for complex systems. Nevertheless, we feel that the curse of dimensionality that plagues instance-based methods is also an issue for particle filtering in high-dimensional spaces. Therefore, we are currently exploring approaches that address this concern. Two ideas seem particularly useful in this setting.

- Combinations of sampling with exact inference — an extension of Rao-Blackwellisation (Doucet, Godsill and Andrieu 2000); this idea was applied successfully to a robot localization and mapping task in Murphy and Russell (2001: this volume).

- Approximate belief state representations that exploit the weak interaction in the domain structure, as was done in (Boyen and Koller 1998, Boyen and Koller 1999) for discrete DBNs using non-sampling-based methods.

We believe that a combination of these techniques will provide us with the tools to reason about the kinds of complex hybrid systems encountered in many real-world applications.

Acknowledgements

We would like to thank Dragomir Angelov, Xavier Boyen, and Raya Fratkina for their help with this project. This research was supported by ARO under the MURI program "Integrated Approach to Intelligent Systems", grant number DAAH04-96-1-0341, by an ONR Young Investigator Award grant number N00014-99-1-0464, by the Powell Foundation, and by the Sloan Foundation.

22

In-Situ Ellipsometry Solutions Using Sequential Monte Carlo

Alan D. Marrs

22.1 Introduction

This chapter presents an example of an actual industrial application in which Sequential Monte Carlo (SMC) methods have led to significant progress toward the goal of on-line control of growing semiconductor composition. This chapter will differ from the majority of contributions to this book in that it presents an example application in which SMC has been used to solve a real problem in industrial process monitoring. By necessity we shall not dwell upon such issues as whether to use SIS, SIR, ASIR, Resample-move, Residual-resample, kernel-smoothing or any of the myriad of mechanisms for "herding" our set of posterior samples from one time step to the next. Similarly, the chapter does not seek to introduce the latest "jolly-wheeze" for implementing a Bayesian recursive filter using Sequential Monte Carlo. The chapter describes how the power of vanilla SMC methods can be used to overcome the analytical problems inherent in highly nonlinear physical measurement models and that, even with a relatively small number of particles, adequate results can be obtained.

The chapter continues with a brief description of the application which is followed by a description of the resulting state-space model. Lastly, composition estimation results for a laboratory grown complex Silicon-Germanium (Si/Ge) semiconductor crystal are presented and compared with composition estimates made using secondary ion mass spectroscopy (SIMS).

22.2 Application background

Silicon (Si) technology is revolutionising the design of both civil and military systems through the production of complex integrated circuits at low unit cost; for examples see (Hewish and Gourley 1999). The electronic and

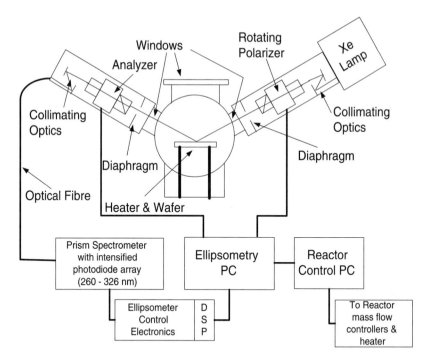

Figure 22.1. Schematic cross-section of the spectroscopic ellipsometry system integrated on the Si/Ge Si reactor.

optical properties of Si/Ge devices are sensitive to the composition, thickness and strain-state of the alloys, and if the additional gains of lower power and higher speed are to be realised in application specific integrated circuits (ASICs), these layer parameters must be accurately controlled in manufacture. The key to success of the control system will be its ability to perform in-situ monitoring of layer composition in real-time as the alloy is grown.

Real-time spectroscopic ellipsometry (RTSE) is currently used for quantitative determination of composition and thickness of growing semiconductor alloys. A schematic diagram of a RTSE system is shown in Figure 22.2. Ellipsometry is a surface analytical technique used to determine the optical properties and morphology of a surface from measurements of the change in polarisation state of light reflected by it (parameterised by the ellipsometric polarisation angles Ψ and Δ). The forward model, based upon the physics of the problem, is well understood. Given the optical properties of the alloy layers, it is possible to calculate the corresponding polarisation angles. For in-situ determination of alloy composition and thickness, the inverse problem must be solved; that is, given the polarisation angles, what are the alloy composition and thickness? As with most inverse problems this is ill defined. Inversion of the forward model is analytically intractable owing to the highly nonlinear nature of the physical model. In composition estimation, the problem is further complicated by the need to define the optical properties of an alloy layer with a given composition and thickness. The current, widely accepted, approach within the semiconductor fabrication industry is to develop a look-up table for optical properties, given composition and thickness, and to use this table in conjunction with a nonlinear model-fitting algorithm to iteratively converge upon a solution (Pickering 1998). In essence, a proposed alloy composition and thickness are used to define the layer optical properties, which are then used to calculate a corresponding value for the polarisation angles. These values are compared with the measured polarisation angles and the error between predicted and measured angles used to direct the search through the parameter space of composition and thickness. The main weaknesses of this technique are the long time lag between real time data acquisition, and computation of film optical properties and thickness. In addition, because each measurement is treated in isolation, the lack of continuity between successive parameter estimates discards one of the most useful features of the data, namely successive measurements are made of a smoothly varying process.

22.3 State-space model

The problem of composition estimation, if viewed as a state-space model, could be considered equivalent to that of target tracking. The target state being the alloy composition and thickness, rather than position and velocity, as is more common in tracking problems. Clearly, as the alloy grows, the composition and thickness of the layer grown in a fixed time period (e.g., 10 seconds) varies smoothly during the growth process. By analogy with target tracking, the current composition and thickness are related to previous composition and thickness as they follow a smooth trajectory through the parameter space in a manner similar to the position and velocity of a moving aircraft.

More generally, we are concerned with the discrete time estimation problem. The state vector, $z_t \in \Re^n$ is assumed to evolve according to the following system model

$$z_{t+1} = f_t(z_t, \epsilon_t), \tag{22.3.1}$$

where $f_t : \Re^n \times \Re^m \to \Re^n$ is the system transition function and $\epsilon_t \in \Re^m$ is a white noise sequence independent of past and current states. The pdf of ϵ_t is assumed known. At discrete times, measurements $y_t \in \Re^p$ become available. These measurements are related to the state vector via the observation equation

$$y_t = h_t(z_t, v_t), \tag{22.3.2}$$

where $h_t : \Re^n \times \Re^r \to \Re^p$ is the measurement function and $v_t \in \Re^r$ is another white noise sequence of known pdf, independent of past and present states and the system noise. It is assumed that the initial pdf $p(z_0 \mid D_0) \equiv p(z_0)$ of the state vector is available together with the functional forms f_i and h_i for $i = 1, \ldots, t$. The available information at time step t is the set of measurements $D_t = \{y_i : i = 1, \ldots, t\}$.

The requirement is to construct the pdf of the current state z_t, given all the available information: $p(z_t \mid D_t)$. In principle, this pdf may be obtained recursively in two stages; prediction and update. Suppose that the required pdf $p(z_{t-1} \mid D_{t-1})$ at time step $t-1$ is available. Then using the system model it is possible to obtain the prior pdf of the state at time step k:

$$p(z_t \mid D_{t-1}) = \int p(z_t \mid z_{t-1})p(z_{t-1} \mid D_{t-1})dz_{t-1}. \qquad (22.3.3)$$

Here the probabilistic model of the state evolution, $p(z_t \mid z_{t-1})$, which is a Markov model, is defined by the system equation (22.3.1) and the known statistics of ϵ_{t-1}.

Then at time step t a measurement y_t becomes available and may be used to update the prior via Bayes rule:

$$p(z_t \mid D_t) = \frac{p(y_t \mid z_t)p(z_t \mid D_{t-1})}{p(y_t \mid D_{t-1})}, \qquad (22.3.4)$$

where the normalising denominator is given by

$$p(y_t \mid D_{t-1}) = \int p(y_t \mid z_t)p(z_t \mid D_{t-1})dz_k. \qquad (22.3.5)$$

The conditional pdf of y_t given z_t, $p(y_t \mid z_t)$, is defined by the measurement model (22.3.2) and the known statistics of v_t. In the update equation (22.3.4), the measurement y_t is used to modify the predicted prior from the previous time step to obtain the required posterior of the state.

The recurrence relations (22.3.3) and (22.3.4) constitute the formal solution to the Bayesian recursive estimation problem.

The two components necessary to carry out inference of a dynamic system are the measurement model and the system evolution model. The corresponding models for the semiconductor composition estimation problem are described in the following subsections.

22.3.1 Ellipsometry measurement model

The physical model for ellipsometry is based upon standard thin film optics theory (Azzam and Bashara 1987). In particular, the Si/Ge crystal is approximated as a two-layer dielectric structure comprising a surface layer on top of a pseudo-substrate layer. The surface layer is that layer of material for which an estimate of composition is desired, while the pseudo-substrate represents all material underneath this surface layer.

Linearly polarised light incident upon the surface of the crystal undergoes a change in polarisation characterised by the ellipsometric polarisation angles Ψ and Δ. These angles are a function of the dielectric properties of the crystal layers, and, if the refractive indices N of the dielectric layers are known, then the ellipsometric polarisation angles for incident light of wavelength λ ($\Psi^\lambda,\Delta^\lambda$) may be estimated from (in this section $j^2 = -1$)

$$\tan(\Psi^\lambda)\exp(j\Delta^\lambda) = \frac{R_p^\lambda}{R_s^\lambda} = \frac{S_{21_p}^\lambda}{S_{11_p}^\lambda} \times \frac{S_{11_s}^\lambda}{S_{21_s}^\lambda}, \qquad (22.3.6)$$

where the reflection coefficients are defined in terms of the elements of the s and p polarisation structure matrices $\mathbf{S}$, which are in turn defined by combining interface matrices ($\mathbf{I}$) and layer matrices ($\mathbf{L}$);

$$\mathbf{S}^\lambda = \mathbf{I}_{gas/surface}^\lambda \mathbf{L}_{surface}^\lambda \mathbf{I}_{surface/pseudo}^\lambda. \qquad (22.3.7)$$

The interface matrices are defined using the interface Fresnel reflection (r) and transmission (t) coefficients for s and p polarisation

$$\mathbf{I}_{a/b}^\lambda = \frac{1}{t_{a/b}^\lambda}\begin{pmatrix} 1 & r_{a/b}^\lambda \\ r_{a/b}^\lambda & 1 \end{pmatrix}, \qquad (22.3.8)$$

where the Fresnel coefficients for the interface between layer a and layer b are calculated from the layer refractive indices N^λ at that wavelength and the incident angle ϕ_a^λ and refraction angle ϕ_b^λ

$$r_{a/b_p}^\lambda = \frac{N_b^\lambda \cos\phi_a^\lambda - N_a^\lambda \cos\phi_b^\lambda}{N_b^\lambda \cos\phi_a^\lambda + N_a^\lambda \cos\phi_b^\lambda} \qquad (22.3.9)$$

$$r_{a/b_s}^\lambda = \frac{N_a^\lambda \cos\phi_a^\lambda - N_b^\lambda \cos\phi_b^\lambda}{N_b^\lambda \cos\phi_a^\lambda + N_a^\lambda \cos\phi_b^\lambda} \qquad (22.3.10)$$

$$t_{a/b_p}^\lambda = \frac{2N_a^\lambda \cos\phi_a^\lambda}{N_b^\lambda \cos\phi_a^\lambda + N_a^\lambda \cos\phi_b^\lambda} \qquad (22.3.11)$$

$$t_{a/b_s}^\lambda = \frac{2N_a^\lambda \cos\phi_a^\lambda}{N_a^\lambda \cos\phi_a^\lambda + N_b^\lambda \cos\phi_b^\lambda}. \qquad (22.3.12)$$

The layer matrices $\mathbf{L}^\lambda$ represent the effect of propagation of light of wavelength λ through a homogeneous layer of refractive index N^λ and thickness d

$$L_a^\lambda = \begin{pmatrix} e^{j\frac{2\pi d N^\lambda}{\lambda}\cos\phi_a^\lambda} & 0 \\ 0 & e^{-j\frac{2\pi d N}{\lambda}\cos\phi_a^\lambda} \end{pmatrix}. \qquad (22.3.13)$$

Equation 22.3.6 only partially fulfills the requirements of a measurement model. In fact, only one element of the state, namely layer depth d, appears explicitly in this model via equation 22.3.13. The use of ellipsometric measurements to monitor composition is only possible through the ability

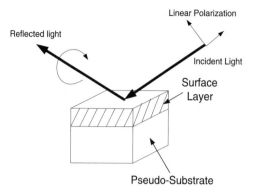

Figure 22.2. Overview of the ellipsometry measurement process.

to relate the alloy fraction of a Si/Ge layer, x, to its refractive index N^λ. This is achieved through the recording of a set of material reference files wherein a set of reference structures are grown in the laboratory and analysed off-line. The material reference files then consist of a set of optical properties for a range of alloy fractions. The optical properties for intermediate values of alloy fraction can be derived using established interpolation methods (Snyder, Woollam, Alterovitz and Johs 1990).

The reader may have noticed that a feature of the layer model is that the current ellipsometry measurement is a function of the current state and all previous states, thus violating the Markov property. This would present a problem when using a filtering approach but for the introduction of the pseudo-substrate model. The pseudo-substrate model (Pickering 1998) is a widely accepted means of ensuring that, as the alloy grows, its corresponding layer model does not need to grow in complexity. Recognizing that the ellipsometry measurement provides information about the optical properties of the structure, the ellipsometric angles can be used to calculate the pseudo-dielectric coefficients for the structure. Using these pseudo-dielectric coefficients, the structure present a time $t-1$ can be used as a pseudo-substrate on which the next layer of material will grow between time $t-1$ and t. In this way, the growing structure can always be represented by a two-layer model and the dependence on previous states is subsumed into the pseudo-substrate. A diagrammatic overview of the ellipsometry measurement process is shown in Figure 22.3.1.

The measurements of Ψ^λ and Δ^λ are taken over a range of discrete wavelengths yielding a set of spectra, $\Psi = [\Psi^{\lambda_1}, ..., \Psi^{\lambda_m}]^T$ and $\Delta = [\Delta^{\lambda_1}, ..., \Delta^{\lambda_m}]^T$. The final novel feature of the ellipsometry measurement model lies in the fact that the measurement noise is wavelength dependent such that $v_t = [v^{\lambda_1}, ..., v^{\lambda_m}]^T$.

22.3.2   System evolution model

The simplest form of system evolution model is the Markov random walk, wherein the state at time t is simply a random perturbation of the state at time $t - 1$, i.e.,

$$z_t = z_{t-1} + \epsilon_{t-1} \tag{22.3.14}$$

where $\epsilon_{t-1} \in \Re^n$ is an i.i.d sequence independent of the initial state.

In many applications, such a simple random walk model is restrictive and fails to encompass all the state evolution behaviour of the process being studied. Often, in tracking applications, the state evolution model is expanded to include multiple models, each capturing a particular aspect of the state evolution behaviour. In the context of the particle filter, it is easy to expand the state evolution model to include a number of models and augment the state vector to include an indicator variable (k) which denotes which state evolution model was used to perform prediction for that state sample (Semerdjiev et al. 1998) (see also McGinnity and Irwin (2001: this volume)).

To represent the possibility that the system evolution could move from one regime to another (i.e., a different evolution model becoming active), a switching model can be included. In this model each sample does not have to remain faithful to the evolution model denoted by the indicator variable. At each prediction step, there can be a finite probability of switching to another evolution model. The switching matrix M contains elements which represent these switching probabilities; thus, element $M_{i,j}$ is the probability that a sample that has previously evolved owing to model i will switch to model j. This switching matrix represents our prior knowledge about the state evolution regimes of the system of interest. If our prior knowledge changes during the course of filtering, the contents of the switching matrix can be updated to reflect this change in knowledge.

To monitor semiconductor growth, three system evolution regimes were identified: increasing alloy fraction, constant alloy fraction, and decreasing alloy fraction. The system models were

$$\mathbf{z}_t = \mathbf{z}_{t-1} + \epsilon_{t-1}, \tag{22.3.15}$$

where $\mathbf{z}_t = [x_t, d_t, k_t]^T$ and $\epsilon_t = [\epsilon_{x_t}^k, \epsilon_{d_t}^k, 0]^T$ are the state and system noise vectors respectively. The system noise is dependent upon the current growth regime and encompasses the following models

	Increasing x $k = 1$	Constant x $k = 2$	Decreasing x $k = 3$
ϵ_x^k	$Gamma(\alpha_{x_+}, \beta_{x_+})$	$Normal(0, \sigma_{x_c}^2)$	$-Gamma(\alpha_{x_-}, \beta_{x_-})$
ϵ_d^k	$Normal(\mu_{d_+}, \sigma_{x_+}^2)$	$Normal(0, \sigma_{d_c}^2)$	$Normal(\mu_{d_-}, \sigma_{d_-}^2)$

The model parameters $(\alpha_{x+}, \beta_{x+}, \mu_{d+}, \sigma_{d+}, \sigma_{x_c}, \mu_{d_c}, \sigma_{d_c}, \alpha_{x-}, \beta_{x-}, \mu_{d-}, \sigma_{d-})$ being predefined based upon prior knowledge of the growth apparatus and could be learned off-line.

Switching System Evolution model

(i) For $i = 1, ..., N$

- $k_{t-1} = z_{t-1}^{(i)}[3]$
- For $j = 1, ..., \mathsf{Length}(M[k_{t-1}])$

$$m[j] = \sum_{p=1}^{j} M[k_{t-1}, p] \qquad (22.3.16)$$

- Sample $u \sim U_{[0,1]}$
- Find j s.t. $m[j-1] < u \le m[j+1]$; $k_t = j$
- $z_t = z_{t-1} + \epsilon_{t-1}^{k_t}$

The switching matrix was initially set to be

$$M = \begin{pmatrix} 0.7 & 0.3 & 0.0 \\ 0.2 & 0.6 & 0.2 \\ 0.0 & 0.3 & 0.7 \end{pmatrix}, \qquad (22.3.17)$$

where $i/j = 1, 2, 3$ denotes the increasing, constant, and decreasing system models respectively. This attempts to model the dynamics of the process. For example, a decreasing or increasing alloy fraction is expected to continue that trend with some small probability of moving into a constant growth regime, but the lags in the process make a transition from increasing to decreasing unlikely over the time period between measurements.

During growth, this switching matrix may be modified to reflect our prior knowledge about the growth conditions, for example after a change in gas flow rates into the growth chamber has been made. To illustrate, the switching evolution model procedure the algorithm is detailed below.

Finally, since each sample has an associated indicator for the particular growth regime, a posterior estimate of which regime dominated the growth can be obtained and used to qualitatively assess the performance of the growth equipment.

22.3.3 Particle filter

This section describes the particle filter algorithm used to carry out recursive Bayesian estimation of Si/Ge composition. The filter is a generic SIR filter as described in (Gordon et al. 1993), but with some enhancements. A detailed description of the main steps in the algorithm follows.

Particle Filter for Si/Ge composition tracking

(i) At $t = 0$

- For $i = 1, ..., N$, draw samples z_0^i from the state prior $p(z_0)$.

(ii) For $t = 1, 2, ...$

Bayesian importance sampling step

- For $i = 1, ..., N$, sample:

$$\hat{z}_t^{(i)} \sim q(z_t | z_{t-1}^{(i)}, D_t) \tag{22.3.18}$$

and set:

$$\hat{z}_t^{(i)} \triangleq (z_{t-1}^{(i)}, \hat{z}_t^{(i)}) \tag{22.3.19}$$

- For $i = 1, ..., N$, evaluate the importance weights up to a normalising constant:

$$w_t^{(i)} = w_{t-1}^{(i)} \frac{p(D_t | \hat{z}_t^{(i)}) p(\hat{z}_t^{(i)} | z_{t-1}^{(i)})}{q(\hat{z}_t^{(i)} | z_{t-1}^{(i)}, D_t)} \tag{22.3.20}$$

- For $i = 1, ..., N$, normalise the importance weights

$$\tilde{w}_t^{(i)} = w_t^{(i)} \left[\sum_{j=1}^{N} w_t^{(j)} \right]^{-1} \tag{22.3.21}$$

If $N_{eff} \leq$ Threshold And $N_{eff} > 0$ Resampling step:

- For $i = 1, ..., N$, set $z_t^{(i)} = \hat{z}_t^{(j)}$ where $j = k$ with probability $w_t^{(k)}$.

Else: *(Lost Track)* MCMC step: $\hat{z}_{t-1}^{(i)} = \hat{z}_{t-1}^{(i)}$ For $i = 1, ..., N$ or time runs out

- Sample $u \sim U_{[0,1]}$.
- Sample the proposal candidate $\hat{z}_t^{*(i)} \sim p(\hat{z}_t^{*(i)} | \hat{z}_{t-1}^{(i)}) = N(\hat{z}_{t-1}^{(i)}, Q_t)$
- If $u \leq \min \left\{ 1, \frac{p(D_t | \hat{z}_t^{*(i)})}{p(D_t | \hat{z}_{t-1}^{(i)})} \right\}$, Accept $\hat{z}_t^{(i)} = \hat{z}_t^{*(i)}$ Else, Reject.

For $i = 1, ..., N$, set $w_t^{(i)} = \tilde{w}_t^{(i)} = \frac{1}{N}$.

In this application, the importance sampling proposal distribution was taken to be $q(z_t | z_{t-1}^{(i)}, D_t) = p(z_t | z_{t-1})$.

One important feature of the particle filter for Si/Ge composition tracking is the low number of samples used because of the expense of evaluating the likelihood and the need for on-line estimates. With a small number of samples it becomes imperative that the number of effective samples be monitored closely and effective measures be available to regain the track. A possible indicator of track quality is the Number of Effective samples

(N_{eff}) (Liu and Chen 1995)

$$N_{eff} = \frac{1}{\sum_{i=1}^{N} \left(\tilde{w}_t^{(i)} \right)^2},$$ (22.3.22)

which essentially says that the empirical variance of the normalised importance weights may be used as a measure of track quality. Once lost track has been declared a Markov Chain Monte Carlo (MCMC) move step is used to try and "herd" the particles to a region of state-space with higher probability (Gilks and Berzuini 1999). Since there is a limited amount of time in which to carry out this MCMC update before the next measurement arrives, it may be possible that not all particles have been considered for updating by the time the next measurement must be processed. In this case, before carrying out the MCMC step, the prior predictive particles are copied to the posterior particles, which are updated in turn. As the next measurement arrives, updating stops, and the current set of posterior samples used to process it. The advantage of using an MCMC step is that the proposal distribution $p(\hat{z}_t^{*(i)} | \hat{z}_t^{(i)})$ can be very different from that used in the importance sampling step, making much larger particle moves possible.

22.4   Results

In this section, we present some results for the laboratory growth of Si/Ge on a Si substrate via chemical vapour deposition. The structure comprised three layers with alloy fraction stepping down for each layer. The rate of growth was approximately one mono-layer ($\sim 1\overset{\circ}{A}$) per second. Since the measurement model is essentially a bulk model, and is only applicable to layers that are at least a few mono-layers thick ($\sim 10\overset{\circ}{A}$), which determines the time-scale within which the particle filter must perform a single update. In this example, 200 samples were used in the particle filter. The evaluation of the ellipsometry sensor model accounted for the bulk of the computational burden, masking any gains to be obtained using efficient sampling schemes.

The resulting state estimate from the particle filter is shown in Figure 22.4. The solid line is the mean posterior estimate formed by taking the posterior mean of the composition x for each incremental layer and estimate of total depth formed by summing the posterior mean estimates of d for all layers grown up to that time. The dotted lines use the same posterior mean estimate for x, but sum the posterior estimates of the $\pm 90\%$ confidence limits for d to form a measure of the uncertainty in the estimate of total depth. For comparison, the results of independent off-line analysis using secondary ion mass spectroscopy (SIMS) are included as the dashed line. SIMS analysis is a destructive procedure and the smoothing of the SIMS

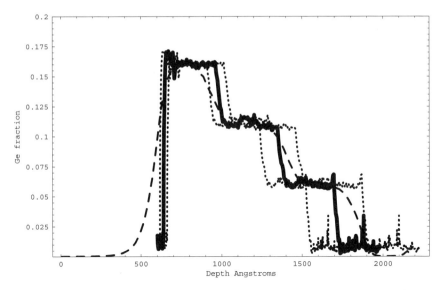

Figure 22.3. Estimates of alloy fraction and layer depth for SiGe structure (Solid line is posterior mean from filter, dashed line is result from SIMS analysis of structure).

profile is characteristic of it. The stepped layer structure is evident in both the particle filter and SIMSs estimates, with the particle filter estimate of alloy fraction x in good agreement with the SIMS analysis. The depth estimate shows poorer agreement. However, the discrepancy is well within the error bounds for both the particle filter and SIMS (typically 5% error in the SIMS depth estimates).

The corresponding estimates of model probability are shown in Figure 22.4. The estimated model probabilities show some variation owing mainly to the small number of samples. However, these estimates clearly show the regions of increasing and decreasing alloy fraction.

22.5   Conclusion

This chapter presented an example of the use of particle filters for solving a complex on-line tracking problem. The particle filter removes the need for gross approximations to the measurement and system evolution models and assumptions of Gaussian noise, enabling sequential inference to be carried out on the most complex problems.

The results demonstrate that using a particle filter inference of semiconductor composition can be performed in real-time, yielding results similar to those obtained by off-line characterisation methods.

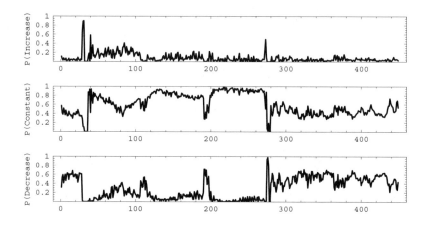

Figure 22.4. Posterior probabilities of system evolution models.

The main advantages of using the particle filter lie in its use of the state probability density function as the answer to the inference problem. This pdf enables point estimates of the composition, and realistic error measures for those estimates to be derived. In addition, the filtering approach recognises the dynamic nature of the growth process and reflects the relation between current composition and past composition. Existing approaches to composition estimation based upon nonlinear model fitting methods treat each measurement in isolation, so the resulting composition estimates do not reflect the continuous growth process and relation between successive measurements.

The drawbacks of the particle filtering approach lie in the use of random samples to approximate the state pdf. Although generally this is not a limitation because the approximation is almost sure to converge to the true pdf as the number of samples approaches infinity, in a real-time application, wherein evaluating the measurement model is computationally expensive, only a small number of samples can be used. However, as the results show, even with as few as 200 samples, composition estimates similar to SIMS analysis can be obtained. The limiting factor in carrying out on-line estimation is the time interval between successive measurements (in this case, 10 seconds), which dictates the average thickness for which composition estimates are desired. If estimates for thicker layers were adequate for the application, the corresponding time interval would increase, permitting the pdf samples to be expanded. Furthermore, as computational resources continue to improve in speed, the number of samples can be increased.

22.6   Acknowledgments

I would like to thank my colleague Neil Gordon for his many useful discussions on the finer aspects of particle filters. Dave Robbins, Chris Pickering, John Russell and John Glasper of DERA Electronics sector are also thanked for their help in growing the crystals, providing the data and offering advice on the practical and theoretical aspects of spectroscopic ellipsometry. This work is published with the permission of DERA on behalf of the Controller of Her Majesty's Stationery Office.

23

Manoeuvring Target Tracking Using a Multiple-Model Bootstrap Filter

Shaun McGinnity
George W. Irwin

23.1   Introduction

The Kalman filter (Sorensen 1966) is the optimal solution to the Bayesian estimation problem for a given linear, stochastic, state-space system with additive Gaussian noise sources of known statistics. The solution is possible since all densities in the system are Gaussian, and remain so under system transformation, and thus closed-form expressions for these are always available in terms of their first and second moments. However, if the model used by the filter does not match the actual system dynamics, the filter will tend to diverge, such that the actual errors fall outside the range predicted by the filter's estimate of the error covariance. By introducing some degree of uncertainty into the filter, small deviations from the actual system can be tracked. This also ensures that the filter gain does not fall to zero, and thus measurements are always used to update the state estimate. The practicality of this approach for large model mismatch problems is limited however.

The tracking of a target that can perform manoeuvres poses a more significant problem, since here the system can switch between different modes of operation, and the filter may then only be accurate for one particular mode. A manoeuvre, whose magnitude and onset time are usually unknown, generally lasts for only a very short period of time compared with the total path of the target. The target trajectory is therefore mainly constant velocity, straight-line motion interspersed with short periods of acceleration. Although optimal for non-manoeuvring periods, a constant velocity filter is clearly not suitable for tracking during manoeuvring periods and generally the filter estimates will diverge. Conversely, a filter matched to the actual motion of the target during manoeuvres will not achieve good performance during non-manoeuvre periods. To accurately match the actual target dynamics throughout the engagement, some form of adaptive filter model is therefore required.

Multiple-model approaches to the manoeuvring target tracking problem operate several Kalman filters in parallel. Each filter is matched to a different target motion, and the state estimate is then a weighted sum of the estimate from each filter, with the weighting determined by the posterior probability of each hypothetical model. The method is decision-free since the weightings are determined as the probability of each model being correct given the current measurement. However, since it is assumed that the target can switch from one model to another with a certain probability, even in the linear, Gaussian case, the optimal solution is impractical owing to the exponential growth with time in the number of possible branches, and therefore Kalman filters, required. Maintaining a constant number of Kalman filters inevitably incurs a loss in accuracy of the system probability distributions, and therefore a loss in optimality. Of these sub-optimal methods, some discard the most unlikely branches based on a *maximum a posteriori* approach (Andersson 1985, Averbuch, Itzikowitz and Kapon 1991). Others carry out hypothesis merging by approximating several possible branches with a single Gaussian density of equal first and second moments. The Generalised Pseudo Bayes (GPB) algorithms (Ackerson and Fu 1970, Bruckner, Scott and Rea 1973) carry out merging after the measurement update step. Recently the Interacting Multiple Model (IMM) algorithm (Bar-Shalom, Chang and Blom 1985, Blom and Bar-Shalom 1988, Mazor, Averbuch, Bar-Shalom and Dayan 1998) has been more commonly used because it gives similar performance to GPB, but, by merging after the hypothesis branching step, a lower complexity and computational load is achieved.

Given that the optimal solution to the multiple-model problem generates complex probability densities, and can be regarded as a special case of the more general nonlinear and non-Gaussian estimation problem, it is natural that Bayesian estimation methods have generated a considerable degree of interest in the field of multiple-model manoeuvring target tracking. In these methods, numerical approximations to the system probability density functions replace analytical solutions which are generally not available. Sorensen (Sorensen 1974) has classified these as either local or global. The density functions formed using local approaches, of which the extended Kalman filter is the most common, are accurate at a single point in the state-space. They are therefore very simple to implement. Global approaches are considerably more complex, since they attempt to give an accurate approximation over the complete region of significance of each density function. Usually a grid is defined over this region and either the value of the density function (Alspach and Sorenson 1972, Bucy and Senne 1971, Doyle and Harris 1996, Kramer and Sorensen 1988) or the Bayesian integrations (Gunther, Beard, Wilson, Oliphant and Stirling

1997, Wang and Klein 1978) are approximated at each grid point. Even for a single model problem these approaches are generally very numerically and computationally intensive. Their application to multiple-model problems is therefore largely impractical.

The bootstrap filter (Gordon et al. 1993, Gordon, Salmond and Ewing 1995) carries out approximate Bayesian estimation by predicting and updating a set of samples representing the system probability density functions. The samples tend to concentrate in regions of high probability, capturing the essential properties of the densities and, because no grid is applied, the evolution of the samples is computationally very simple. Thus this approach is very attractive for recursive Bayesian estimation of nonlinear and non-Gaussian systems, and therefore multiple model systems also.

This chapter describes a novel extension of the bootstrap filter for application to the multiple-model estimation problem (McGinnity 1998). It is shown that the optimal multiple-model solution can be obtained with a constant number of filters by taking the general Bayesian approach, and using more complex non-Gaussian densities. Hypothesis branching is approximated by resampling from the system switching probability distribution function, in a similar manner to the bootstrap resampling step. Further, since the bootstrap filter is a general, recursive, Bayesian estimator, the multiple-model extension is also directly applicable to nonlinear and non-Gaussian multiple-model cases. Similar applications of multiple models and bootstrap filtering can be found in (Semerdjiev et al. 1998, Isard and Blake 1998b).

The paper begins by describing the optimal multiple-model estimator followed by the sub-optimal IMM algorithm. The extension of the bootstrap filter to multiple-models is then given. A comparative example illustrates the benefits of the bootstrapping approach in approximating the branched densities of the multiple-model system, for both a Gaussian and a non-Gaussian system. Lastly, the application of these filters to a benchmark target tracking scenario is illustrated, using both a linear and nonlinear representation.

23.2   Optimal multiple-model solution

The general parameterised state-space system is given by:

$$\begin{aligned}
\mathbf{x}(k) &= f_\alpha(\mathbf{x}(k-1), \mathbf{u}_\alpha(k-1), \mathbf{w}_\alpha(k-1)) \\
\mathbf{y}(k) &= h_\alpha(\mathbf{x}(k), \mathbf{v}_\alpha(k)),
\end{aligned} \tag{23.2.1}$$

where f_α and h_α are the parameterised state transition and measurement functions, $\mathbf{x}(k)$, $\mathbf{u}(k)$, $\mathbf{y}(k)$ are the state, input and measurement vectors, $\mathbf{w}(k)$, $\mathbf{v}(k)$ are the process and measurement noise vectors of known statistics. The model index parameter α can take on any one of N values: $\alpha = 1, \ldots, N$.

As in the single-model case, the optimal Bayesian minimum error variance estimate of the state is the expected value of its posterior density. Beginning at time k with N hypotheses, the posterior density, including the parameter α, becomes

$$p(\mathbf{x}(k)|Y^k) = \sum_{j=1}^{N} p(\mathbf{x}(k)|\alpha(k) = j, Y^k) \cdot p(\alpha(k) = j|Y^k), \qquad (23.2.2)$$

where Y^k is the set of all measurements $\{\mathbf{y}(1), \ldots, \mathbf{y}(k)\}$ and $p(\alpha(k) = j|Y^k)$ is the posterior probability that the actual model index at time k is j. Hence, the optimal Bayesian estimate is a weighted sum of parameter-conditioned estimates with the weightings determined by the posterior probability of each parameter.

It is assumed that the model index, α, is governed by an underlying Markov process such that the conditioning parameter can branch at the next time-step with probability

$$p(\alpha(k+1) = i|\alpha(k) = j) = h_{ij}, i = 1, \ldots, N, j = 1, \ldots, N, \qquad (23.2.3)$$

Thus, for N candidate hypotheses at time k, N^2 are possible at time $k+1$:

$$\{\alpha(k+1) = i, \alpha(k) = j\}, i = 1, \ldots, N, j = 1, \ldots, N. \qquad (23.2.4)$$

The subsequent posterior density, $p(\mathbf{x}(k+1)|Y^{k+1})$, can be formed in two ways. Firstly, N^2 model-conditioned posterior densities can be estimated from each of the branched prior densities such that the posterior is the sum of N^2 posteriors conditioned on the branching history:

$$p(\mathbf{x}(k+1)|Y^{k+1}) = \sum_{i=1}^{N} \sum_{j=1}^{N} p(\mathbf{x}(k+1)|\alpha(k+1) = i, \alpha(k) = j, Y^{k+1})$$
$$\cdot p(\alpha(k+1) = i, \alpha(k) = j|Y^{k+1}). \qquad (23.2.5)$$

Here it is clear that an exponential increase in the number of filters with time is required, making this approach impractical even for small values of N.

The second method involves merging the branched prior densities together into N model conditioned prior densities:

$$p(\mathbf{x}(k)|\alpha(k+1) = i, Y^k) = \sum_{j=1}^{N} p(\mathbf{x}(k)|\alpha(k+1) = i, \alpha(k) = j, Y^k)$$
$$\cdot p(\alpha(k) = j|\alpha(k+1) = i, Y^k).$$

Since $\mathbf{x}(k)$ is independent of $\alpha(k+1)$, the branching simplifies to a weighting of the model conditioned posterior densities:

$$p(\mathbf{x}(k)|\alpha(k+1) = i, Y^k) = \sum_{j=1}^{N} p(\mathbf{x}(k)|\alpha(k) = j, Y^k)$$
$$\cdot p(\alpha(k) = j|\alpha(k+1) = i, Y^k). \qquad (23.2.6)$$

The weighting term is determined using Bayes' expansion:

$$p(\alpha(k)= j|\alpha(k+1)= i, Y^k) = \frac{p(\alpha(k+1)= i|\alpha(k)= j, Y^k)p(\alpha(k) = j|Y^k)}{p(\alpha(k+1)= i|Y^k)}$$

$$= \frac{h_{ij}p(\alpha(k)= j|Y^k)}{p(\alpha(k+1)= i|Y^k)}, \qquad (23.2.7)$$

where $p(\alpha(k+1) = i|Y^k)$ is a normalising term.

Although the densities produced by this merging are generally quite complex, the posterior density at $k+1$ is given by the sum of just N model conditioned posteriors:

$$p(\mathbf{x}(k+1)|Y^{k+1})=\sum_{i=1}^{N}p(\mathbf{x}(k+1)|\alpha(k+1)= i, Y^{k+1})p(\alpha(k+1)= i|Y^{k+1}).$$

$$(23.2.8)$$

Thus, the latter approach requires a constant number of estimators and, from a computational perspective, is preferred to the former, (23.2.5).

23.3 The IMM algorithm

When each of the N hypothetical models in the general parameterised state-space system is linear with Gaussian disturbances, the model conditioned posterior densities at time k can be found using Kalman filters. The posterior density, $p(\mathbf{x}(k)|Y^k)$, is then a weighted Gaussian sum. After branching, however, the merged densities given by (23.2) will also be Gaussian sums and therefore the Kalman filter can no longer be used to estimate the model conditioned posterior density required in (23.2.8).

In order to maintain both the computational load and the number of densities stored at a constant level, but also to continue to use the Kalman filter, the IMM algorithm approximates the merged branched prior densities at $k+1$ by a single Gaussian of matched first and second moments.

The posteriors at time k for this linear, Gaussian case are given as

$$p(\mathbf{x}(k)|\alpha(k) = j, Y^k) = N(\hat{\mathbf{x}}_j(k|k), P_j(k|k)), \qquad (23.3.9)$$

where $N(\hat{\mathbf{x}}, P)$ denotes a Gaussian density of mean $\hat{\mathbf{x}}$ and covariance P.

The merged densities, (23.2.6), are then Gaussian sums. Reducing these to a single Gaussian density, with equal mean and covariance, gives

$$p(\mathbf{x}(k)|\alpha(k+1) = i, Y^k) \approx N(\hat{\mathbf{x}}_i^m(k|k), P_i^m(k|k)), \qquad (23.3.10)$$

where

$$\hat{\mathbf{x}}_i^m(k|k) = \sum_{j=1}^{N} \hat{\mathbf{x}}_j(k|k) \cdot \frac{h_{ij}p(\alpha(k) = j|Y^k)}{p(\alpha(k+1) = i|Y^k)} \qquad (23.3.11)$$

and

$$P_i^m(k|k) = \sum_{j=1}^{N} [P_j(k|k) + (\hat{x}_i^m(k|k) - \hat{x}_j(k|k))(\hat{x}_i^m(k|k) - \hat{x}_j(k|k))^T]$$

$$\cdot \frac{h_{ij}p(\alpha(k) = j|Y^k)}{p(\alpha(k+1) = i|Y^k)} \quad (23.3.12)$$

The number of Kalman filters now required is maintained at N. Although sub-optimal, because the densities are always reduced to Gaussian representations, this algorithm has been widely used in different areas of manoeuvring target tracking and has been shown to function well at a modest computational load (Mazor et al. 1998).

For nonlinear systems or systems with non-Gaussian disturbances, a further approximation is introduced into the IMM algorithm because the linear, Gaussian assumptions of the Kalman filter are no longer valid. The posterior densities, and consequently the merged priors, are therefore only locally accurate and do not reflect the actual system densities.

23.4   Multiple model bootstrap filter

Since the bootstrap filter is not restricted to Gaussian densities, it is ideally suited to estimate the optimal multiple model posterior densities using the more computationally efficient density merging technique given by (23.2) and (23.2.8). Thus, the number of estimators can be maintained equal to the number of models N, at the expense of an increase in the complexity of the densities, and therefore the estimator, required.

The extension of the bootstrap filter to multiple models is developed by considering a single set of samples, augmented by an index vector representing the model parameter, $\alpha(k)$. Thus, the posterior densities at k, $p(\mathbf{x}(k)|\alpha(k), Y^k)$, are given by the set of N_s samples:

$$\{\mathbf{X}(k|k), A(k|k)\} = \{\mathbf{X}_l(k|k), A_l(k|k), l = 1, \ldots, N_s\}, \quad (23.4.13)$$

where $\mathbf{X}_l(k|k)$ is the sample from the state and $A_l(k|k) \in 1, \ldots, N$ represents the index of the model which generated the sample $\mathbf{X}_l(k|k)$.

The model-conditioned posterior densities, $p(\mathbf{x}(k)|\alpha(k) = j, Y^k)$, are then approximated by the subset of these samples given by

$$\{\mathbf{X}(k|k), A(k|k) = j\} \quad (23.4.14)$$

and the posterior model probabilities are approximately equal to the proportion of samples in the index set $A(k|k)$ from each model:

$$p(\alpha(k) = j|Y^k) \approx \frac{n(A(k|k) = j)}{N_s}, \quad (23.4.15)$$

where $n(A(k|k) = j)$ denotes the number of samples in $A(k|k)$ equal to j.

Extending this idea, samples from the branched prior densities at $k+1$, $p(\mathbf{x}(k)|\alpha(k), \alpha(k+1), Y^k)$, are formed by augmenting the posterior samples with a further vector, $A(k+1|k)$. The values of $A(k+1|k)$ are generated by applying the model switching Markov chain to each individual sample in the vector $A(k|k)$. Thus, if $A_l(k|k) = j$, then $A_l(k+1|k)$ will be set to i with probability h_{ij}.

The actual values of $A(k+1|k)$ are determined in a similar manner to the likelihood sampling step of the bootstrap filter. Hence, if $A_l(k|k) = j$, $A_l(k+1|k)$ is selected as m where

$$\sum_{i=1}^{m-1} h_{ij} \leq s_l \leq \sum_{i=1}^{m} h_{ij} \qquad (23.4.16)$$

and s_l is a uniformly distributed number in the range $(0, 1]$. The sum, $\sum_{i=1}^{m} h_{ij}$, represents the probability distribution function of the Markov chain, $P(\alpha(k+1) \leq m|\alpha(k) = j)$, $m = 1, \dots, N$.

Each branched prior density, $p(\mathbf{x}(k)|\alpha(k) = j, \alpha(k+1) = i, Y^k)$, is then approximated by the samples:

$$\{\mathbf{X}(k|k), A(k|k) = j, A(k+1|k) = i\}. \qquad (23.4.17)$$

Samples from the merged prior densities are formed by discarding the vector $A(k|k)$ and, therefore, $p(\mathbf{x}(k)|\alpha(k+1) = i, Y^k)$ is approximated by the set:

$$\{\mathbf{X}(k|k), A(k+1|k) = i\}. \qquad (23.4.18)$$

Samples from the prediction density $p(\mathbf{x}(k+1)|\alpha(k+1) = i, Y^k)$ are formed by transforming each sample, $\mathbf{X}_l(k|k)$, using the model indexed by $A_l(k+1|k)$. Hence the set of samples is now given as $\{\mathbf{X}(k+1|k), A(k+1|k) = i\}$, where

$$\mathbf{X}_l(k+1|k) = f_\alpha(\mathbf{X}_l(k|k), \mathbf{u}_\alpha(k), \mathbf{W}_l(k))|_{\alpha = A_l(k+1|k)} \qquad (23.4.19)$$

and $\mathbf{W}_l(k)$ is a sample drawn from the process noise distribution.

The number of samples in this set must be large enough to ensure that the region of overlap between the prediction and likelihood is sufficient to give an adequate representation of the posterior probability density function, since, if the overlap is small, only a fraction of the points will have a high likelihood and the bootstrap sampling-and-replace strategy will tend to concentrate on these few points. A suitable solution is to repeat the prediction $N^*(\geq 10)$ times such that the process noise ensures the samples are sufficiently distinct to give an accurate posterior.

The set of samples, $\{\mathbf{X}(k+1|k+1), A(k+1|k+1)\}$, representing the posterior density are then chosen from the prediction samples using the normal bootstrap resampling technique.

Resampling is carried out based only on the likelihoods of the prediction set, $\mathbf{X}(k+1|k)$, since the individual model weightings are already captured by the proportion of samples in the index vector from each model.

Therefore, samples in the posterior densities will automatically be weighted not just towards high likelihoods, but also towards models with a high proportion of samples in the prediction density.

The approach is structurally similar to that taken by the IMM, but with a more general Bayesian estimator compared to a Kalman filter. Further, in IMM only the first and second moments of the branched densities are matched, rather than forming the true density. Using Sorensen's classification, the IMM algorithm can therefore be viewed as a local approximation, whereas this one is global. Finally, the bootstrap approach has not assumed any linear or Gaussian properties on the system, making it a much more generally applicable technique.

23.4.1 Example

For illustration, the merging approaches of the IMM and bootstrap methods are compared for a scalar two-model case. First, Gaussian priors are considered:

$$\begin{aligned}p(x(k)|\alpha(k)=1,Y^k) &= N(-2,1)\\ p(x(k)|\alpha(k)=2,Y^k) &= N(2,2).\end{aligned}\tag{23.4.20}$$

The prior model probabilities are $[0.3,0.7]^T$ and the Markovian probability transition matrix, H, is given by

$$H=\begin{bmatrix}0.8 & 0.2\\ 0.2 & 0.8\end{bmatrix}.\tag{23.4.21}$$

The merged densities approximated by IMM are Gaussian with the same mean and variance as the true merged densities:

$$\begin{aligned}p(x(k)|\alpha(k+1)=1,Y^k) &= N(-0.53,5.09)\\ p(x(k)|\alpha(k+1)=2,Y^k) &= N(1.61,3.30).\end{aligned}\tag{23.4.22}$$

Using a total sample size of 5000, samples from the prior densities given in (23.4.20) were approximated using $0.3\times 5000=1500$ and $0.7\times 5000=3500$ samples respectively.

After applying the bootstrap density mixing scheme, 1906 samples represented $p(x(k)|\alpha(k+1)=1,Y^k)$ and 3094 represented $p(x(k)|\alpha(k+1)=2,Y^k)$. Thus,

$$\begin{aligned}p(\alpha(k+1)=1|Y^k) &\approx 0.3812\\ p(\alpha(k+1)=2|Y^k) &\approx 0.6188,\end{aligned}\tag{23.4.23}$$

which are very close to the true values of 0.3800 and 0.6200 respectively. The means were determined as -0.52 and 1.59 and the variances as 5.33 and 3.42, which are again very close to the actual values.

Figure 23.1 illustrates the density approximations formed with IMM and the bootstrap method using Kernel Density Estimation (KDE) (Silverman

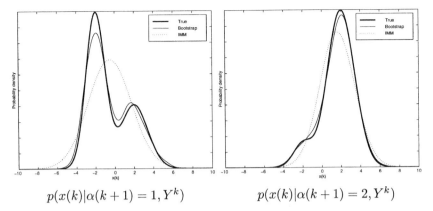

$$p(x(k)|\alpha(k+1) = 1, Y^k) \qquad\qquad p(x(k)|\alpha(k+1) = 2, Y^k)$$

Figure 23.1. Bootstrap and IMM approximations to merged densities, equation (23.4.20).

1986). In this case, the bootstrap method forms a much better approximation to the shape of the merged densities than does IMM. The errors present in the bootstrap approach are a combination of the sampling method and the KDE techniques applied.

A more likely scenario for nonlinear estimation is that the densities will be non-Gaussian. To show this the merging methods are illustrated for the same example, but with

$$
\begin{aligned}
p(x(k)|\alpha(k+1) = 1, Y^k) &= uniform(-3, -1) \\
p(x(k)|\alpha(k+1) = 2, Y^k) &= uniform(0.5, 3.5).
\end{aligned}
\tag{23.4.24}
$$

In a nonlinear estimation scenario, the IMM algorithm would be using an EKF and would therefore approximate these densities as Gaussian. Hence, the best estimate would have the same mean and covariance and mixing these gives

$$
\begin{aligned}
p_{IMM}(x(k)|\alpha(k+1) = 1, Y^k) &= N(-2, 0.33) \\
p_{IMM}(x(k)|\alpha(k+1) = 2, Y^k) &= N(2, 0.75).
\end{aligned}
\tag{23.4.25}
$$

The bootstrap approach, however, would not require such an approximation, and therefore the prior densities can be realised using samples from a uniform distribution.

Figure 23.2 illustrates the density approximations formed after merging using both the IMM and the bootstrap methods, using the same initial probabilities and Markov transition matrix as before. The merged samples for the bootstrap method fall into two distinct regions and the density approximations are realised by assuming a uniform density across each. Here, the errors in the IMM technique are much more evident.

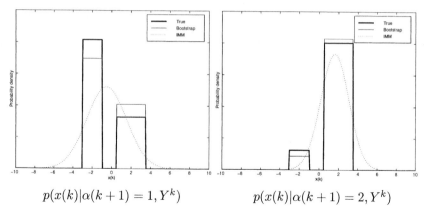

$$p(x(k)|\alpha(k+1) = 1, Y^k) \qquad p(x(k)|\alpha(k+1) = 2, Y^k)$$

Figure 23.2. Bootstrap and IMM approximations to merged densities in a nonlinear scenario, equation (23.4.24).

23.5   Target tracking examples

23.5.1   Target scenarios

A benchmark target scenario, based on a simplification of the problem of the beam pointing control of a phased array radar for manoeuvring targets (Blair and Watson 1994), was selected to illustrate the multiple-model bootstrap tracker. The target motion is representative of an incoming anti-ship missile. Ignoring motion in the z-plane, the target starts at a speed of $0.3kms^{-1}$ and a heading of $75degrees$ in the x-y plane. After 3 seconds, the motor ignites and accelerates the missile along the same heading to a speed of $0.335kms^{-1}$. The target then begins a series of lateral acceleration manoeuvres of magnitude approximately $7gkms^{-2}$. On completing these turns, the target flies in a constant heading for $10s$, after which it begins a further lateral acceleration manoeuvre of magnitude $3gkms^{-2}$. Figure 23.3 shows the complete target trajectory. In all tests, the sample time, T, was taken as $1.0s$. These results have been reported in (McGinnity and Irwin 1998).

23.5.2   Linear, Gaussian tests

The first set of simulations compares the new multiple-model bootstrap estimator and the IMM tracker using linear models driven with Gaussian noise. Two types of models were employed. The first assumed constant velocity, non-manoeuvring motion in the Cartesian frame with small deviations in velocity, $v(t)$, on each axis approximated by zero-mean, Gaussian white noise, $w(t)$, of variance q_{cv}:

$$\dot{v}(t) = w(t) \qquad (23.5.26)$$

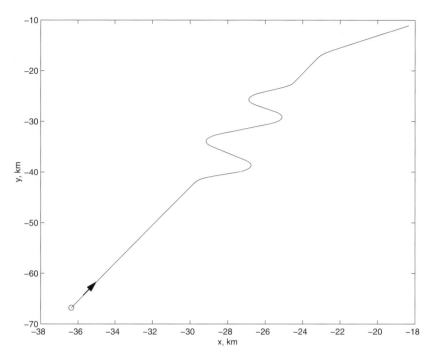

Figure 23.3. Test trajectory.

The discrete-time model for each axis is then

$$\mathbf{x}(k) = A\mathbf{x}(k-1) + \mathbf{w}(k-1)$$
$$A = \begin{bmatrix} 1 & T \\ 0 & 1 \end{bmatrix}, Q_{cv} = q_{cv} \begin{bmatrix} \frac{T^3}{3} & \frac{T^2}{2} \\ \frac{T^2}{2} & T \end{bmatrix}, \quad (23.5.27)$$

where the state $\mathbf{x}(k)$ denotes target position and velocity, A is the state transition matrix and $\mathbf{w}(k)$ is a zero-mean, Gaussian white noise sequence of covariance Q_{cv}.

The second model, designed to track manoeuvring motion, assumed constant acceleration motion on each Cartesian axis, with small deviations in acceleration, $a(t)$, approximated by zero-mean, Gaussian white noise, $w(t)$, of variance q_{ca}.

The discrete-time model for this case is also given by (23.5.27), but the state $\mathbf{x}(k)$ now denotes target position, velocity and acceleration on each axis and $\mathbf{w}(k)$ is a zero-mean, Gaussian, white noise sequence of covariance

Q_{ca}. Thus,

$$A = \begin{bmatrix} 1 & T & \frac{T^2}{2} \\ 0 & 1 & T \\ 0 & 0 & 1 \end{bmatrix}, Q_{ca} = q_{ca} \begin{bmatrix} \frac{T^5}{20} & \frac{T^4}{8} & \frac{T^3}{6} \\ \frac{T^4}{8} & \frac{T^3}{3} & \frac{T^2}{2} \\ \frac{T^3}{6} & \frac{T^2}{2} & T \end{bmatrix}. \quad (23.5.28)$$

In simulation studies a single constant-velocity model, with $\alpha = 1$ and $q_{cv} = 10^{-6}(kms^{-2})^2$, and two constant-acceleration models were used. The first, $\alpha = 2$, used a high disturbance covariance, $q_{ca} = 10^{-3}(kms^{-3})^2$, to enable changes in acceleration to be followed. The other, $\alpha = 3$, had a low disturbance covariance, $q_{ca} = 10^{-6}(kms^{-3})^2$, to enable roughly constant-acceleration motion to be tracked.

To enable good tracking and smooth switching from one model to another, it is prudent to ensure that the constant-acceleration model with high process noise is used when changes in acceleration are detected, since this provides the greatest flexibility in this case. Hence, the system could not switch directly from the constant-velocity model to the low covariance, constant-acceleration model, or vice versa. The Markov transition probability matrix was defined as

$$H = \begin{bmatrix} 0.95 & 0.33 & 0.0 \\ 0.05 & 0.34 & 0.05 \\ 0.0 & 0.33 & 0.95 \end{bmatrix} \quad (23.5.29)$$

and the initial probabilities were set to

$$p(\alpha(0)) = \begin{bmatrix} 0.6 & 0.2 & 0.2 \end{bmatrix}^T. \quad (23.5.30)$$

In all scenarios, measurements $z^x(k)$, $z^y(k)$ were taken of the Cartesian position of the target, corrupted by white, Gaussian noise of zero-mean and variance $R = 0.01km^2$ on each axis. Separate filters were implemented for each axis and, for the bootstrap filter, the number of sample points, N_s, was set to 5000 and the prediction multiplication factor N^* equal to 10. This gave a mean execution time on a Pentium Pro 200 PC of $0.42s$ per iteration, which is significantly less than the sample time.

The initial estimate of the state was taken as the true initial state corrupted by an additional random error given by the initial error covariance matrix

$$P(0) = \begin{bmatrix} R & \frac{R}{T} & 0 \\ \frac{R}{T} & 2\frac{R}{T^2} & 0 \\ 0 & 0 & 10^{-6} \end{bmatrix}. \quad (23.5.31)$$

For comparison purposes, the Normalised Position Error(NPE), defined as the ratio of the mean-square position estimation error to the mean-square measurement error over S simulations,

$$NPE(k) = \sqrt{\frac{\sum_{i=1}^{S}[(x_i(k) - \hat{x}_i(k))^2 + (y_i(k) - \hat{y}_i(k))^2]}{\sum_{i=1}^{S}[(x_i(k) - z_i^x(k))^2 + (y_i(k) - z_i^y(k))^2]}} \qquad (23.5.32)$$

was applied. An NPE value below unity signifies good estimation.

Figure 23.4 plots the variation in NPE for both the multiple-model bootstrap filter and the IMM algorithm over 50 simulations using the linear models. Here the bootstrap filter improves on the IMM during the non-manoeuvring sections, signifying that it forms a better representation of the system densities in this case. During the manoeuvres the performance of each algorithm is similar. This may be expected because the linear, constant-acceleration model does not accurately reflect the actual target motion during this period and therefore neither filter can settle on the correct model.

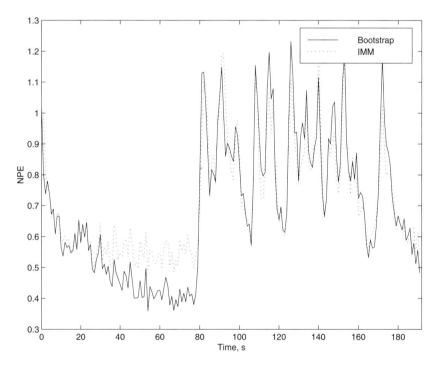

Figure 23.4. Variation in NPE using IMM and multiple model bootstrap filter - linear, Gaussian tests.

23.5.3   Polar simulation results

To illustrate the performance of the nonlinear, multiple-model bootstrap filter, the example target was also tracked in polar co-ordinates. Here the state is defined as: position(r), bearing (θ), speed (V), heading(φ), longitudinal acceleration(U_{long}) and lateral acceleration(U_{lat}). A strategy similar to the linear case was used. Thus the model set consisted of a constant-velocity model and several constant lateral and longitudinal acceleration models assuming different process noise levels. The continuous-time equations of motion for constant velocity motion are

$$\begin{aligned}
\dot{r} &= V\cos\left(\varphi - \theta\right)\\
\dot{\theta} &= (V\sin\left(\varphi - \theta\right))/r\\
\dot{V} &= 0\\
\dot{\varphi} &= 0,
\end{aligned}$$
(23.5.33)

and for constant longitudinal and lateral acceleration motion

$$\begin{aligned}
\dot{r} &= V\cos\left(\varphi - \theta\right)\\
\dot{\theta} &= (V\sin\left(\varphi - \theta\right))/r\\
\dot{V} &= U_{long}\\
\dot{\varphi} &= U_{lat}/V\\
\dot{U}_{long} &= 0\\
\dot{U}_{lat} &= 0.
\end{aligned}$$
(23.5.34)

In the constant-velocity system, small changes in the speed and heading were modelled by longitudinal and lateral acceleration jerks, approximated as bursts of process noise, acting at the start of each sample period. Thus,

$$\begin{aligned}
V(t) &= V(t) + w_{long}(kT)\delta(t - kT)\\
\varphi(t) &= \varphi(t) + \tfrac{w_{lat}(kT)}{V(t)}\delta(t - kT),
\end{aligned}$$
(23.5.35)

where k denotes the sample time, $w_{long}(kT)$ and $w_{lat}(kT)$ represent longitudinal and lateral acceleration jerks, and $\delta(t - kT)$ is the impulse function.

Similarly, in the constant acceleration system, the jerk approximation is:

$$\begin{aligned}
U_{long}(t) &= U_{long}(t) + w_{long}(kT)\delta(t - kT)\\
U_{lat}(t) &= U_{lat}(t) + w_{lat}(kT)\delta(t - kT).
\end{aligned}$$
(23.5.36)

Discretised versions of these models, using first-order differences, were applied in the multiple-model bootstrap filter. For comparison the IMM algorithm using extended Kalman filters was also implemented.

The first model employed, $\alpha = 1$, was constant-velocity. The second, $\alpha = 2$, assumed constant-acceleration motion with large lateral acceleration variance and small longitudinal acceleration variance to deal with lateral acceleration manoeuvres. The third, $\alpha = 3$, was based on constant-acceleration motion with large longitudinal acceleration variance and small lateral acceleration variance to handle adaption to longitudinal acceleration manoeuvres. To enable tracking of manoeuvres after adaption, the final model, $\alpha = 4$, assumed constant-acceleration motion with lateral and longitudinal acceleration variances, both of which small. Table 1 lists the variance values employed, with those of the manoeuvring models selected so that the standard deviation was approximately equal to the expected maximum size of manoeuvre. Equation (23.5.37) details the Markov switching matrix employed.

Model, α	Longitudinal acceleration jerk variance	Lateral acceleration jerk variance
Constant velocity, 1	$1.0 \times 10^{-10}(kms^{-2})^2$	$1.0 \times 10^{-10}(kms^{-2})^2$
Constant acceleration, 2	$1.0 \times 10^{-10}(kms^{-3})^2$	$5.0 \times 10^{-3}(kms^{-3})^2$
Constant acceleration, 3	$1.0 \times 10^{-5}(kms^{-3})^2$	$1.0 \times 10^{-10}(kms^{-3})^2$
Constant acceleration, 4	$1.0 \times 10^{-8}(kms^{-3})^2$	$1.0 \times 10^{-8}(kms^{-3})^2$

Table 23.1. Model parameters for the polar simulations.

$$H = \begin{bmatrix} 0.9 & 0.34 & 0.34 & 0.00 \\ 0.05 & 0.33 & 0.0 & 0.1 \\ 0.05 & 0.0 & 0.33 & 0.1 \\ 0.0 & 0.33 & 0.33 & 0.8 \end{bmatrix} \tag{23.5.37}$$

Thus, the constant-velocity model could switch to either of the adaptive manoeuvring models, but with low probability. The constant-acceleration motion could also switch to the adaptive models, but with a slightly higher probability. The adaptive models had an equal probability of remaining in the same mode, reverting to constant-velocity or maintaining constant-acceleration motion.

The initial probabilities were set to

$$p(\alpha(0)) = \begin{bmatrix} 0.5 & 0.1 & 0.2 & 0.2 \end{bmatrix}^T. \tag{23.5.38}$$

Measurements were taken of the target position (range and bearing) with additive zero-mean, Gaussian white measurement noise of covariance:

$$R_r = 4.0 \times 10^{-4} km^2, R_\theta = 5.0 \times 10^{-5} rad^2. \tag{23.5.39}$$

To ensure good initialisation, the error covariance of the initial state estimate was set to a diagonal matrix with: $\sigma_r^2 = R_r$, $\sigma_\theta^2 = R_\theta$, $\sigma_V^2 = 1.0 \times 10^{-4}(kms^{-1})^2$, $\sigma_\varphi^2 = 1.5 \times 10^{-5}(rads^{-1})^2$, and $\sigma_{U_{long}}^2 = \sigma_{U_{lat}}^2 =$

$1.0 \times 10^{-6}(kms^{-2})^2$. The mean was then set to the true mean plus an additive random element generated from this covariance matrix.

Again, the multiple-model bootstrap filter used a total of 5000 sample points, with the multiplicative factor N_s equal to 10. This gave a mean execution time of $0.95s$ which again is smaller than the sample time.

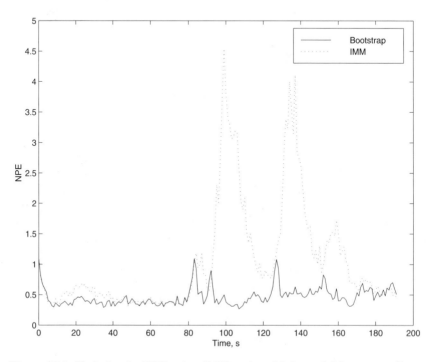

Figure 23.5. Variation in NPE using IMM and multiple model bootstrap filter - polar tests.

Converting the polar measurements to Cartesian allows the NPE for each filter to be plotted, Figure 23.5. Here the IMM algorithm performs poorly and was found to diverge in approximately 40% of simulations.

Using one of the divergent cases as an example allows the superior tracking of the bootstrap filter to be illustrated. The first lateral manoeuvre begins at $t = 78s$. Therefore, for comparison, the bootstrap filter is initialised using the IMM estimate at $t = 76s$. For a single run, Figure 23.6 plots the corresponding model weightings from each filter over the period $t = 76s$ to $t = 100s$.

At $t = 93s$ the Cartesian velocity densities (transformed from the polar estimates) from each model are shown in figure 23.7. The bootstrap estimates are shown as dots representing samples, the IMM estimates as

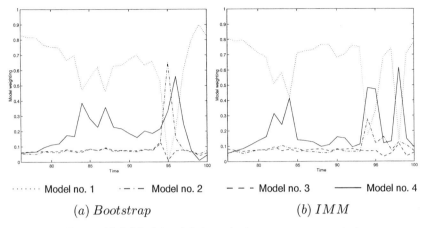

(a) *Bootstrap*   (b) *IMM*

Figure 23.6. Model weightings during manoeuvre period.

Gaussian densities plotted at ±2 standard deviations. At this instant, it is clear that the IMM and bootstrap approximations are quite similar.

At the next sample time, however, a distinct difference is evident. A split has occurred in the bootstrap samples between those of high likelihood, close to the true velocity, and those samples more densely distributed close to the previous estimate. The weighting also shifts towards the manoeuvring models. The IMM algorithm cannot deal with these non-Gaussian densities and maintains its Gaussian estimates close to the erroneous values. This continues at subsequent sample times and causes the filter to eventually diverge. With the bootstrap filter, the number of samples in the correct region increases and the manoeuvre is tracked successfully.

23.6   Conclusions

The use of general Bayesian bootstrapping estimation techniques within a multiple-model target-tracking framework has been discussed. For linear, Gaussian systems the optimal solution to the multiple-model problem requires an exponentially growing number of Kalman filters. Bayesian techniques, however, are not restricted to linear, Gaussian scenarios. With an associated increase in the complexity of the system probability density functions, it has been shown that the optimal solution can be found using the same number of Bayesian estimators as models.

Just as the Bayesian estimation techniques for single models are not restricted to linear, Gaussian systems (as in the Kalman filter), so the multiple-model extensions of these techniques are applicable without alteration to nonlinear and non-Gaussian problems.

An extension of the bootstrap filter to multiple-models has been proposed. Each sample is indexed to a particular hypothetical model. To approximate the model branching of the optimal multiple-model solution, the Markovian switching is applied to each sample.

The superior accuracy of the branched densities obtained by the bootstrap approach, compared with the linear, Gaussian IMM algorithm was illustrated by branching two Gaussian and two non-Gaussian densities.

The new approach has been illustrated using a standard target-tracking scenario. Using linear, Gaussian systems the multiple-model bootstrap filter showed generally performance superior to the IMM algorithm during non-manoeuvre periods and comparable performance during manoeuvres.

The multiple-model bootstrap filter was then investigated within a nonlinear framework. In this case it showed good tracking performance whereas the IMM algorithm using EKFs displayed very large estimation errors and poor adaption to manoeuvres.

In conclusion the multiple-model bootstrap filter has been shown to be a very powerful tool for multiple-model linear and nonlinear filtering which, especially in nonlinear situations, gives superior performance to the IMM algorithm. Most importantly, although computationally intensive, in both cases the bootstrap filter execution time was lower than the sample time.

23.7   Acknowledgments

The authors wish to acknowledge the support of the Department of Education for Northern Ireland and SMS Ltd. in helping us complete this work.

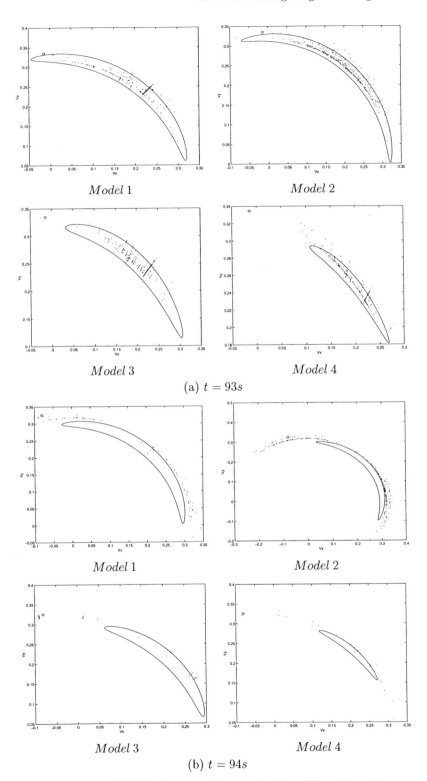

(a) $t = 93s$

(b) $t = 94s$

Figure 23.7. Velocity density estimates of each model - polar tests.

24

Rao-Blackwellised Particle Filtering for Dynamic Bayesian Networks

Kevin Murphy
Stuart Russell

24.1 Introduction

Particle filtering in high dimensional state-spaces can be inefficient because a large number of samples is needed to represent the posterior. A standard technique to increase the efficiency of sampling techniques is to reduce the size of the state space by marginalizing out some of the variables analytically; this is called Rao-Blackwellisation (Casella and Robert 1996). Combining these two techniques results in Rao-Blackwellised particle filtering (RBPF) (Doucet 1998, Doucet, de Freitas, Murphy and Russell 2000). In this chapter, we explain RBPF, discuss when it can be used, and give a detailed example of its application to the problem of map learning for a mobile robot, which has a very large ($\sim 2^{100}$) discrete state space.

In Bayesian filtering, the goal is to compute the distribution $P(Z_t|y_{1:t})$, where Z_t is the hidden state at time t, and $y_{1:t} = (y_1, \dots, y_t)$ is all the evidence up to time t. We can compute this quantity recursively (on-line) by applying Bayes' rule sequentially. However, in general, the required integrals cannot be computed in closed form. Particle filtering (PF) therefore approximates the posterior using sequential importance sampling. Please see the chapter by Doucet, de Freitas and Gordon (2001: this volume) for a more detailed introduction to standard PF.

Suppose we partition the state-space Z_t into two sub-spaces, R_t and X_t. Then, by the chain rule of probability, we can write

$$P(X_{1:t}, R_{1:t}|y_{1:t}) = P(X_{1:t}|R_{1:t}, y_{1:t})P(R_{1:t}|y_{1:t}).$$

If we can update $P(X_t|R_{1:t}, y_{1:t})$ analytically and efficiently, then we only need to sample $P(R_{1:t}|y_{1:t})$ using the particle filter. Since we are now sampling in a smaller space, in general we will need far fewer particles to reach the same accuracy as standard PF. This is the key idea behind RBPF.

In RBPF, each particle maintains not just a sample from $P(R_{1:t}|y_{1:t})$, which we shall denote by $r_{1:t}^{(i)}$, but also a *parametric* representation of the distribution $P(X_t|r_{1:t}^{(i)}, y_{1:t})$, which we shall denote by $\alpha_t^{(i)}$. (The parametric representation might be a mean vector and a covariance matrix, for instance.) The R_t samples are updated as in standard PF, and then the X_t distributions are updated using an exact filter, conditional on R_t. The overall algorithm is shown in Figure 24.1.

In Section 24.2, we discuss when this algorithm can be usefully applied, which is best described using the language of dynamic Bayesian networks. In Section 24.3, we discuss in detail how to compute the equations used by the algorithm. Finally, in Section 24.4, we discuss the application of RBPF to map learning.

24.2   RBPF in general

We can most easily characterise the models to which RBPF may be efficiently applied by representing them graphically as dynamic Bayesian networks (DBNs), which we now briefly explain. Bayesian networks (Pearl 1988, Cowell, Dawid, Lauritzen and Spiegelhalter 1999) are directed acyclic graphs, in which nodes represent random variables, and the lack of arcs represents conditional independencies. Dynamic Bayesian networks extend Bayesian networks to probability distributions that evolve over time. As in a state-space model, we must specify the transition model, $P(Z_t|Z_{t-1})$, the observation model, $P(Y_t|Z_t)$, and the prior, $P(Z_1)$. We use a two-slice graph to represent the conditional independence relationships in $P(Z_t|Z_{t-1})$. In addition to the graph structure, we must specify the conditional probability distribution (CPD) of each node given its parents. For a more detailed introduction to DBNs, please see Koller and Lerner (2001: this volume).

RBPF can be applied to any DBN that can be made topologically equivalent to the model shown in Figure 24.2, clustering nodes together if necessary. However, to apply the algorithm in Figure 24.1, we must compute the term

$$P(R_t|r_{1:t-1}^{(i)}, y_{1:t-1}) = \sum_{x_{t-1}} P(R_t|r_{1:t-1}^{(i)}, y_{1:t-1}, x_{t-1})P(x_{t-1}|r_{1:t-1}^{(i)}, y_{1:t-1})$$

$$= \sum_{x_{t-1}} P(R_t|r_{t-1}^{(i)}, x_{t-1})P(x_{t-1}|r_{1:t-1}^{(i)}, y_{1:t-1}). \quad (24.2.1)$$

If X_{t-1} is a vector in $\{1, \ldots, k\}^m$, i.e., a tuple of m discrete k-valued random variables, then this equation takes $O(k^m)$ time to compute, which is usually unacceptably high, especially since it must be computed once per particle. Likewise, if X_{t-1} is a vector in $\mathbb{R}^m$, the required integra-

Generic RBPF

(i) Sequential importance sampling step

- For $i = 1, \ldots, N$, sample

$$\left(\widehat{r}_t^{(i)}\right) \sim q(r_t; r_{1:t-1}^{(i)}, y_{1:t})$$

and set

$$\left(\widehat{r}_{1:t}^{(i)}\right) \triangleq \left(\widehat{r}_t^{(i)}, r_{1:t-1}^{(i)}\right)$$

- For $i = 1, \ldots, N$, evaluate the importance weights up to a normalising constant:

$$w_t^{(i)} = \frac{p\left(y_t \mid y_{1:t-1}, \widehat{r}_{1:t}^{(i)}\right) p\left(\widehat{r}_t^{(i)} \mid \widehat{r}_{1:t-1}^{(i)}, y_{1:t-1}\right)}{q\left(\widehat{r}_t; \widehat{r}_{1:t-1}^{(i)}, y_{1:t}\right)}$$

- For $i = 1, \ldots, N$, normalise the importance weights:

$$\widetilde{w}_t^{(i)} = w_t^{(i)} \left[\sum_{j=1}^{N} w_t^{(j)}\right]^{-1}$$

(ii) Selection step

- Resample N samples from $(\widehat{r}_{1:t}^{(i)})$ according to the importance distribution $\widetilde{w}_t^{(i)}$ to obtain N random samples $(r_{1:t}^{(i)})$ approximately distributed according to $p(r_{1:t}^{(i)} | y_{1:t})$.

(iii) Exact step

- Update $\alpha_t^{(i)}$ given $\alpha_{t-1}^{(i)}$, $r_t^{(i)}$, $r_{t-1}^{(i)}$, and y_t.

Figure 24.1. Generic Rao-Blackwellised particle filtering algorithm. If we replace references to r_t with $z_t = (x_t, r_t)$ and omit step 3, the result is a regular (non Rao-Blackwellised) particle filter. Notice that if we are only interested in filtering, we do not need to store the full trajectory $r_{1:t}^{(i)}$ in each particle, just its most recent component, $r_t^{(i)}$, since we are updating $\alpha_t^{(i)}$ on-line. ($\alpha_t^{(i)}$ is a parametric representation of $P(X_t | y_{1:t}, r_{1:t}^{(i)})$.) This figure is adapted from (Doucet, de Freitas, Murphy and Russell 2000).

tion often cannot be computed in closed form. Hence, it is common to assume that there is no arc from X_{t-1} to R_t, so that we can eliminate the marginalization over X_{t-1}. In other words, the equation becomes

$$P(R_t | r_{1:t-1}^{(i)}, y_{1:t-1}) = P(R_t | r_{t-1}^{(i)}).$$

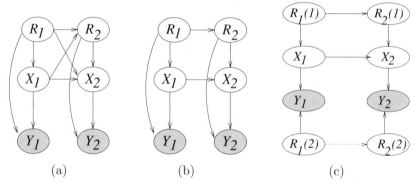

Figure 24.2. (a) The canonical DBN to which RBPF can be applied. The shaded Y_t nodes denote observations. (b) A simplification in which we have removed the cross arcs. The R_t nodes are called the roots, and the X_t nodes the leaves. (c) We have partitioned the root node into two components, $R_t(1)$, which affects the dynamics of X_t, and $R_t(2)$, which affects the observation model Y_t. The dashed arc indicates an optional temporal dependence between the measurement model failures.

If we consider a single time slice, there are now no arcs entering R_t, so we refer to it as a root. Similarly, X_t will be called a leaf. It is also common to assume there is no arc from R_{t-1} to X_t, although this does not change the algorithm in any significant way.[1] After eliminating both of these inter-slice cross-arcs, we end up with the DBN shown in Figure 24.2(b).

Given the simplified model of Figure 24.2(b), we are left with two tasks: to sample the root nodes efficiently, and to update the leaf distributions efficiently, given that the roots have known values (and are therefore effectively disconnected from the graph). Because the roots can be sampled using a standard particle filter, many of the techniques discussed in this book are directly applicable. Hence we focus on the latter issue here.

A simple example of the leaves being efficiently updated occurs when the CPDs for X_t and Y_t are linear-Gaussian, in which case the model is called a conditionally linear-Gaussian state-space model. We can compute $P(X_t|r_{1:t}, y_{1:t})$ recursively using a Kalman filter. If R_t is discrete, the model is called a jump Markov linear system (JMLS), or switching state-space model (Chen and Liu 2000, Doucet et al. 1999).

The JMLS can be generalised in many ways. Some of these generalizations are still amenable to RBPF, but others require further approximations. A simple generalization that can still be handled by RBPF is to split the root node into two independent components, as in Figure 24.2(c). $R_t(1)$ is a parent of X_t, and can be used to model discontinuous

---

[1] Specifically, it just means we only have to condition on $r_t^{(i)}$, instead of both $r_t^{(i)}$ and $r_{t-1}^{(i)}$, when updating $\alpha_t^{(i)}$ in step 3 of the algorithm in Figure 24.1.

changes in state (c.f. McGinnity and Irwin (2001: this volume)). $R_t(2)$ is a parent of Y_t, and can be used to model outliers in the observations. (Another case in which we might have multiple root nodes is when we are doing on-line parameter estimation: in the Bayesian approach, the parameters are just additional random variables (nodes in the graph) that can be sampled using particle filtering.)

A second generalization of the JMLS is to put back the arc from X_{t-1} to R_t. This can be used to model the fact that the jumps depend on the system's (continuous) state, instead of occurring spontaneously. Unfortunately, equation 24.2.1 becomes difficult to compute in this case, since we have a hidden continuous parent connected to a (hidden) discrete child. One possible approach would be to use the variational approximation discussed in (Murphy 1999). Another would be to compute a Monte Carlo approximation of the sum, by sampling values of X_{t-1} from α_{t-1}.

A third generalization of the JMLS is to allow the system dynamics to be nonlinear, even conditioned on R_t. This makes it hard to integrate out X_t exactly. We could apply an approximate inference technique, such as the extended Kalman filter, or the unscented filter (Julier and Uhlmann 1997), but we would no longer be doing strict Rao-Blackwellisation. (In particular, these approximations may diverge.) So, the only known sound approach in this case is to use standard PF (perhaps with clever proposals).

24.2.1 How do we choose which nodes to sample?

Given an arbitrary DBN with a potentially complex topology, which nodes should be considered as roots, and which as leaves? If we disallow arcs from X_{t-1} to R_t for the reasons discussed above, the answer is fairly straightforward. We define the set of nodes that are eligible to be roots, $\mathcal{R}$, to be all the nodes either having no parents, or whose parents are also in the set $\mathcal{R}$, possibly shifted back in time. We initialise $\mathcal{R}$ to be all the nodes $Z_t(i)$ that have no parents, or whose only parent is $Z_{t-1}(i)$. We then add to $\mathcal{R}$ all the nodes, $Z_t(i)$, whose parents are in $\mathcal{R} \cup \{Z_{t-1}(i)\}$. For example, in the DBN in Figure 24.3, we start with $\mathcal{R}_1$ containing all the nodes in dotted boxes, to get

$$\mathcal{R}_2 = \{R1o, R2o, RF1o, RF2o, MF1o, MF2o, MF12\}.$$

The idea is to keep growing the set $\mathcal{R}$ until the set of remaining variables, $\mathcal{X}_i \stackrel{\text{def}}{=} \mathcal{S} \setminus \mathcal{R}_i$, can be updated exactly and efficiently, where $\mathcal{S}$ represents all the nodes in a single slice of the DBN, and $\mathcal{R}_i$ represents the root set at iteration i of the above root-growing algorithm. In the two-tank case, $\mathcal{R}_2$ is locally maximal: there is no single node that can be added such that the desired closure property could be maintained. The next valid root set is $\mathcal{R}_3 = \mathcal{S}$, which corresponds to sampling all the variables, as in standard PF.

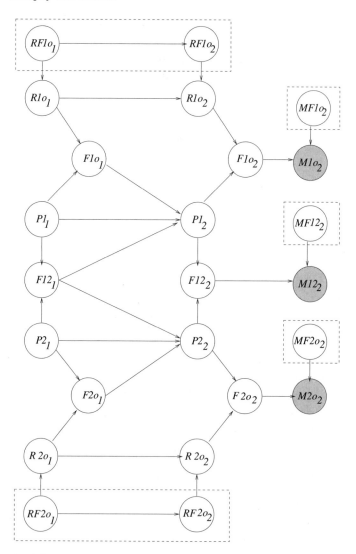

Figure 24.3. The two-tank DBN, adapted from Koller and Lerner (2001: this volume). We have omitted the $RF12$ and $R12$ nodes to keep the figure simple. RF standards for resistance failure, and represents a permanent change in the resistivity of a pipe. MF stands for measurement failure, and represents a temporary error in the sensor. The MF and observed (shaded) nodes are only shown for slice 2, for simplicity, since they are transient variables, i.e., not connected across time-slices.

Now suppose the set of non-boxed variables $\mathcal{X}_1$ had *linear*-Gaussian dynamics. In this case, we could sample the discrete root nodes $\mathcal{R}_1$ and apply the Kalman filter to the continuous nodes $\mathcal{X}_1$, as we discussed before. Expanding from $\mathcal{R}_1$ to $\mathcal{R}_2$, while legitimate, would probably not be very helpful, since $R1o$ and $R2o$ are jointly Gaussian with $\mathcal{X}_2$ by assumption, and hence can be marginalised out efficiently.

When all the (hidden) nodes in a DBN are discrete, we can always carry out exact inference in closed form, using the HMM filter or the junction tree algorithm (Smyth et al. 1997, Cowell et al. 1999). The problem is that the complexity is generally $O(k^n)$, where n is the number of hidden nodes, and k is the number of values each node can take on. (We shall see an example of this in Section 24.4.) As before, we should keep growing $\mathcal{R}$ until $\mathcal{X}$ becomes small enough that exact inference becomes computationally tractable.[2] Of course, the larger $\mathcal{R}$, the more samples we shall need, so this tradeoff must be made carefully. We hope to examine this issue in the future. In this chapter, we assume that the set $\mathcal{R}$ has been pre-specified.

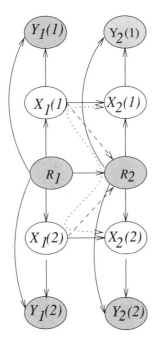

Figure 24.4. We have partitioned the leaves and observations into two components. The dashed arcs entering the root node induce correlation between the leaves, as indicated by the dotted path of influence, which passes through the root nodes, which are shown shaded, because they are instantiated.

---

[2]For example, applying the greedy root-growing algorithm to the BAT DBN (Figure 6 in the chapter by Koller and Lerner) gives the following results, where we use the notation $\Delta_i \stackrel{\text{def}}{=} \mathcal{R}_i \setminus \mathcal{R}_{i-1}$ to represent the extra nodes added at the ith iteration. $\mathcal{R}_1 = \{$LeftClr, RightClr, EngStatus, BYdotDiff, SensorValid, Bclr$\}$, $\Delta_2 = \{$ FYdotDiff, BXdot, BcloseFast, Fclr $\}$, $\Delta_3 = \{$ FcloseSlow $\}$, $\Delta_4 = \{$ FrontBackStatus $\}$. To continue growing, we would have to consider additions of two or more parents simultaneously, etc.

In some cases, the non-root nodes might be conditionally independent given the roots, as in Figure 24.4. We will see an example of this in Section 24.4. The advantage is that we can update each leaf independently, conditioned on the root, which is exponentially more efficient than updating the leaves jointly. Notice, however, that, if we had arcs from the leaves entering the root, as shown by the dotted lines in Figure 24.4, we would no longer be able to update the leaves independently. This is because R_t would act like a common observed child node, inducing correlation amongst its parents, a phenomenon called *explaining away* (Pearl 1988).

24.3   The RBPF algorithm in detail

We now explain the RBPF algorithm, which is sketched in Figure 24.1, in more detail. We assume that there are no arcs from X_{t-1} to R_t, and for simplicity, that there are also no arcs from R_{t-1} to X_t, so the generic structure is isomorphic to Figure 24.2(b).

As we mentioned in the introduction, the belief state, $P(X_t, R_{1:t}|y_{1:t})$, is represented by a set of N weighted particles. The marginal distribution on the root nodes is approximated as

$$P(r_{1:t}|y_{1:t}) \approx \sum_{i=1}^{N} w_t^i \delta(r_{1:t}^{(i)}, r_{1:t}),$$

where w_t^i is the weight of the ith particle, and $\delta(x, y) = 1$ if $x = y$ and is 0 otherwise. The marginal on the leaf is approximated as

$$
\begin{aligned}
P(X_t|y_{1:t}) &= \sum_{r_{1:t}} P(X_t|r_{1:t}, y_{1:t}) P(r_{1:t}|y_{1:t}) \\
&\approx \sum_{r_{1:t}} P(X_t|r_{1:t}, y_{1:t}) \sum_{i=1}^{N} w_t^i \delta(r_{1:t}^{(i)}, r_{1:t}) \\
&= \sum_{i=1}^{N} w_t^i P(X_t|r_{1:t}^{(i)}, y_{1:t}).
\end{aligned}
$$

For notational simplicity, we will assume that all nodes are discrete, so we can represent each leaf marginal as a vector in $[0,1]^k$ which sums to 1, where k is the number of possible values of the node. In addition, since the nodes are discrete, all the CPDs can be expressed as tables:

$$
\begin{aligned}
\pi_R(r) &\overset{\text{def}}{=} P(R_1 = r) \\
T_R(r'; r) &\overset{\text{def}}{=} P(R_t = r|R_{t-1} = r') \\
T_X(x', r; x) &\overset{\text{def}}{=} P(X_t = x|R_t = r, X_{t-1} = x') \\
O_Y(x, r; y) &\overset{\text{def}}{=} P(Y_t = y|X_t = x, R_t = r).
\end{aligned}
$$

Since the leaf nodes are discrete, we can update their distributions, conditional on having sampled R_t, using the HMM filter, as follows. First, we compute the one step-ahead prediction:

$$
\begin{aligned}
\alpha_{t|t-1}^{(i)}(x) &\overset{\text{def}}{=} P(X_t = x | y_{1:t-1}, r_{1:t}^{(i)}) \\
&= \sum_{x'} P(X_t = x | X_{t-1} = x', r_t^{(i)}) P(X_{t-1} = x' | y_{1:t-1}, r_{1:t-1}^{(i)}) \\
&= \sum_{x'} T_X(x', r_t^{(i)}; x) \alpha_{t-1}^{(i)}(x').
\end{aligned}
$$

Then we do Bayesian updating:

$$
\begin{aligned}
\alpha_t^{(i)}(x) &\overset{\text{def}}{=} P(X_t = x | y_{1:t}, r_{1:t}^{(i)}) \\
&= (1/C_t^{(i)}) P(y_t | X_t = x, r_t^{(i)}) P(X_t = x | y_{1:t-1}, r_{1:t}^{(i)}) \\
&= (1/C_t^{(i)}) O_Y(x, r_t^{(i)}; y_t) \alpha_{t|t-1}^{(i)}(x),
\end{aligned}
$$

where the denominator is equal to the likelihood

$$
C_t^{(i)} \overset{\text{def}}{=} P(y_t | y_{1:t-1}, r_{1:t}^{(i)}) = \sum_x O_Y(x, r_t^{(i)}; y_t) \alpha_{t|t-1}^{(i)}(x). \qquad (24.3.2)
$$

If we have L conditionally independent leaves, as in Figure 24.4, we apply these equations to each leaf separately. Each such update will result in a local likelihood term, $C_t(j)^{(i)}$, like the one above. The overall likelihood then becomes a product of the local likelihoods:

$$
\begin{aligned}
P(y_t | y_{1:t-1}, r_{1:t}^{(i)}) &= \sum_{x_1, \dots, x_L} \prod_{j=1}^{L} \left[P(y_t(j), X_t(j) = x_j | r_{1:t}^{(i)}, y_{1:t-1}) \right] \\
&= \prod_{j=1}^{L} \left[\sum_x P(y_t(j) | X_t(j) = x, r_{1:t}^{(i)}) P(X_t(j) = x | r_{1:t-1}^{(i)}, y_{1:t-1}) \right] \\
&= \prod_{j=1}^{L} C_t(j)^{(i)}.
\end{aligned}
$$

All that remains is to specify how to do the following standard PF steps.

- Sample new values of the roots.
- Compute the weight of a particle.
- Resample the particles.

We shall now explain these steps in detail. We drop the i superscript for brevity.

As discussed in the chapter by Doucet, de Freitas and Gordon (2001: this volume), in sequential importance sampling, if we sample from the proposal distribution $q(R_t; r_{1:t-1}, y_{1:t})$, we must assign the particle weight equal to the ratio between the true posterior and the proposal density:

$$w_t = \frac{P(y_t|y_{1:t-1}, r_{1:t})P(R_t|r_{t-1})}{q(R_t; r_{1:t-1}, y_{1:t})}.$$

The simplest case is if we sample from the prior, $q(R_t) = P(R_t|r_{t-1})$. In this case, the weight is simply the likelihood computed in Equation 24.3.2.

The optimal proposal distribution, in the sense of minimizing the variance of the importance weights (Doucet et al. 1999), is given by

$$P(R_t|r_{1:t-1}, y_{1:t}) = \frac{P(y_t|y_{1:t-1}, r_{1:t})P(R_t|r_{t-1})}{P(y_t|y_{1:t-1}, r_{1:t})},$$

where the denominator is the one-step-ahead likelihood

$$P(y_t|y_{1:t-1}, r_{1:t}) = \sum_{r=1}^{|R_t|} P(y_t|y_{1:t-1}, r_{1:t-1}, R_t = r)P(R_t = r|r_{t-1}).$$

This requires computing the likelihood $|R_t|$ times, which can be quite expensive. Whether the computational expense is worthwhile depends on the relative reliability of the observations and the transition prior: if the prior is weak (diffuse) and the observation likelihood is strong (sharply peaked), many particles may be proposed in a part of the state-space that has low likelihood, which is wasteful. In this case, it might be worthwhile to take the most recent evidence, y_t, into account before proposing. See Pitt and Shephard (2001: this volume) for a more detailed discussion.

Finally, given a set of particles and weights, we can resample a fresh set using any of the standard methods, such as residual resampling, discussed in Doucet, de Freitas, and Gordon (2001: this volume).

24.4   Application: concurrent localisation and map learning for a mobile robot

In this section, we discuss an application of RBPF to a highly simplified version of the problem of map learning for mobile robots (Murphy 2000). The application of standard (non-RB) PF to the problem of robot localisation (finding the robot's position given a known map) using real robots is discussed in Thrun et al. (2001: this volume).

Consider a robot that can move on a discrete, two-dimensional grid. The goal is to learn the color of each grid cell, which, for instance, can be either black or white. The difficulty is that the color sensors are not perfect (they may accidently flip bits with probability p_o), nor are the motors (the robot may fail to move in the desired direction with probability p_a, owing e.g., to wheel slippage). Consequently, it is easy for the robot to get lost. And when the robot is lost, it does not know what part of the map to update.

(Note that, if the robot always knew its location, e.g., by using GPS, map learning would be easy; unfortunately, GPS does not work indoors, nor is it accurate enough.)

The optimal Bayesian solution to this problem is to maintain a belief state over both the location of the robot, $L_t \in \{1, \dots, N_L\}$, and the color of each grid cell, $M_t(l) \in \{0,1\}$, $l = 1, \dots, N_L$, where N_L is the number of cells. For notational simplicity, in this subsection we shall assume that there are only two colors; the technique generalises easily.

We assume that the color of the cells can change, to represent the fact that the environment can be dynamic. For example, in Section 24.4.2, we use four colors that represent whether a cell is unoccupied, or contains a wall, or an open door, or a closed door, and we allow doors to change between open and closed. In this case, M_t is like an occupancy grid (Moravec and Elfes 1985), which is a simple kind of map. For simplicity, we assume that the colors of the cells change independently, but with an identical distribution, specified by the matrix $T_M(c; c') \stackrel{\text{def}}{=} P(M_t(l) = c' \mid M_{t-1}(l) = c)$. If this is an identity matrix, it means that the colors do not change.[3]

The observation model is that the robot sees the color of the grid cell at its current location, corrupted by noise:

$$P(y_t|m_1, \dots, m_{N_L}, L_t = l) \stackrel{\text{def}}{=} B(m_l; y_t) = \begin{cases} 1 - p_o & \text{if } y_t = m_l \\ p_o & \text{if } y_t \neq m_l, \end{cases}$$

where p_o is the probability that a color gets misread, and B is the 2×2 observation matrix with $1 - p_o$ on the diagonal and p_o off the diagonal. A more realistic model would capture the fact that the robot can see the color of neighboring cells as well. This is not hard to do, but for notational simplicity, we shall stick to the single-cell model for now.

Let us assume for now that the robot is in a one-dimensional grid world, so it can only move left or right, depending on the control input, U_t. The robot moves in the desired direction with probability $1 - p_a$, and otherwise stays put. In addition, it cannot move off the edges of the grid. Algebraically, this becomes

$$P(L_t = l'|L_{t-1} = l, U_t = \rightarrow) = \begin{cases} 1 - p_a & \text{if } l' = l + 1 \text{ and } 1 \leq l' < N_L \\ p_a & \text{if } l = l' \text{ and } 1 \leq l' < N_L \\ 1 & \text{if } l = l' = N_L \\ 0 & \text{otherwise.} \end{cases}$$

---

[3]If the colors do not change, and if we are satisfied with learning a maximum likelihood estimate of the map, instead of a full posterior, we can treat the $M(i)$s as fixed parameters and use EM to learn them (Thrun, Burgard and Fox 1998). However, doing this on-line results in learning a map that bears little resemblance to the true mode of the posterior (Murphy 2000).

The equation for $P(L_t = l'|L_{t-1} = l, U_t = \leftarrow)$ is analogous. In Section 24.4.2, we shall consider the case of a robot moving in two dimensions. In that case, the motion model will be slightly more complex.

Finally, we need to specify the prior. The robot is assumed to know its initial location, so the prior $P(L_1)$ is a delta function. If the robot did not know its initial location, there would be no well defined origin to the map; in other words, the map is relative to the initial location. Also, the robot has a uniform prior over the colors of the cells. However, it knows the size of the environment, so the state-space has fixed size.

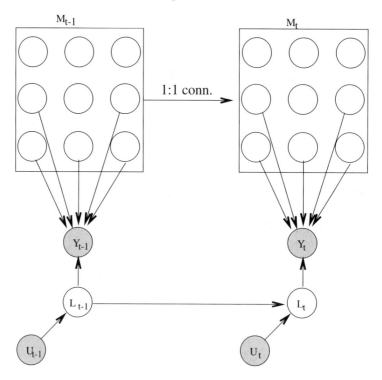

Figure 24.5. The DBN used in the map-learning problem. $M_t(l)$ represents the color of grid cell l at time t, L_t represents the robot's location, Y_t the current observation, and U_t is the current input (control).

The DBN we are using is shown in Figure 24.5. This kind of topology (ignoring the observed input nodes) is called a factorial HMM (Ghahramani and Jordan 1997). Inference in these models is computationally intractable. To be precise, if there are n chains, each of which can take k possible values, then the belief state has size $O(k^n)$: all the chains become coupled because Y_t is a common observed child. Exact inference, using the junction tree algorithm (Smyth et al. 1997, Cowell et al. 1999), takes $O(nk^{n+1})$ operations per time step. In general, if each node in chain j can take on

k_j values, then the belief state has size $S = \prod_{j=1}^{n} k_j$, and exact inference takes $O(S \sum_{j=1}^{n} k_j)$ operations per time step. In our application, there are N_L chains with 2 possible values each, and one chain with N_L possible values, so we need $3N_L^2 2^{N_L}$ operations per time step. In Section 24.4, we use a 10×10 grid, so this requires $O(2^{100})$ operations per time step for exact inference.

In our case, however, the observation model has the crucial property that the observation Y_t only depends on a single map element of $\mathbf{M}_t$ once the location L_t is known. Notice that this conditional independence property is not obvious from the structure of the graph, but is implicit in Y_t's CPD. (See also (Boutilier, Friedman, Goldszmidt and Koller 1996).) Consequently, we can rewrite the model of Figure 24.5 to take the form of Figure 24.2(a), as follows. L_t corresponds to the root R_t, and the map cells $M_t(j)$ are correspond to the leaves $X_t(j)$. (We ignore the U_t nodes for simplicity.) The transition model for the root is the motion model of the robot, and the transition model of the leaves is the matrix T_M, defined to be independent of L_t. Finally, the observation nodes are as follows. We define

$$P(Y_t(j) = m' | M_t(j) = m, L_t = l) = \begin{cases} B(m; m') & \text{if } j = l \\ 1/2 & \text{if } j \neq l, \end{cases}$$

where $j, l \in \{1, \dots, N_L\}$ and $m, m' \in \{0, 1\}$. The parent node L_t is acting like a switch, and ensures that the observation model only gives information about $M_t(l)$, where $L_t = l$ is the robot's current location; for all other cells, the observation model is non-informative. It is now straightforward to apply RBPF to this model.

24.4.1   Results on a one-dimensional grid

To evaluate the effectiveness of this algorithm, we first applied it to a problem which was sufficiently small (8 cells) that we could compute the ground truth using exact inference. In particular, consider the one-dimensional grid shown in Figure 24.6. We have $N_L = 8$, so exact inference takes about 50,000 operations per time step (!). For simplicity, we fixed the control policy as follows: the robot starts at the left, moves to the end, and then returns home. Suppose there are no sensor errors, but there is a single "slippage" error at $t = 4$. We summarise this below, where U_t represents the input (control action) at time t.

t	1	2	3	4	5	6	7	8	9	10	11	12	13	14	15	16
L_t	1	2	3	4	4	5	6	7	8	7	6	5	4	3	2	1
Y_t	0	1	0	1	1	0	1	0	1	0	1	0	1	0	1	0
U_t	-	$\rightarrow$	$\rightarrow$	$\rightarrow$	$\rightarrow$	$\rightarrow$	$\rightarrow$	$\rightarrow$	$\leftarrow$	$\leftarrow$	$\leftarrow$	$\leftarrow$	$\leftarrow$	$\leftarrow$	$\leftarrow$	$\leftarrow$

Figure 24.6. A one-dimensional grid world.

To study the effect of this sequence, we used exact inference to compute $P(L_t|y_{1:t})$ and $P(\mathbf{M}_t|y_{1:t})$: see Figure 24.7. At each time step, the robot thinks it is moving one step to the right, but the uncertainty gradually increases. However, as the robot returns to familiar territory, it is able to better localise itself, and hence the map becomes sharper, even in distant cells. Notice that this effect only occurs because we are modelling the correlation between cells. (See also the stochastic map representation of (Smith, Self and Cheeseman 1988).) For a more detailed interpretation of this example, see (Murphy 2000).

In Figure 24.8, we show the results obtained using RBPF. We see that it approximates the exact solution very closely, using only 50 particles. The results shown are for a particular random number seed; other seeds produce qualitatively very similar results, indicating that 50 particles are in fact sufficient in this case. Obviously, as we increase the number of particles, the error and variance decrease, but the running time increases linearly.

The question of how many particles to use is a difficult one: it depends both on the noise parameters and the structure of the environment. (If every cell has a unique color, localization, and hence map learning, is easy.) Since

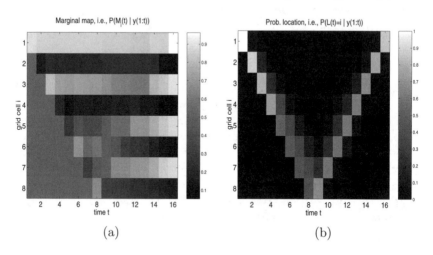

Figure 24.7. Results of exact inference on the 1D grid world. (a) A plot of $P(\mathbf{M}_t(i) = 1|y_{1:t})$, where i is the vertical axis and t is the horizontal axis; lighter cells are more likely to be color 1 (white). (b) A plot of $P(L_t = i|y_{1:t})$, i.e., the estimated location of the robot at each time step.

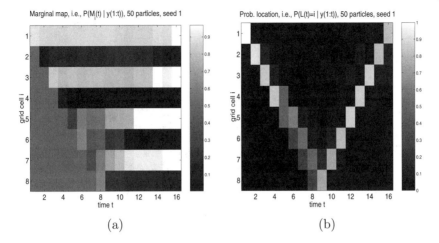

(a)   (b)

Figure 24.8. Results of the RBPF algorithm on the 1D grid world using 50 particles.

we are sampling trajectories, the number of hypotheses grows exponentially with time, as in a jump Markov linear system. In the worst case, the number of particles needed may depend on the length of the longest cycle in the environment, since this determines how long it might take for the robot to return to familiar territory and kill off some of the hypotheses (since a uniform prior on the map cannot be used to determine L_t when the robot is visiting places for the first time). In the above example, the robot was able to localise itself quite accurately when it reached the end of the corridor, since it knew that this corresponded to cell 8. In general, we may need to use a clever exploration policy, such as the one we discuss in the next section, to control the number of particles required.

For comparison, we also tried the Boyen-Koller (BK) algorithm (Boyen and Koller 1998), which is another popular approximate inference algorithm for discrete DBNs. In its simplest, fully factorised form, BK represents the belief state as a product of marginals:

$$P(\mathbf{M}_t, L_t | y_{1:t}) = P(L_t | y_{1:t}) \prod_{j=1}^{N_L} P(M_t(j) | y_{1:t}).$$

The results of using BK are shown in Figure 24.9. As we can see, it functions very poorly in this case, because it ignores correlation between the cells. Of course, it is possible to use products of pairwise or higher-order marginals for tightly coupled sets of variables. Unfortunately, there is no natural subset of variables to use in this case, since all the grid cells are potentially correlated.

514   Murphy and Russell

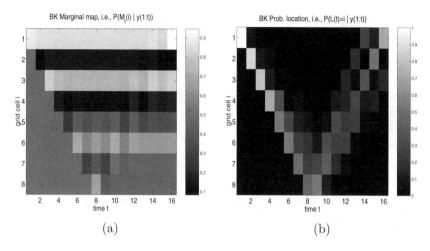

Figure 24.9. Results of the BK algorithm on the 1D grid world.

24.4.2   Results on a two-dimensional grid

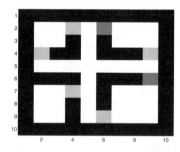

Figure 24.10. A simple 2D grid world. Grey cells denote doors (which are either open or closed), black denotes walls, and white denotes free space.

We now consider the 10×10 grid world in Figure 24.10. We use four colors, which represent closed doors, open doors, walls, and free space. Doors can toggle between open and closed independently with probability $p_c = 0.1$, but the other colors remain static; hence the cell transition matrix T_M is

$$\begin{pmatrix} 1-p_c & p_c & 0 & 0 \\ p_c & 1-p_c & 0 & 0 \\ 0 & 0 & 1 & 0 \\ 0 & 0 & 0 & 1 \end{pmatrix}.$$

The robot observes a 3×3 neighborhood centered on its current location. The total probability that each pixel gets misclassified is $p_o = 0.1$. The robot can move north, east, south or west; there is a $p_a = 0.1$ chance it will accidently move in a direction perpendicular to the one specified by U_t.

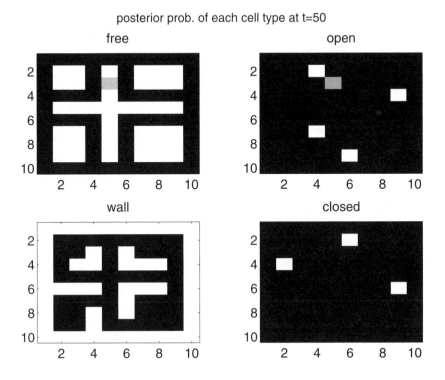

Figure 24.11. Results of RBPF on the 2D grid world using 200 particles.

We use a control policy that alternates between exploring new territory when the robot is confident of its location, and returning to familiar territory to re-localise itself when the entropy of $P(L_t|y_{1:t})$ becomes too high, as in (Fox et al. 1998); see (Murphy 2000) for details.

The results of applying the RBPF algorithm to this problem, using 200 particles, are shown in Figure 24.11. We see that by time 50, it has learnt an almost perfect map, even though the state-space has size 2^{100}.

24.5   Conclusions and future work

We have shown how to exploit tractable substructure in certain kinds of DBNs by combining particle filtering with exact inference. In the future, we hope to find more applications for this technique, and to extend the algorithm to the batch (off-line) case.

25

Particles and Mixtures for Tracking and Guidance

David Salmond
Neil Gordon

25.1 Introduction

The guidance algorithm is the central decision and control element of a missile system. It is responsible for taking data from all available missile sensors, together with targeting information and generating a guidance demand. This is usually in the form of an acceleration demand that is passed to the autopilot.

Broadly speaking, guidance algorithms are composed of two elements: estimation and control. The role of the estimator is to construct the probability density function (pdf) of the missile-target states required by the control law. The control element takes this pdf and translates it into a guidance demand. In current systems, the estimator is provided by a Kalman filter.

Optimal control theory is concerned with finding a control strategy that minimises a cost function that defines the objective. For the guidance problem, the objective is to effect an intercept with the required target, and therefore a suitable cost function is based on the missile-target miss distance. The modern subject of optimal control theory may be taken as starting with the work of (Pontryagin 1957, Bellman 1961), upon which much of the development of modern deterministic and stochastic control theory is based. In this chapter, missile guidance is presented as a stochastic control problem. Limitations of the solution of the mathematically tractable Linear-Quadratic-Gaussian (LQG) special case are then discussed. A particle based control law is then proposed and demonstrated in a simulation example involving guiding against a target with corruption from intermittent spurious objects. The problem is further complicated by the presence of finite sensor resolution, clutter and state dependent classification measurements.

25.1.1   Guidance as a stochastic control problem

Problems where the system states are not exactly known and the system to be controlled may be subject to random disturbances require stochastic control theory. For guidance problems we have an (uncertain) model for the system evolution, a model describing the relation between the measurements and the system state, and a utility (cost) function that details the gains (losses) of particular control options relative to the objectives of the engagement. This clearly fits into the stochastic control framework (Bayard 1991, Bertsekas 1987, Bertsekas 1995). (This framework is also referred to as reinforcement learning in the artificial intelligence literature.)

The system is assumed to evolve according to the discrete time dynamic model

$$x_{k+1} = f_k(x_k, u_k, w_k), \qquad (25.1.1)$$

where x_k is the state vector of the system, f_k is the (assumed known) model of the system dynamics, u_k is the control demand and w_k is the system noise (of assumed known pdf). The state vector contains all terms necessary to describe the system. For example, in a guidance application the state vector would typically include terms such as relative position and velocity.

Measurements are assumed to become available at timestep t_k and they are related to the system state by

$$z_k = h_k(x_k, e_k), \qquad (25.1.2)$$

where z_k is the measurement vector, h_k is the (assumed known) measurement function and e_k is the measurement noise (of assumed known pdf).

The cost function describing the engagement is taken to be

$$J = E\left(g_N(x_N) + \sum_{i=1}^{N-1} g_i(x_i, u_i) \right), \qquad (25.1.3)$$

where g_N is a function representing the terminal cost (usually a function of miss distance in guidance applications) and $\{g_i : i = 1, \dots, N-1\}$ are costs on the method of getting there (for example they could favour trajectories requiring minimal control effort). The cost is a random variable and so the optimal control is chosen on the basis of the expected value. This expectation is taken with respect to all unknown random quantities. These include the random noise sequences $\{w_i, e_i : i = 1, 2, ..., N\}$, the initial state of the system (as specified by a prior pdf) and the response of the observed object in light of the strategy adopted. The optimal choice will not only take into account the data currently available, but also information about what data are likely to become available by following a given policy.

Optimisation problems of this kind are usually solved by backward iteration from the terminal time step (assuming this is not also a random variable). If the final point is time step t_N then the control selected at time step t_{N-1} will be

$$\min_{u_{N-1}} E\Big(g_{N-1} + g_N \mid D_{N-1}\Big), \tag{25.1.4}$$

where $D_{N-1} = \{z_1, \ldots z_{N-1}\}$, all the data received up to time step t_{N-1}. The expectation is taken with respect to the distribution of all unknown quantities. This process then iterates backwards through alternating minimisations and expectations to give the control choice at time step t_k as the solution of

$$\min_{u_k} E\Big\{g_k + \min_{u_{k+1}} E\Big\{g_{k+1} + \ldots$$
$$+ \min_{u_{N-1}} E\Big\{g_{N-1} + g_N \mid D^{N-1}\Big\} \ldots \mid D^{k+1}\Big\} \mid D^k\Big\} \tag{25.1.5}$$

The solution of this optimisation problem will generally be a trade-off between the conflicting requirements of learning about the unknown states and achieving the terminal objective. Apart from the (restrictive but important) special case of linear-quadratic-Gaussian (LQG) assumptions for the models describing the system, (25.1.5) is not in general soluble.

A further complication arises because the above development assumes that the target response is independent of the chosen control. Clearly, this is unrealistic, and instead the joint strategies of the pursuer and pursued should be considered. This leads us into the requirement for a game-theoretic framework to describe the interaction. In the development below, we shall avoid this complication by assuming that the target does not respond to the pursuer's behaviour.

The solution is to find a control strategy of the form $u_k = u_k(D_k, U_{k-1})$, where $U_k = (u_1, \ldots, u_k)$, that minimises the cost function J under the constraint $u_k \in \overline{U}_k$ (some admissible set of controls).

25.1.2 Information state

In the above statement of the control problem, the requirement is to define a control law u_k that maps past and current measurements D_k and past controls U_{k-1} onto the set of admissible controls. The number of elements in the set (D_k, U_{k-1}) increases with k and so also does the complexity of the control law. Thus, there is a strong motivation to find a reduced set of parameters, or at least a more convenient representation of fixed dimension, that summarises the necessary information for control purposes. Such an equivalent representation is known as a sufficient statistic or an information state (see (Bertsekas 1987, Bar-Shalom 1981)). For the control problem, the most useful sufficient statistic is the conditional pdf of the system state, $p(x_k \mid D_k, U_{k-1})$. Thus, the control problem can be effectively restated: instead of requiring a mapping from (D_k, U_{k-1}) to the admissible controls, an equivalent problem is to find a mapping from $p(x_k \mid D_k, U_{k-1})$ to the admissible controls. The proof that this conditional pdf is an information state relies on the assumption that x_k is an incompletely observed Markov process.

The use of the conditional pdf as an information state provides a convenient bridge to Bayesian state estimation. Thus, for most problems of interest the optimal stochastic controller may be viewed as two distinct elements.

- A recursive Bayesian estimator that constructs $p(x_k \mid D_k, U_{k-1})$ from z_k, u_{k-1} and from the previous pdf $p(x_{k-1} \mid D_{k-1}, U_{k-2})$

- A control law that produces u_k as a function of $p(x_k \mid D_k, U_{k-1})$ - with knowledge of the future system and measurement models

Apart from being supplied with the deterministic input u_{k-1}, the estimator is quite separate from the control law. This partitioning of the stochastic control problem is the starting point for the derivation of many sub-optimal controllers. However, it should be noted that, although the conditional pdf is equivalent to (D_k, U_{k-1}) for control purposes, it does not necessarily provide a representation of reduced dimension. Only for certain special cases is a significant reduction in dimension possible. By far the most important of these is the Linear-Gaussian (LG) class of problem, where (25.1.1,25.1.2) are linear models with Gaussian disturbances. Here, the Bayesian estimator reduces to Kalman filter and the required conditional pdf is a Gaussian which can be precisely described by a mean and covariance.

25.1.3   Dynamic programming and the dual effect

The optimal solution to the (discrete) stochastic control problem may, in principle, be obtained by dynamic programming. This concept was introduced by (Bellman 1961). It may be shown (Aoki 1967, Feldbaum 1965, Bertsekas 1987) that the optimal control is the solution to the stochastic Bellman equation:

$$J_k(D_k, U_{k-1}) = \min_{u_k} E\left(g_k(x_k, u_k) + J_{k+1}(D_{k+1}, U_k) \mid D_k, U_{k-1} \right)$$

$$(25.1.6)$$

for $k < N$, with end condition

$$J_N(D_N, U_{N-1}) = E\left(g_N(x_N) \mid D_N, U_{N-1} \right).$$ $\quad (25.1.7)$

Here, $J_k(D_k, U_{k-1})$ is the expected cost from time t_k to the end given the values D_k and U_{k-1}. (25.1.6) is a statement of the principal of optimality (see (Bellman 1961)) for this problem. The two terms on the right hand side of (25.1.6) show how the cost is built up recursively. The first term g_k is the cost incurred for the current stage k, and the second term is the "cost-to-go" from t_{k+1} to t_N. Thus, the optimal control is found by minimising over the current system state and the most probable future behaviour of the system.

From above, it can be seen that, although the control is causal and is a function of the values of the past and current measurements, the optimal control takes into account that future measurements will become available (the future measurements are effectively averaged out in (25.1.6), conditional upon the present information). This can lead to a most interesting phenomenon known as "dual-effect" (Feldbaum 1965, Bar-Shalom and Tse 1974). Here the optimal control attempts actively to learn about the system with the ultimate objective of improving the control performance. This is exhibited in deliberate "probing" by the control that does not help directly to achieve the control objective. A system for which the control has no effect on the learning process is said to be "neutral". The converse of probing is "caution". Here the control takes account of the uncertainty in the state, typically by moderating the magnitude of the demand (relative to the equivalent deterministic problem). Caution is sometimes exhibited by a "hedging" type of behaviour. The solution to (25.1.6) achieves the optimal balance between active probing to reduce the uncertainty, caution due to uncertainty and deliberate control to realise the objective (see (Bar-Shalom 1981)). Clearly, the dual effect phenomenon plays no part in full information systems where the state vector is known (as in the deterministic case), and control effort is solely devoted to achieving the objective. Missile guidance schemes involving dual control are reported in (Birmiwal and Bar-Shalom 1984, Hull, Speyer and Burris 1990).

Dynamic programming provides a valuable insight into the nature of optimal stochastic control, but, from a practical standpoint, the procedure is usually infeasible. The control part of the problem is bedeviled by the "curse of dimensionality" (Bellman 1961) and compounded by the difficulty of obtaining the conditional expectations in (25.1.6).

25.1.4 Separability and certainty equivalence

A stochastic control problem is said to possess the certainty equivalence property if the optimal control law is of the same functional form as that of the optimal feedback control law for the equivalent deterministic system. The optimal stochastic control law is obtained using the expected value of the state in place of the actual state, which is not available for the stochastic problem. Thus, if $u_k = U_k^D(x_k)$ is the optimal control for the deterministic system and the stochastic system has the certainty equivalence property, then the optimal control is given by $u_k = u_k^D(\hat{m}_k)$, where $\hat{m}_k = E(x_k \mid D_k, U_{k-1})$, the expected value of the system state given all current information. Thus, at each time step a deterministic optimisation problem based on $\hat{m}_k$ and the expected values of all stochastic parameters in the system model must be solved (unless it is possible to obtain a closed form expression for $u_k^D(.)$ in advance). By far the most important class of problem for which the certainty equivalence property holds is the linear-quadratic-Gaussian (LQG) class. In this case $\hat{m}_k$ may be obtained from a Kalman filter and the optimal control law is a linear function of $\hat{m}_k$ (the optimal state regulator).

The certainty equivalence class is a subset of the wider (but somewhat less interesting) class of separable problems. For separable problems, the optimal control is of the form $u_k = u_k(\hat{m}_k)$, where the function $u_k(.)$ is not necessarily the same as $u_k^D(.)$. This is the usual definition of separable (Bar-Shalom and Tse 1974, Bertsekas 1987) although the term is sometimes used loosely.

25.1.5 Sub-optimal strategies

As is clear from the above discussion, optimal solutions to the stochastic control problem are only available for a few (admittedly very useful) special cases, notably the class of LQG problems. Thus, sub-optimal strategies are of great interest. Possibly the most common approach is to approximate a problem to the LQG framework (by linearisation) and then apply this control law. Typically, this will involve the extended Kalman filter. Although this solution is sub-optimal, for mild deviations from LQG assumptions, the result is often quite satisfactory. For more gross deviations, such as significant nonlinearity, measurement association uncertainty or non-quadratic cost functions, the LQG type solution is unlikely to be appropriate.

It has been suggested (Bar-Shalom 1981) that sub-optimal strategies for stochastic control can be conveniently divided into two broad classes, feedback algorithms and closed-loop algorithms.

Feedback control

Feedback control strategies depend on the information that is currently available (such as $p(x_k \mid D_k, U_{k-1})$ or (D_k, U_{k-1})), but ignore the possibility that future measurements may become available. That is, prior statistical descriptions of future data are not used: no future feedback is expected. However, information on the future behaviour of the system (via the system model and system noise statistics) may be used. This type of strategy precludes the possibility of any probing and the system merely learns passively as measurements are received.

The application of certainty equivalence is an example of feedback control. Another feedback strategy is the so-called open-loop feedback controller (OLFC) (also known as the open-loop optimal feedback (OLOF) controller), which explicitly takes account of current uncertainty in x_k and the future disturbances $w_k, w_{k+1}, \ldots, w_{N-1}$. Thus, OLOF control takes account of the full information state $p(x_k \mid D_k, U_{k-1})$ rather than just its mean. For the linear-Gaussian problem, the OLOF strategy can take account of the covariance P_k of the state estimate, as well as the mean $\hat{m}_k$ (Tanner 1995). It has been shown (Bertsekas 1987) that, unlike the certainty equivalence control, the average performance of the OLOF control law is at least as good as that of the open-loop-optimal (OLO) control (i.e., assuming no measurements at all). Notice that because OLOF ignores future measurements, it tends to produce an overcautious control as the expected increase in uncertainty owing to system noise is not balanced by information from new measurements.

Closed-loop control

These strategies make use of current information and take at least some account of the fact that future measurements may be available. This type of strategy may result in active probing by the control. Sub-optimal strategies of the closed-loop type include limited look ahead policies where the control makes use of the statistical description of the next N_L measurements that may be received. Other sub-optimal strategies include somewhat ad-hoc control excursions to improve the observability of the system. In general, closed-loop strategies are more difficult to obtain than feedback policies.

25.2   Derivation of control laws from particles

25.2.1   Certainty equivalence control

Apart from completely open loop control (which ignores all measurements and so is useless for guidance applications), the certainty equivalence assumption is the most straightforward route to a control with possibly useful performance. As indicated in Section 25.1.4, the certainty equivalence assumption is that all uncertain quantities assume their expected values, thereby converting the stochastic problem into a deterministic one. Thus, at time step k, the problem is to find the sequence of future controls $u_k, u_{k+1}, \ldots, u_{N-1}$ that minimises the cost function

$$J_k^{CE}(D_k, U_{k-1}) = \min \left[g_N(\overline{m}_N) + \sum_{i=k}^{N-1} g_i(\overline{m}_i, u_i) \right], \qquad (25.2.8)$$

where $\overline{m}_{i+1} = f_i(\overline{m}_i, u_i, \overline{w}_i)$ for $i \geq k$ with $\overline{m}_k = \hat{m}_k = E(x_k \mid D_k, U_{k-1})$ and $\overline{w}_i = E(w_i)$. The expected value $\hat{m}_k$ may be approximated from the sample set.

Hence, using the system model as indicated above, the cost function may be evaluated for any control sequence, so at each time step a deterministic optimisation must be carried out. Alternatively, for certain forms of f_i and

g_i, it may be possible to obtain a closed form solution for u_k as a function of $\hat{m}_k$ (especially for the LQ problem).

Notice that the above scheme makes poor use of the available information on the uncertainty in x_k contained in the sample set. For most (non-LQ type) problems, this information is relevant. Also, for certainty equivalence control, it may be possible to obtain a reasonable (adequate) estimate of $\hat{m}_k$ from some other, more simple, estimation scheme. A variation on this procedure might be to use a location estimate other than the mean – for example it might be preferable to base the certainty equivalence control on the MAP estimate of x_k, especially if the pdf is highly multi-modal.

25.2.2 A scheme based on open-loop feedback control

This scheme is based on a simple and direct evaluation of the desired cost function with the OLOF (or OLFC) approximation – it may be viewed as an extension of the certainty equivalence method. As explained in Section 25.1.4, the OLOF approximation ignores the possibility of future measurements at each time step when a control decision is obtained (and so avoids the possibility of a dual control strategy). Thus, at each time step k, the problem is to find the sequence of future controls $u_k, u_{k+1}, \dots, u_{N-1}$ that minimises the cost function

$$J_k^{OLOF}(D_k, U_{k-1}) = \min_{u_k, u_{k+1}, \dots, u_{N-1}} E\left(J \mid D_k, U_{k-1}\right) \qquad (25.2.9)$$

within any constraints on the control sequence. Here the expectation in equation (25.2.9) is over the current and future uncertainty in the state given (D_k, U_{k-1}), but not averaging over the contribution of future measurements, which are ignored. Thus,

$$J_k^{OLOF}(D_k, U_{k-1}) = \min_{u_k, u_{k+1}, \dots, u_{N-1}} \left(\int g_N(x_N)p(x_N \mid D_k, U_{N-1})dx_N + \sum_{i=k}^{N-1} \int g_i(x_i, u_i)p(x_i \mid D_k, U_{i-1})dx_i \right), \quad (25.2.10)$$

where $p(x_i \mid D_k, U_{i-1})$ for $i > k$ is the predicted pdf of this state given the information available at time t_k. If the estimation side of the problem is to be tackled via sequential Monte Carlo methods, then an approximation of the predicted pdf is available as $\{x_{i|k}(i) : i = 1, \dots, N_s\}$, where N_s is the size of the representative sample set. Thus, for any future sequence of controls $\{u_k, u_{k+1}, \dots, u_{N-1}\}$, the integrals in (25.2.10) may be approximately evaluated to give

$$J_k^{OLOF}(D_k, U_{k-1}) = \min_{u_k, u_{k+1}, \dots, u_{N-1}} \sum_{i=1}^{N_s} \left[g_N(x_{N|k}(i)) + \sum_{l=k}^{N-1} g_l(x_{l|k}(i), u_l) \right].$$
$$(25.2.11)$$

Thus, in principle, it is possible to search the control space to find the control sequence that minimises the cost function $J_k^{OLOF}(D_k, U_{k-1})$: this would be the OLOF control. In general, such a procedure would be computationally prohibitive, but may be tenable for certain special cases, such as when the number of control options is limited. Also note that this scheme could be applied with any set of cost functions g_k. The control law applied to the guidance example in the next section is a problem specific approximation to OLOF control.

25.3   Guidance in the presence of intermittent spurious objects and clutter

25.3.1   Introduction

The potential advantages of sample based guidance schemes are likely to be most apparent for distinctly non-Gaussian problems. This is so for the example of guidance in the presence of intermittent spurious objects (such as decoys) and clutter presented here. This example was introduced from a tracking perspective in (Salmond, Fisher and Gordon 1998, Gordon, Marrs and Salmond 2001) where results based on the hybrid bootstrap filter were given. In this section, the analysis is extended with guidance solutions based on the control law suggested above.

25.3.2   Problem formulation

The nonlinearity of this guidance example owes to the presence of an intermittent interfering object in close proximity to the required target. The decoys are spawned in close proximity to the required target at random times and persist for random periods. Only one such source is present at any instant and its birth/death process is modelled by a Markov process. The sensor (e.g. radar) receives measurements from both the required target and the decoy (if present). However, the resolution of the sensor is finite, so when in close proximity, the target and spurious object may generate only a single common return. Also, the sensor may produce random spurious measurements (clutter). Associated with each measurement is a classification flag derived within the sensor signal processing. This classification is imperfect and only produces a correct result with some known probability. The performance of the classifier depends upon the aspect of the target subtended to the sensor (i.e., it is state dependent). The problem is to generate a guidance command in spite of all the interfering measurements.

Here we consider an application of particle filters to this problem. It was noted by (Avitzour 1995) that particle filters can be applied directly to problems involving measurement association uncertainty, provided the measurement likelihood can be specified (also see (Gordon 1997)). The main advantage of this approach is that measurement association hypotheses from previous time steps do not need to be explicitly stored - they are contained implicitly in the sample set. In this case, sub-optimality owes to the number of samples being finite.

25.3.3 Simulation example

Scenario

Both objects T and D (if it exists) are assumed to obey the common second order linear Cartesian model. The X-coordinate for this model is

$$\left.\begin{array}{l} x_{ik+1} = x_{ik} + \Delta t \dot{x}_{ik} + (\Delta t^2/2) w_{iXk} \\ \dot{x}_{ik+1} = \dot{x}_{ik} + \Delta t w_{iXk}, \end{array}\right\} \qquad (25.3.12)$$

for i=T and D. The Y coordinate is similar. The driving noise sequences w_{TXk}, w_{TYk}, w_{DXk} and w_{DYk} are independent Gaussian random processes of variance q. Notice that, since the models for T and D are the same in this example, trajectory characteristics cannot be used to distinguish between T and D. If the dynamics of T and D were known to be different and this could be modelled (as allowed in the general formulation), this would aid the discrimination process. The state vector for the system is

$$X_k = (x_T, \dot{x}_T, y_T, \dot{y}_T, x_D, \dot{x}_D, y_D, \dot{y}_D, \gamma)_k \,.$$

In the example below, the system driving noise variance q is set to 0.015^2 and the time step Δt=0.25. Also, the probability of D being born at time k+1 given that it does not exist at k is p_{01} =Pr$\{\gamma_{k+1} = 1 \mid \gamma_k=0\}$=0.1, and the probability of it dying is p_{10}=Pr$\{\gamma_{k+1} = 0 \mid \gamma_k=1\}$=0.2. Thus, the average lifetime of a D object is $5\Delta t$, and the average interval between the death of one D object and the birth of another is $10\Delta t$.

Measurements of range and bearing $(z_{ik} = (z_{irk}, z_{i\theta k}))$ are taken from a sensor platform. Thus, for measurements originating from T or D, the nonlinear measurement function is defined by

$$z_{iRk} = r_{\lambda(i)k} + v_{iRk} \quad \text{and} \quad z_{i\theta k} = \theta_{\lambda(i)k} + v_{i\theta k}, \qquad (25.3.13)$$

where for resolved objects (λ(i)=T or D)

$$r_{\lambda(i)k} = \sqrt{x^2_{\lambda(i)k} + y^2_{\lambda(i)k}} \quad \text{and} \quad \theta_{\lambda(i)k} = \tan^{-1}\left(\frac{y_{\lambda(i)k}}{x_{\lambda(i)k}}\right), \qquad (25.3.14)$$

and similarly for the case of an unresolved composite measurement (λ(i)=J) using the centroid of the individual states. The measurement errors v_{irk} and $v_{i\theta k}$ are independent, zero mean, Gaussian processes of variance σ_r^2 and σ_θ^2 , respectively (for λ(i)=T, D or J). T and D may be resolved provided they do not fall into the same range/bearing resolution cell (Δr, $\Delta\theta$), i.e.

$$P_{res}(x_T, x_D) = \begin{cases} 0 & \text{if } |r_T - r_D| < \Delta r \text{ and } |\theta_T - \theta_D| < \Delta\theta \\ 1 & \text{otherwise.} \end{cases}$$

$$(25.3.15)$$

For the example results presented below, $\sigma_r=0.01$, $\Delta r=0.05$, $\sigma_\theta=0.01$ radians and $\Delta\theta=0.05$ radians. Notice that the measurement error is substantially less than the sensor resolution, a common feature of radar systems. Also, the detection probabilities for T, D and the composite return are all 0.99 (i.e., $P_{TD} = P_{DD} = P_{JD} = 0.99$).

Clutter measurements are uniformly distributed in range and bearing over the field of view of the sensor. In the simulation experiments, an acceptance gate in the measurement space is defined to reject any measurements that clearly originate from clutter and do not assist in the estimation of the state vector. This gate is defined by the maximum extent of the predicted T and D position samples plus four standard deviations of the measurement error. In the simulation experiment below, the clutter density is high: the average number of clutter returns in a $\Delta r \times \Delta\theta$ resolution cell is 0.25 and the average number of returns in a $\sigma_r \times \sigma_\theta$ cell is 0.01.

Associated with each measurement is a discrete classification flag that takes the values T, D or C. The classifier performance of the sensor against the target is a function of the target aspect presented. Assuming that the target's axis is directed along its velocity vector, the classification performance is a function of $\psi = \theta_{heading} - \theta_T$ (where $-180° < \psi < 180°$), where θ_T is the sightline angle between the sensor and the target with respect to the X-axis, and

$$\theta_{heading} = \tan^{-1}(\dot{y}_T/\dot{x}_T) \qquad (25.3.16)$$

is the target heading relative to the X-axis. The classification probabilities for the simulation example are given in Table 5.1. Thus, it is assumed

Origin of actual measurement	Classifier output		
	T	D	C
T for $|\psi| \in (10°, 170°)$	0.60	0.30	0.10
T for $|\psi| \notin (10°, 170°)$	0.45	0.45	0.10
D	0.30	0.60	0.10
J	0.45	0.45	0.10
C	0.15	0.15	0.70

Table 25.1. Classification probabilities.

that for D and clutter, performance is independent of aspect. Notice that when the subtended target aspect is within 10° of the target's axis, the

classifier output is equally likely to be T or D. At other aspects, performance is useful. Also the classifier has a 10% chance of mistaking the T,
D or J for clutter, and correctly recognises clutter with a probability of
70%. The composite return J is equally likely to be classified as T or D
- there is no classification J in this example. The state dependence of the
target classification performance considerably complicates the estimation
problem. The accuracy of the state information affects the degree to which
the filter can rely on the classifications, while, conversely, it is possible to
learn about the direction of the target velocity vector from the sequence
of classifications. The particle filter is able to accommodate (and exploit)
this state dependence.

At k=0, $\gamma \equiv 0$ so that only T is present. The prior distribution of the
position of T is Gaussian with mean (x_{T0}, y_{T0}) and covariance

$$
\begin{pmatrix} \sigma^2_{TX0} & 0 \\ 0 & \sigma^2_{TY0} \end{pmatrix}.
$$

The prior distribution of the initial target velocity is defined in terms of
direction and magnitude, the direction being uniformly distributed over
$[0, 2\pi)$ and the magnitude being uniformly distributed over $[0, V_{Tmax}]$.
In the example below, $\bar{x}_{T0} = 0.1, \bar{y}_{T0} = 1.9, \sigma_{TX0} = \sigma_{TY0} =$
0.25 and $V_{T\,max} = 0.05$. The prior distribution of the position and velocity
of each D at birth is assumed to Gaussian with mean x_T and a diagonal
covariance matrix. In the example below, the standard deviation of the x
and y positions is $0.03(= 3\sigma_r = 0.6\Delta r)$ and the standard deviations of the
x and y velocity components is $0.01(= 0.2V_{Tmax})$. Thus, D type objects
are generated in close proximity and with similar velocities to the target
(relative to the problem parameters).

25.3.4   Guidance results

In this section we extend the previous example to the case of a guided
interceptor that carries the sensor. The interceptor is assumed to move
at a constant speed V_M in the same plane as the target. The heading of
the pursuer is controlled by a guidance demand u_k (turn rate) updated at
every time step using the output of the estimator. So, the heading ϕ of the
interceptor is given by

$$
\phi_{k+1} = \phi_k + u_k \Delta t, \tag{25.3.17}
$$

where $| u_k | < \Delta\phi_{MAX} = a_{MAX}/V_M$, a_{MAX} being the maximum lateral
acceleration capability of the interceptor (which is assumed to respond
instantly to the demand – i.e., is lag free).

Cost function

For guidance interception problems, it is often convenient to specify a cost function in terms of "zero effort miss" (ZEM) (Lawrence 1998). The ZEM m_k at time t_k is the miss distance that would occur if the pursuer were to continue on a constant velocity course defined by its current position and velocity - i.e., $u_j = 0$ for $j > k$. Miss distance is the Euclidean distance between the interceptor and the target at the point of closest approach. Nulling the ZEM is equivalent to putting the interceptor on a collision course. An approximate expression for the ZEM in terms of the current pursuer and target state is

$$m(x) = \frac{(x_T - x_M)(\dot{y}_T - \dot{y}_M) - (y_T - y_M)(\dot{x}_T - \dot{x}_M)}{\sqrt{(\dot{x}_T - \dot{x}_M)^2 + (\dot{y}_T - \dot{y}_M)^2}}. \qquad (25.3.18)$$

Here, $(x_T, y_T, \dot{x}_T, \dot{y}_T)$ are from the state vector of the filter and $(x_M, y_M, \dot{x}_M, \dot{y}_M)$ are the known position and velocity of the pursuer. The pursuer can attempt to counter the ZEM by generating a "divert" via a change in the heading. If the pursuer manoeuvres at a constant change of heading rate u and if the period from the current time to the point of closest approach is T, then a divert of approximately $uV_M T^2/2$ will be produced to counter the ZEM. Unfortunately, T is not known, but may be approximated by range divided by closing speed. Thus, an approximate expression for the divert in terms of the control u and the current pursuer and target states is

$$d(x, u) = \frac{1}{2} V_M u \frac{(x_T - x_M)^2 + (y_T - y_M)^2}{(\dot{x}_T - \dot{x}_M)^2 + (\dot{y}_T - \dot{y}_M)^2}. \qquad (25.3.19)$$

If we simplify the problem by restricting the choice of future controls to a constant turn rate, i.e., $u_j = u_k$ for $j > k$, then the optimisation requirement becomes: at each time step k, find the admissible control u_k to minimise the expected cost

$$J(u_k) = E\left[g\big(m(x_k) - d(x_k, u_k)\big) \mid D_k\right] \qquad (25.3.20)$$

(compare with equation (25.2.9)). Notice that this is not a true OLOF approximation because future uncertainty in the target state (owing to manoeuvres) is ignored. Using the N_s samples from the particle filter, in a similar fashion to the development in Section 25.2.2, we obtain the approximation

$$J(u_k) = \sum_{i=1}^{N_s} g\big(m(x_{k|k}(i)) - d(x_{k|k}(i), u_k)\big). \qquad (25.3.21)$$

The form of the cost function $g(.)$ requires careful consideration for this multiple object scenario. In particular, the familiar quadratic cost function is inappropriate because it always attempts to drive the system towards the

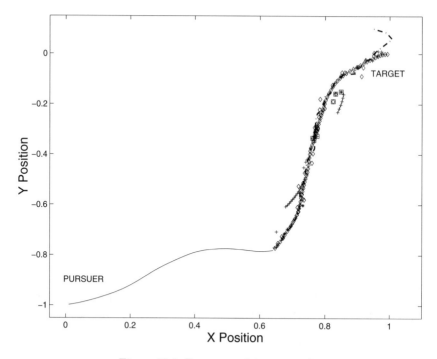

Figure 25.1. Pursuer and target tracks

expected value of the state. For a scenario in which the pdf of the state will frequently be multi-modal, this is highly undesirable. The problem owes to the unbounded nature of the quadratic cost function, which means that probability mass of the target distribution remote from accessible positions will still significantly affect the guidance demand. It is therefore more appropriate to employ a bounded cost function, and in this case we have used an inverse Gaussian function of the form

$$g(\tilde{m}) = 1 - exp\left(-\frac{\tilde{m}^2}{\sigma^2}\right), \qquad (25.3.22)$$

where the parameter σ determines the extent of the "well" in the cost (Tanner 1995). With this cost function, the penalty for missing any target probability mass is essentially constant (unity) if the miss distance exceeds 3σ. This is a more accurate description of what is usually required in an interception problem and it avoids the excessive hedging that can occur with a quadratic cost and a multi-model target pdf. Instead, there is a smooth transition from a hedging strategy in the early phases of guidance to a firm decision on a dominant mode as range closes. Notice that, for a given value of u, evaluation of the expected cost $J(u)$ via (25.3.18-25.3.21) and the sample set from the particle filter is straightforward - as it would be for any form of the cost function g. For our guidance example, where the control u is a scalar and can only take values over a finite interval, a near-optimal control can be simply obtained by evaluating (25.3.21) for a grid of u in the admissible interval.

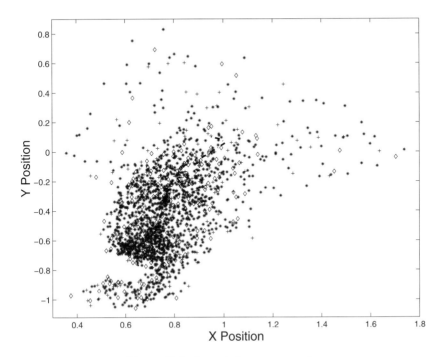

Figure 25.2. Accepted measurements.

In Figure 25.1, the actual paths of the target, pursuer and the intermittent objects are shown as continuous lines – the target line joining diamonds and the intermittent objects joining crosses. There are two D type paths (alive at different times) in this example. The pursuer path is a result of applying the guidance strategy described in the previous section. The parameters of the pursuer are: constant speed $V_M = 0.04$ and maximum lateral acceleration $a_{MAX} = 0.01$, so the maximum heading change per time step $\Delta\phi_{MAX} = 0.025$. The cost function parameter is set to $\sigma = 0.1$ and the control at each time step is obtained from (25.3.18) over a grid of 100 values. It can be seen that the pursuer achieves a satisfactory intercept of the target. The guidance law has successfully accommodated the uncertainty resulting from interfering secondary objects and dense clutter. To illustrate the difficulty of this scenario due to clutter interference, Figure 25.2 shows the full set of accepted measurements throughout the 70 time step scenario. Figure 25.3 shows the number accepted at each time step. Ignoring the other complicating features of the problem, this level of association uncertainty alone is completely beyond standard analytic ap-

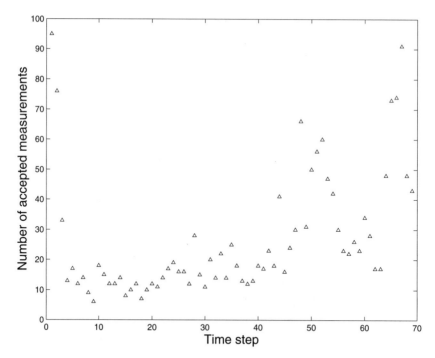

Figure 25.3. Number of accepted measurements.

proaches to the filtering problem, whereas the particle filter has been able to cope.

Acknowledgements

This research was sponsored by the UK MOD TG3 and TG9.
©Crown copyright 2000
Published with the permission of DERA on behalf of HMSO.

26

Monte Carlo Techniques for Automated Target Recognition

Anuj Srivastava
Aaron D. Lanterman
Ulf Grenander
Marc Loizeaux
Michael I. Miller

26.1 Introduction

Motivated by the extraordinary performance of the human visual system, automated recognition of targets from their remotely sensed images has become an active area of research. In particular, Bayesian techniques based on statistical models of system components such as targets, sensors, and clutter have emerged. Since the probability models associated with such physical systems are complicated, general closed-form analytical solutions are ruled out, and computational approaches become vital. This chapter explores a family of computational solutions applied to finding the unknowns associated with Automated Target Recognition (ATR) problems. In general ATR, remote sensors (such as visual or infrared cameras, or microwave or laser radars) observe a scene containing a number of targets, either moving or stationary. These sensors produce measurements that are analyzed by computer algorithms to *detect, track* and *recognise* the targets of interest in that scene. (Dudgen and Lacoss 1993) and (Srivastava, Miller and Grenander 1999) offer more detailed introductions.

A fundamental issue in ATR is the so-called variability of target pose. To illustrate, consider a regular hand-held camera taking images of a car. The images are vastly different depending on the relative orientation and the distance between the camera and the car. This underlines a major difficulty in the design of ATR algorithms: how can we mathematically model the variation in the images (pixel-values) owing to the variation in target pose? The task is further complicated by imperfections in the sensor and the presence of structured clutter in the scene, which often obscures the targets. Following Grenander's formulation of general pattern theory (Grenander 1993), we have separated the two components. On one hand, we have developed detailed models for handling the variation in target occurrences in three-dimensions. On the other hand, using physical considerations, we have statistically modelled the sensor-transformations that map three-dimensional targets into their two-dimensional images.

For modelling the target occurrences, we start with a three-dimensional rigid CAD model (a triangulated surface, for example) for each possible target we expect to find in the scene. These CAD models are called *templates*. The targets occur in a scene at arbitrary positions and orientations, with respect to the sensor, these occurrences are generated by the action of rotation and translation groups on the appropriate templates. Please refer to (Boothby 1986) for a review of group action and geometry of these groups. In this scheme, given an observed image, the target recognition task reduces to: (i) estimating the target pose and location, and (ii) selecting the most likely template.

Target orientation is represented by the elements of the special orthogonal group, $SO(3) = \{O \in \mathbb{R}^{3\times3} : O^\dagger O = I, \det(O) = 1\}$, and translation by the elements of $\mathbb{R}^3$. Together, their joint action is through the special Euclidean group $SE(3)$, the semi-direct product of $SO(3)$ and $\mathbb{R}^3$. The target type is assumed to be an element α of the discrete set $\mathcal{A}$. For instance, for the ground-based scenarios explored in this chapter, $\mathcal{A} = \{$M60-tank, T62-tank, M2-APC$\}$. If k targets are present, for some positive integer k, the total pose variation is an element of $SE(3)^k$. Since the number of targets k is unknown a-priori, the parameter space is generalised to include all values of k, i.e., define $\mathcal{X} = \bigcup_{k=0}^{\infty} \mathcal{X}_k$, where $\mathcal{X}_k = (SE(3) \times \mathcal{A})^k$; each $\mathcal{X}_k$ is then a subspace of $\mathcal{X}$. In the special case of targets located on the ground, we assume that the earth is flat, and the targets may be represented by elements of $\bigcup_{k=0}^{\infty}(SE(2) \times \mathcal{A})^k$.

Although there may be many other nuisance parameters, such as the locations of light sources (in video images) and the targets' heat profiles (in infrared images), this chapter will focus only on pose variation. The other variations can also be modelled as group actions on the templates in a similar way as pose and location variations. ATR algorithms are provided with either a single image or multiple images (a movie) to carry out parameter estimation and target recognition. In the case of a time-sequence of images, it is possible that there may be a relative motion between the sensor and the target: either one or both are moving, or in other words, the nuisance parameters are changing in time. For inference at an observation time n, the representation space extends to $\mathcal{X}^n$, $n = \{1, 2, \ldots, N\}$, where N is the total number of observed images. Shown in Figure 26.1 are three images of a stationary tank observed by a FLIR (forward looking infrared) camera mounted on a moving missile.

In our representation, any object present in the scene, that is not a target of interest is a source of *clutter*. Probability models for background clutter

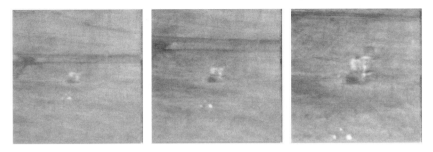

Figure 26.1. Sample FLIR sequence (courtesy Dr. Richard Sims, AMCOM).

play an important role in image understanding. An important question is: should the clutter be analyzed in three-dimensions by physically modelling sources of clutter, or should it be analyzed in two-dimensions by considering the pixels directly? The first approach is fundamental but prohibitive, while the second approach seems reasonable, but works with reduced knowledge. Taking the second approach, some simplified clutter models are developed in the paper (Grenander and Srivastava 2000). In this chapter, however, we avoid clutter models by assuming the background to be known completely.

This chapter considers two cases: (i) ATR from a single image, and (ii) ATR from multiple images having relative motion between the sensor and the targets. The second case deals with estimating nuisance parameters and target recognition in a sequential manner as multiple images are included in the inferences. The algorithms can be adapted to real-time implementation by considering conditionally optimal estimates and related conditional error bounds. Improvement in algorithmic performance, as more images are included in the inference, will be demonstrated.

26.1.1   The Bayesian posterior

To represent and analyze our knowledge about the observed scenes, we take a Bayesian approach by establishing a posterior probability on the representation space. The posterior has contributions from two quantities: a prior probability denoting our *a-priori* information about the targets and their behavior, and the sensor data likelihood expressing the probability of observed images for a given set of targets and their poses.

First, we consider the posterior on $\mathcal{X}$ (for ATR from a single image). For each subspace $\mathcal{X}_k$ (where $\mathcal{X}_k$ is a Lie group) a product of Haar measure on $SE(3)^k$ and a counting measure on $\mathcal{A}^k$, denoted γ_k, is chosen as the base measure with respect to which probability densities are defined. Let $\pi_{0,k}$ be a prior density on $\mathcal{X}_k$ and L be the image-likelihood. L is a function of a target configuration s and comes from the probability density of observing a given image conditioned on a value of s. Image synthesis models for various sensors and the resulting expressions for $L(s)$ are given later. Given k, the posterior probability measure on $\mathcal{X}_k$ is, for a measurable set $A \subset \mathcal{X}_k$,

$$\mu_k(A) = \frac{\int_A \pi_k(s)\gamma_k(ds)}{\int_{\mathcal{X}_k} \pi_k(s)\gamma_k(ds)},$$

where $\pi_k(s) \propto \pi_{0,k}(s)L(s)$ is the posterior density. For later convenience, we define $H_k \propto -\log(\pi_k)$ as the posterior energy on the subspace $\mathcal{X}_k$. Now, for all measurable sets $A \subset \mathcal{X}$, define the posterior probability measure μ, on the whole $\mathcal{X}$, according to

$$\mu(A) = \sum_{k=0}^{\infty} p(k) \int_{A \cap \mathcal{X}_k} \frac{1}{\mathcal{Z}} e^{-H_k(s)} \gamma_k(ds), \qquad (26.1.1)$$

with the normaliser $\mathcal{Z} = \sum_{k=0}^{\infty} p(k) \int_{\mathcal{X}_k} e^{-H_k(s)} \gamma_k(ds)$, and where $\sum_k p(k) = 1$.

Next, we extend this construction to $\mathcal{X}^N$ (for ATR from multiple images). For an observation time $n \in \{1, 2, \ldots, N\}$, define $\mu^{(n)}$ to be the posterior distribution of $A \subset \mathcal{X}^n$ measurable according to

$$\mu^{(n)}(A) = \sum_{k=0}^{\infty} p(k) \int_{A \cap \mathcal{X}^n} \frac{1}{\mathcal{Z}(n)} e^{-H_K^{(n)}(s)} \gamma_k^{(n)}(ds), \qquad (26.1.2)$$

where $H_k^{(n)} \propto -\log(\pi_k^{(n)})$ is the energy associated with the posterior $\pi_k^{(n)} \propto \pi_{0,k}^{(n)}(s)L^{(n)}(s)$ on $\mathcal{X}^n$ given n observations. $\pi_{0,k}^{(n)}(s)$ is the prior density function on $\mathcal{X}^n$ and, for example, can come from Newtonian dynamics governing the relative target/sensor motion, as described in (Miller, Srivastava and Grenander 1995). For real-time applications, with focus on speedier algorithms, $\pi_{0,k}^{(n)}$ can often be written in a recursive form in n. If the inference algorithm is iterative, requiring initial conditions, the solutions from $\mu^{(n-1)}$ can be used to initialise the procedure to generate inferences from $\mu^{(n)}$ using *sequential Monte Carlo* sampling. Shown in Figure 26.2 are two examples of conditional prior densities given all previous observations. For the case of airplane tracking in $I\!R^3$, the left figure shows high prior probability candidates (in broken lines), for extending the airplane track, conditioned on the previously estimated track (bold line). For the case of estimating tank orientation in $SO(2)$, the right figure displays a prior density on tank orientations, computed for the sampled points in $SO(2)$, conditioned on previously observed images.

26.1.2   Inference engines

As in traditional estimation-theoretic contexts, inferences are generated through statistics under the posterior and/or solutions with high posterior probabilities. Unlike the processing strategies for optimal detection and recognition in classical linear estimation problems, there is a nonlinear relation between the target parameters and the image synthesis models. The

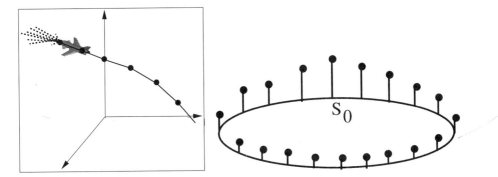

Figure 26.2. Depiction of prior densities on $I\!R^3$ (left) and $SO(2)$ (right) conditioned on the previous inferences.

posterior μ (or $\mu^{(n)}$) is often too complicated to analytically derive classical estimators such as conditional means and variances. Hence, taking a numerical approach, we generate these estimators empirically by sampling the posterior in such a way that averaging the samples asymptotically equals the conditional expectations. In addition to computing estimated target representations, we shall also characterise estimation performance by computing lower bounds on the expected estimation errors. For a specific choice of the distance function to specify the estimation errors, these bounds will serve as prognostic and diagnostic tools for algorithm evaluation.

The main tool presented here is a Markov process having two distinct components: (i) sample-path continuous diffusions generated by solving stochastic differential equations on the subgroups $\mathcal{X}_k$, and (ii) discrete jumps from one subspace to another to cover all of $\mathcal{X}$. This construction extends the *jump-diffusion* processes, originally described in (Grenander and Miller 1994) for vector spaces such as $I\!R^n$, to Lie groups having curved geometry (non-vector spaces). The diffusions search over the subspaces (connected sets), of a fixed dimension, while the jumps search over the discrete parameters such as the number of targets present (k) and their labels (α). This construction extends to sampling from $\mu^{(n)}$ on $\mathcal{X}^n$ by having a higher dimensional process, each component of which traverses the corresponding $\mathcal{X}$. For simplicity of presentation, we restrict to the single image ATR for the most part.

As described in (Geman and Geman 1984, Geman and Hwang 1987), for a fixed k, a diffusion process which samples from a given probability density π_k can be constructed using Langevin's stochastic differential equation (SDE)

$$dX(t) = -\nabla H_k(X(t))dt + \sqrt{2}dW(t), \tag{26.1.3}$$

where $W(t)$ is a standard Wiener process with appropriate dimensions. Given certain regularity conditions on H_k, π_k is the unique stationary probability density of this Markov process $X(t)$. However, Langevin's SDE is valid only for vector spaces; for curved Lie groups (such as $\mathcal{X}_k$), this equation has to be modified according to the underlying geometry. To illustrate, consider the curved component of $\mathcal{X}_k$, namely $SO(3)^k$. From the differential geometry of $SO(3)$, we specify the tangent vector space at each point on $SO(3)$. The process $X(t)$ is allowed to move only along the tangential (gradient) directions, resulting in a flow, which stays on $SO(3)^k$. Adding a stochastic component, similar to equation 26.1.3, results in a stochastic gradient flow, or a diffusion, on that subspace.

For extending the inference to the complete space, $\mathcal{X}$, a family of discrete moves called *jumps*, which take the process from one subspace $\mathcal{X}_k$ to another $\mathcal{X}_{k'}$, are added. The two components, jumps and diffusions, are combined together to form a jump-diffusion process $X(t)$ in such a way that it satisfies the ergodic result: the empirical averages of functions evaluated on the sample paths converge to their expectations under the posterior μ. This is a type of Markov chain Monte Carlo (MCMC) procedure for generating statistics from the posterior (conditional means, covariances, and others) when the posterior is highly nonlinear in the parameter space. To make it rigorous, we have studied equilibrium measures and ergodicity properties. Refer to (Srivastava 1996) for details. By selecting the infinitesimal drifts and jump probabilities appropriately, μ becomes the unique stationary measure of the jump-diffusion Markov process $X(t)$. From this, the ergodic result follows:

$$\frac{1}{T}\int_0^T f(X(t))dt \overset{T\to\infty}{\Rightarrow} \int_{\mathcal{X}} f(s)\mu(ds), \qquad (26.1.4)$$

for all bounded, measurable functions f defined on $\mathcal{X}$. f is determined by the cost function chosen for optimization. For example, consider the problem of estimating orientation (in $SO(2)$) of a single ground-target from one image. Let μ be the posterior on $SO(2)$ and for $s_1, s_2 \in SO(2)$, define $d(s_1, s_2) = \|s_1 - s_2\|_2^2$, where $\|\cdot\|_2$ denotes the Frobenious norm of a matrix. Then, we can compute an optimal estimate (in the MMSE sense) according to

$$\hat{s} = \underset{s_1 \in SO(2)}{\operatorname{argmin}} \int_{SO(2)} d(s_1, s_2)^2 \mu(ds_2). \qquad (26.1.5)$$

A detailed formulation and an algorithm to compute $\hat{s}$ is presented in (Grenander, Miller and Srivastava 1998). Furthermore, it is shown that the error associated with $\hat{s}$, called the Hilbert-Schmidt bound, is least among all possible orientation estimators. The Hilbert-Schmidt bound (HSB) involves a joint expectation over the parameter space and the observations, $HSB = \mathcal{E}(d(\hat{s}, s)^2)$, and is computed by particularizing the above-mentioned sampling techniques, as described in (Grenander et al. 1998).

In the case of ATR from image sequences, either the parameter space $(\mathcal{X}^n)$ or the posterior distribution $(\mu^{(n)})$ or both are changing in time. The inference, which constitutes conditional mean estimates, or simply high probability points at a particular time, corresponds to the posterior distribution on the appropriate parameter space at that time. Let $s_1^{(n)}, s_2^{(n)}, \ldots, s_j^{(n)}$ be the samples generated from $\mu^{(n)}$ and $\hat{s}^{(n)}$ be the resulting optimal MMSE estimate. If the prior distribution allows a recursive structure in n, then it is possible to simply map the samples/inferences at time n into the samples/inferences at times $(n+1)$ without resorting to the full sampling procedure at each sample time. The performance bounds (HSB) associated with these optimal estimates can also be propagated forward. In more general situations, where the complete sampling procedure is necessary, the samples $s_1^{(n)}, s_2^{(n)}, \ldots, s_j^{(n)}$ or inferences $\hat{s}^{(n)}$ can be used to initialise the Markov process for generating samples/inferences at time $(n+1)$.

A vast literature on MCMC based algorithms for sampling from given probability distributions is available; see, for example, (Green 1994, Green 1995), (Meyn and Tweedie 1993), (Roberts and Tweedie September, 1995) and many more. The book edited by Gilks, Richardson, and Spiegelhalter addresses these algorithms from a practical standpoint (Gilks et al. 1996); see also the papers (Besag and Green 1993, Smith and Roberts 1993). (Green 1994) has applied these techniques to pixel-based image analysis and object discovery. (Reno and Booth 1998) have constructed jump-diffusion algorithms to recognise 2-D targets in visual aerial images. Comprehensive ergodic results associated with discrete Markov chains and continuous Markov processes are proven in (Doss, Sethuraman and Athreya 1996), (Meyn and Tweedie 1993) and (Athreya and Ney 1972). This chapter employs results from (Grenander and Miller 1994, Srivastava, Grenander, Jensen and Miller December, 1999) to develop empirical approaches for ATR.

26.2   Jump-diffusion sampling

Our goal is to construct a process taking values in $\mathcal{X}$ (or $\mathcal{X}^n$) such that it samples from the given posterior probability measure μ (or $\mu^{(n)}$). For convenience we will focus on sampling from μ on $\mathcal{X}$ with the extension to $\mu^{(n)}$ being similar. In the latter case (ATR from image sequences), the task is to sample from $\mu^{(n)}$ on $\mathcal{X}^n$ for each n. The samples generated at n^{th}-stage are often used to either initialise the sampling process of the $(n+1)^{th}$-stage or sometimes directly modified to generate the samples for the $(n+1)^{th}$-stage. The latter procedure is feasible if there is a convenient, recursive structure on $\mu^{(n)}$ in terms of n. In that case, the samples from $\mu^{(n)}$ can simply be projected and scaled to obtain samples from $\mu^{(n+1)}$ (using ideas from *importance sampling*). However, in a general problem this structure may not be present and the solution relies on sampling from $\mu^{(n)}$ for each n in its entirety.

The sampling process $X(t)$ has two components: a sample path continuous diffusion component and a discrete jump component. At random times, called the *jump times*, the process transitions from one point in $\mathcal{X}$ to another, according to some pre-defined transition probability. The jumps can be to nearby points or points further away in $\mathcal{X}$, depending upon the choice of the transition probability. In between the jumps, the process follows a stochastic differential equation, of appropriate dimension, to generate diffusions in the current subspace. The diffusions are sample path continuous processes which travel short distances in small times. The choice of jump times, transition probabilities, and diffusion parameters (means and variances) depends on the posterior μ and the nature of the application. The diffusions contribute to the Bayesian estimation of pose/location parameters, whereas the jumps discover discrete characteristics of a scene, namely the target numbers and their labels. To illustrate, we first construct diffusion processes on Lie groups $(\mathcal{X}_k)$ of fixed dimension (k) and fixed labels (α) and later define a jump component that changes k and α.

26.2.1 Diffusion Processes

It is well-known that for (flat) vector spaces, the diffusions are generated as solutions of Langevin's SDE, equation 26.1.3, see for example (Geman and Geman 1984, Geman and Hwang 1987). This equation can also be viewed as a stochastic gradient equation with infinitesimal mean given by the gradient of the posterior energy $(-\nabla H_k(X(t)))$ and a constant infinitesimal variance $(\sqrt{2})$. We extend that idea to write stochastic differential equations on Lie groups $\mathcal{X}_k$ such that their solutions are diffusions with appropriate ergodic properties. The analogous stochastic differential equation on a Lie group is given by

$$dX(t) = -\sum_{i=1}^{n}(Y_{i,X(t)}H_k)Y_{i,X(t)}dt + \sum_{i=1}^{n}Y_{i,X(t)} \circ dW_i(t), \qquad (26.2.6)$$

where the Y_i's are orthonormal, smooth (tangent) vector fields on $\mathcal{X}_k$, n is the dimension of the group, and W_i's are real-valued, independent, standard Wiener processes.

Consider a simple case of sampling via diffusions on $SO(3)$ and let H_1 be the posterior energy defined on $SO(3)$. Let $Y_{1,e}$, $Y_{2,e}$, and $Y_{3,e}$ be the three orthonormal bases of 3×3 skew-symmetric matrices. For any point $s \in SO(3)$, $Y_{i,s} = sY_{i,e}$, $i = 1,2,3$, form an orthonormal bases for the vector space tangent to $SO(3)$ at s. In this notation, for the energy function $H_1 : SO(3) \mapsto \mathbb{R}_+$, $(Y_{i,s}H_1) \in \mathbb{R}$, $i = 1,2,3$, denotes the directional derivative of the posterior energy H_1 in the direction of Y_i (at point s).

The composite vector, $\sum_{i=1}^{3}(Y_{i,s}H_1)Y_{i,s}$ is called the gradient vector of H_1 at s. It is the direction of maximum change in the value of H_1 at s. A deterministic gradient process, $X(t)$, is a solution of the equation, $\frac{dX(t)}{dt} = -\sum_{i=1}^{3}(Y_{i,X(t)}H_1)Y_{i,X(t)}$. In other words, $X(t)$ is a curve in $SO(3)$ such that the tangent at any point along the curve is equal to the (negative) gradient vector of H_1 at that point. Adding random perturbations uniformly to all basis-directions in the tangent plane, according to equation 26.2.6, makes it a stochastic gradient process or a diffusion.

To establish the ergodic properties of this $X(t)$, we utilise tools from operator semi-group theory. A Markov process is generally characterised by an initial condition and a transition probability function. Equivalently, the behavior of a Markov process can also be characterised by its infinitesimal generator or backward operator (Ethier and Kurtz 1986). (Srivastava 1996) shows that for the above choice of diffusion parameters (means and variances), the semi-group operator of $X(t)$ satisfies the conditions for π_k to be the unique stationary probability density of $X(t)$. This leads to the ergodic result (Eqn. 26.1.4) for sampling from π_1 on $\mathcal{X}_1$. A simple extension to higher-dimensions, $SO(3)^k$, or product spaces $SE(3)^k$, for arbitrary k, follows.

This construction provides a sampling method on the orientation/location variabilities associated with a fixed number of targets with fixed labels. The next step extends this construction by adding discrete moves which can change the number of targets, and their labels, to sample from the posterior over the whole space $\mathcal{X}$.

26.2.2 Jump processes

As described in (Gikhman and Skorokhod 1969), $X(t)$ is a Markov jump process with jump measure $q(s, A)$, if $X(t)$ is a Markov process having a transition probability $P_t(s, A)$ and satisfies: (i) $X(t)$ is continuous in probability, i.e., for all $A \subset \mathcal{X}$ and $s \in \mathcal{X}$,

$$\lim_{t \downarrow 0} P_t(s, A) = 1_A(s), \qquad (26.2.7)$$

and (ii) $q(s, \cdot)$ is a bounded measure $q(s, \cdot)$ defined by

$$\lim_{t \downarrow 0} \frac{P_t(s, A) - 1_A(s)}{t} = q(s, A), \qquad \text{uniformly in } s, A. \qquad (26.2.8)$$

For any point $s_1 \in \mathcal{X}$, the jump intensity $q(s_1)$ is defined as the integral: $\int_{\mathcal{X}/\{s_1\}} q(s_1, \gamma(ds_2))$.

Unlike diffusions, which are sample path continuous processes with probability one, jump processes can move in ϵ time to neighborhoods far from the initial state with probability:

$$P_\epsilon(s, A) = 1_A(s) + q(s, A/\{s\})\epsilon + o(\epsilon).$$

Given a jump measure, q, the transition probability of the Markov process can be calculated as:

$$Q(s_1, \gamma(ds_2)) = \begin{cases} \frac{q(s_1, \gamma(ds_2))}{q(s_1)} & \text{for } q(s_1) \neq 0 \\ 1_{ds_2}(s_1) & \text{for } q(s_1) = 0, \end{cases} \quad (26.2.9)$$

with $\int_{X/\{s_1\}} Q(s_1, \gamma(ds_2)) = 1$. $Q(s_1, \cdot)$ represents the transition probability distribution from the state $\{s_1\}$ to any measurable set.

To completely define a jump process, two specifications are necessary, jump times and jump transition probabilities. We choose the process to jump at Poisson times by subordinating the process to a Poisson Markov chain. That is, the jumps are performed at random times separated by independent, exponential (with mean 1) times. Secondly, we have to specify the jump intensity q and the resulting transition probability Q for the ATR problem. In other words, for all $s \in \mathcal{X}$, we have to specify the set of points to which the process can transition from s (denoted by $\mathcal{T}^1(s)$), and with what probabilities. There are several possible jump moves for changing the target number k and the target types α. We restrict the set of allowed jump moves to keep them simple. In our scheme, the jumps are designed such that they manipulate only one target at a time. For example, the algorithm can add a target in the scene (incrementing k), remove a target from the scene (decrementing k), or change the target type associated with a particular target, keeping k fixed. For more sophisticated procedures, and perhaps better convergence rates, other jump types can also be incorporated, although we have not explored that topic any further. For the current state s, $Q(s, \cdot)$ puts a non-zero measure only on those sets falling within the range of allowed jump moves. There are two additional conditions imposed on the jump component of $X(t)$:

(i) for a chosen metric on $\mathcal{X}$, the separation between a point $s_1 \in \mathcal{X}$ and any other point in the support of $Q(s_1, \cdot)$ must be bounded.

(ii) For any two subspaces $\mathcal{X}_{k_1}$ and $\mathcal{X}_{k_2}$, it must be possible to transition from one to another in a finite number of steps.

These properties ensure that the process does not move, in probability, arbitrarily far from its current state in one jump, but still can travel from one subspace to another in a reasonable time.

Similar to the diffusion component, the ergodic properties of a jump process can also be analyzed via its infinitesimal generator. For the desired ergodic behavior, the generator must satisfy the stationarity condition, also well known as the *detailed balance* condition: the transition parameters q and Q must satisfy

$$q(s_1)\mu(ds_1) = \int_{\mathcal{X}} q(s_2)Q(s_2, ds_1)\mu(ds_2). \quad (26.2.10)$$

There are several ways to choose the jump measure q, for the ATR problem, such that the above requirements are satisfied. We briefly illustrate one such approach based on a modified version of the Metropolis-Hastings algorithm, as described in (Lanterman, Miller and Snyder 1997). A basic Metropolis-Hastings algorithm is characterised by two entities: (i) a *proposal probability* measure to generate candidate states, and (ii) an *acceptance probability* $a(s_1, s_2)$ which compares the current state s_1 with the candidate state s_2 and probabilistically decides whether to accept the transition to s_2. At jump times, a candidate $s_2 \in \mathcal{X}$ is drawn from the *proposal density* $r(s_1, s_2)$ (a density in s_2 parameterised by s_1); the *acceptance probability* is then computed:

$$a(s_1, s_2) = \min \left\{ \frac{\pi(s_2) r(s_2, s_1)}{\pi(s_1) r(s_1, s_2)}, 1 \right\}. \qquad (26.2.11)$$

Here $\pi(s)$ is the posterior density on the subspace to which s belongs. The proposal is accepted with the probability $a(s_1, s_2)$ and rejected with the probability $1 - a(s_1, s_2)$. This scheme corresponds to choosing the jump measure

$$q(s_1, \gamma(ds_2)) = \min \left\{ \frac{\pi(s_2) r(s_2, s_1)}{\pi(s_1) r(s_1, s_2)}, 1 \right\} r(s_1, s_2) \gamma(ds_2). \qquad (26.2.12)$$

A wide variety of proposal densities may be used, satisfying the basic condition that $r(s_1, s_2) > 0$ for $s_2 \in \mathcal{T}^1(s_1)$ and $r(s_1, s_2) = 0$ for $s_2 \notin \mathcal{T}^1(s_1)$. Special cases of the general Metropolis-Hastings jump-diffusion scheme described here have been reported previously. If $r(s_1, s_2) = r(s_2, s_1)$, we have $a(s_1, s_2) = \min\{\pi(s_2)/\pi(s_1), 1\}$, which corresponds to the traditional Metropolis algorithm (Metropolis et al. 1953). In a special situation it may be possible to have an informative prior, which is relatively easy to sample from. Then, it is beneficial to choose

$$r(s_1, s_2) = \frac{p(k) \pi_{0,k}(s_2)}{\int_{\mathcal{T}^1(s_1)} p(k) \pi_{0,k}(s') \gamma(ds')} \qquad (26.2.13)$$

for $s_2 \in \mathcal{T}^1(s_1)$, which corresponds to drawing candidates from the prior and accepting and rejecting based on the likelihood. This is one example of *importance-sampling*, sampling from one probability (prior) and modifying the samples appropriately to generate samples from another probability (posterior). This approach is effective in the ground-to-air tracking studies of (Miller et al. 1995) because drawing from the prior on a target's motion dynamics is efficient. In the single image ATR scenarios, the uniform prior on positions and orientations is not informative; it provides little help in locating new targets or determining their orientations. Therefore, the approach will be to propose the jump moves according to the prior $p(\cdot)$ on the number of targets, and draw the proposal from the posterior density over the space that can be reached via the chosen move type.

26.2.3 Jump-diffusion algorithm

Combining the two elements, jumps and diffusions, we can now specify a complete jump-diffusion algorithm. The jump-diffusion process $\{X(t), t \geq 0\}$ jumping at random times $t_1, t_2, \ldots$ is constructed by the following algorithm.

Algorithm 1. *Let $i = 0, t_0 = 0$, and $X(0) = X_0 \in \mathcal{X}$ be any initial condition.*

(i) *Generate a sample u of an exponential random variable with constant mean 1.*

(ii) *Follow the stochastic differential equation generating a diffusion process in the subspace determined by $X(t_i) \in \mathcal{X}_k$ for the time interval $t \in [t_i, t_{i+1})$, $t_{i+1} = t_i + u$:*

$$dX(t) = -\sum_{i=1}^{n}(Y_{i,X(t)}H_k)Y_{i,X(t)}dt + \sum_{i=1}^{n}Y_{i,X(t)} \circ dW_i(t).$$

(26.2.14)

of $\mathcal{X}_k$.

(iii) *At $t = t_{i+1}$, perform a jump move, from $X(t_{i+1})$ to an element of $\mathcal{T}^1(X(t_{i+1}))$, according to the transition probability measure given by $Q(X(t_{i+1}), \cdot)$.*

(iv) *Set $i \leftarrow i + 1$, go to step 1.*

To characterise this Markov jump-diffusion process, we invoke the composite infinitesimal generator formed as the sum of the jump and diffusion generators. It has been shown in (Srivastava 1996) that if the two individual generators satisfy the stationarity condition then the composite generator also satisfies that condition. Therefore, this algorithm results in a Markov jump-diffusion process $X(t)$ whose unique stationary measure is given by the posterior μ. Furthermore, for any bounded measurable function f on $\mathcal{X}$ $\lim_{T\to\infty} \frac{1}{T}\int_0^T f(X(t))dt \to \int_{\mathcal{X}} f(s)d\mu(s)$. In the case of ATR from image sequences this algorithm can be extended to sampling from $\mu^{(n)}$ on $\mathcal{X}^n$ for each n.

26.3   Sensor models

A sensor (or multiple sensors) observing a scene gives rise to observations that can be in the form of images, sampled signal waveforms, or matched filter outputs. The characteristics of these observations are dependent upon the sensor mechanism which captures the image, and often its mode of operation. This mechanism can be thought of as a mapping T from the target space $\mathcal{I}$ to an observation space $\mathcal{I}^{\mathcal{D}}$, $T : \mathcal{I} \to \mathcal{I}^{\mathcal{D}}$. $\mathcal{I}$ is the set of all possible occurrences of the targets, generated as the action of all possible transformations on all the templates. Let $sI \in \mathcal{I}$ denote a three-dimensional scene containing targets at poses specified by the elements of $s \in \mathcal{X}$. Then, for $sI \in \mathcal{I}$, let TsI be the ideal (noiseless) image of these targets at the parameters determined by $s \in \mathcal{X}$. T generally constitutes nonlinear transformations such as projections, obscurations, and accumulations. For statistical target recognition and improved performance, the statistical behavior of T has to be included in the algorithms. It is often difficult to derive analytical forms of T for various sensors. In that case, T can be incorporated in the algorithms using high-quality simulators such as PRISM, IRMA, XPATCH, etc., or by storing real images of the targets of interest at lots of pre-determined orientations.

Orthographic and perspective projection schemes are illustrated in the left panels (top and bottom, respectively) in Figure 26.3. The top-left panel shows an orthographic projection imaging of a truck and the bottom-left shows the perspective projection scaling of orthographic images.

(i) **Video Sensor Model**. Video images (visual spectrum) are generated by projecting light rays reflected from the target surface onto the focal plane of the camera. We have exploited SGI graphics environments to simulate the projection mechanism T for video cameras. In addition to the projection mechanism, the sensor has random noise ε whose characteristics depend upon the sensor and its mode of operation. For additive noise models, the image obtained from the sensor is modelled by the equation

$$I^{\mathcal{D}} = TsI + \varepsilon \in \mathcal{I}^{\mathcal{D}}, \quad s \in \mathcal{X}. \tag{26.3.15}$$

Assuming i.i.d. Gaussian noise, with the standard deviation σ at each pixel, the likelihood of observing an image $I^{\mathcal{D}}$ given the parameters s is

$$L(s) = \frac{1}{(2\pi\sigma^{2d})^{1/2}} \exp\{-\frac{1}{2\sigma^2}||I^{\mathcal{D}} - TsI||_2^2\}, \quad d = \dim(\mathcal{I}^{\mathcal{D}}), \tag{26.3.16}$$

where $s \in \mathcal{X}_k$ for some k. Shown in the middle panels of Figure 26.3 are the ideal image TsI (top-middle panel) and the noisy image $I^{\mathcal{D}}$ (bottom-middle panel) of a tank.

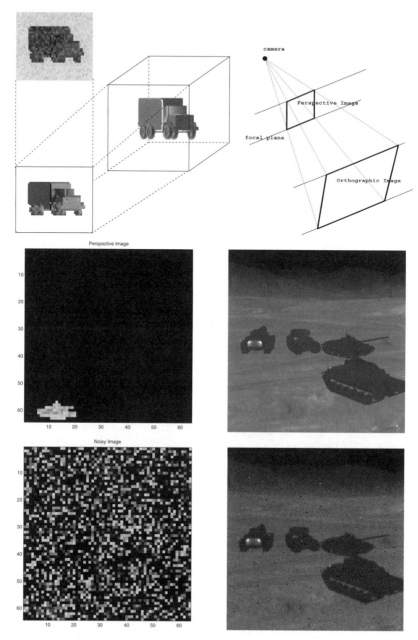

Figure 26.3. Top left: orthographic projection model. Top right: perspective scaling model. Middle left: an ideal video tank image. Middle right: an ideal FLIR image. Bottom left: a noisy video image. Bottom right: a noisy FLIR image.

(ii) **Infrared (FLIR) Model**. FLIR cameras sense the infrared radiation emitted by targets, which depends on their thermodynamic states, shapes, and surface emissivities. To simulate the FLIR signature generation, we employ radiance for M60 and T62 tanks and M62 as predicated by the PRISM software (Michigan Technological University 1987) developed by Michigan Technological University's Keweenaw Research Center (data courtesy Dr. A. Curran, Michigan Tech/KRC) and currently marketed by ThermoAnalytics. For simplicity, we assume that target facets radiate known intensities. Of course, in general, targets will radiate vastly different intensities depending on their thermal state, which is a function of their operating conditions, weather, and so forth. We also assume that the ground and the sky each radiate constant known intensities. These assumptions can be relaxed by incorporating thermodynamic variation into the representation as described in (Cooper, Lanterman, Joshi and Miller 1996, Cooper, Grenander, Miller and Srivastava 1997, Lanterman 1998).

Following the CCD camera model of (Snyder, Hammoud and White 1993), the camera introduces blurring according to

$$TsI(\mathbf{y}) = \sum_{\mathbf{x}} q(\mathbf{y}|\mathbf{x})TsI_1(\mathbf{x}), \qquad (26.3.17)$$

where $\mathbf{x}$, $\mathbf{y}$ are pixel locations, $TsI_1(\cdot)$ is the ideal observed intensity and $q(\mathbf{y}|\mathbf{x})$ is the camera point spread function, normalised so that $\sum_{\mathbf{y} \in \mathcal{Y}} q(\mathbf{y}|\mathbf{x}) = 1$. The data $I^D(\mathbf{y})$ collected by a CCD camera is Poisson distributed with mean $TsI(\mathbf{y})$. The Poisson log-likelihood of the data I^D given I_1 is

$$L(s) = -\sum_{\mathbf{y} \in \mathcal{Y}} TsI(\mathbf{y}) + \sum_{\mathbf{y} \in \mathcal{Y}} \ln[TsI(\mathbf{y})]I^D(\mathbf{y}). \qquad (26.3.18)$$

The right panels of Figure 26.3 show the ideal image (top-right) blurred with a Gaussian point-spread function and corrupted with Poisson noise and uniformly distributed dead and saturated pixels (bottom-right panel). Although the likelihood model used here does not explicitly incorporate dead and saturated pixels, they have little effect on the performance of the algorithm.

26.4   Experiments

Now we present applications of jump-diffusion algorithms in ATR for two cases: (i) single image ATR, and (ii) multiple image ATR.

(i) **Single Image**. For a single image ATR experiment, we have allowed an unknown number of targets located on the ground at unknown

positions and orientations. The implementation details may be found
in (Lanterman et al. 1997). In the sampling process, the types of
jumps allowed are, in one step: (i) addition of a target to the scene,
(ii) removal of a target from the scene, and (iii) change one target
to another (for example, M60 to T62). Move (iii) can also involve
large changes in orientation. In between the random jump times, the
process follows diffusions on the space $SE(2)^k$, where k is the number of targets currently discovered. The initial condition for $X(t)$ is
assumed to be an empty scene, i.e., $k = 0$.

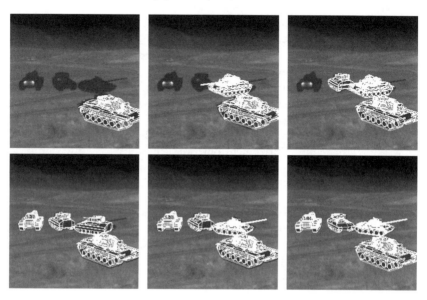

Figure 26.4. Iterations 1, 3, 24, 87, 88, and 103 of a jump-diffusion process for
FLIR data.

Figure 26.4 shows a few selected states of a sample path of the process sampling from the posterior conditioned on the image data (I^D)
shown in Figure 26.3 (bottom-right panel). Each picture depicts a
rendering of the scene, corresponding to $X(t) \in \mathcal{X}$, superimposed on
the underlying true ideal image.

Because we started the process with an empty scene, the algorithm
tries to add a target on the first iteration. The M60 on the right
appears larger than the other targets since it is closer to the sensor.
The algorithm finds it first since it can "explain" the largest amount
of data pixels. In iteration 3, it mistakes the T62 for an M60. It
does this since it has not yet found the adjacent M2 and is trying
to explain some of the M2's pixels using the barrel of the M60. This
demonstrates the importance of moves that allow changes of type. In
iteration 24, the algorithm finds the M2, but the hypothesis is facing

the wrong direction. Although the diffusions may refine orientation estimates, they are impractical for making large orientation changes, suggesting the necessity of a jump move for making drastic changes in orientation. Notice the diffusions have found the correct placement of the M60. The algorithm continues to propose adding targets, but these proposals are rejected because the data does not support a fifth target.

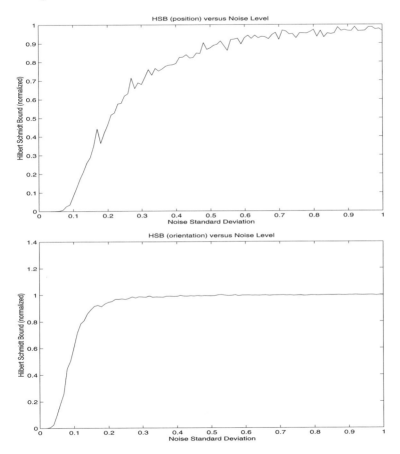

Figure 26.5. HSB as a function of noise level. Top: position component of HSB. Bottom: orientation component of HSB.

In addition to estimating optimal target representations, the Bayesian formulation also results in lower bounds (HSB) on expected estimation errors. As a simple illustration, shown in Figure 26.5 is the variation of HSB versus the sensor-noise level for estimating orientation $(SO(2))$ and position $(I\!R^2)$ of a M60 tank from a single image (of the type shown in the lower-middle panel of Figure 26.3). The left plot shows the position component of HSB and the right plot shows the orientation component of HSB, as functions of the noise level. At low noise levels the HSB value is zero, implying that perfect pose/location estimation is possible. At higher noise levels the error in estimation steadily increases until it reaches its maximum value. Further details can be found in (Loizeaux, Srivastava and Miller 1999).

(ii) **Image Sequence.** Next, we describe an experiment in estimating pose and location of a stationary target, using a sequence of images taken from a sensor moving towards the target. The target type is assumed known for a single target so the model-selection is not included in this example. Shown in Figure 26.6 is a sequence of three (noiseless) FLIR images of an M60 tank, starting from a distant sensor-location ($n = 1$) and moving closer to the M60 at each increment in n. The posterior distribution is now a function of n; Markov chain is employed to sample from the posterior for each value of n. In these imaging applications, often the posterior, at each time, is concentrated in small subsets and the important task is to track these subsets from one n to another. For details please refer to (Srivastava 2000).

Simulating noisy images (using the additive, Gaussian-noise model), for different noise levels, we estimate tank position and orientation and analyze estimation performance (HSB) as a function of noise level and n. HSB variation against sensor-noise and n is plotted in Figure

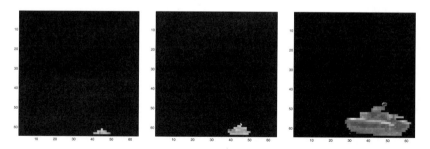

Figure 26.6. ATR from an image sequence: these pictures show (ideal) images of an M60 at successive estimates of pose and location. Moving from left to right, the sensor is closing in on the target.

26.7. Each curve represents the estimation error bound using images accumulated up to that time, i.e. for inference from $\mu^{(n)}$. The left picture shows the error bounds on position estimation while the right picture is for orientation estimation.

These bounds can be used to analyze procedures with regard to their estimation performances. In applications such as missile aim-point selection, where the sensor generates a sequence of images while closing in on the target, the comprehensive estimation uses all previous im-

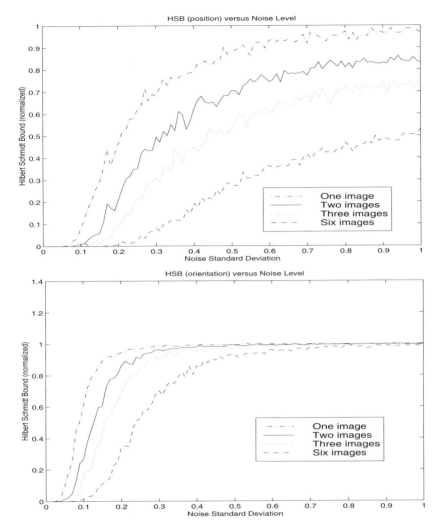

Figure 26.7. Curves showing improved performance as the number of images (n) increases. Left: position component of HSB. Right: orientation component of HSB.

ages. How much accuracy do we lose if only the last image is used? In response, we present Figure 26.8, which shows three HSB curves. The first uses an image from the first sensor position only. The second uses the image taken from the second sensor position only, ignoring the first image, which was used to steer the camera towards the target. The third curve is based on both images, and shows the gain when previous images are utilised. Again, the left picture is for position estimation and the right picture is for orientation estimation.

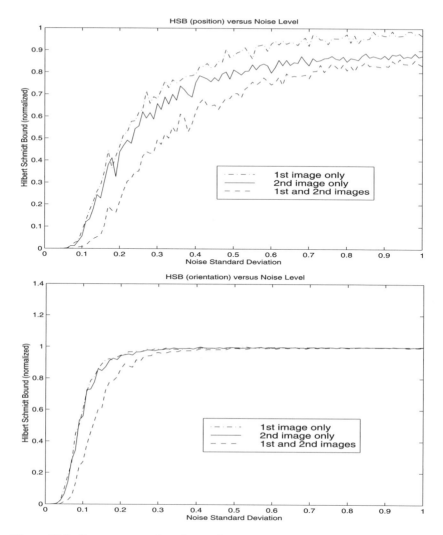

Figure 26.8. Curves comparing the performance when previous images are/are not included in the inference. Left: HSB due to position. Right: HSB due to orientation.

26.5   Acknowledgments

This research was supported by ARO DAAH04-95-0494, ONR N00014-95-0095, ARO-MURI DAAH04-96-1-0445, ARO DAA-G55-98-1-0102, and NSF-9871196. A. Srivastava and A.D. Lanterman are currently supported by the contracts DAAD19-99-0267 and DARPA F49620-98-1-0498, respectively.

Bibliography

Ackerson, G. A. and Fu, K. S. (1970).On state estimation in switching environments, *IEEE Transactions on Automatic Control* **15**(1): 10–17.

Aguilar, O. and West, M. (2000).Bayesian dynamic factor models and portfolio allocation, *Journal of Business and Economic Statistics*.

Alspach, D. L. and Sorenson, H. W. (1972).Nonlinear Bayesian estimation using Gaussian sum approximations, *IEEE Transactions on Automatic Control* **AC-17**(4): 439–448.

Anderson, B. D. and Moore, J. B. (1979).*Optimal Filtering*, Prentice-Hall, New Jersey.

Andersson, P. (1985).Adaptive forgetting in recursive identification through multiple models, *International Journal of Control* **42**(5): 1175–1193.

Andrade Netto, M. L., Gimeno, L. and Mendes, M. J. (1978).On the optimal and suboptimal nonlinear filtering problem for discrete-time systems, *IEEE Transactions on Automatic Control* **23**: 1062–1067.

Andrieu, C., de Freitas, J. F. G. and Doucet, A. (1999a).Sequential Bayesian estimation and model selection applied to neural networks, *Technical Report CUED/F-INFENG/TR 341*, Cambridge University Engineering Department.

Andrieu, C., de Freitas, J. F. G. and Doucet, A. (1999b).Sequential MCMC for Bayesian model selection, *IEEE Higher Order Statistics Workshop*, Caesarea, Israel, pp. 130–134.

Aoki, M. (1967).*Optimisation of stochastic sytstems*, Academic Press.

Athreya, K. B. and Ney, P. E. (1972).*Branching Processes*, Springer-Verlag.

Averbuch, A., Itzikowitz, S. and Kapon, T. (1991).Radar target tracking - Viterbi versus IMMA, *IEEE Transactions on Aerospace and Electronic Systems* **27**(3): 550–563.

Avitzour, D. (1995).Stochastic simulation Bayesian approach to multitarget tracking, *IEE Proceedings-Radar Sonar and Navigation* **142**(2): 41–44.

Azzam, R. M. A. and Bashara, N. M. (1987).*Ellipsometry and Polarized Light*, North-Holland Physics Publishing.

Baker, J. (1987).Reducing bias and inefficiency in the selection algorithm, *International Conference on Genetic Algorithms and Their Applications*.

Bally, V. and Talay, D. (1996).The law of the Euler scheme for stochastic differential equations: 2. Approximation of the density, Monte Carlo methods and applications, *Probab. Th. Related Fields* **2**: 93–128.

Bar-Shalom, Y. (1981).Stochastic dynamic programming: caution and probing, *IEEE Transactions on Automatic Control* **AC-26**(5): 1184–1195.

Bar-Shalom, Y. and Fortman, T. E. (1987).*Tracking and Data Association*, Academic Press, New York.

Bar-Shalom, Y. and Tse, E. (1974).Dual effect, certainty equivalence and separation in stochastic control, *IEEE Transactions on Automatic Control* **AC-19**(5): 494–500.

Bar-Shalom, Y., Chang, K. C. and Blom, H. A. P. (1985).Tracking a maneuvering target using input estimation versus the interacting multiple model algorithm, *IEEE Transactions on Aerospace and Electronic Systems* **25**(2): 296–300.

Bascle, B. and Blake, A. (1998).Separability of pose and expression in facial tracking and animation, *Proceedings of the 6th International Conference on Computer Vision*, pp. 323–328.

Baumberg, A. and Hogg, D. (1994).Learning flexible models from image sequences, *in* J.-O. Eklundh (ed.), *Proceedings of the 3rd European Conference on Computer Vision*, Springer-Verlag, pp. 299–308.

Baumberg, A. and Hogg, D. (1995a).An adaptive eigenshape model, *Proceedings of the British Machine Vision Conference* pp. 87–96.

Baumberg, A. and Hogg, D. (1995b).Generating spatiotemporal models from examples, *Proceedings of the British Machine Vision Conference*, Vol. 2, pp. 413–422.

Bayard, D. S. (1991).A forward method for optimal stochastic nonlinear and adaptive control, *IEEE Transactions on Automatic Control* **36**(9): 1046–1055.

Bellman, R. (1961).*Adaptive control processes: a guided tour*, Princeton University Press.

Bentley, J. L. (1980).Multidimensional divide and conquer, *Communications of the ACM* **23**(4): 214–229.

Bergman, N. (1999).*Recursive Bayesian Estimation: Navigation and Tracking Applications*, PhD thesis, Department of Electrical Engineering.Linköping Studies in Science and Technology. Dissertations No. 579.

Bergman, N., Ljung, L. and Gustafsson, F. (1999).Terrain navigation using Bayesian statistics, *IEEE Control Systems Magazine* **19**(3): 33–40.

Bernardo, J. M. and Smith, A. F. M. (1994).*Bayesian Theory*, Wiley Series in Applied Probability and Statistics.

Bertsekas, D. P. (1987).*Dynamic programming: deterministic and stochastic models*, Prentice-Hall.

Bertsekas, D. P. (1995).*Dynamic programming and optimal control (Volumes 1 and 2)*, Athena Scientific.

Berzuini, C., Best, N. G., Gilks, W. R. and Larizza, C. (1997).Dynamic conditional independence models and Markov Chain Monte Carlo methods, *Journal of the American Statistical Association* **92**(440): 1403–1412.

Besag, J. and Green, P. J. (1993).Spatial statistics and bayesian computation, *Journal of the Royal Statistical Society B* **55**: 25–38.

Beymer, D. and Poggio, T. (1995).Face recognition from one example view, *Proceedings of the 5th International Conference on Computer Vision*, pp. 500–507.

Bickel, P. J., Ritov, Y. and Ryden, T. (1998).Asymptotic normality of the maximum-likelihood estimator for general hidden Markov models, *Annals of Statistics* **26**(4): 1614–1635.

Billio, M., Monfort, A. and Robert, C. P. (1998).The simulated likelihood ratio (SLR) method, *Technical report*, University Cá Foscari of Venice.

Birmiwal, K. and Bar-Shalom, Y. (1984).Dual control guidance for simultaneous identification and interception, *Automatica* **20**(6): 737–749.

Bishop, C. M. (1995).*Neural Networks for Pattern Recognition*, Oxford University Press.

Blair, W. D. and Watson, G. A. (1994).IMM algorithm for solution to benchmark problem for tracking maneuvering targets, *Proceedings of the SPIE* **2221**: 476–488.

Blake, A. and Isard, M. (1998).*Active contours*, Springer-Verlag.

Blake, A. and Isard, M. A. (1994).3D position, attitude and shape input using video tracking of hands and lips, *Proceedings of Siggraph*, ACM, pp. 185–192.

Blake, A., Bascle, B., Isard, M. and MacCormick, J. (1998).Statistical models of visual shape and motion, *Philosophical Transactions of the Royal Society A.* **356**: 1283–1302.

Blake, A., Isard, M. A. and Reynard, D. (1995).Learning to track the visual motion of contours, *Journal of Artificial Intelligence* **78**: 101–134.

Blom, H. A. P. and Bar-Shalom, Y. (1988).The interacting multiple model algorithm for systems with Markovian switching coefficients, *IEEE Transactions on Automatic Control* **AC–33**(8): 780–783.

Bobick, A. F. and Wilson, A. D. (1995).A state-based technique for the summarisation and recognition of gesture, *Proceedings of the 5th International Conference on Computer Vision*, pp. 382–388.

Bollerslev, T., Engle, R. F. and Nelson, D. B. (1994).ARCH models, *in* R. F. Engle and D. McFadden (eds), *The Handbook of Econometrics, Volume 4*, North-Holland, Amsterdam, pp. 2959–3038.

Bølviken, E., Glöckner, F. and Stenseth, N. C. (1999).Estimating parameters in autoregressive processes of low order from noisy, binary data, manuscript under preparation.

Boothby, W. M. (1986).*An Introduction to Differential Manifolds and Riemannian Geometry*, Academic Press, Inc.

Borenstein, J., Everett, B. and Feng, L. (1996).*Navigating Mobile Robots: Systems and Techniques*, A K Peters, Ltd., Wellesley, MA.

Borobsky, B. Z. and Zakai, M. (1975).A lower bound on the estimation error for Markov processes, *IEEE Transactions on Automatic Control* **20**(6): 785–788.

Boutilier, C., Friedman, N., Goldszmidt, M. and Koller, D. (1996).Context-specific independence in Bayesian networks, *Proceedings of the 12th Conference on Uncertainty in AI*, Morgan Kauffmann, pp. 115–123.

Boyen, X. and Koller, D. (1998).Tractable inference for complex stochastic processes, *Proceedings of the Fourteenth Annual Conference on Uncertainty in AI (UAI)*, pp. 33–42.

Boyen, X. and Koller, D. (1999).Exploiting the architecture of dynamic systems, *Proceedings of the Sixteenth National Conference on Artificial Intelligence (AAAI)*, pp. 313–320.

Brindle, A. (1981).*Genetic algorithm for function optimization*, PhD thesis, University of Alberta.

Bruckner, J. M., Scott, H. R. W. and Rea, G. R. (1973).Analysis of multimodal systems, *IEEE Transactions on Aerospace and Electronic Systems* **9**(6): 883–888.

Bucy, R. S. and Senne, K. D. (1971).Digital synthesis of Nonlinear Filters, *Automatica* **7**: 287–298.

Burgard, W., Fox, D., Hennig, D. and Schmidt, T. (1996).Estimating the absolute position of a mobile robot using position probability grids, *Proceedings of the Thirteenth National Conference on Artificial Intelligence*, AAAI, AAAI Press/MIT Press, Menlo Park.

Campbell, L. W. and Bobick, A. F. (1995).Recognition of human body motion using phase space constraints, *Proceedings of the 5th International Conference on Computer Vision*, pp. 624–630.

Carlin, B. P., Polson, N. G. and Stoffer, D. S. (1992).A Monte Carlo approach to nonnormal and nonlinear state-space modeling, *Journal of the American Statistical Association* **87**(418): 493–500.

Carpenter, J., Clifford, P. and Fearnhead, P. (1999a).Building robust simulation-based filters for evolving data sets, unpublished, Department of Statistics, Oxford University.

Carpenter, J., Clifford, P. and Fearnhead, P. (1999b).An improved particle filter for nonlinear problems, *IEE Proceedings-Radar Sonar and Navigation* **146**(1): 2–7.

Carter, C. K. and Kohn, R. (1994).On Gibbs sampling for state space models, *Biometrika* **81**(3): 541–553.

Carter, C. K. and Kohn, R. (1996).Markov-Chain Monte-Carlo in conditionally Gaussian state-space models, *Biometrika* **83**(3): 589–601.

Casella, G. and Robert, C. P. (1996).Rao-Blackwellisation of sampling schemes, *Biometrika* **83**(1): 81–94.

Cavers, J. K. (1991).An analysis of pilot symbol assisted modulation for rayleigh fading channels, *IEEE Transactions on Vehicle Technology* **40**: 686–693.

Chan, K. S. and Ledolter, J. (1995).Monte Carlo EM estimation for time series models involving counts, *Journal of the American Statistical Association* **90**(429): 242–252.

Chen, R. and Liu, J. S. (2000).The mixture Kalman filter, *Journal of the Royal Statistical Society, Ser. B*.

Clapp, T. C. (2000).*Statistical Methods in the Processing of Communications Data*, PhD thesis, Cambridge University Engineering Department.

Clapp, T. C. and Godsill, S. J. (1999).Fixed-lag smoothing using sequential importance sampling, *in* J. M. Bernardo, J. O. Berger, A. P. Dawid and A. F. M. Smith (eds), *Bayesian Statistics 6*, Oxford University Press, Oxford, pp. 743–52.

Cleveland, W. S. and Devlin, S. J. (1988).Locally weighted regression: An approach to regression analysis by local fitting, *Journal of the American Statistical Association* **83**: 596–610.

Collings, I. B. and Moore, J. B. (1994).An HMM approach to adaptive demodulation of QAM signals in fading channels, *International Journal of Adaptive Control and Signal Processing* **8**(5): 457–474.

Collings, I. B. and Moore, J. B. (1995).An adaptive hidden Markov model approach to FM and M-Ary DPSK demodulation in noisy fading channels, *Signal Processing* **47**(1): 71–84.

Cooper, M. L., Grenander, U., Miller, M. I. and Srivastava, A. (1997).Accommodating geometric and thermodynamic variability for forward-looking infrared sensors, *Proceedings of the SPIE, Algorithms for Synthetic Aperture Radar IV*, Vol. 3070.

Cooper, M. L., Lanterman, A. D., Joshi, S. C. and Miller, M. I. (1996).Representing the variation of thermodynamic state via principal components analysis, *Proceedings of the Third Workshop On Conventional Weapon ATR*.

Cootes, T. F., Taylor, C. J., Lanitis, A., Cooper, D. H. and Graham, J. (1993).Building and using flexible models incorporating grey-level information, *Proceedings of the 4th International Conference on Computer Vision*, pp. 242–246.

Cover, T. and Thomas, J. (1991).*Elements of Information Theory*, Wiley.

Cowell, R. G., Dawid, A. P., Lauritzen, S. L. and Spiegelhalter, D. J. (1999).*Probabilistic Networks and Expert Systems*, Springer-Verlag.

Cox, I. J. (1991).Blanche—an experiment in guidance and navigation of an autonomous robot vehicle, *IEEE Transactions on Robotics and Automation* **7**(2): 193–204.

Cox, I. J. and Wilfong, G. T. (eds) (1990).*Autonomous Robot Vehicles*, Springer-Verlag.

Crisan, D. and Doucet, A. (2000).Convergence of sequential Monte Carlo methods, *Technical Report 381*, CUED-F-INFENG.

Crisan, D. and Grunwald, M. (1999).Large deviation comparison of branching algorithms versus resampling algorithms: Application to discrete time stochastic filtering, *Technical Report 1999-9*, Cambridge University Statistical Laboratory.

Crisan, D. and Lyons, T. (1999).Minimal entropy approximations and optimal algorithms for the filtering problem, Preprint.

Crisan, D. and Lyons, T. J. (1997).Nonlinear filtering and measure–valued processes, *Probability Theory and Related Fields* **109**(2): 217–244.

Crisan, D., Del Moral, P. and Lyons, T. (1999).Discrete filtering using branching and interacting particle systems, *Markov Processes and Related Fields* **5**(3): 293–318.

Cybenko, G. (1989).Approximation by superpositions of a sigmoidal function, *Mathematics of Control, Signals, and Systems* **2**(4): 303–314.

Dagum, P. and Galper, A. (1995).Forecasting sleep apnea with dynamic network models, *Proc. Ninth Conference on Uncertainty in Artificial Intelligence (UAI)*, pp. 64–71.

Davis, L. D. (1991).*Handbook of Genetic Algorithms*, Van Nostrand Reinhold, New York.

de Freitas, J. F. G. (1999).*Bayesian Methods for Neural Networks*, PhD thesis, University of Cambridge, Cambridge, UK.

de Freitas, J. F. G., Macleod, I. M. and Maltz, J. S. (1999).Neural networks for pneumatic actuator fault detection, *Transactions of the South African IEE* **90**: 28–34.

de Freitas, J. F. G., Niranjan, M., Gee, A. H. and Doucet, A. (2000).Sequential Monte Carlo methods to train neural network models, *Neural Computation* **12**(4): 955–993.

Dean, T. and Kanazawa, K. (1989).A model for reasoning about persistence and causation, *Computational Intelligence* **5**(3): 142–150.

Dean, T. L. and Boddy, M. (1988).An analysis of time-dependent planning, *Proceedings of the Seventh National Conference on Artificial Intelligence AAAI-92*, AAAI, AAAI Press/The MIT Press, Menlo Park, CA, pp. 49–54.

DeGroot, M. (1970).*Optimal Statistical Decisions*, McGraw-Hill, New York, NY.

Del Moral, P. (1996).Nonlinear filtering: interacting particle solution, *Markov Processes and Related Fields* **2**(4): 555–580.

Del Moral, P. (1998).Measure valued processes and interacting particle systems, application to non linear filtering problems, *Annals of Applied Probability* **8**: 438–495.

Del Moral, P. and Guionnet, A. (1998a).Large deviation for interacting particle systems. Application to nonlinear filtering problems, *Stochastic Processes and their Applications* **78**(1): 69–95.

Del Moral, P. and Guionnet, A. (1998b).On the stability of measure valued processes. Applications to nonlinear filtering and interacting particle systems, *Publication du Laboratoire de Statistique et Probabilités 3-98*, Université Paul Sabatier, Toulouse.

Del Moral, P. and Guionnet, A. (1999).A central limit theorem for non linear filtering using interacting particle systems, *Annals of Applied Probability* **9**: 275–297.

Del Moral, P. and Jacod, J. (1998).The Monte-Carlo method for filtering with discrete time observations, *Publications 453 du Laboratoire de Probabilités Paris 6*.

Del Moral, P. and Jacod, J. (1999).The Monte-Carlo method for filtering with discrete time observations: Central limit theorems, *Publications 515 du Laboratoire de Probabilités Paris 6*.

Del Moral, P. and Ledoux, M. (2000).Convergence of empirical processes for interacting particle systems with applications to non-linear filtering, *Journal of Theoretical Probability*.

Del Moral, P. and Miclo, L. (2000).Branching and interacting particle systems approximations of Feynman-Kac formulae with applications to non linear filtering, *in* M. Emery, M. Ledoux and M. Yor (eds), *Séminaire de Probabilités XXXIV*, Lecture Notes in Mathematics, Springer-Verlag, pp. 1–145.

Dellaert, F., Burgard, W., Fox, D. and Thrun, S. (1999).Using the condensation algorithm for robust, vision-based mobile robot localization, *Proceedings of the IEEE International Conference on Computer Vision and Pattern Recognition*, IEEE, Fort Collins, CO.

Dellaert, F., Fox, D., Burgard, W. and Thrun, S. (1999).Monte Carlo localization for mobile robots, *Proceedings of the IEEE International Conference on Robotics and Automation (ICRA)*.

Dellaert, F., Thorpe, C. and Thrun, S. (1999).Mosaicing a large number of widely dispersed, noisy, and distorted images: A Bayesian approach, *Technical Report CMU-RI-TR-99-34*, Carnegie Mellon University, Pittsburgh, PA.

Delyon, B., Lavielle, M. and Moulines, E. (1999).Convergence of a stochastic approximation version of the EM algorithm, *Annals of Statistics* pp. 94–128.

Dempster, A. P., Laird, N. M. and Rubin, D. B. (1977).Maximum likelihood from incomplete data via EM algorithm, *Journal of the Royal Statistical Society Series B-Methodological* **39**(1): 1–38.

Devroye, L. (1986).*Non–Uniform Random Variate Generation*, Springer–Verlag.

Devroye, L. (1987).*A Course on Density Estimation*, Vol. 14 of *Progress in Probability and Statistics*, Birkhäuser, Boston.

Devroye, L. and Györfi, L. (1985).*Nonparametric Density Estimation : The L_1 View*, Wiley series in Probability and Mathematical Statistics, John Wiley & Sons, New York.

Diaconis, P. and Shahshahani, M. (1987).The subgroup algorithm for generating uniform random variables, *Probability in the Engineering and Informational Sciences* **1**: 15–32.

Diebold, F. X. and Nerlove, M. (1989).The dynamics of exchange rate volatility: a multivariate latent factor ARCH model, *Journal of Applied Econometrics* **4**: 1–21.

Djurić, P. M. (1999).Monitoring and selection of dynamic models by Monte Carlo sampling, *IEEE Higher Order Statistics Workshop*, Caesarea, Israel, pp. 191–194.

Doerschuk, P. C. (1995).Cramér-Rao bounds for discrete-time nonlinear filtering problems, *IEEE Transactions on Automatic Control* **40**(8): 1465–1469.

Doss, H., Sethuraman, J. and Athreya, K. B. (1996).On the convergence of the Markov chain simulation, *Annals of Statistics* **24**: 69–100.

Doucet, A. (1998).On sequential simulation-based methods for Bayesian filtering, *Technical Report CUED/F-INFENG/TR 310*, Department of Engineering, Cambridge University.

Doucet, A., Andrieu, C. and Fitzgerald, W. J. (1998).Bayesian estimation of HMM via Monte Carlo methods, *IEEE Workshop on Neural Networks and Signal Processing*, Cambridge.

Doucet, A., Barat, E. and Duvaut, P. (1995).Monte Carlo approach to recursive Bayesian state estimation, *Proceedings of the IEEE Signal Processing/Athos Workshop on Higher Order Statistics*, Girona, Spain, pp. 12–14.

Doucet, A., de Freitas, J. F. G., Murphy, K. and Russell, S. (2000).Rao Black-wellised particle filtering for dynamic Bayesian networks, *Uncertainty in Artificial Intelligence*.

Doucet, A., Godsill, S. and Andrieu, C. (2000).On sequential Monte Carlo sampling methods for Bayesian filtering, *Statistics and Computing* **10**(3): 197–208.

Doucet, A., Gordon, N. J. and Krishnamurthy, V. (1999).Particle filters for state estimation of jump Markov linear systems, *Technical Report CUED/F-INFENG/TR 359*, Cambridge University Engineering Department.

Doyle, R. S. and Harris, C. J. (1996).Multi-sensor data fusion for helicopter guidance using neuro-fuzzy estimation algorithms, *The Aeronautical Journal* **100**(996): 241–251.

Du, J. and Vucetic, B. (1991).New 16QAM trellis codes for fading channels, *Electronics Letters* **27**(7): 1009–1010.

Duda, R. O. and Hart, P. E. (1973).*Pattern Classification and Scene Analysis*, John Wiley & Sons.

Dudgen, D. E. and Lacoss, R. T. (1993).An overview of automatic target recognition, *MIT Lincoln Laboratory Journal* **6**(1): 3–10.

Durbin, J. and Koopman, S. J. (1997).Monte Carlo maximum likelihood estimation for non-Gaussian state space models, *Biometrika* **84**: 1403–1412.

Engelson, S. and McDermott, D. (1992).Error correction in mobile robot map learning, *Proceedings of the 1992 IEEE International Conference on Robotics and Automation*, Nice, France, pp. 2555–2560.

Ethier, S. N. and Kurtz, T. G. (1986).*Markov Processes*, John Wiley & Sons.

Fahrmeir, L. (1992).Posterior mode estimation by extended Kalman filtering for multivariate dynamic generalized linear model, *Journal of the American Statistical Association* **87**(418): 501–509.

Fahrmeir, L. and Kaufmann, H. (1991).On Kalman filtering, posterior mode estimation and Fisher-scoring in dynamic exponential family regression, *Metrika* **38**: 37–60.

Feldbaum, A. A. (1965).*Optimal control systems*, Academic Press.

Fernyhough, J. H., Cohn, A. G. and Hogg, D. C. (1996).Generation of semantic regions from image sequences, *Proceedings of the 4th European Conference on Computer Vision*, pp. 475–484.

Fisher, N. I. (1993).*Statistical Analysis of Circular Data*, Cambridge University Press, Cambridge.

Fox, D., Burgard, W. and Thrun, S. (1998).Active Markov localization for mobile robots, *Robotics and Autonomous Systems* **25**: 195–207.

Fox, D., Burgard, W. and Thrun, S. (1999).Markov localization for mobile robots in dynamic environments, *Journal of Artificial Intelligence Research (JAIR)* **11**: 391–427.

Fox, D., Burgard, W., Dellaert, F. and Thrun, S. (1999).Monte Carlo localization: Efficient position estimation for mobile robots, *Proceedings of the National Conference on Artificial Intelligence (AAAI)*, AAAI, Orlando, FL.

Fox, D., Burgard, W., Kruppa, H. and Thrun, S. (2000).A probabilistic approach to collaborative multi-robot localization, *Autonomous Robots*.

Frühwirth-Schnatter, S. (1994).Applied state space modelling of non-Gaussian time series using integration-based Kalman filtering, *Statistics and Computing* **4**: 259–269.

Fukuda, T., Ito, S., Oota, N., Arai, F., Abe, Y., Tanake, K. and Tanaka, Y. (1993).Navigation system based on ceiling landmark recognition for autonomous mobile robot, *Proceedings of the International Conference on Industrial Electronics Control and Instrumentation (IECON'93)*, Vol. 1, pp. 1466 – 1471.

Fung, R. and Del Favero, B. (1994).Backward simulation in Bayesian networks, *Proceedings of the Conference on Uncertainty in AI (UAI)*.

Galdos, J. I. (1980).A Cramér-Rao bound for discrete-time nonlinear filtering problems, *IEEE Transactions on Automatic Control* **25**(1): 117–119.

Gelb, A. (ed.) (1974).*Applied Optimal Estimation*, MIT Press, Cambridge, MA.

Geman, S. and Geman, D. (1984).Stochastic relaxation, Gibbs distributions and the Bayesian restoration of images., *IEEE Transactions on Pattern Analysis and Machine Intelligence* **6**(6): 721–741.

Geman, S. and Hwang, C.-R. (1987).Diffusions for global optimization, *SIAM Journal of Control and Optimization* **24**(24): 1031–1043.

Georghiades, C. N. and Han, J. C. (1997).Sequence estimation in the presence of random parameters via the EM algorithm, *IEEE Transactions on Communications* **45**(3): 300–308.

Gerlach, R., Carter, C. and Kohn, R. (1999).Diagnostics for time series analysis, *Journal Time Series Analysis* **21**: 309–330.

Gersch, W. and Kitagawa, G. (1988).Smoothness priors in time series, *Bayesian Analysis of Time Series and Dynamic Models*, New York, pp. 431–476.

Gertsman, M. J. and Lodge, J. H. (1997).Symbol-by-symbol MAP demodulation of CPM and PSK signals on rayleigh flat-fading channels, *IEEE Transactions on Communications* **45**(7): 788–799.

Geweke, J. (1989).Bayesian inference in econometric models using Monte Carlo integration, *Econometrica* **24**: 1317–1399.

Geweke, J. F. and Zhou, G. (1996).Measuring the pricing error of the arbitrage pricing theory, *The Review of Financial Studies* **9**: 557–587.

Geyer, C. J. (1993).Estimating normalizing constants and reweighting mixtures in Markov chain Monte Carlo, *Technical report*, University of Minnesota.

Ghahramani, Z. and Jordan, M. I. (1997).Factorial hidden Markov models, *Machine Learning* **29**(2–3): 245–273.

Gikhman, I. I. and Skorokhod, A. V. (1969).*Introduction to the Theory of Random Processes*, Saunders Company.

Gilks, W. R. and Berzuini, C. (1999).Following a moving target – Monte Carlo inference for dynamic Bayesian models, *submitted to Journal of the Royal Statistical Society, Series B.*

Gilks, W. R., Richardson, S. and Spiegelhalter, D. J. (eds) (1996).*Markov Chain Monte Carlo in Practice*, Chapman and Hall, Suffolk.

Gill, R. D. and Levit, B. Y. (1995).Application of the van Trees inequality: a Bayesian Cramér-Rao bound, *Bernoulli* **1**(1/2): 59–79.

Givens, G. H. and Raftery, A. E. (1996).Local adaptive importance sampling for multivariate densities with strong nonlinear relationships, *Journal of the American Statistical Association* **91**: 132–141.

Godsill, S. J. (1997).Bayesian enhancement of speech and audio signals which can be modelled as ARMA processes, *International Statistical Review* **65**(1): 1–21.

Godsill, S. J. and Rayner, P. J. W. (1998a).*Digital Audio Restoration: A Statistical Model-Based Approach*, Springer-Verlag.

Godsill, S. J. and Rayner, P. J. W. (1998b).Robust reconstruction and analysis of autoregressive signals in impulsive noise using the Gibbs sampler, *IEEE Transactions on Speech and Audio Processing* **6**(4): 352–372.

Godsill, S. J., Doucet, A. and West, M. (2000).Methodology for Monte Carlo smoothing with application to time-varying autoregressions, *Proceedings of the International Symposium on Frontiers of Time Series Modelling.*Institute of Statistical Mathematics, Tokyo.

Goldberg, D. E. (1989).*Genetic Algorithms in Search, Optimization and Machine Learning*, Addison-Wesley.

Goodwin, C. G. and Sin, K. S. (1984).*Adaptive filtering prediction and control*, Prentice-Hall.

Gordon, N. (1993).*Bayesian methods for tracking*, PhD thesis, University of London.

Gordon, N. (1997).A hybrid bootstrap filter for target tracking in clutter, *IEEE Transactions on Aerospace and Electronic Systems* **33**: 353–358.

Gordon, N. J., Marrs, A. D. and Salmond, D. J. (2001).Sequential analysis of nonlinear dynamic systems using particles and mixtures, *in* W. J. Fitzgerald, A. Walden, P. C. Young and R. L. Smith (eds), *Nonlinear and Nonstationary Signal Processing*, Cambridge University Press.

Gordon, N. J., Salmond, D. J. and Ewing, C. (1995).Bayesian state estimation for tracking and guidance using the bootstrap filter, *Journal of Guidance, Control and Dynamics* **18**(6): 1434–1443.

Gordon, N. J., Salmond, D. J. and Smith, A. F. M. (1993).Novel approach to nonlinear/non-Gaussian Bayesian state estimation, *IEE Proceedings-F* **140**(2): 107–113.

Grassberger, P. (1997).Pruned-enriched Rosenbluth method: simulations of q polymers of chain length up to 1000000., *Physical Review E* **56**: 3682–93.

Green, P. J. (1994).Markov chain monte carlo in image analysis, *Markov chain Monte Carlo in Practice.*

Green, P. J. (1995).Reversible jump Markov chain Monte Carlo computation and Bayesian model determination, *Biometrika* **82**(4): 711–32.

Grenander, U. (1976–1981).*Lectures in Pattern Theory I, II and III*, Springer-Verlag.

Grenander, U. (1993).*General Pattern Theory*, Oxford University Press.

Grenander, U. and Miller, M. I. (1994).Representations of knowledge in complex systems, *Journal of the Royal Statistical Society* **56**(3): 549–603.

Grenander, U. and Srivastava, A. (2000).Probability models for background clutter in natural images, *Monograph of Department of Statistics, Florida State University*.

Grenander, U., Chow, Y. and Keenan, D. M. (1991).*HANDS. A Pattern Theoretical Study of Biological Shapes*, Springer-Verlag.

Grenander, U., Miller, M. I. and Srivastava, A. (1998).Hilbert-Schmidt lower bounds for estimators on matrix Lie groups for ATR, *IEEE Transactions on Pattern Analysis an Machine Intelligence* **20**(8): 790–802.

Gunther, J., Beard, R., Wilson, J., Oliphant, T. and Stirling, W. (1997).Fast nonlinear filtering via Galerkins method, *American Control Conference*, Albuquerque, USA, pp. 2815–2819.

Haeb, R. and Meyr, H. (1989).A systematic approach to carrier recovery and detection of digitally phase modulated signals on fading channels, *IEEE Transactions on Communications* **COM-37**: 748–759.

Hammersley, J. M. and Morton, K. W. (1954).Poor man's Monte Carlo, *Journal of the Royal Statistical Society B* **16**: 23–38.

Handschin, J. E. and Mayne, D. Q. (1969).Monte Carlo techniques to estimate the conditional expectation in multi-stage non-linear filtering, *International Journal of Control* **9**(5): 547–559.

Harrison, P. J. and Stevens, C. F. (1976).Bayesian forecasting (with discussion), *Journal of the Royal Statistical Society, Series B* **38**: 205–247.

Harvey, A. C., Ruiz, E. and Sentana, E. (1992).Unobserved component time series models with ARCH disturbances, *Journal of Econometrics* **52**: 129–158.

Hastings, W. K. (1970).Monte Carlo sampling methods using Markov chains and their applications, *Biometrika* **57**: 97–109.

Hesterberg, T. (1995).Weighted average importance sampling and defensive mixture distributions, *Technometrics* **37**(2): 185–194.

Hewish, M. and Gourley, S. R. (1999).Ultra-wideband technology opens up new horizons, *Jane's International Defence Review* p. 20.

Higuchi, T. (1997).Monte Carlo filter using the genetic algorithm operators, *Journal of Statistical Computation and Simulation* **59**(1): 1–23.

Higuchi, T. (1999).Applications of quasi-periodic oscillation models to seasonal small count time series, *Computational Statistics & Data Analysis* **30**: 281–301.

Hinkel, R. and Knieriemen, T. (1988).Environment perception with a laser radar in a fast moving robot, *Proceedings of Symposium on Robot Control*, Karlsruhe, Germany, pp. 68.1–68.7.

Holland, J. H. (1975).*Adaption in natural and artificial systems*, University of Michigan Press, Ann Arbor.

Holmes, C. C. and Mallick, B. K. (1998).Bayesian radial basis functions of variable dimension, *Neural Computation* **10**(5): 1217–1233.

Holst, U. and Lindgren, G. (1991).Recursive estimation in mixture models with Markov regime, *IEEE Transactions on Information Theory* **37**: 1683–1690.

Horn, L. J., Showalter, M. R. and Russell, C. T. (1996).Detection and behavior of pan wakes in Saturn's ring, *Icarus* **124**: 643–676.

Huang, T., Koller, D., Malik, J., Ogasawara, G., Rao, B., Russell, S. J. and Weber, J. (1994).Automatic symbolic traffic scene analysis using belief networks, *Proceedings of the Twelfth National Conference on Artificial Intelligence (AAAI)*, pp. 966–972.

Huang, X. D., Arika, Y. and Jack, M. A. (1990).*Hidden Markov Models for Speech Recognition*, Edinburgh University Press.

Hull, D. G., Speyer, J. L. and Burris, D. B. (1990).Linear-quadratic guidance law for dual-control of homing missiles, *Journal of Guidance, Control and Dynamics* **13**(1): 137–144.

Hull, J. and White, A. (1987).The pricing of options on assets with stochastic volatilities, *Journal of Finance* **42**: 281–300.

Hürzeler, M. (1998).*Statistical Methods for General State–Space Models*, Ph.D. Thesis, Department of Mathematics, ETH Zürich, Zürich.

Hürzeler, M. and Künsch, H. R. (1998).Monte Carlo approximations for general state space models, *Journal of Computational and Graphical Statistics* **7**(2): 175–193.

Ikoma, N. (1996).Estimation of time varying peak of power spectrum based on non-Gaussian nonlinear state space modeling, *Journal of Signal Processing* **49**: 85–95.

Irwing, M., Cox, N. and Kong, A. (1994).Sequential imputation for multilocus linkage analysis, *Proceedings of the National Academy of Science, USA* **91**: 11684–11688.

Isard, M. and Blake, A. (1996).Contour tracking by stochastic propagation of conditional density, *European Conference on Computer Vision*, Cambridge, UK, pp. 343–356.

Isard, M. and Blake, A. (1998a).Condensation – conditional density propagation for visual tracking, *International Journal of Computer Vision* **28**(1): 5–28.

Isard, M. and Blake, A. (1998b).A mixed-state condensation tracker with automatic model switching, *Proceedings of the 6th International Conference on Computer Vision*, pp. 107–112.

Jacquier, E., Polson, N. G. and Rossi, P. E. (1994).Bayesian analysis of stochastic volatility models (with discussion), *Journal of Business and Economic Statistics* **12**: 371–417.

Jazwinski, A. H. (1970).*Stochastic Processes and Filtering Theory*, Academic Press.

Jensen, F. V., Kjærulff, U., Olesen, K. G. and Pedersen, J. (1989).An expert system for control of waste water treatment – a pilot project, *Technical report*, Judex Datasystemer A/S, Aalborg, Denmark.

Julier, S. J. and Uhlmann, J. K. (1997).A new extension of the Kalman filter to nonlinear systems, *Proc. of AeroSense: The 11th International Symposium on Aerospace/Defence Sensing, Simulation and Controls, Orlando, Florida*, Vol. Multi Sensor Fusion, Tracking and Resource Management II.

Kaelbling, L. P., Cassandra, A. R. and Kurien, J. A. (1996).Acting under uncertainty: Discrete Bayesian models for mobile-robot navigation, *Proceedings of the IEEE/RSJ International Conference on Intelligent Robots and Systems*.

Kalman, R. E. (1960).A new approach to linear filtering and prediction problems, *Journal of Basic Engineering* **82**: 35–45.

Kam, P. Y. and Ching, H. M. (1992).Sequence estimation over the slow nonselective rayleigh fading channel with diversity reception and its application to viterbi decoding, *IEEE J.Select. Areas Commun.* **COM-10**(3): 562–570.

Kanazawa, K., Koller, D. and Russell, S. J. (1995).Stochastic simulation algorithms for dynamic probabilistic networks, *Proceedings of the Eleventh Annual Conference on Uncertainty in AI (UAI '95)*, pp. 346–351.

Kashiwagi, N. and Yanagimoto, T. (1992).Smoothing serial count data through a state-space model, *Biometrics* **48**: 1187–1194.

Kendall, M. G. and Stuart, A. (1979).*The advanced theory of statistics, vol 2, inference and relationship*, Charles Griffing and Co Ltd, London.

Kerr, T. (1989).Status of CR-like lower bounds for nonlinear filtering, *IEEE Transactions on Aerospace and Electronic Systems* **25**: 590–601.

Kim, S., Shephard, N. and Chib, S. (1998).Stochastic volatility: Likelihood inference and comparison with ARCH models, *Review of Economic Studies* **65**: 361–393.

King, M., Sentana, E. and Wadhwani, S. (1994).Volatility and links between national stock markets, *Econometrica* **62**: 901–933.

Kitagawa and Sato (2000).Nonlinear state space model approach to financial time series with time-varying variance, *Proceedings of the Hong Kong International Workshop on Statistics in Finance*, Imperial College Press, London.

Kitagawa, G. (1987).Non-Gaussian state-space modeling of nonstationary time series, *Journal of the American Statistical Association* **82**(400): 1032–1063.

Kitagawa, G. (1991).A nonlinear smoothing method for time series analysis, *Statistica Sinica* **1**(2): 371–388.

Kitagawa, G. (1996).Monte Carlo filter and smoother for non-Gaussian nonlinear state space models, *Journal of Computational and Graphical Statistics* **5**(1): 1–25.

Kitagawa, G. (1998).Self-organising state space model, *Journal of the American Statistical Association* **93**(443): 1203–1215.

Kitagawa, G. and Gersch, W. (1985).A smoothness priors time varying AR coefficient modeling of nonstationary time series, *IEEE Transactions on Automatic Control* **AC-30**: 48–56.

Kitagawa, G. and Gersch, W. (1996).*Smoothness Priors Analysis of Time Series*, Vol. 116 of *Lecture Notes In Statistics*, Springer-Verlag.

Knill, D. C., Kersten, D. and Yuille, A. L. (1996).A Bayesian formulation of visual perception, *in* D. C. Knill and W. Richard (eds), *Perception as Bayesian inference*, Cambridge University Press, pp. 1–22.

Koenderink, J. J. and van Doorn, A. J. (1991).Affine structure from motion, *Journal of the Optical Society of America A.* **8**(2): 337–385.

Koenig, S. and Simmons, R. (1996).Passive distance learning for robot navigation, *in* L. Saitta (ed.), *Proceedings of the Thirteenth International Conference on Machine Learning*.

Koller, D. and Fratkina, R. (1998).Using learning for approximation in stochastic processes, *Proceedings of the International Conference on Machine Learning (ICML)*.

Koller, D., Lerner, U. and Angelov, D. (1999).A general algorithm for approximate inference and its application to hybrid Bayes nets, *Proceedings of the Fifteenth Annual Conference on Uncertainty in AI (UAI)*, pp. 324–333.

Koller, D., Weber, J. and Malik, J. (1994).Robust multiple car tracking with occlusion reasoning, *Proceedings of the 3rd European Conference on Computer Vision*, Springer-Verlag, pp. 189–196.

Kong, A., Liu, J. S. and Wong, W. H. (1994).Sequential imputations and Bayesian missing data problems, *Journal of the American Statistical Association* **89**(425): 278–288.

Kortenkamp, D., Bonasso, R. P. and Murphy, R. (eds) (1998).*AI-based Mobile Robots: Case studies of successful robot systems*, MIT Press.

Kotecha, J. H. and Djurić, P. M. (2000).Sequential Monte Carlo sampling detector for Rayleigh fast-fading channels, *Proceedings of IEEE International Conference on Acoustics, Speech, and Signal Processing*, Istanbul, Turkey.

Kramer, S. C. and Sorensen, H. W. (1988).Recursive Bayesian estimation using piece-wise constant approximations, *Automatica* **24**(6): 789–801.

Kremer, K. and Binder, K. (1988).Monte Carlo simulation of lattice models for macromolecules, *Computer Physics Reports*.

Künsch, H. R. (2000).State space and hidden Markov models, *in* C. Klüppelberg (ed.), *Complex Stochastic Systems*, Chapman and Hall.

Lanitis, A., Taylor, C. J. and Cootes, T. F. (1995).A unified approach to coding and interpreting face images, *Proceedings of the 5th International Conference on Computer Vision*, pp. 368–373.

Lanterman, A. (1998).*Modeling Clutter and Target Signatures for Pattern-Theoretic Understanding of Infrared Scenes*, D. Sc. Thesis, Washington University, St. Louis, Missouri.

Lanterman, A., Miller, M. and Snyder, D. (1997).General Metropolis-Hastings jump diffusions for automatic target recognition in infrared scenes, *Optical Engineering* **36**(4): 1123–1137.

Lawrence, R. V. (1998).Interceptor line of sight rate steering: necessary conditions for a direct hit, *Journal of Guidance, Control and Dynamics* **21**(3): 471–476.

Le Gland, F., Musso, C. and Oudjane, N. (1998).An analysis of regularized interacting particle methods for nonlinear filtering, *Proceedings of the IEEE European Workshop on Computer–Intensive Methods in Control and Data Processing, Prague*, pp. 167–174.

Lenser, S. and Veloso, M. (2000).Sensor resetting localization for poorly modelled mobile robots, *Proceedings of the IEEE International Conference on Robotics and Automation (ICRA)*, IEEE, San Francisco, CA.

Leonard, J., Durrant-Whyte, H. F. and Cox, I. J. (1992).Dynamic map building for an autonomous mobile robot, *International Journal of Robotics Research* **11**(4): 89–96.

Liu, J. S. (1996a).Metropolized independent sampling with comparisons to rejection sampling and importance sampling, *Statistics and Computing* **6**(2): 113–119.

Liu, J. S. (1996b).Nonparametric hierarchical Bayes via sequential imputations, *Annals of Statistics* **24**(3): 911–930.

Liu, J. S. and Chen, R. (1995).Blind deconvolution via sequential imputations, *Journal of the American Statistical Association* **90**(430): 567–576.

Liu, J. S. and Chen, R. (1998).Sequential Monte Carlo methods for dynamic systems, *Journal of the American Statistical Association* **93**(443): 1032–1044.

Liu, J. S., Chen, R. and Wong, W. H. (1998).Rejection control and sequential importance sampling, *Journal of the American Statistical Association* **93**(443): 1022–1031.

Liu, J. S., Wong, W. H. and Kong, A. (1994).Covariance structure of the Gibbs sampler with applications to the comparisons of estimators and augmentation schemes, *Biometrika* **81**(1): 27–40.

Liu, Y. and Blostein, S. D. (1995).Identification of frequency non-selective fading channels using decision feedback and adaptive linear prediction, *IEEE Transactions on Communications* **43**: 1484–1492.

Ljung, L. (1979).Asymptotic behavior of the extended Kalman filter as a parameter estimator for linear systems, *IEEE Transactions on Automatic Control* **AC-24**(1): 36–50.

Ljung, L. (1987).*System identification: theory for the user*, Prentice-Hall.

Lodge, J. H. and Moher, M. L. (1990).Maximum likelihood sequence estimation of CPM signals transmitted over rayleigh flat-fading channels, *IEEE Transactions on Communications* **38**: 787–794.

Loizeaux, M., Srivastava, A. and Miller, M. I. (1999).Pose/location estimation of ground targets, *Proceedings of SPIE Conference on Signal processing, sensor fusion, and target recognition*, Orlando, FL, pp. 140–151.

MacCormick, J. (1999).Partitioned sampling, *Accepted to ICCV'99*.

MacCormick, J. and Blake, A. (1998).A probabilistic contour discriminant for object localisation, *Proceedings of the 6th International Conference on Computer Vision*, pp. 390–395.

MacEachern, S. N., Clyde, M. and Liu, J. S. (1999).Sequential importance sampling for nonparametric Bayes models: the next generation, *Canadian Journal of Statistics* **27**(2): 251–267.

Mardia, K. V., Ghali, N. M., Howes, M., Hainsworth, T. J. and Sheehy, N. (1993).Techniques for online gesture recognition, *Journal of Image and Vision Computing* **11**(5): 283–294.

Marshall, A. W. (1956).The use of multi-stage sampling schemes in Monte Carlo computations, *in* M. Meyer (ed.), *Symposium on Monte Carlo Methods*, Wiley, New York, pp. 123–140.

Mazor, E., Averbuch, A., Bar-Shalom, Y. and Dayan, J. (1998).Interacting multiple model methods in target tracking: a survey, *IEEE Transactions on Aerospace and Electronic Systems* **34**(1): 103–123.

McGinnity, S. (1998).*Nonlinear estimation techniques for target tracking*, PhD thesis, The Queen's University of Belfast, Department of Electrical and Electronic Engineering.

McGinnity, S. and Irwin, G. W. (1998).A multiple model bootsrap filter for manoeuvring target tracking, *Submitted to IEEE Transactions on Aerospace and Electronic Systems*.

Meirovitch, H. (1985).The scanning method with a mean-field parameter: Computer-simulation study of critical exponents of self-avoiding walks on a square lattice, *Macromolecules* **18**(3): 563–569.

Melvin, D. G. (1996).A comparison of statistical and connectionist techniques for liver transplant monitoring, *Technical Report CUED/F-INFENG/TR 282*, Cambridge University Engineering Department.

Metropolis, N., Rosenbluth, A. W., Rosenbluth, M. N., Teller, A. H. and Teller, E. (1953).Equations of state calculations by fast computing machines, *Journal of Chemical Physics* **21**: 1087–1091.

Meyn, S. P. and Tweedie, R. L. (1993).*Markov chains and stochastic stability*, Springer-Verlag.

Michigan Technological University (1987).*Prism 3.1 User's Manual*, Keweenaw Research Center.

Miller, M. I., Srivastava, A. and Grenander, U. (1995).Conditional-expectation estimation via jump-diffusion processes in multiple target tracking/recognition, *IEEE Transactions on Signal Processing* **43**(11): 2678–2690.

Monahan, J. and Genz, A. (1997).Spherical-radial integration rules for Bayesian computation, *Journal of the American Statistical Association* **92**(438): 664–674.

Moore, A. W. (1990).*Efficient Memory-based Learning for Robot Control*, PhD thesis, Trinity Hall College, Cambridge University, England.

Moore, A. W., Schneider, J. and Deng, K. (1997).Efficient locally weighted polynomial regression predictions, *Proceedings of the International Conference on Machine Learning (ICML)*.

Moravec, H. and Elfes, A. (1985).High resolution maps from wide angle sonar, *Proceedings of the IEEE International Conference on Robotics & Automation (ICRA)*, pp. 116–121.

Mosterman, P. J. and Biswas, G. (1999).Diagnosis of continuous valued systems in transient operating regions, *IEEE Transactions on Systems, Man, and Cybernetics, Part A* **29**(6): 554–565.

Mumford, D. (1996).Pattern theory: a unifying perspective, *in* D. C. Knill and W. Richard (eds), *Perception as Bayesian inference*, Cambridge University Press, pp. 25–62.

Murphy, K. P. (1999).A variational approximation for Bayesian networks with discrete and continuous latent variables, *Proceedings of the 14th Conference on Uncertainty in AI*, Morgan Kauffmann, pp. 457–466.

Murphy, K. P. (2000).Bayesian map learning in dynamic environments, *Advances in Neural Information Processing Systems 12*, MIT Press, pp. 1015–1021.

Murray, D. W. and Basu, A. (1994).Motion tracking with an active camera, *IEEE Transactions on Pattern Analysis and Machine Intelligence* **16**(5): 449–459.

Musso, C. and Oudjane, N. (1998).Regularisation schemes for branching particle systems as a numerical solving method of the nonlinear filtering problem, *Proceedings of the Irish Signals Systems Conference, Dublin*.

Nardone, S. C. and Aidala, V. J. (1981).Observability criteria for bearings–only target motion analysis, *IEEE Transactions on Aerospace and Electronic Systems* **AES–17**(2): 162–166.

Narendra, K. S. and Parthasarathy, K. (1990).Identification and control of dynamical systems using neural networks, *IEEE Transactions on Neural Networks* **1**: 4–27.

Neal, R. M. (1998).Annealed importance sampling, *Technical Report 9805*, Department of Statistics, University of Toronto.

Nicholson, A. E. and Brady, J. M. (1994).Dynamic belief networks for discrete monitoring, *IEEE Transactions on Systems, Man and Cybernetics* **24**(11): 1593–1610.

North, B. and Blake, A. (1997).Using expectation-maximisation to learn dynamical models from visual data, *Proceedings of the British Machine Vision Conference*, pp. 669–678.

Nourbakhsh, I., Powers, R. and Birchfield, S. (1995).DERVISH, an office-navigating robot, *AI Magazine* **16**(2): 53–60.

Oehlert, G. W. (1998).Faster adaptive importance sampling in low dimensions, *Journal of Computational and Graphical Statistics* **7**: 158–174.

O'Hagan, A. (1994).*Kendall's Advanced Theory of Statistics: Bayesian Inference, Volume 2B*, Halsted Press, New York.

Omohundro, S. M. (1991).Bumptrees for efficient function, constraint, and classification learning, *in* R. P. Lippmann, J. E. Moody and D. S. Touretzky (eds), *Advances in Neural Information Processing Systems 3*, Morgan Kaufmann.

Oudjane, N. and Musso, C. (1999).Multiple model particle filter, *Actes du 17ème Colloque GRETSI, Vannes*, pp. 681–684.

Parthasarathy, P. (1967).*Probability Measures on Metric Spaces*, Academic Press, New York.

Pearl, J. (1988).*Probabilistic Reasoning in Intelligent Systems*, Morgan Kaufmann.

Penny, W. D., Roberts, S. J., Curran, E. and Stokes, M. (1999).EEG-based communication: a pattern recognition approach, To appear in *IEEE Transactions on Rehabilitation Engineering*.

Pickering, C. (1998).Complementary in-situ and post-deposition diagnostics of thin film semiconductor structures, *Thin Solid Films* pp. 313–314.

Pitt, M. and Shephard, N. (1999a).Analysis of time varying covariances: A factor stochastic volatility approach, *in* J. O. Berger, J. M. Bernardo, A. P. Dawid and A. F. M. Smith (eds), *Bayesian Statistics 5*, Oxford University Press, pp. 547–570.

Pitt, M. K. and Shephard, N. (1999b).Filtering via simulation: Auxiliary particle filters, *Journal of the American Statistical Association* **94**(446): 590–599.

Pitt, M. K. and Shephard, N. (1999c).Time varying covariances: a factor stochastic volatility approach (with discussion), *in* J. M. Bernardo, J. O. Berger, A. P. Dawid and A. F. M. Smith (eds), *Bayesian Statistics 6*, Oxford University Press, pp. 547–570.

Poggio, T. and Girosi, F. (1990).Networks for approximation and learning, *Proceedings of the IEEE* **78**(9): 1481–1497.

Pole, A. (1988).Transfer response models: A numerical approach, *in* J. M. Bernardo, M. H. DeGroot, D. V. Lindley and A. F. M. Smith (eds), *Bayesian Statistics 3*, Oxford University Press, pp. 733–746.

Pole, A. and West, M. (1988).Efficient numerical integration in dynamic models, *Technical Report Warwick Research Report 136*, Department of Statistics, University of Warwick, Coventry.

Pole, A. and West, M. (1990).Efficient Bayesian learning in non-linear dynamic models, *Journal of Forecasting* **9**(2): 119–136.

Pole, A., West, M. and Harrison, P. J. (1988).Non-normal and non-linear dynamic Bayesian modelling, *in* J. C. Spall (ed.), *Bayesian Analysis of Time Series and Dynamic Models*, Marcel Dekker, New York, pp. 167–198.

Pontryagin, L. S. (1957).Some mathematical properties arising in connection with the theory of optimal automatic control systems, *Proceedings of the Conference on Basic Problems in Automatic Control and Regulation*, Academy of Sciences, Moscow.

Prado, R., West, M. and Krystal, A. D. (1999).Multi-channel EEG analyses via dynamic regression models with time-varying lag/lead structure, *ISDS Discussion paper 99-05, Duke University*.

Press, W. H., Teukolsky, S. A., Vetterling, W. T. and Flannery, B. P. (1992).*Numerical Recipes in C - The Arts of Scientific Computing*, 2nd edn, Cambridge University Press.

Priestley, M. B. (1981).*Spectral Analysis and Time Series*, Probability and Mathematical Statistics, Academic Press, Inc.

Proakis, J. G. (1995).*Digital Communications*, 3rd edn, McGraw-Hill.

Punskaya, E., Andrieu, C., Doucet, A. and Fitzgerald, W. J. (2000).Particle filtering for demodulation in fading channels, *Technical Report 384*, CUED-F-INFENG.

Puskorius, G. V. and Feldkamp, L. A. (1994).Neurocontrol of nonlinear dynamical systems with Kalman filter trained recurrent networks, *IEEE Transactions on Neural Networks* **5**(2): 279–297.

Rabiner, L. and Bing-Hwang, J. (1993).*Fundamentals of speech recognition*, Prentice-Hall.

Rabiner, L. and Juang, B. (1986).An introduction to hidden Markov models, *IEEE Acoustics, Speech & Signal Processing Magazine* **3**(1): 4–16.

Rajan, J. J., Rayner, P. J. W. and Godsill, S. J. (1997).A Bayesian approach to parameter estimation and interpolation of time-varying autoregressive processes using the Gibbs sampler, *IEE Proceedings on Vision, Image and Signal Processing*.

Rao, C. R. (1973).*Linear Statistical Inference and Its Applications*, John Wiley & Sons, New York.

Rencken, W. D. (1993).Concurrent localisation and map building for mobile robots using ultrasonic sensors, *Proceedings of the IEEE/RSJ International Conference on Intelligent Robots and Systems*, Yokohama, Japan, pp. 2129–2197.

Reno, A. and Booth, D. (1998).Deformable models for object recognition in aerial image, *F. Sadjadi (ed.), Automatic Target Recognition VIII* **SPIE Proc. 3371**: 322–333.

Reynard, D., Wildenberg, A. P., Blake, A. and Marchant, J. (1996).Learning dynamics of complex motions from image sequences, *Proceedings of the 4th European Conference on Computer Vision*, Cambridge, England, pp. 357–368.

Ripley, B. D. (1996).*Pattern Recognition and Neural Networks*, Cambridge University Press.

Ripley, B. D. and Sutherland, A. L. (1990).Finding spiral structures in images of galaxies, *Philosophical Transactions of the Royal Society of London A* **332**(1627): 477–485.

Robert, C. P. and Casella, G. (1999).*Monte Carlo Statistical Methods*, Springer-Verlag.

Roberts, G. O. and Tweedie, R. L. (September, 1995).Exponential convergence of Langevin's diffusions and their discrete approximations, *MCMC website*.

Rosenbluth, M. N. and Rosenbluth, A. W. (1955).Monte Carlo calculation of the average extension of molecular chains, *Journal of Chemical Physics* **23**: 356–359.

Rowe, S. M. and Blake, A. (1996).Statistical mosaics for tracking, *Journal of Image and Vision Computing* **14**: 549–564.

Royama, T. (1992).*Analytical Population Dynamics*, Vol. 10 of *Population and community biology series*, Chapman & Hall, London.

Rubin, D. B. (1987a).*Multiple imputation for nonresponse in surveys*, John Wiley & Sons, New York.

Rubin, D. B. (1987b).A noniterative sampling/importance resampling alternative to the data augmentation algorithm for creating a few imputations when fractions of missing information are modest: the SIR algorithm, *Journal of the American Statistical Association*.

Rubin, D. B. (1988).Using the SIR algorithm to simulate posterior distributions, *in* J. M. Bernardo, M. H. DeGroot, D. V. Lindley and A. F. M. Smith (eds), *Bayesian Statistics 3*, Oxford University Press, pp. 395–402.

Rydén, T. (1995).On recursive estimation for hidden Markov models, *Stochast. Proc. Applic.* **66**: 79–96.

Sage, A. P. and Melsa, J. L. (1971).*Estimation Theory with Applications to Communication and Control*, McGraw-Hill.

Salmond, D. J. (1990).Mixture reduction for target tracking in clutter, *SPIE Vol. 1305, Signal and Data Processing of Small Targets*, pp. 434–445.

Salmond, D. J., Fisher, D. and Gordon, N. J. (1998).Tracking in the presence of intermittent spurious objects and clutter, *in* O. E. Drummond (ed.), *Signal and Data Processing of Small Targets 1998*, Vol. 3373, SPIE, pp. 460–474.

Sampei, S. and Sunaga, T. (1993).Rayleigh fading compensation for QAM in land mobile communications, *IEEE Transactions on Vehicle Technology* **42**: 137–147.

Schnatter, S. (1992).Integration-based Kalman-filtering for a dynamic generalized linear trend model, *Computational Statistics & Data Analysis* **13**(13): 447–459.

Semerdjiev, T., Jilkov, V. and Angelove, D. (1998).Target tracking using Monte Carlo simulation, *Mathematics and Computers in Simulation* **47**: 441–447.

Seymour, J. P. and Fitz, M. P. (1995).Near-optimal symbol-by-symbol detection schemes for flat rayleigh fading, *IEEE Transactions on Communications* **43**(2/3/4): 1525–1533.

Shachter, R. and Kenley, C. (1989).Gaussian influence diagrams, *Management Science* **35**: 527–550.

Shan, L. and Doerschuk, P. C. (1997).Performance bounds for nonlinear filters, *IEEE Transactions on Aerospace and Electronic Systems* **33**(1): 316–318.

Shephard, N. (1994).Partial non-Gaussian state space, *Biometrika* **81**: 115–131.

Shephard, N. and Pitt, M. K. (1997).Likelihood analysis of non-Gaussian measurement time series, *Biometrika* **84**: 653–67.

Shiryaev, A. N. (1996).*Probability*, Graduate Texts in Mathematics, 2nd edn, Springer-Verlag.

Silverman, B. W. (1986).*Density Estimation for Statistics and Data Analysis*, Vol. 26 of *Monographs on Statistics and Applied Probability*, Chapman & Hall, London.

Simmons, R. and Koenig, S. (1995).Probabilistic robot navigation in partially observable environments, *Proceedings of IJCAI-95*, IJCAI, Inc., Montreal, Canada, pp. 1080–1087.

Simmons, R., Goodwin, R., Haigh, K., Koenig, S. and O'Sullivan, J. (1997).A layered architecture for office delivery robots, *Proceedings of the First International Conference on Autonomous Agents*, Marina del Rey, CA.

Smith, A. and Roberts, G. (1993).Bayesian computation via the Gibbs sampler and related Markov chain Monte Carlo methods, *J. Royal Statistical Society B* **55**(1): 3–37.

Smith, A. F. M. and Gelfand, A. E. (1992).Bayesian statistics without tears: a sampling–resampling perspective, *American Statistician* **46**(4): 84–88.

Smith, A. F. M. and West, M. (1983).Monitoring renal transplants: An application of the multi-process Kalman filter, *Biometrics* **39**: 867–878.

Smith, R., Self, M. and Cheeseman, P. (1988).Estimating uncertain spatial relationships in robotics, *in* Lemmer and Kanal (eds), *Uncertainty in Artificial Intelligence*, Vol. 2, Elsevier, pp. 435–461.

Smyth, P., Heckerman, D. and Jordan, M. I. (1997).Probabilistic independence networks for hidden Markov probability models, *Neural Computation* **9**(2): 227–269.

Snyder, D. L., Hammoud, A. M. and White, R. L. (1993).Image recovery from data acquired with a charge-coupled-device camera, *Journal of the Optical Society of America A* **10**(5): 1014–1023.

Snyder, P. G., Woollam, J. A., Alterovitz, S. A. and Johs, B. (1990).Modelling $Al_xGa_{1-x}As$ optical constants as functions of composition, *Journal of Applied Physics.* **68**(11): 5925–5926.

Solo, V. (1980).Some aspect of recursive parameter estimation, *International Journal of Control* **32**: 395–410.

Sorensen, H. W. (1966).Kalman filtering techniques, *Advanced Control Systems* **3**: 219–292.

Sorensen, H. W. (1974).On the development of practical nonlinear filters, *Information Sciences* **7**: 253–270.

Sorenson, H. W. and Alspach, D. L. (1971).Recursive Bayesian estimation using Gaussian sums, *Automatica* **7**: 465–479.

Srivastava, A. (1996).*Inferences on Transformation Groups Generating Patterns on Rigid Motions*, D. Sc. Thesis, Washington University, St. Louis, Missouri.

Srivastava, A. (2000).Bayesian filtering for tracking pose and location of rigid targets, *Proceedings of SPIE*.

Srivastava, A., Grenander, U., Jensen, G. R. and Miller, M. I. (December, 1999).Jump-diffusion Markov processes on orthogonal groups for object recognition, *accepted for publication by Journal of Statistical Planning and Inference*.

Srivastava, A., Miller, M. I. and Grenander, U. (1999).Bayesian automated target recognition, *Handbook of Image and Video Processing*.

Stavropoulos, P. and Titterington, D. M. (1998).Computational Bayesian filtering, *Technical report 98-13, Department of Statistics, University of Glasgow*.

Stenseth, N. C., Bjørnstad, O. N. and Falck, W. (1996).Is spacing behaviour coupled with predation causing the microtine density cycle? A synthesis of current process-oriented and pattern-oriented studies, *Proceedings of the Royal Society of London, B* **263**: 1423–1435.

Stenseth, N. C., Bjørnstad, O. N. and Saitoh, T. (1996).A gradient from stable to cyclic populations of Clethrionomys rufocanus in Hokkaido, *Proceedings of the Royal Society of London, B* **263**: 1117–1126.

Stenseth, N. C., Falck, W., Bjørnstad, O. N. and Krebs, C. J. (1997).Population regulation in snowshoe hare and Canadian lynx: Asymmetric food web configurations between hare and lynx, *Proceedings of National Academy of Science, Washington* **94**: 5147–5152.

Stenseth, N. C., Falck, W., Chain, K. S., Bjørnstad, O. N., Donoghue, M. O. and Tong, H. (1998).From ecological patterns to ecological processes: Phase- and density-dependencies in the Canadian lynx cycle, *Proceedings of National Academy of Science, Washington* **95**: 15430–15435.

Storvik, G. (1994).A Bayesian approach to dynamic contours through stochastic sampling and simulated annealing, *IEEE Transactions on Pattern Analysis and Machine Intelligence* **16**(10): 976–986.

Stuber, G. L. (1996).*Principles of Mobile Communication*, Kluwer.

Sullivan, J., Blake, A., Isard, M. and MacCormick, J. (1999).Bayesian correlation, *Accepted to ICCV'99*.

Sutherland, A. I. and Titterington, D. M. (1994).Bayesian analysis of image sequences, *Technical report 94-3, Department of Statistics, University of Glasgow*.

Tanizaki, H. (1993).*Nonlinear Filters*, Springer-Verlag, New York.

Tanner, G. L. (1995).Missile control against multiple targets using non-quadratic cost functions, *American Control Conference*, Vol. 1, pp. 515–519.

Tanner, M. A. (1993).*Tools for Statistical Inference*, 2nd edn, Springer Verlag.

Taylor, J. H. (1979).Cramer-Rao estimation error lower bound analysis for non-linear systems with unknown deterministic variables, *IEEE Transactions on Automatic Control* **24**(2): 343–344.

Thrun, . (1998).Bayesian landmark learning for mobile robot localization, *Machine Learning*.

Thrun, S., Bennewitz, M., Burgard, W., Cremers, A. B., Dellaert, F., Fox, D., Hähnel, D., Rosenberg, C., Roy, N., Schulte, J. and Schulz, D. (1999).MINERVA: A second generation mobile tour-guide robot, *Proceedings of the IEEE International Conference on Robotics and Automation (ICRA)*.

Thrun, S., Burgard, W. and Fox, D. (1998).A probabilistic approach to concurrent mapping and localization for mobile robots, *Machine Learning* **31**: 29–53.

Thrun, S., Fox, D. and Burgard, W. (2000).Monte Carlo localization with mixture proposal distribution, *Proceedings of the AAAI National Conference on Artificial Intelligence*, AAAI, Austin, TX.

Tichavský, P., Muravchik, C. and Nehorai, A. (1998).Posterior Cramér-Rao bounds for discrete-time nonlinear filtering, *IEEE Transactions on Signal Processing* **46**(5): 1386–1396.

Tierney, L. (1994).Markov chains for exploring posterior distributions, *Annals of Statistics* **22**: 1701–1728.

Titterington, D. M. (1973).A method of extremum adaptation, *Journal of the Institute of Mathematics and its Applications* **11**: 297–315.

Titterington, D. M., Smith, A. F. M. and Makov, U. E. (1985).*Statistical Analysis of Finite Mixture Distributions*, John Wiley & Sons, New York.

Tong, H. (1990).*Non-linear Time Series: A Dynamical System Approach*, Oxford statistical science series, Oxford University Press.

Torrance, J. M. and Hanzo, L. (1995).Comparative study of pilot symbol assisted modem schemes, *Proceedings of Radio Receivers and Associated Systems*, Bath, England, pp. 36–41.

Ullman, S. and Basri, R. (1991).Recognition by linear combinations of models, *IEEE Transactions on Pattern Analysis and Machine Intelligence* **13**(10): 992–1006.

van der Merwe, R., Doucet, A., de Freitas, J. F. G. and Wan, E. (2000).The unscented particle filter, *Technical Report CUED/F-INFENG/TR 380*, Cambridge University Engineering Department.

Van Trees, H. L. (1968).*Detection, Estimation and Modulation Theory*, John Wiley & Sons, New York.

Vermaak, J., Andrieu, C., Doucet, A. and Godsill, S. J. (1999).On-line Bayesian modelling and enhancement of speech signals, *Technical Report CUED/F-INFENG/TR 361*, Cambridge University Engineering Department.

Vetter, T. and Poggio, T. (1996).Image synthesis from a single example image, *Proceedings of the 4th European Conference on Computer Vision*, Cambridge, England, pp. 652–659.

Vidoni, P. (1999).Exponential family state space models based on a conjugate latent process, *Journal of the Royal Statistical Society* **B 61**: 213–221.

Vitetta, G. M. and Taylor, D. P. (1995).Maximum likelihood decoding of uncoded and coded PSK signal sequences transmitted over rayleigh flat-fading channels, *IEEE Transactions on Communications* **43**(11): 2750–2578.

von Neumann, J. (1951).Various techniques used in connection with random digits, *National Bureau of Standards Applied Mathematics Series* **12**: 36–38.

Wall, F. T. and Erpenbeck, J. (1959).New method for the statistical computation of polymer dimensions, *Journal of Chemical Physics* **30**(3): 634–37.

Wang, A. H. and Klein, R. L. (1978).Optimal quadrature formula nonlinear estimators, *Information Sciences* **16**: 169–184.

Wei, G. C. G. and Tanner, M. A. (1990).A Monte Carlo implementation of the EM algorithm and the poor man's data augmentation algorithms, *Journal of the American Statistical Association* **85**: 699–704.

Weiß, G., Wetzler, C. and von Puttkamer, E. (1994).Keeping track of position and orientation of moving indoor systems by correlation of range-finder scans, *Proceedings of the International Conference on Intelligent Robots and Systems*, pp. 595–601.

West, M. (1986).Bayesian model monitoring, *Journal of the Royal Statistical Society (Ser. B)* **48**: 70–78.

West, M. (1993a).Approximating posterior distributions by mixtures, *Journal of Royal Statistical Society* **55**: 409–422.

West, M. (1993b).Mixture models, Monte Carlo, Bayesian updating and dynamic models, *in* J. H. Newton (ed.), *Computing Science and Statistics: Proceedings of the 24th Symposium on the Interface*, Interface Foundation of North America, Fairfax Station, Virginia, pp. 325–333.

West, M. and Harrison, J. (1997).*Bayesian forecasting and dynamic models*, Springer Series in Statistics, 2nd edn, Springer-Verlag, New York.

West, M. and Harrison, P. J. (1986).Monitoring and adaptation in Bayesian forecasting models, *Journal of the American Statistical Association* **81**: 741–750.

West, M. and Harrison, P. J. (1989).Subjective intervention in formal models, *Journal of Forecasting* **8**: 33–53.

West, M., Harrison, P. J. and Migon, H. S. (1985).Dynamic generalized linear models and Bayesian forecasting (with discussion), *Journal of the American Statistical Association* **80**: 73–97.

West, M., Prado, R. and Krystal, A. (1999).Evaluation and comparison of EEG traces: Latent structure in non-stationary time series, *Journal of the Acoustical Society America* **94**: 1083–1095.

Whitley, D. (1994).A genetic algorithm tutorial, *Statistics and Computing* **4**: 63–85.

Williams, D. (1991).*Probability with Martingales*, Cambridge University Press.

Zeger, S. L. (1988).A regression model for time series of counts, *Biometrika* **75**: 621–629.

Zilberstein, S. and Russell, S. (1995).Approximate reasoning using anytime algorithms, *in* S. Natarajan (ed.), *Imprecise and Approximate Computation*, Kluwer.

Zweig, G. and Russell, S. J. (1998).Speech recognition with dynamic Bayesian networks, *Proceedings of the Fifteenth National Conference on Artificial Intelligence (AAAI)*, pp. 173–180.

Index